PIEZOELECTRIC MATERIALS

PIEZOELECTRIC MATERIALS
Design and Applications

edited by
N. B. Singh
Dev Kumar Mahato

JENNY STANFORD
PUBLISHING

Published by

Jenny Stanford Publishing Pte. Ltd.
101 Thomson Road
#06-01, United Square
Singapore 307591

Email: editorial@jennystanford.com
Web: www.jennystanford.com

British Library Cataloguing-in-Publication Data
A catalogue record for this book is available from the British Library.

Piezoelectric Materials: Design and Applications

ISBN 978-981-5129-40-3 (Hardcover)
ISBN 978-1-003-59897-8 (eBook)

Contents

Preface

Piezoelectric materials have played a transformative role in science and technology for over a century. These materials, with their ability to convert mechanical energy into electrical energy and vice versa, have applications in fields such as energy harvesting, electronics, sensors, and medical devices. The unique properties of these materials gave birth to innovative technologies. This book deals with the science, engineering, and practical applications of piezoelectric materials in detail. It discusses these materials from their fundamental principles to their cutting-edge applications. The book comprises 20 chapters that deal with different aspects of piezoelectric materials. It showcases various techniques to engineer piezoelectric materials to customize these materials for specific applications. It provides ample information to equip materials scientists, engineers, or students with the knowledge and tools to create and manipulate piezoelectric materials for their projects and innovations. Furthermore, it explores a wide array of real-world applications of piezoelectric materials. From sensors that detect subtle movements to energy-harvesting devices that convert vibrations into electricity and from piezoelectric actuators that enable precise motion control to medical devices that improve healthcare outcomes, the potential applications of piezoelectric materials are vast and ever-expanding. The book is intended for a broad audience—from those with a deep scientific background to those seeking a practical understanding of piezoelectric materials. Each chapter is structured to guide the readers progressively, building upon the foundational concepts and culminating in advanced topics and applications. It is not just a book—it is an invitation to explore the limitless potential of piezoelectric materials and join the exciting journey of discovery and innovation. We hope that the knowledge and insights shared in this book will inspire the readers to embark on adventures in the world of piezoelectricity and, in doing so, contribute to the evolution of science and technology.

Prof. N. B. Singh
RDC, Sharda University, Greater Noida, India
Dr Dev Kumar Mahato
Department of Physics, NIT, Patna, India
Summer 2024

Chapter 1

Introduction to Piezoelectricity

Narender Singh and N. B. Singh
Department of Chemistry and Biochemistry, Sharda University, Greater Noida, India
n.b.singh@sharda.ac.in

Piezoelectricity is a fascinating and scientifically significant phenomenon that lies at the heart of various modern technologies. This chapter provides an overview of the fundamental principles of piezoelectricity, its historical development, and its applications in various fields. The phenomenon of piezoelectricity involves the generation of an electric charge in certain materials when subjected to mechanical stress and, conversely, the deformation of these materials when subjected to an electric field. The historical evolution of piezoelectricity is given. This chapter discusses synthetic methods and characterization of piezoelectric materials.

1.1 Introduction

Piezoelectric materials are a class of materials that exhibit a unique property: they can generate an electric charge in response

Piezoelectric Materials: Design and Applications
Edited by N. B. Singh and Dev Kumar Mahato

ISBN 978-981-5129-40-3 (Hardcover), 978-1-003-59897-8 (eBook)
www.jennystanford.com

to mechanical stress or deform when an electric field is applied to them. Due to this property, they are used for various applications where the conversion of mechanical energy into electrical energy and vice versa is required. Some common examples of piezoelectric materials include Quartz, Piezoceramics, Piezoelectric polymers, Gallium phosphate, Rochelle salt (potassium sodium tartrate), and others. Piezoelectric materials find applications in numerous sectors, including electronics, consumer electronics, mechanical engineering, robotics, environmental monitoring, civil engineering and infrastructure monitoring, nanotechnology, acoustic and audio, automotive, medical, energy harvesting, aerospace, and defense [1, 2]. The direct piezoelectric effect refers to the phenomenon where these materials generate an electric charge in proportion to the applied mechanical stress. Conversely, when subjected to an electric field, they experience strain proportional to the applied field, known as the inverse piezoelectric effect [1, 2]. Despite their valuable properties, piezoelectric materials have some limitations. These include their operating temperature range, mechanical strength, and frequency response. In conclusion, piezoelectric materials are unique substances with remarkable properties that make them invaluable in a wide array of applications [3, 4]. In this chapter, various aspects of piezoelectric materials, their properties, applications, and limitations have been explored.

1.2 History of Piezoelectric Materials

Beginning around the turn of the 18th century, scientists began to discover and study piezoelectricity as an intentional physical phenomenon. During that time, the Dutch imported tourmaline, a rare stone from the East Indies, to Europe [5]. Tourmaline [6] has an unusual property: when heated, it attracts other substances, such as ashes. Nearly 50 years later, Swedish physician and botanist Carl Linnaeus, best known for creating biological nomenclature, speculated that this occurrence might be related to electricity. Indeed, within a decade, German physicist Franz Aepinus established that this "peculiar feature" was indeed related to electricity. Later, this phenomenon was termed "pyroelectricity" – the capacity of a substance to produce a transient voltage when heated or cooled. The discovery of piezoelectricity, primarily investigated by French researchers, followed the discovery of pyroelectricity. René-Just

Haüy and Antoine César Becquerel performed experiments based on their knowledge of pyroelectricity to test Charles-Augustin de Coulomb's hypothesis that electric charge might be produced by pressure. However, they were not successful in their efforts.

The direct piezoelectric effect was eventually discovered by Pierre and Jacques Curie in 1880, building on the theoretical basis of pyroelectricity [7, 8]. They observed that by applying specific mechanical stress to crystals of tourmaline, quartz, cane sugar, and Rochelle salt (also known as Seignette's salt), charge at certain locations on their surfaces could be produced. The term "piezoelectricity," derived from the Greek words "piezein," meaning to press or squeeze, and "elektron," meaning amber (a substance that attracts other materials when rubbed), was coined by the German physicist Wilhelm G. Hankel to describe this phenomenon. The inverse piezoelectric effect was predicted by French-Luxembourgish physicist Gabriel Lippmann a year later and experimentally confirmed by the Curie brothers in the same year.

1.3 Classification of Piezoelectric Materials

Classification of piezoelectric materials is given in Fig. 1.1 [9–14].

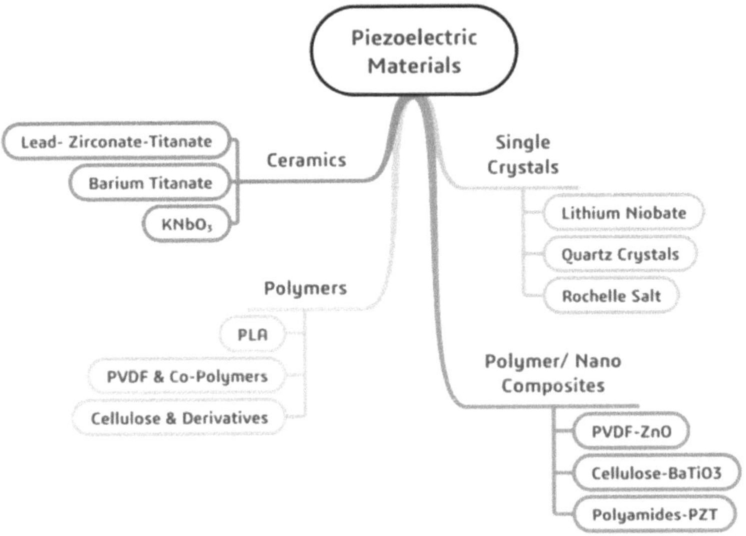

Figure 1.1 Classification of piezoelectric materials.

1.4 Piezoelectric Single Crystals

Unlike piezoelectric ceramics, which are polycrystalline materials composed of many small crystals, piezoelectric single crystals consist of a continuous, single crystal structure. The piezoelectric effect arises due to the asymmetrical arrangement of atoms within the crystal lattice. When an external force is applied to the crystal, it causes a displacement of the charged ions, resulting in a separation of positive and negative charges along certain crystallographic planes. This separation of charges leads to the generation of an electric potential across the crystal. Some of the well-known piezoelectric single crystals include: Quartz (SiO_2), Lithium niobate ($LiNbO_3$), Lead zirconate titanate (PZT), Potassium niobate ($KNbO_3$), Barium titanate ($BaTiO_3$), etc. Piezoelectric single crystals have several advantages over piezoelectric ceramics, such as higher piezoelectric coefficients, improved electromechanical coupling coefficients, and better temperature stability. These properties make them suitable for precision applications where high sensitivity and performance are required [15–17].

1.5 Piezoelectric Ceramics

Most common ceramics are polycrystalline, which means they may be considered as a collection of many little single crystals called "crystallites" or "grains." Ceramics can be viewed as microscopic grains that are baked together to form a solid material [18, 19]. This happens during a process known as sintering. Figure 1.2 depicts the grains in barium titanate ceramics as an example.

Ceramics are further divided into the following two categories:

1. Traditional ceramics (Fig. 1.3) [20, 21]
2. Technical ceramics (Fig. 1.4) [22, 23]

Technical ceramics are available in a variety of forms, including tiles used in the Space Shuttle program, ballistic protective plates, biomedical implants, disk brakes, and mechanical bearings. The basic materials utilized in these ceramics are frequently high-grade powders rather than clay as in conventional ceramics [24, 25].

Figure 1.2 Optical microscopy image of $BaTiO_3$ ceramics obtained by sintering process.

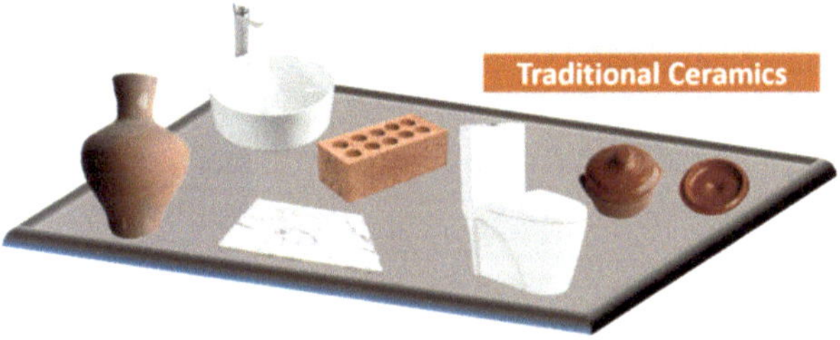

Figure 1.3 Representation of traditional ceramics.

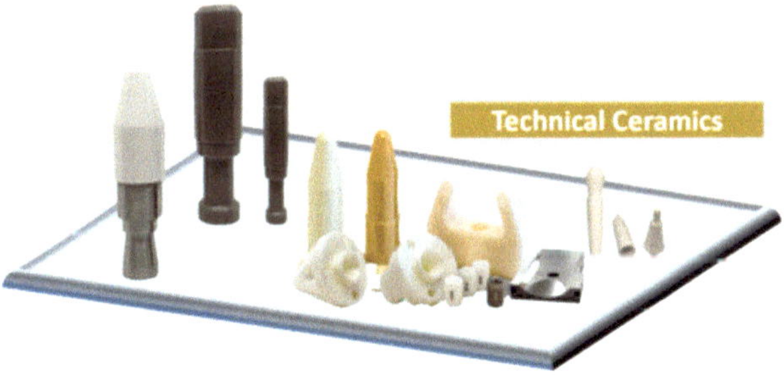

Figure 1.4 Representation of technical ceramics.

The class of technical ceramics includes electroceramics, which are ceramic materials principally employed for their electrical capabilities. The composition and crystal structure of these ceramics determine their distinctive electrical, magnetic, or optical capabilities. Electroceramics [26] encompasses a wide range of materials:

- Dielectric ceramics are utilized as insulators or in temperature stable capacitors.
- Electronically conducting oxides, such as indium tin oxide (ITO), are employed to serve as transparent conducting materials in displays and OLEDs [27].
- Magnetic ceramics, such as spinel ferrites, hexaferrites, and garnets. These are commonly used in the electronics sector as permanent magnetic materials and in transformers.
- Ion conducting materials, such as yttria stabilized zirconia (YSZ), are used in solid-state fuel cells [28].

Barium titanate and lead zirconium titanate (PZT) are the most often utilized piezoelectric ceramics [29]. Both materials form in the so-called perovskite crystal structure, as do many other ferroelectric ceramics.

1.6 Piezoelectric Polymers

Piezoelectric polymer materials have sparked considerable interest due to their capacity to convert vibrational energy into electrical energy in a controlled environment. Polymer piezoelectric materials have a high degree of flexibility, high mechanical strength, low production cost, high biocompatibility, and high chemical resistance [30]. Polyvinylidene fluoride (PVDF) and its copolymer have the highest piezoelectric coefficients and exhibit promising results when blended with various inorganic fillers [31]. Polymers have antioxidative qualities in addition to mechanical, chemical, and hydrolytic properties [32]. High thermal and electrical actuation, as well as surface charge alteration due to poling, have sparked interest in polymer-based piezoelectric materials [33]. PVDF is primarily composed of non-polar (α), polar (β, γ, σ), and ε-phases [34], the latter of which can be adjusted to high piezo-coefficient values

[35]. Other polymers, such as trifluoroethylene (TrFE), employed alongside PVDF, show better electromechanical responses because of β-phase and chain mobility [36]. For pressure sensors, PVDF-TrFE demonstrated 40 times better sensitivity and strength [37]. Table 1.1 shows the comparison of properties of PVDF with other materials.

Polymer-based piezoelectric materials can enable energy harvesting concepts that do not require external power from various environmental sources. They can be wearable and flexible, which allows them to be employed in human gesture identification techniques [38], real-time monitoring devices [39], health evaluations [40], electronic textile cardiograms [41], and other applications.

Table 1.1 PVDF property comparison against other piezoelectric materials [42–44]

Factors	Soft PZT	$BaTiO_3$	PVDF Film	Copolymer
density [10^3 kg/m^3]	8.0	5.7	1.78	1.82
dielectric constant ε_r [–]	1800–4000	1700	12	8
d_{33} [10^{-12} C/N]	395-600	78	-30	-38
k_{31} [-] at 1 [kHz]	0.30–0.40	0.21	0.12	0.20
g_{33} [10^{-3} Vm/N]	21-25	5	-330	-540
acoustic impedance [10^6 kg/m^2 s]	30	30	2.7	2.7

1.7 Piezoelectric Composites

In the realm of electronic materials used in devices, meeting the required combination of desirable qualities can often be challenging to achieve with a single-phase material. In some cases, the inherent limitations of materials can be overcome by creating a composite product that involves mixing two or more classes of materials together [45]. This composite approach allows for the integration of different properties from each constituent material, resulting in a product that possesses a unique combination of traits. The focus of the investigation lies in composites consisting of a polymer matrix containing piezoelectric particles or fibers [46]. These composites, known as piezocomposites, are categorized based on the "Connectivity" between the different phases, as depicted in Fig. 1.5. When analyzing a two-phase piezoelectric composite, each phase's connection is identified using two integers. For instance, if a phase is interconnected in all three directions (x, y, and z), it exhibits "3"-dimensional connectivity. Conversely, if the other phase is only connected in the z-direction, it displays "1"-dimensional connectivity [47]. Consequently, the composite is designated as a 1-3 composite, indicating the connectivity pattern of its constituent phases.

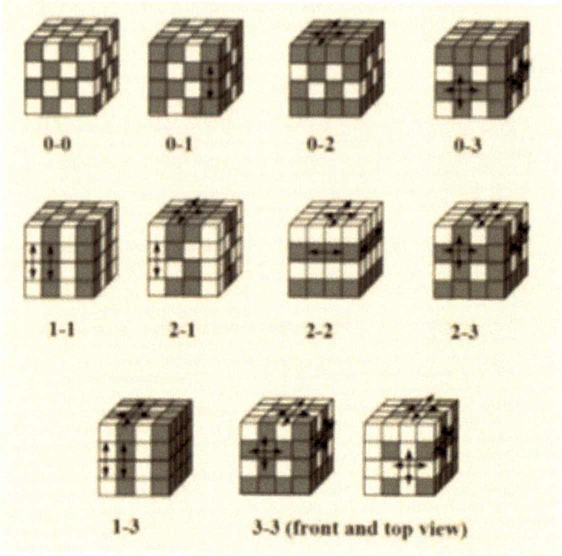

Figure 1.5 Schematic layout of multiple piezocomposites with varying connectivity [47].

Composite materials [48] facilitate production, design, and study of many mechanisms, particularly when particle or grain size is reduced. The nano or submicron grain size of composite material results in low yield loss in thin thickness layered composites, boosting reliability and mechanical strength [49].

Three steps can be used to understand composites in piezoelectric materials research, as given below:

1. The first step is to find materials with the highest piezoelectric coefficient, electromechanical coupling, big bandwidth with high sensitivity, and source level.

2. The second step consists of increasing Curie temperature, decreasing heat shunts, maintaining the electric field, and increasing mechanical strength.
3. The third step entails enhancing the quality factor (Q) and electromechanical coefficients for high- and low-frequency applications.

Piezocomposites combine the desirable features of both polymers and ceramics, resulting in a material with high flexibility (as seen in polymers) and improved piezoelectric coefficients and coupling (as seen in ceramics) [50]. This combination allows to produce high-quality composites that can be tailored to suit a wide range of applications. The customization of the composite's properties can be achieved by modifying various critical aspects, such as phases, connectivity, and volume ratios. One of the simplest methods to customize the attributes of piezocomposites is by changing the volume fraction. By altering the volume fraction of piezoelectric, dielectric, or other physical properties, strong sum or combination effects can be observed [51–53]. The presence of a surrounding polymer phase in the composite induces stopband resonances, which absorb the transverse piezoelectric coefficients and planar electromechanical coupling factors, resulting in a high kt/kp ratio. This, in turn, significantly enhances the directivity of the input energy through the composite and minimizes potential cross-talk issues [54, 55]. Moreover, the composite structure optimizes the use of ferroelectric ceramics' or crystals' maximum k33 coupling when employing a 1-3 connectivity or k33 coupling with a 2-2 connectivity, respectively. This design approach surpasses the kt of monolithic materials, leading to superior performance [56]. In summary, piezocomposites offer a wide range of possibilities for tailoring their properties, making them highly versatile for various applications. Their combination of flexibility and enhanced piezoelectric coefficients and coupling makes them ideal candidates for multiple uses where precise control of energy and reduced cross-talk are essential.

1.8 Synthesis of Piezoelectric Materials

There are a number of methods for the preparation of piezoelectric materials (Fig. 1.6) [57]. A brief description of the methods is described.

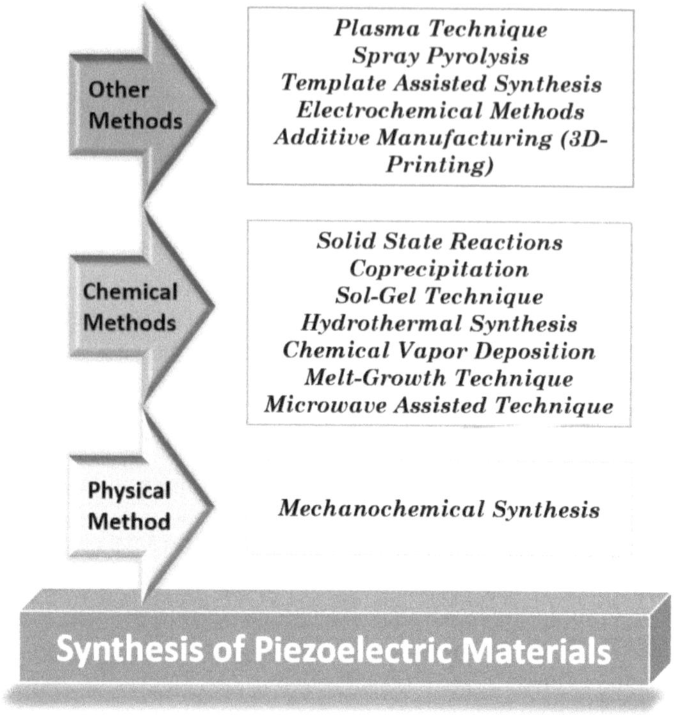

Figure 1.6 Different methods for the preparation of piezoelectric materials.

Physical Methods:

Mechanochemical Synthesis: This technique involves the mechanical activation of precursor materials for chemical reactions and leads to the formation of piezoelectric materials. The process takes place due to grinding or high-energy ball milling.

Chemical Methods:

Solid-State Reaction: This is one of the most traditional methods for synthesizing piezoelectric materials. It involves the direct reaction of solid-state precursors to form the desired piezoelectric compound. The precursors are mixed thoroughly in stoichiometric proportions and then heated at elevated temperatures. The high temperature leads to the formation of the piezoelectric material.

Taking an example of the production of $Pb(Zr_xTi_{1-x})O_3$ powders, the raw powders PbO, ZrO_2, and TiO_2 are weighed in proper proportions, combined, and calcined for 1–2 h at 800–900°C [58]. $BaTiO_3$ is prepared by heating equimolar amounts of BaO and TiO_2 raw materials. $BaCO_3$ powder is generally preferred over BaO because high purity of BaO is costlier and chemically less reactive.

Coprecipitation: In this technique, suitable metal salts are mixed, and a precipitating agent is added to the solution to obtain a gelatinous precipitate. The precipitate is then thermally treated to form piezoelectric materials with controlled composition and particle size distribution.

Sol-Gel Technique: The sol-gel method is an important technique to synthesize various materials, including piezoelectric ones. It involves the hydrolysis and condensation of metal alkoxides or metal chlorides to form a sol (a colloidal suspension). This sol is then subjected to gelation and further heat treatment to remove the solvent and organic components. The resulting gel is heated at a higher temperature to obtain the piezoelectric material.

Hydrothermal Synthesis: Hydrothermal synthesis is performed in an autoclave or pressure vessel at elevated temperatures and pressures. It involves the reaction of precursors in an aqueous solution under hydrothermal conditions, promoting the formation of the desired piezoelectric crystals. Different steps involved are shown in Fig. 1.7.

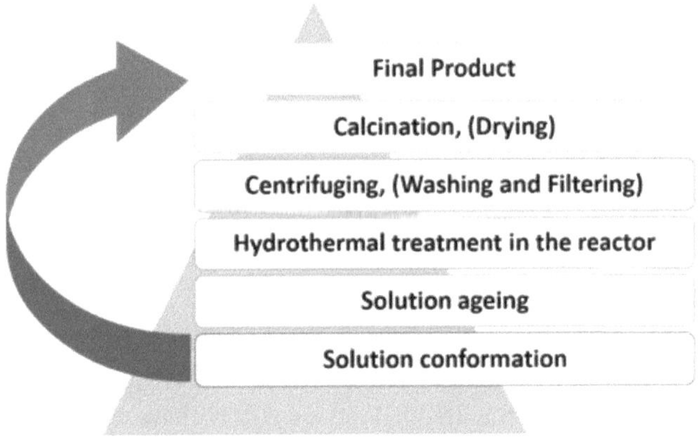

Figure 1.7 Steps showing for hydrothermal process.

Chemical Vapor Deposition (CVD): The CVD technique involves the decomposition of gaseous precursor compounds on a substrate surface at elevated temperatures. The reaction forms a thin film of the desired material with a controlled composition and crystal structure.

Melt Growth Technique: For some piezoelectric materials, such as lead-based compounds like lead zirconate titanate (PZT), the melt growth technique is employed. In this method, the precursors are melted together at high temperatures and then slowly cooled to promote the formation of piezoelectric materials.

Microwave-Assisted Synthesis: Microwave irradiation can be employed to accelerate chemical reactions and enhance the synthesis of piezoelectric powders.

Other Methods:

The above techniques are commonly used for the preparation of piezoelectric materials, but there are some emerging methods that show promise in the formation of piezoelectric materials as given below.

Plasma Synthesis: Plasma chemical synthesis utilizes high-temperature plasma to synthesize piezoelectric powders with enhanced purity and tailored properties.

Spray Pyrolysis: In this method, precursor solutions are atomized into fine droplets, which undergo pyrolysis to form piezoelectric materials. Spray pyrolysis offers good control over particle size and composition.

Template-Assisted Synthesis: Template-assisted methods, such as template sol-gel and template-assisted crystal growth, involve using templates or substrates with predefined patterns or structures. These templates guide the growth of piezoelectric materials.

Electrochemical Methods: Electrochemical methods, such as electroplating and electrophoretic deposition, are employed for the synthesis of thin films and coatings of piezoelectric materials. These techniques involve the use of electric fields to drive the deposition of piezoelectric materials onto conductive substrates.

Additive Manufacturing (3D Printing): With advancements in additive manufacturing techniques, researchers have started exploring 3D printing methods to create complex piezoelectric structures. By combining piezoelectric materials with suitable polymers, it is possible to fabricate intricate piezoelectric devices and systems.

1.9 Characterization of Piezoelectric Materials

In order to understand the properties and behavior of piezoelectric materials, the characterization of piezoelectric materials is important [59]. The different methods used are discussed below.

Piezoelectric Constant Measurement: Piezoelectric constant measurement is a process used to determine the piezoelectric properties of a material. Piezoelectric materials can generate an electric charge in response to mechanical stress (direct piezoelectric effect) or deform when subjected to an electric field (inverse piezoelectric effect). The piezoelectric constant, also known as the piezoelectric coefficient, quantifies this relationship between mechanical stress and electric charge.

The piezoelectric constant is represented by different symbols depending on the specific property being measured:

1. d-Constant (Charge Sensitivity): This constant represents the generated charge (in Coulombs) per unit of mechanical stress applied (in Newtons) and is denoted by "d." The d-constant is the most common and widely measured piezoelectric constant.
2. g-Constant (Voltage Sensitivity): This constant represents the deformation (strain) experienced by the material per unit of applied electric field and is denoted by "g." It is also known as the voltage coefficient.
3. e-Constant (Piezoelectric Strain Constant): This constant represents the generated strain in the material per unit of applied electric field and is denoted by "e."

It is important to note that the measurement process may differ depending on the specific application and the properties being evaluated. Moreover, measuring piezoelectric constants can be a complex task, and researchers often follow standardized testing procedures to ensure reliable and reproducible results.

Resonance Method: This method involves measuring the resonance frequency and impedance of a piezoelectric sample. By applying an AC voltage at various frequencies and monitoring the mechanical response, the resonant frequency and impedance can be determined. This information helps in calculating important parameters such as mechanical quality factor (Q_m) and electromechanical coupling coefficient (kt).

Damping Factor Measurement: Damping factor measurement involves studying the energy dissipation in the material when subjected to mechanical vibrations. The damping factor is crucial in determining the stability and efficiency of piezoelectric devices.

Ferroelectric Domain Imaging: This method allows visualizing the ferroelectric domains within a piezoelectric material. Techniques such as piezoresponse force microscopy (PFM) and transmission electron microscopy (TEM) can be used for domain imaging.

Electromechanical Resonance Analysis: Electromechanical resonance analysis of piezoelectric materials involves the study of the behavior of these materials when subjected to an alternating electric field and mechanical stress or vibration. To analyze the resonance behavior of piezoelectric materials, the material's response to different frequencies of the applied electrical and mechanical stimuli is investigated. The resonance analysis results are crucial for designing and optimizing piezoelectric devices for specific applications. Resonance frequencies can be matched with operational frequencies in devices to achieve maximum efficiency and sensitivity. The electromechanical resonance behavior of piezoelectric materials can be affected by various factors, such as material composition, geometry, temperature, and mechanical boundary conditions.

Thermal Characterization: Understanding how a piezoelectric material responds to changes in temperature is essential for many applications. Thermal methods like differential scanning calorimetry (DSC) and thermomechanical analysis (TMA) can be employed for this purpose.

Dielectric Constant: The dielectric constant, also known as the relative permittivity, is an important parameter for characterizing piezoelectric materials [60]. When dealing with piezoelectric materials, one typically encounters two different dielectric constants: the "static" or "low-frequency" dielectric constant (ε_0) and the "high-frequency" dielectric constant (ε_r). Static or low-frequency dielectric constant (ε_0) characterizes the ability of the material to store electric charge when subjected to an external electric field. High-frequency dielectric constant (ε_r) is important when the material is used in high-frequency applications, such as piezoelectric transducers or sensors operating at radio frequencies (RF) or ultrasound frequencies. The dielectric constant of a piezoelectric material depends on various factors, including its crystal structure, composition, temperature, and frequency of operation. Measuring the dielectric constant and loss tangent can provide valuable insights into the material's performance. Piezoelectric materials are also ferroelectric, so their dielectric properties play a significant role in their behavior.

Electrical Impedance Spectroscopy: This method involves sweeping a range of frequencies and measuring the impedance response of the piezoelectric material. It helps in determining electrical behavior and identifying any anomalies.

Mechanical Testing: Standard mechanical tests like tensile testing and compressive testing can be performed to assess the mechanical properties of piezoelectric materials, including their Young's modulus and Poisson's ratio.

Dynamic Testing: Dynamic testing involves applying dynamic forces or vibrations to the material and studying its response. This helps in understanding the dynamic behavior and mechanical reliability of the material.

1.10 Lead-Free Piezoelectric Materials

Lead zirconate titanates ($Pb(Zr,Ti)O_3$), commonly referred to as PZT, are the most widely used piezoelectric materials for various applications, such as actuators, sensors, and transducers, because of their exceptional piezoelectric properties [61]. Lead oxide present in these materials is known to pose health risks as it accumulates in organisms, causing detrimental effects on the brain and nervous system. Thus, disposal of these materials such as recycling of consumer products containing PZT in cars, smart systems, and sound generators is a problem. Despite these problems, the advantages of PZT cannot be ignored. In critical fields like medical imaging, the use of PZT enables high signal resolution, facilitating precise diagnoses and ultimately saving numerous lives. Because of the toxicity of lead, researches are going on lead-free piezoelectric materials.

An alternative to PZT piezoceramics, lead-free perovskite ferroelectric oxides are now known as environmentally friendly piezoelectric materials. There are many lead-free perovskite ferroelectric oxides, which can replace PZT effectively [62]. Three main groups have been considered as potential replacements for PZT in various applications: (i) Sodium potassium niobate ($(K,Na)NbO_3$)-based solid solutions, with $(K_{0.5}Na_{0.5})NbO_3$ (KNN) being the most common representative, (ii) sodium bismuth titanate ($Na_{1/2}Bi_{1/2}TiO_3$, NBT)-based piezoelectrics and related perovskite solid-solution derivatives like NBT-BT-KNN, and (iii) modified barium titanate ($BaTiO_3$, BT), specifically barium calcium zirconate titanate ($(Ba,Ca)(Zr,Ti)O_3$, BCZT). BCZT exhibits exceptional piezoelectric properties suitable for actuation but is limited to a narrow temperature range below its Curie temperature (below 100°C) to maintain its performance. Additionally, there are other lead-free candidates of interest, such as the multiferroic and high-temperature piezoelectric $BiFeO_3$ and its modified formulations like $BiFeO_3$-BT.

1.11 Applications of Piezoelectric Materials

Piezoelectric materials have a wide range of applications across various industries due to their unique ability to convert mechanical energy into electrical energy and vice versa. The unique properties of piezoelectric materials make them invaluable in many cutting-edge technologies and practical applications, and ongoing research continues to explore new uses and improve existing applications. Some of the applications are given in Fig. 1.8.

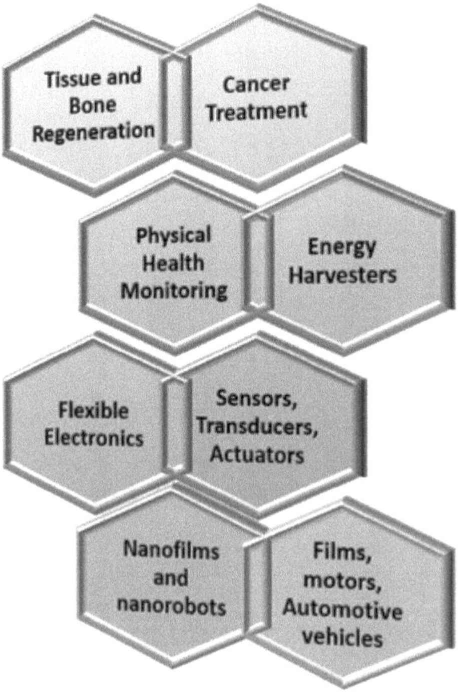

Figure 1.8 Applications of piezoelectric materials.

Some important applications are elaborated below.

Sensors and Transducers: Piezoelectric materials are commonly used in sensors and transducers to detect and convert physical parameters like pressure, force, acceleration, and strain into electrical signals. As shown in Fig. 1.9, the piezoelectric sensor has applications in many diverse fields due to advancements in micro- and nano-scale fabrication techniques. A number of piezoelectric sensor applications are known. The study of piezoelectric sensors in pavement suggested applications for electric vehicle charging in the future due to its distributed nature of electricity generation [63].

Ultrasound Technology: Piezoelectric transducers are essential components in ultrasound machines. They generate and detect ultrasound waves, enabling medical imaging, non-destructive testing, and various other applications like distance measurements in robotics and object detection.

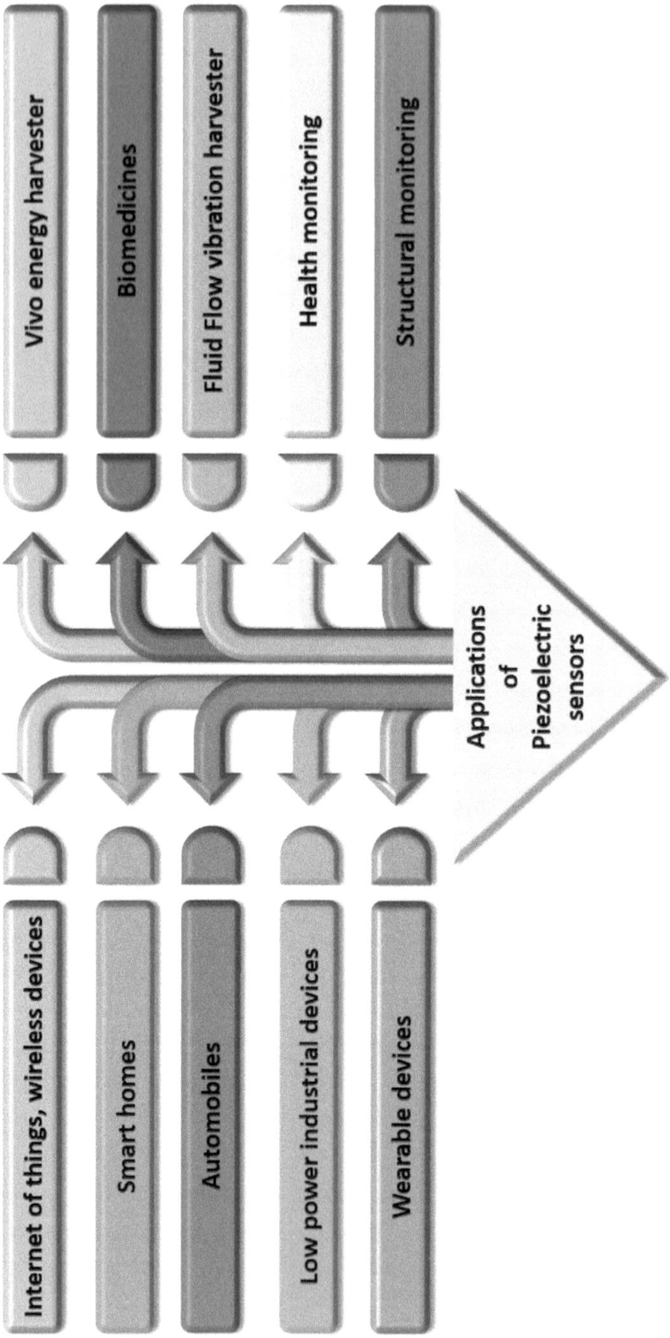

Figure 1.9 Applications of piezoelectric sensors.

Energy Harvesting: Piezoelectric materials can be used to scavenge energy from mechanical vibrations, such as those from machinery, footsteps, or traffic. This energy harvesting can power low-energy devices or be stored for later use.

Piezoelectric Actuators: Piezoelectric actuators can convert electrical signals into mechanical motion with high precision and responsiveness. They find applications in micro-positioning systems, nano-manipulation, adaptive optics, and microvalves.

Inkjet Printers: Piezoelectric actuators are used in inkjet printers to eject ink droplets from the printhead onto the paper. They provide fine control over droplet size and placement, resulting in high-quality printing.

Acoustic Devices: Piezoelectric materials are used in various acoustic applications, such as piezoelectric speakers, buzzers, and microphones.

Structural Health Monitoring: By embedding piezoelectric sensors in structures like bridges, buildings, and aircraft, it is possible to monitor their health and detect potential damages or defects.

Piezoelectric Igniters: They are used in gas appliances like gas stoves and grills to create sparks for igniting the gas.

Vibration Damping: Piezoelectric materials can be used to actively control vibrations in structures or machinery, reducing unwanted oscillations and noise.

Sonar and Underwater Applications: In underwater sonar systems, piezoelectric transducers are used to generate and detect sound waves for navigation, communication, and underwater mapping [64].

Biomedical Applications: Piezoelectric materials are used in medical devices such as ultrasonic scalpels, ultrasound imaging probes, and bone growth stimulators. These flexible piezoelectric materials have growing applications in implantable and wearable biomedical devices [65].

Non-Destructive Testing (NDT): In NDT applications, piezoelectric transducers are used to generate ultrasonic waves that can be used to inspect and evaluate the integrity of materials and structures [66].

1.12 Conclusions

Piezoelectric materials are a source of energy and can be used in various applications such as sensors, machines, gadgets, etc. using normal manufacturing procedures. These piezoelectric materials have improved flexibility, high mechanical strength, fast responding sensors (within seconds), modified films, monitoring, and assessment for human health, and regenerating damaged tissue and bone. Now lead-free piezoelectric materials are also gaining a lot of attention as they are environment friendly and have low risk to human health. In this chapter, the classification, synthesis, characterization, and applications of piezoelectric materials have been discussed. All types of piezoelectric materials are creating new possibilities for advancing future growth.

References

1. Muganda JM, Jansen B, Homburg E, van de Burgt Y, den Toonder J. Influence function measurement technique using the direct and indirect piezoelectric effect for surface shape control in adaptive systems. *IEEE Transactions on Automation Science and Engineering.* **19**(2), 994, 2021.
2. Zhao C, Kong D. An indirect comparison quasi-static calibration method for piezoelectric pressure sensors based on an inverse model. *Measurement.* **159**, 107778, 2020.
3. Sekhar MC, Veena E, Kumar NS, Naidu KC, Mallikarjuna A, Basha DB. A review on piezoelectric materials and their applications. *Crystal Research and Technology.* **58**(2), 2200130, 2023.
4. Behera A, Behera A. Piezoelectric materials. In *Advanced Materials: An Introduction to Modern Materials Science, Springer,* Cham. 43, 2022.
5. Mason WP. Piezoelectricity, its history and applications. *The Journal of the Acoustical Society of America.* **70**(6), 1561, 1981.
6. Lang SB. The history of pyroelectricity: From ancient Greece to space missions. *Ferroelectrics.* **230**(1), 99, 1999.
7. Koptsik VA, Rez IS. Pierre Curie's works in the field of crystal physics (on the one-hundredth anniversary of the discovery of the piezoelectric effect). *Soviet Physics Uspekhi.* **24**(5), 426, 1981.
8. Heywang W, Thomann H. Tailoring of piezoelectric ceramics. *Annual Review of Materials Science.* **14**(1), 27, 1984.

9. Koruza J, Liu H, Höfling M, Zhang MH, Veber P. (K,Na)NbO_3-based piezoelectric single crystals: Growth methods, properties, and applications. *Journal of Materials Research.* **35**(8), 990, 2020.
10. Guo Q, Li F, Xia F, Wang P, Gao X, Hao H, Liu H, Sun H, Zhang S. Piezoelectric ceramics with high piezoelectricity and broad temperature usage range. *Journal of Materiomics.* **7**(4), 683, 2021.
11. Smith M, Kar-Narayan S. Piezoelectric polymers: Theory, challenges and opportunities. *International Materials Reviews.* **67**(1), 65, 2022.
12. Maity K, Mandal D. Piezoelectric polymers and composites for multifunctional materials. In *Advanced Lightweight Multifunctional Materials*, Woodhead Publishing. 239, 2021.
13. Kabakov P, Kim T, Cheng Z, Jiang X, Zhang S. The versatility of piezoelectric composites. *Annual Review of Materials Research.* **53**, 165, 2023.
14. Su Y, Li W, Cheng X, Zhou Y, Yang S, Zhang X, Chen C, Yang T, Pan H, Xie G, Chen G. High-performance piezoelectric composites via β phase programming. *Nature Communications.* **13**(1), 4867, 2022.
15. Verma S, Arya S, Gupta V, Khosla A. Highly stable self-charging piezoelectric (Rochelle salt) driven supercapacitor based on Ni nanowires. *Chemical Engineering Journal.* **424**, 130567, 2021.
16. Aramaki M, Yoshimura T, Murakami S, Satoh K, Fujimura N. Demonstration of high-performance piezoelectric MEMS vibration energy harvester using $BiFeO_3$ film with improved electromechanical coupling factor. *Sensors and Actuators A: Physical.* **291**, 167, 2019.
17. Zhang S, Yu F. Piezoelectric materials for high temperature sensors. *Journal of the American Ceramic Society.* **94**(10), 153, 2011.
18. Chen C, Wang X, Wang Y, Yang D, Yao F, Zhang W, Wang B, Sewvandi GA, Yang D, Hu D. Additive manufacturing of piezoelectric materials. *Advanced Functional Materials.* **30**(52), 2005141, 2020.
19. Pardo L. Piezoelectric ceramic materials for power ultrasonic transducers. In *Power Ultrasonics*, Woodhead Publishing. 65, 2023.
20. Usher TD, Cousins KR, Zhang R, Ducharme S. The promise of piezoelectric polymers. *Polymer International.* **67**(7), 790, 2018.
21. Shi Y, Yang W, Ge G, Wu S, Shen B, Zhai J. Improved ferroelectric properties of $BiFeO_3$-based piezoelectric ceramics through morphotropic phase boundary construction. *Ceramics International.* **46**(10), 15991, 2020.
22. Hao J, Li W, Zhai J, Chen H. Progress in high-strain perovskite piezoelectric ceramics. *Materials Science and Engineering: R: Reports.* **135**, 1, 2019.

23. Khaliq J, Deutz DB, Frescas JA, Vollenberg P, Hoeks T, van der Zwaag S, Groen P. Effect of the piezoelectric ceramic filler dielectric constant on the piezoelectric properties of PZT-epoxy composites. *Ceramics International.* **43**(2), 2774, 2017.

24. Eltouby P, Shyha I, Li C, Khaliq J. Factors affecting the piezoelectric performance of ceramic-polymer composites: A comprehensive review. *Ceramics International.* **47**(13), 17813, 2021.

25. Pradhan SK, Kumar A, Kour P, Pandey R, Kumar P, Kar M, Sinha AN. Piezoelectric and mechanical properties of PVDF-PZT composite. *Ferroelectrics.* **558**(1), 59, 2020.

26. Baraskar BG, Kambale RC, James AR, Mahesh ML, Ramana CV, Kolekar YD. Ferroelectric, piezoelectric and electrostrictive properties of Sn^{4+}-modified $Ba_{0.7}Ca_{0.3}TiO_3$ lead-free electroceramics. *Journal of the American Ceramic Society.* **100**(12), 5755, 2017.

27. Kandhan S, Krishnan P, Jagan R, Aravindhan S, Srinivasan S, Gunasekaran S. Structural, optical and piezoelectric investigation on new Brucinium Chlorate di-hydrate NLO single crystal for optoelectronic, piezo-sensor, transducer and OLED applications. *Optical Materials.* **84**, 556, 2018.

28. Marcaud G, Serna S, Panaghiotis K, Alonso-Ramos C, Le Roux X, Berciano M, Maroutian T, Agnus G, Aubert P, Jollivet A, Ruiz-Caridad A. Third-order nonlinear optical susceptibility of crystalline oxide yttria-stabilized zirconia. *Photonics Research.* **8**(2), 110, 2020.

29. Liao WQ, Zhao D, Tang YY, Zhang Y, Li PF, Shi PP, Chen XG, You YM, Xiong RG. A molecular perovskite solid solution with piezoelectricity stronger than lead zirconate titanate. *Science.* **363**(6432), 1206, 2019.

30. Hu D, Yao M, Fan Y, Ma C, Fan M, Liu M. Strategies to achieve high performance piezoelectric nanogenerators. *Nano Energy.* **55**, 288, 2019.

31. Wu L, Jin Z, Liu Y, Ning H, Liu X, Alamusi, Hu N. Recent advances in the preparation of PVDF-based piezoelectric materials. *Nanotechnology Reviews.* **11**(1), 1386, 2022.

32. Haddadi SA, Ghaderi S, Amini M, Ramazani SA. Mechanical and piezoelectric characterizations of electrospun PVDF-nanosilica fibrous scaffolds for biomedical applications. *Materials Today: Proceedings.* **5**(7), 15710, 2018.

33. Sengupta P, Ghosh A, Bose N, Mukherjee S, Roy Chowdhury A, Datta P. A comparative assessment of poly(vinylidene fluoride)/conducting polymer electrospun nanofiber membranes for biomedical applications. *Journal of Applied Polymer Science.* **137**(37), 49115, 2020.

34. Sappati KK, Bhadra S. Piezoelectric polymer and paper substrates: A review. *Sensors.* **18**(11), 3605, 2018.

35. Gong S, Zhang B, Zhang J, Wang ZL, Ren K. Biocompatible poly(lactic acid)-based hybrid piezoelectric and electret nanogenerator for electronic skin applications. *Advanced Functional Materials.* **30**(14), 1908724, 2020.

36. Meng N, Zhu X, Mao R, Reece MJ, Bilotti E. Nanoscale interfacial electroactivity in PVDF/PVDF-TrFE blended films with enhanced dielectric and ferroelectric properties. *Journal of Materials Chemistry C.* **5**(13), 3296, 2017.

37. Ali F, Raza W, Li X, Gul H, Kim KH. Piezoelectric energy harvesters for biomedical applications. *Nano Energy.* **57**, 879, 2019.

38. Maity K, Mondal A, Saha MC. Cellulose nanocrystal-based all-3D-printed pyro-piezoelectric nanogenerator for hybrid energy harvesting and self-powered cardiorespiratory monitoring toward the human–machine interface. *ACS Applied Materials & Interfaces.* **15**(11), 13956, 2023.

39. Ji Z, Zhang M. Highly sensitive and stretchable piezoelectric strain sensor enabled wearable devices for real-time monitoring of respiratory and heartbeat simultaneously. *Nanotechnology and Precision Engineering (NPE).* **5**(1), 013002, 2022.

40. Okano T, Izumi S, Katsuura T, Kawaguchi H, Yoshimoto M. Multimodal cardiovascular information monitor using piezoelectric transducers for wearable healthcare. *Journal of Signal Processing Systems.* **91**, 1053, 2019.

41. Azimi S, Golabchi A, Nekookar A, Rabbani S, Amiri MH, Asadi K, Abolhasani MM. Self-powered cardiac pacemaker by piezoelectric polymer nanogenerator implant. *Nano Energy.* **83,** 105781, 2021.

42. Khaliq J, Deutz DB, Frescas JA, Vollenberg P, Hoeks T, van der Zwaag S, Groen P. Effect of the piezoelectric ceramic filler dielectric constant on the piezoelectric properties of PZT-epoxy composites. *Ceramics International.* **43**(2), 2774, 2017.

43. Zhang S, Tong W, Wang J, Wang W, Wang Z, Zhang Y. Modified sepiolite/PVDF-HFP composite film with enhanced piezoelectric and dielectric properties. *Journal of Applied Polymer Science.* **137**(9), 48412, 2020.

44. Zhang X, Xia W, Liu J, Zhao M, Li M, Xing J. PVDF-based and its copolymer-based piezoelectric composites: Preparation methods and applications. *Journal of Electronic Materials.* **51**(10), 5528, 2022.

45. Uchino K. The development of piezoelectric materials and the new perspective. In *Advanced Piezoelectric Materials*, Woodhead Publishing. 1, 2017.

46. Mallik N, Ray MC. Effective coefficients of piezoelectric fiber-reinforced composites. *AIAA Journal*. **41**(4), 704, 2003.

47. James NK. Piezoelectric and dielectric properties of polymer-ceramic composites for sensors. 2015. https://doi.org/10.4233/uuid:7923c925-d1a9-4ad7-aa58-3d6376dcc210.

48. Chen J, Qiu Q, Han Y, Lau D. Piezoelectric materials for sustainable building structures: Fundamentals and applications. *Renewable and Sustainable Energy Reviews*. **101**, 14, 2019.

49. Ponnamma D, Chamakh MM, Deshmukh K, Basheer Ahamed M, Erturk A, Sharma P, Al-Maadeed MA. Ceramic-based polymer nanocomposites as piezoelectric materials. In *Smart Polymer Nanocomposites: Energy Harvesting, Self-Healing and Shape Memory Applications*. Springer, Cham. 77, 2017.

50. Khanbareh H, de Boom K, Schelen B, Scharff RB, Wang CC, van der Zwaag S, Groen P. Large area and flexible micro-porous piezoelectric materials for soft robotic skin. *Sensors and Actuators A: Physical*. **263**, 554, 2017.

51. Nayak S, Sahoo B, Chaki TK, Khastgir D. Development of polyurethane–titania nanocomposites as dielectric and piezoelectric material. *RSC Advances*. **3**(8), 2620, 2013.

52. Lin XJ, Zhou KC, Zhang XY, Zhang D. Development, modeling and application of piezoelectric fiber composites. *Transactions of Nonferrous Metals Society of China*. **23**(1), 98, 2013.

53. Wang D, Chen JS. Progress on the applications of piezoelectric materials in sensors. *Materials Science Forum*. **848**, 749, 2016.

54. Xie L, Wang G, Jiang C, Yu F, Zhao X. Properties and applications of flexible poly(vinylidene fluoride)-based piezoelectric materials. *Crystals*. **11**(6), 644, 2021.

55. Jones DJ, Prasad SE, Wallace JB. Piezoelectric materials and their applications. *Key Engineering Materials*. **122**, 71, 1996.

56. Liu J, Tian H, Lucas E, Raja AS, Lihachev G, Wang RN, He J, Liu T, Anderson MH, Weng W, Bhave SA. Monolithic piezoelectric control of soliton microcombs. *Nature*. **583**(7816), 385, 2020.

57. Miclea C. Preparation of piezoelectric nanoparticles. In *Piezoelectric Nanomaterials for Biomedical Applications*, Springer Berlin Heidelberg. 29, 2012.

58. Huang X. Sintering kinetics and properties of highly pure lead zirconate titanate ceramics. *Universitaet Bayreuth (Germany)*, 2009.

59. Kraśny MJ, Bowen CR. A system for characterisation of piezoelectric materials and associated electronics for vibration powered energy harvesting devices. *Measurement.* 168,108285, 2021.

60. Hu J, Fan H, Wu S, Tang L, Qin L, Luo W. Characterization of temperature dependence of dielectric, elastic and piezoelectric properties of $BaTiO_3$ ceramics. *Ceramics International.* 48, 25741–25746, 2022

61. Maeder Demartin M, Damjanovic D, Setter N. Lead free piezoelectric materials. *Journal of Electroceramics*, **13**, 385–392, 2004.

62. Malič B, Otoničar M, Radan K, Koruza J. Lead-free piezoelectric ceramics. *Encyclopedia of Materials: Technical Ceramics and Glasses*, **3**, 358–368, 2021. doi:10.1016/ B978-0-12-803581-8.12131-9

63. Sawane M, Prasad M. MEMS piezoelectric sensor for self-powered devices: A review. *Materials Science in Semiconductor Processing*, **158**, 107324, 2023.

64. Yeon A, Yeo HG, Roh Y, Kim K, Seo HS, Choi H. A piezoelectric micro-electro-mechanical system vector sensor with a mushroom-shaped proof mass for a dipole beam pattern. *Sensors and Actuators A: Physical.* **332**, 113129, 2021.

65. Roy A, Dwari D, Ram MK, Datta P. Piezoelectric nanomaterials for biomedical applications. In *Food, Medical, and Environmental Applications of Nanomaterials*, Elsevier. 355, 2022.

66. Arnau A, editor. *Piezoelectric Transducers and Applications,* Springer Berlin, Heidelberg. 2004.

Chapter 2

Manufacturing Methods for Piezoelectric Ceramic Materials

Anamol Gautam,[a] Bhasker Pratap Choudhary,[a] Mirgender Kumar,[b] and Gaurav Singh[c]

[a]*Department of Applied Sciences, Chandigarh Engineering College Jhanjeri, Mohali, India*

[b]*Department of Electronics Engineering, Chandigarh Engineering College Jhanjeri, Mohali, India*

[c]*Department of Chemistry, Anurag University, Hyderabad, Telangana, India*

pratapbhasker6@gmail.com

Piezoelectric materials, renowned for their pivotal role in precision and acoustic engineering, have become increasingly sought-after in various sectors such as electrical, energy, and biomedical industries. This growing demand has spurred researchers to delve into fresh materials and device combinations, while engineers tirelessly strive to advance existing technologies. Given the extensive scope of this field, it is imperative to provide a comprehensive overview of the diverse facets of piezoelectric materials. This chapter delves into a range of topics, starting with the fabrication processes for these

Piezoelectric Materials: Design and Applications
Edited by N. B. Singh and Dev Kumar Mahato

ISBN 978-981-5129-40-3 (Hardcover), 978-1-003-59897-8 (eBook)
www.jennystanford.com

materials. It explores the objectives behind the material design, including the consideration of solid solution compositions and dopants. Additionally, it scrutinizes the production of piezoelectric ceramic devices through a two-step process: the creation of ceramic powders and the subsequent sintering of shaped structures. To ensure the reproducibility of advanced device characteristics, wet chemical preparation methods are employed for ceramic powder production. Furthermore, this chapter delves into innovative synthesis techniques and the development of thin films, shedding light on popular device designs like multi-layers, bimorphs, and other composite types. It also examines the fundamental impact of particle size and film thickness on ferroelectricity, exploring their relationship with micro-/nanotechnologies. In terms of preparation methods for piezoelectric ceramics, this chapter offers a comparative discussion encompassing solid-state reaction routes, ball milling processes, and co-precipitation. Additionally, it elucidates alternative processing routes, such as spark plasma sintering and hot pressing, known for achieving high ceramic density in lead-free piezoelectric ceramics, albeit at a higher cost. Recognizing the limitations of these expensive techniques for mass production, this chapter explores cost-effective, rapid, and environmentally friendly alternatives. Specifically, it delves into the ultrasonic irradiation processing method and the polymer-modified ceramic processing method, highlighting their suitability for broader applications.

2.1 Introduction

The key to achieving the desired properties in nanostructured piezoelectric materials lies in controlling factors like particle size, dimensionality, and composition. These materials are essentially constructed from nanocrystals, serving as the building blocks, making the initial combination of high-purity, extremely fine, and clustered-free piezoelectric oxide nanocrystals through precise control over particle size, shape, and chemical makeup a critical first step in the process of creating piezoelectric ceramics as the desired characteristics.

There are two primary challenges in synthesizing piezoelectric oxide nanocrystals: firstly, ensuring precise control over composition and particle size and, secondly, managing the arrangement and distribution of these nanoscale building blocks inside the final piezoelectric ceramic compound of nanoscale. For many decades, researchers have developed numerous methods for crafting piezoelectric oxide nanocrystals and the associated nanostructured piezoelectric materials. In the following sections, we will delve into some important techniques used for preparing piezoelectric oxide nanocrystals.

2.2 Solid-State Reaction Method

The conventional approach for crafting nanocrystals of piezoelectric oxides (like $BaTiO_3$, $PbTiO_3$, $Pb(Zr,Ti)O_3$, $CaTiSiO_3$, metal-free organic compounds such as $C_6H_{12}N_2$-NH_4-Br_3/I_3, N-methyl-N-diazabicyclo-octonium, and others like $SmCoO_3$ and $LaCoO_3$) is the solid-state reaction method [1–6]. This method involves a series of steps to transform starting materials (which can be oxides or a combination of oxides and carbonates) into the desired perovskite phase as shown in Figs. 2.1 and 2.2. The process starts by accurately measuring the precursor compounds in the necessary stoichiometric ratios for the target piezoelectric composition. This includes the A-site cation, B-site cation, and anion X, which can be sourced from oxides, carbonates, nitrates, or other suitable sources, ensuring high purity.

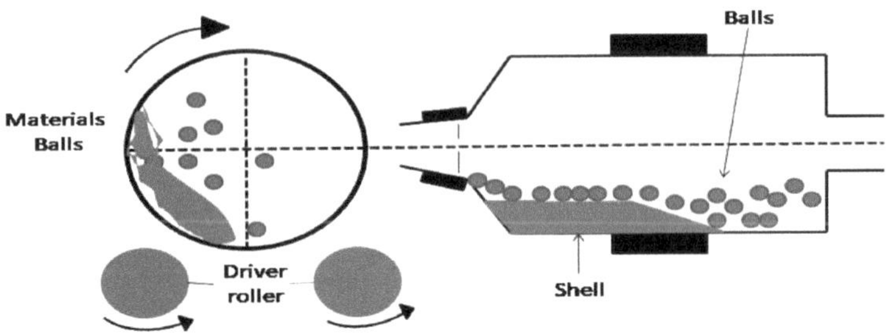

Figure 2.1 Ball mill process for grinding.

The reaction occurs in the presence of air. Precursor powders are mixed thoroughly, initially in a mortar and pestle, and then further ground into fine particles using a ball mill to achieve homogeneity. The process relies on the diffusion of cations from one precursor into another. The next step involves placing the precursor mixture or pellets (prepared with a hydraulic press machine at the appropriate pressure) into a crucible, which is then loaded into a high-temperature furnace. The crucible is heated to the desired synthesis temperature, typically ranging from 900°C to 1600°C, and maintained at that temperature for several hours, allowing solid-state reactions to take place.

During the heating process, the precursor initially reacts among the superficial surface of others to create a surface coating of piezoelectric on discrete precursor grains. As the cations continue to diffuse into other precursor materials, intermediate compounds are formed. With prolonged sintering, these intermediate phases react with the remaining precursor in the core regions of the precursor grains, resulting in the formation of piezoelectric material. After the synthesis period, the furnace and sample are gradually cooled to room temperature to prevent thermal shock and material cracks.

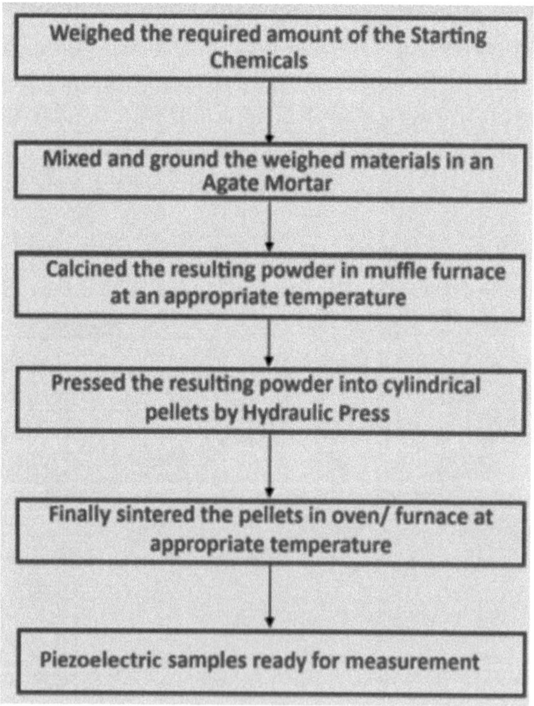

Figure 2.2 The detailed steps for the synthesis of the studied materials by solid-state reaction method.

Calcination at high temperatures produces clustered powder with widespread particle size, necessitating further grinding. However, it's important to note that the milling process can introduce contamination and unwanted defects into the final products. Moreover, achieving the desired homogeneity, stoichiometry, and phases becomes more challenging as the number of components in the ceramic powders increases. Higher temperatures for calcination can accelerate the initial stages of the reaction, but often lead to refining and other agglomerated powders.

2.3 Sol-Gel Synthesis

The sol-gel method has emerged as a widely utilized approach for the synthesis of various compositions of piezoelectric ceramics (e.g., $SrCo_{1-x}Fe_xO_3$, $Sm_{1-x}Ca_xFeO_3$, $LaGaO_3$, $SrLaGaO_3$, $SrFeO_3$, $GdAlO_3$, $CaCu_3Ti_4O_{12}$, etc.), driven by its ability to provide precise control over composition, morphology, and structure [7–12]. There are several key steps (as shown in Fig. 2.3). First, suitable metal precursors, often in the form of metal acetates, nitrates, or halides, are weighed and mixed in the desired stoichiometric ratio. These precursors are then dissolved in a chosen solvent, such as ethanol or methanol, to form a homogeneous solution. To control particle growth and stability, a stabilizer or ligand may be added. The solution is stirred thoroughly to ensure uniformity. Once the precursors, solvents, and organic components are in place, the synthesis process proceeds with rigorous mixing. This step is indispensable to ensure the homogeneity of the solution, preventing the formation of undesirable defects or variations in composition. Achieving a well-mixed solution sets the stage for subsequent phases in the sol-gel process. The sol-gel method is characterized by gelation, which plays a crucial role in the transformation of the precursor solution into a solid piezoelectric material. Gelation is initiated by the formation of metal-oxygen-metal bonds, giving rise to a three-dimensional network within the solution. The next critical step in the sol-gel method is solvent removal through drying. This process results in the formation of a porous metal oxide matrix. Achieving the desired porosity is essential, as it can significantly impact the piezoelectric material's properties, including its surface area, crystallinity, and capacity for ion transport. The final phase in the sol-gel synthesis entails calcination, wherein the dried gel is subjected to elevated temperatures, typically within the range of 300–600°C. The temperature and duration of calcination are carefully controlled to achieve the optimal crystal phase and morphology.

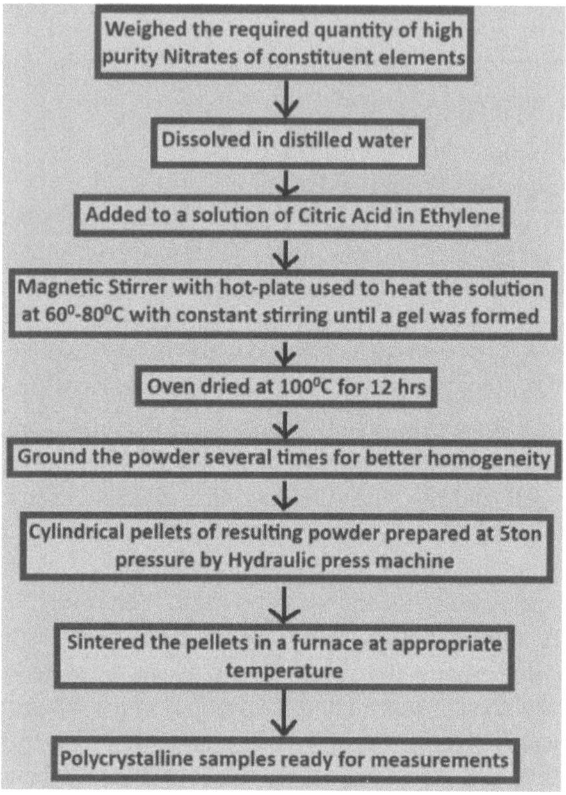

Figure 2.3 The detailed steps for the synthesis of the studied materials by sol-gel method.

2.4 Vapor Deposition (CVD/MOCVD)

The chemical vapor deposition (CVD) and metal-organic chemical vapor deposition (MOCVD) methods have emerged as powerful techniques for synthesizing piezoelectric materials with precise control over film properties, such as $A_{1-x}B_xMnO_3$ (where A = La, Nd, Pr; B = Li, Na, Ca, Sr, Ba, K, Rb), $LaAlO_3$, $SrTiO_3$, $NdGaO_3$, $ZrO_2(Y_2O_3)$, orthorhombic $CuPbSbS_3$, and $P(Zr_{1-x}Ti_x)O_3$, etc. [13–18]. This process involves several critical steps as featured in Fig. 2.4, starting with precursor selection and preparation. Cutting-edge metal acetates, nitrates, or halides are chosen for their compatibility with vapor deposition techniques. These precursors are accurately weighed and mixed along with organic compounds to tailor the piezoelectric composition. Once prepared, these precursors are introduced into specialized delivery systems within CVD/MOCVD reactors. The controlled flow rates and concentrations of precursor gases are crucial for achieving uniform deposition and the desired stoichiometry in the perovskite films. Careful regulation of gas flows is essential to avoid precursor depletion or excess. Researchers employ substrates like single-crystal silicon, glass, or flexible polymeric materials depending on the application. Surface treatments, such as functionalization and plasma cleaning, are applied to promote adhesion and enhance film quality. Temperature control during deposition is meticulously regulated to ensure the growth of high-quality piezoelectric films. Precursor decomposition involves breaking chemical bonds within the precursor molecules, releasing metal and halide ions that react with organic components, leading to piezoelectric crystallization on the substrate surface. In situ monitoring and control systems play a critical role in ensuring deposition quality. Post-deposition treatments, including annealing, are integral to the formation of high-quality perovskite films. Innovative annealing strategies, such as rapid thermal annealing (RTA) and sequential deposition-annealing processes, enhance the crystallinity of formed thin film and reduce defect densities. These techniques contribute to the development of piezoelectric films with superior charge transport properties.

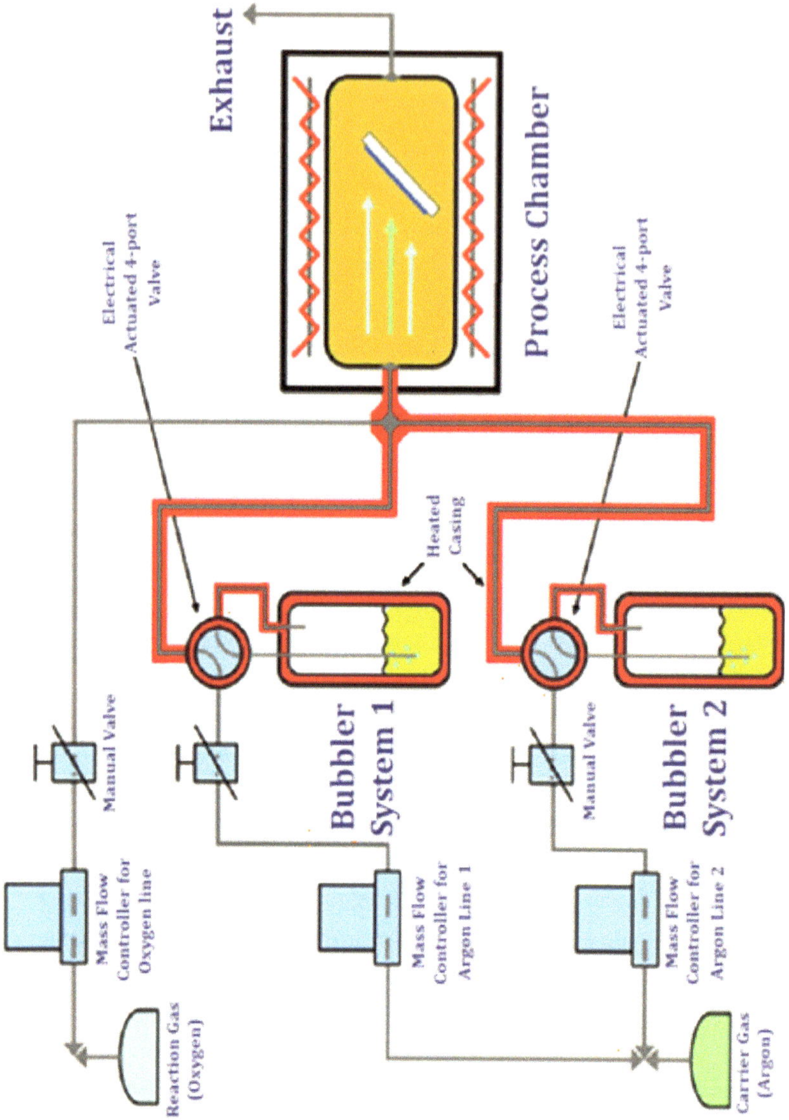

Figure 2.4 Schematic working principle of an atmospheric pressure MOCVD system.

2.5 Hydrothermal Method

Hydrothermal synthesis has emerged as a powerful technique for the controlled preparation of piezoelectric ceramic, e.g., (Pb,Ba)-(Sr,Zr) O_3, $PbZr_xTi_{1-x}O_3$, $BaZr_xTi_{1-x}O_3$, $Pb_xBa_{1-x}TiO_3$, $CaTiO_3$, $SrTiO_3$, and $(Pb_{0.92},La_{0.08})$-$(Zr_{0.65},Ti_{0.35})O_3$, a class of compounds with diverse applications, including catalysis, photovoltaics, and sensors [19–24]. This method leverages high-temperature and high-pressure aqueous environments to facilitate the formation of crystalline perovskite structures. Researchers often utilize metal salts or metal oxides as starting materials, depending on the desired piezoelectric composition. For instance, barium chloride ($BaCl_2$) and titanium dioxide (TiO_2) can serve as precursors for the synthesis of barium titanate ($BaTiO_3$) piezoelectric. These precursors are weighed and mixed in stoichiometric ratios to achieve the desired composition. The next step involves dissolving the prepared precursors in a suitable solvent, typically deionized water or other aqueous solutions. This results in a homogeneous solution containing metal cations and oxygen species. To promote piezoelectric formation, alkaline hydroxide or other pH-adjusting agents may be added. The solution is thoroughly stirred to ensure even distribution of the precursors. The hydrothermal reactor is a critical component of the synthesis setup. It consists of a high-pressure vessel capable of withstanding the elevated temperatures and pressures required for hydrothermal synthesis. The reaction vessel is sealed to prevent the escape of water vapor, creating a closed-loop system that maintains a controlled environment. The reactor is heated to the desired temperature, typically ranging from 100°C to 300°C, depending on the specific piezoelectric compound being synthesized. Higher temperatures can accelerate reaction kinetics, while lower temperatures may be chosen to optimize crystallinity. Pressure is controlled by maintaining a sealed environment within the reactor. The elevated pressure, typically ranging from a few to several hundred atmospheres, facilitates the dissolution of reactants and the formation of piezoelectric crystals. At the desired temperature and pressure, the hydrothermal reaction proceeds. Under these conditions, metal cations from the dissolved precursors react with oxygen species in the solution to form piezoelectric crystals. After the hydrothermal reaction is complete, the reactor is cooled, and the piezoelectric crystals are collected from the reaction solution. The collected crystals may require further processing steps, such as washing, drying, and calcination, to remove residual solvent and enhance crystallinity. Calcination involves heating the crystals at moderate temperatures to remove organic residues and promote crystal growth and phase purity. The flow chart is shown in Fig. 2.5.

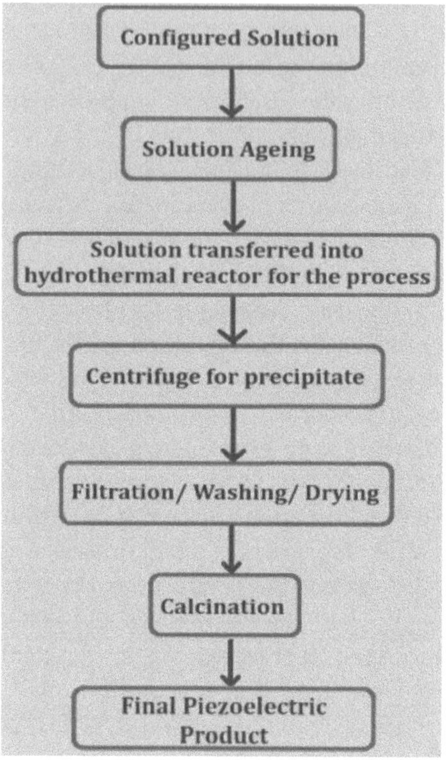

Figure 2.5 Flow chart of hydrothermal synthesis for piezoelectric ceramics.

2.6 Pulsed Laser Deposition

Pulsed Laser Deposition (PLD) stands out for its ability to deposit high-quality thin films of complex oxides with precise control over film properties. This technique relies on the interaction between an intense laser pulse and a target material to produce a thin film on a substrate. The PLD process offers exceptional control over film composition, thickness, and crystalline structure of different compositions of $Ba_{1-x}Sr_xTiO_3$, $Na_{0.5}Bi_{0.5}$-TiO_3, Pb(Zr,Ti) O_3, $Ba(Sb_xSn_{1-x})O_3$, $BaTiO_3$, $SrTiO_3$, and Cu/Pt/Ag/Au deposited on ceramic substrates SrTiO3/$LaAlO_3$/MgO, etc., making it a valuable tool in materials science [25–30].

The first critical step in PLD is the selection of appropriate precursor materials and the preparation of a target. The target is typically composed of the desired piezoelectric material in bulk form. High-purity metal oxides or ceramic powders are carefully mixed in stoichiometric ratios to achieve the desired composition. The mixed powder is then pressed into a dense pellet, which is suitable for the PLD process. In PLD, a high-energy laser pulse is positioned onto the surface of the target material. The laser pulse is typically generated by a pulsed laser source, as a solid-state laser. When the laser pulse strikes the target, it vaporizes a small quantity of material from the target's surface. This vaporized material is transformed into a plasma cloud consisting of ions, electrons, and neutral atoms. The plasma cloud expands rapidly and moves toward the substrate. The substrate, which is carefully positioned at a controlled distance from the target, serves as the deposition surface for the piezoelectric film. The plasma cloud condenses and deposits onto the substrate, forming a thin film with a piezoelectric structure. The kinetic energy of the vaporized species allows for the growth of highly crystalline and dense films. The schematic diagram is presented in Fig. 2.6.

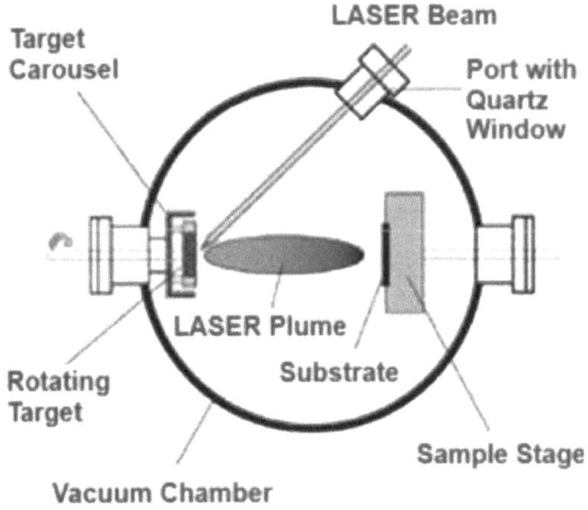

Figure 2.6 Schematic diagram of the PLD system.

PLD offers precise control over several critical parameters to tailor the properties of the deposited piezoelectric film. These parameters include laser energy, pulse duration, deposition rate, substrate temperature, and oxygen partial pressure. By adjusting these parameters, researchers can fine-tune the composition, thickness, crystallinity, and morphology of the piezoelectric film.

2.7 Solvothermal Synthesis

Solvothermal synthesis is a technique used to prepare piezoelectric materials, a class of compounds (R_2NiMnO_6, where R = La, Nd, Sm, etc.; K/Na-NbO_3, ternary piezoelectric $NiTiO_3$, ABO_3, where A = Ba, Sr, Ca, Cs, Pb, etc.; B = Zr, Ti, etc.) with a wide range of applications, including catalysts, photovoltaics, and sensors [31–36]. This method relies on high-temperature and high-pressure solvent environments to facilitate the formation of crystalline perovskite structures as shown in Fig. 2.7. Solvothermal synthesis allows for the precise tuning of piezoelectric properties, making it a valuable tool in materials science. The process of solvothermal synthesis begins with the careful selection and preparation of precursor materials. Researchers often use metal salts or metal halides as starting materials, depending on the desired piezoelectric composition. These precursors are weighed and mixed in stoichiometric ratios to achieve the desired composition. The precursor materials are dissolved in a suitable solvent, as it affects the solubility of the precursors and the growth of piezoelectric crystals. The precursor solution is stirred to ensure an even distribution of the precursors and create a homogeneous mixture.

Solvothermal synthesis relies on precise control of temperature and pressure. A high-pressure reactor is used to create a closed system where the reaction occurs. The reactor is sealed to prevent the escape of solvents and gases. The temperature is elevated to the desired level, typically in the range of 150°C to 300°C, depending on the specific piezoelectric compound being synthesized. Under these conditions, metal cations from the dissolved precursors react with anions from the solvent to form piezoelectric crystals. The controlled environment within the reactor ensures that the reaction occurs homogeneously and results in well-defined crystal structures. The reaction time can be adjusted to control crystal size and morphology.

Longer reaction times often lead to larger crystals with well-defined facets, while shorter times may produce smaller, more irregularly shaped particles. After the solvothermal reaction is complete, the reactor is cooled, and the piezoelectric crystals are collected from the reaction solution. The collected crystals may require further processing steps, such as washing, drying, and calcination, to remove residual solvent and enhance crystallinity. Calcination involves heating the crystals at moderate temperatures to remove organic residues and promote crystal growth and phase purity.

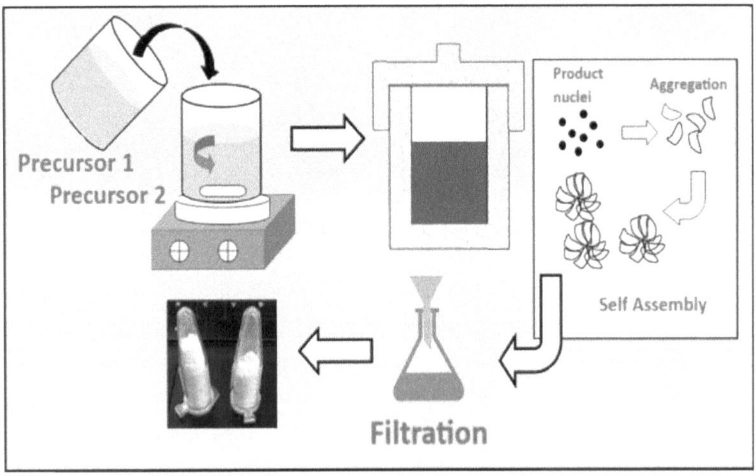

Figure 2.7 Schematic diagram of the solvothermal synthesis.

2.8 Spray Pyrolysis Method

Spray pyrolysis is a chemical deposition technique that involves the formation of a thin film of $LaMnO_{3+\delta}$, $BaTiO_3$, $SrTiO_3$, $La_{0.2}Sr_{0.7}TiO_3$, Ni-Yr-Stabilized-Zr, $SrZrO_3$, $La_xZr_xO_7$, La/Sr-MnO_3, $La_2Ti_2O_7$, etc., through the controlled spraying of precursor solutions onto a heated substrate [37–42]. The process begins with the dissolution of precursor compounds, typically metal salts or metal-organic compounds, in a suitable solvent. The resulting solution is then atomized into fine droplets or an aerosol mist. These droplets are directed onto a heated substrate, where the solvent rapidly evaporates, leaving behind the precursor materials. Subsequent thermal decomposition of the precursors leads to the formation of crystalline piezoelectric films as shown in Fig. 2.8.

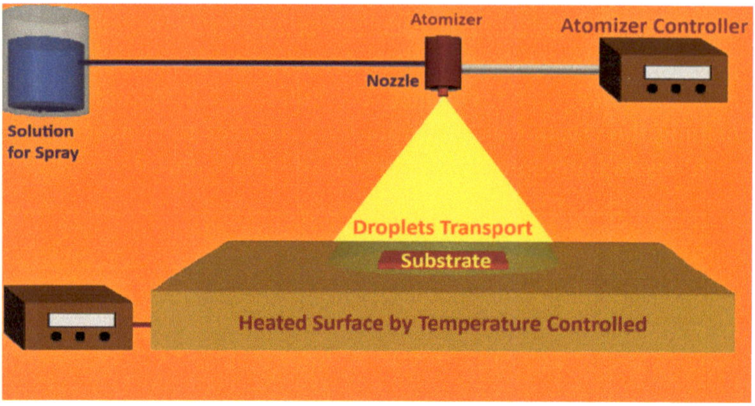

Figure 2.8 Schematic diagram of the spray pyrolysis synthesis.

The primary advantage of the spray pyrolysis method is the ability to precisely control deposition parameters. These parameters include the substrate temperature, spray rate, precursor concentration, and the choice of solvent. Adjusting these variables allows researchers to tailor the properties of the deposited piezoelectric films. For example, controlling the substrate temperature during deposition can influence the crystallinity and grain size of the piezoelectric material.

2.9 In Situ Growth Methods

In situ growth methods are advanced techniques for the synthesis of materials, including piezoelectric compounds, directly within a device or substrate [43–48]. Researchers typically choose metal salts or metal-organic compounds as starting materials, depending on the desired piezoelectric composition as ABX_3, A = organic group or inorganic ion, B = Pb, and X = Cl, Br, or I; $CsPbBr_3$, $Ce_{1\text{-}x}Sm_xO_{1.9}$, $(Pr_{1\text{-}x}Sr_x)_3\text{-}(Fe_yNi_{1\text{-}y})_2O_7$, $Sr_2Fe_{1.3}Ni_{0.2}Mo_{0.5}O_{6\text{-}\delta}$, etc. First, the solution containing metal nitrates and hydroxides, ethylene glycol, citric acid, and carbon was mixed to achieve homogeneity at room temperature.

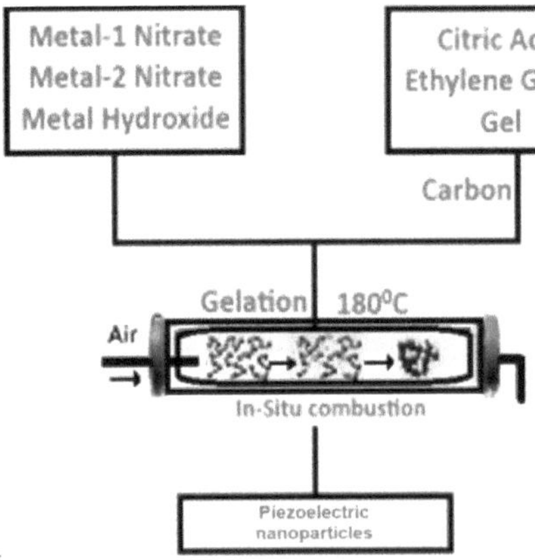

Figure 2.9 Schematic diagram of the in situ growth method.

The solution was then evaporated to form a gel network, which anchored the precursor metal ions. The as-prepared gel was then transferred to a tube furnace at 180°C until the reactants were self-ignited (using intermediate nitrate and oxygen as the oxidizers; and ethylene glycol, citric acid, and carbon as the fuels). The combustion of the mixture is enhanced in the presence of carbon, which allows the decomposition of intermediate nitrate at around 180°C. By using the heat released from self-combustion, the desired piezoelectric nanoparticles could quickly nucleate and grow in situ into the unburned carbon, creating a conductive network. These methods are particularly valuable when precise control over material properties and integration into complex structures or devices is required. A schematic diagram is presented in Fig. 2.9.

References

1. Pan L., Zhu G. (eds.), *Perovskite Materials – Synthesis, Characterisation, Properties, and Applications*. InTech, Croatia, 2016.
2. Saki Z., Byranvand M.M., Taghavinia N., Kedia M., Saliba M., Solution-processed perovskite thin-films: The journey from lab- to large-scale solar cells, *Energy & Environmental Science*, **14**, 5690, 2021.
3. Lyu F., Chen Z., Shi R., Yu J., Lin B.L., Solid phase synthesis of metal-free perovskite crystalline materials. *Journal of Solid-State Chemistry*, **304**, 122548, 2021.

4. Liu R., Xu D., Li S., Zhe L., Yanfeng X., Deyong W., Wenhui S., Solid-state synthesis and properties of $SmCoO_3$, *Frontiers of Chemistry in China*, **1**, 398, 2006.
5. Li F., Yu X., Chen L., Pan H., Xin X., Solid-state synthesis of $LaCoO_3$ perovskite nanocrystals, *Journal of American Ceramic Society*, **85**, 2177, 2004.
6. Zhang H., Zhang J., Ding H., Li Y., Sun S., Ao W., Liang Y., Solid-phase synthesis of perovskite using spent SCR catalyst and calcium carbonate and its application as ceramic opacifier, *Journal of Industrial and Engineering Chemistry*, **114**, 499, 2022.
7. Ahmad A.L., Idrus N.F., Othman M.R., Preparation of perovskite alumina ceramic membrane using sol–gel method, *Journal of Membrane Science*, **262**, 129, 2005.
8. Felix A.M.H., Hernandez T., Parra S.D.L., Ibarra J., Kharisov B., Sol–gel based Pechini method synthesis and characterization of $Sm_{1-x}Ca_xFeO_3$ perovskite $0.1 \le x \le 0.5$, *Powder Technology*, **229**, 290, 2012.
9. Huang K., Feng M., Goodenough J.B., Sol-gel synthesis of a new oxide-ion conductor Sr- and Mg-doped $LaGaO_3$ perovskite, *Journal of American Ceramic Society*, **79**, 1100, 1996.
10. Cizauskaite S., Reichlova V., Nenartaviciene G., Beganskiene A., Pinkas J., Kareiva J., Sol-gel preparation and characterization of perovskite gadolinium aluminates, *Materials Science-Poland*, **25**, 755, 2007.
11. Jaiswal S.K., Kashyap V.K., Kumar J., On the sol–gel synthesis and characterization of strontium ferrite ceramic material, *Materials Research Bulletin*, **47**, 692, 2012.
12. Jin S., Xia H., Zhang Y., Guo J., Xu J., Synthesis of $CaCu_3Ti_4O_{12}$ ceramic via a sol-gel method, *Materials Letters*, **61**, 1404, 2007.
13. Alharbi Y.T., Alam F., Parvez K., Missous M., Lewis D.J., Molecular precursor route to bournonite ($CuPbSbS_3$) thin films and powders, *Inorganic Chemistry*, **60**, 13691, 2021.
14. Mathur S., Thin film deposition by sol-gel and CVD processing of metal-organic precursors. In: Pauleau, Y. (ed.) *Chemical Physics of Thin Film Deposition Processes for Micro and Nanotechnologies*. NATO Science Series by Springer, Dordrecht (NAII Vol. 55), 2002.
15. Bibes M., Gorbenko O., Martínez B., Kaul A., Fontcuberta J., Alkaline-doped manganese perovskite thin films grown by MOCVD, *Journal of Magnetism and Magnetic Materials*, **211**, 47, 2000.
16. Okada M., Takai S., Amemiya M., Tominaga K., Preparation of c-axis-oriented $PbTiO_3$ thin films by MOCVD under reduced pressure, *Japanese Journal of Applied Physics*, **28**, 1030, 1989.

17. Jero P.D., Rebillat F., Kent D.J., Jones J.G., Crystallization of lanthanum hexaluminate from MOCVD precursors, *22nd Annual Conference on Composites, Advanced Ceramics, Materials, and Structures: A: Ceramic Engineering and Science Proceedings*, **19**, 1988.
18. Sakashita Y., Ono T., Segawa H., Tominaga K., Okada M., Preparation and electrical properties of MOCVD-deposited PZT thin films, *Journal of Applied Physics*, **69**, 8352, 1991.
19. Lencka M.M., Riman R.E., Hydrothermal synthesis of perovskite materials: Thermodynamic modeling and experimental verification, *Ferroelectrics*, **151**, 159, 2011.
20. Yang G., Park S.J., Conventional and microwave hydrothermal synthesis and application of functional materials: A Review, *Materials*, **12**, 1177, 2019.
21. Moreira M.L., Paris E.C., Nascimento G.S.D., Longo V.M., Sambrano J.R., Mastelaro V.R., Bernardi M.I.B., Andres J., Varela J.A., Elson L., Structural and optical properties of $CaTiO_3$ perovskite-based materials obtained by microwave-assisted hydrothermal synthesis: An experimental and theoretical insight, *Acta Materialia*, **57**, 5174, 2009.
22. Suchanek W.L., Lencka M.M., Riman R.E., Hydrothermal synthesis of ceramic materials, *Aqueous Systems at Elevated Temperatures and Pressures, Physical Chemistry in Water, Steam and Hydrothermal Solutions*, Elsevier Inc., Piscataway, USA, **717**, 2004.
23. Wendelbo R., Akporiaye D.E., Karlsson A., Plassen M., Olafsen A., Combinatorial hydrothermal synthesis and characterisation of perovskites, *Journal of the European Ceramic Society*, **26**, 849, 2006.
24. Moon J., Kerchner J.A., Krarup H., Adair J.A., Hydrothermal synthesis of ferroelectric perovskites from chemically modified titanium isopropoxide and acetate salts, *Journal of Materials Research*, **14**, 425, 1999.
25. Goux L., Gervais M., Gervais F., Champeaux C., Catherinot A., Pulsed laser deposition of ferroelectric BST thin films on perovskite substrates: An infrared characterization, *International Journal of Inorganic Materials*, **3**, 839, 2001.
26. Cojocaru C.V., Nechache R., Harnagea C., Pignolet A., Rosei F., Nanoscale patterning of functional perovskite-type complex oxides by pulsed laser deposition through a nanostencil, *Applied Surface Science*, **256**, 4777, 2010.
27. Liu Q., Dai J., Liu Z., Zhang X., Zhu G., Ding G., Electrical and optical properties of Sb-doped $BaSnO_3$ epitaxial films grown by pulsed laser deposition, *Journal of Physics D: Applied Physics*, **43**, 455401, 2010.

28. Dragoi C., Cernea M., Trupina L., Lead-free ferroelectric $BaTiO_3$ doped-$(Na_{0.5}Bi_{0.5})TiO_3$ thin films processed by pulsed laser deposition technique, *Applied Surface Science*, **257**, 9600, 2011.

29. Gatea H.A., Khalil S.M., Evaluating the impact of substrate deposition on optical properties of perovskite barium strontium titanate $(Ba_{0.5}Sr_{0.5}TiO_3)$ thin films prepared by pulsed laser deposition technique, *The European Physical Journal D*, **76**, 148, 2022.

30. Francis A.J., Salvador P.A., Crystal orientation and surface morphology of face-centered-cubic metal thin films deposited upon single-crystal ceramic substrates using pulsed laser deposition, *Journal of Materials Research*, **22**, 89, 2007.

31. Modeshia D.R., Walton R.I., Solvothermal synthesis of perovskites and pyrochlores: Crystallisation of functional oxides under mild conditions, *Chemical Society Reviews*, **39**, 4303, 2010.

32. Walton R.I., Perovskite oxides prepared by hydrothermal and solvothermal synthesis: A review of crystallisation, chemistry, and compositions, *Chemistry-A European Journal*, **26**, 9041, 2020.

33. Lu C.H., Naresh N., Kumar P.S., Som S., Microwave-assisted solvothermal synthesis and electrochemical characterizations of ternary perovskite $NiTiO_3$ anode materials for lithium-ion batteries, *Ceramics International*, **45**, 19517, 2019.

34. Caruntu, D., Rostamzadeh T., Costanzo, T., Saman S.P., Gabriel C., Solvothermal synthesis and controlled self-assembly of monodisperse titanium-based perovskite colloidal nanocrystals, *Nanoscale*, **7**, 12955, 2015.

35. Singh J., Kumar A., Solvothermal synthesis dependent structural, morphological and electrochemical behaviour of mesoporous nanorods of Sm_2NiMnO_6, *Ceramics International*, **46**, 11041, 2020.

36. Li L., Miao L., Zhang Z., Pu X., Feng Q., Yanagisawa K., Fan Y., Fan M., Wena P., Hu D., Recent progress in piezoelectric thin film fabrication via the solvothermal process, *Journal of Materials Chemistry A*, **7**, 16046, 2019.

37. Fukui T., Oobuchi T., Ikuhara Y., Ohara S., Kodera K., Synthesis of (La,Sr) MnO_3–YSZ composite particles by spray pyrolysis, *Journal of American Ceramic Society*, **80**, 261, 1997.

38. Burnat D., Heel A., Holzer L., Kata D., Lis J., Graule T., Synthesis and performance of A-site deficient lanthanum-doped strontium titanate by nanoparticle based spray pyrolysis, *Journal of Power Sources*, **201**, 26, 2012.

39. Patil L.A., Suryawanshi D.N., Pathan I.G., Patil D.G., Effect of firing temperature on gas sensing properties of nanocrystalline perovskite $BaTiO_3$ thin films prepared by spray pyrolysis techniques, *Sensors and Actuators B: Chemical*, **195**, 643, 2014.

40. Beyribey B., Kim H., Persky J., Electrochemical characterization of $BaCe_{0.7}Zr_{0.1}Y_{0.16}Zn_{0.04}O_{3-\delta}$ electrolyte synthesized by combustion spray pyrolysis, *Ceramics International*, **47**, 1976, 2021.

41. Kuai L., Kan E., Cao W., Huttula M., Ollikkala S., Ahopelto T., Honkanen A.-P., Huotari S., Wang W., Geng B., Mesoporous $LaMnO_{3+\delta}$ perovskite from spray–pyrolysis with superior performance for oxygen reduction reaction and Zn–air battery, *Nano Energy*, **43**, 81, 2018.

42. Abe Y., Laine R.M., Photocatalytic plate-like $La_2Ti_2O_7$ nanoparticles synthesized via liquid-feed flame spray pyrolysis (LF-FSP) of metalo-organic precursors, *Journal of American Ceramic Society*, **103**, 4832, 2020.

43. Chen H., Zhou L., Fang Z., Wang S., Yang T., Zhu L., Hou X., Wang H., Wang Z.L., Piezoelectric nanogenerator based on in situ growth all-inorganic $CsPbBr_3$ perovskite nanocrystals in PVDF fibers with long-term stability, *Advanced Functional Materials*, **31**, 2011073, 2021.

44. Wang Y., Lei X., Zhang Y., Chen F., Liu T., In-situ growth of metallic nanoparticles on perovskite parent as a hydrogen electrode for solid oxide cells, *Journal of Power Sources*, **405**, 114, 2018.

45. Niu B., Lu C., Yi W., Luo S., Li X., Zhong X., Zhao X., Xu B., In-situ growth of nanoparticles-decorated double perovskite electrode materials for symmetrical solid oxide cells, *Applied Catalysis B: Environmental*, **270**, 118842, 2020.

46. Wang J., Zhou J., Yang J., Neagu D., Fu L., Lian Z., Shin T.H., Wu K., Tailoring the surface of perovskite through in situ growth of Ru/RuO_2 nanoparticles as robust symmetrical electrodes for reversible solid oxide cells, *Advanced Materials Interfaces*, **7**, 2000828, 2020.

47. Zhou J., Shin T.H., Ni C., Chen G., Wu K., Cheng Y., Irvine J.T.S., In situ growth of nanoparticles in layered perovskite $La_{0.8}Sr_{1.2}Fe_{0.9}Co_{0.1}O_{4-\delta}$ as an active and stable electrode for symmetrical solid oxide fuel cells, *Chemistry of Materials*, **28**, 2981, 2016.

48. Neagu D., Tsekouras G., Miller D.N., Menard H., Irvine J.T.S., In situ growth of nanoparticles through control of non-stoichiometry, *Nature Chemistry*, **5**, 916, 2013.

Chapter 3

PZT-Based Piezoelectric Materials: From Synthesis to Functional Properties and Diverse Applications

Sonu Kumar,[a] D. N. Singh,[b] and Dev Kumar Mahato[a]
[a]*Department of Physics, National Institute of Technology, Patna 800 005, India*
[b]*Department of Materials Engineering, Indian Institute of Science, Bangalore 560012, India*
drdevkumar@yahoo.com

This chapter extensively reviews lead-based piezoelectric materials, emphasizing their distinct benefits and vital role in commercial applications. Because of their superior physical qualities, lead-based materials such as PZT and PLZT have become promising materials for the industry's requirements. They have higher piezoelectric constants, which results in improved sensitivity and unique mechanical strength, so they are preferable for sensors and actuators. Furthermore, the availability of sufficient lead resources and efficient manufacturing methods helps with their cost-

Piezoelectric Materials: Design and Applications
Edited by N. B. Singh and Dev Kumar Mahato

ISBN 978-981-5129-40-3 (Hardcover), 978-1-003-59897-8 (eBook)
www.jennystanford.com

effectiveness. Based on extensive research and proven commercially usable, this review shows that lead-based materials continue their exceptional performance and applicability, strengthening their position as the prioritized choice for piezoelectric applications in different fields of science and technology. Also, this chapter will give a gesture synthesis and characterization techniques used for piezoceramic analysis. The emphasis is given to show the structural, dielectric, and ferroelectric response of $(1\text{-}x)PbZr_{0.6}Ti_{0.4}O_3\text{-}xBaTiO_3$ [x = 0.05] (0.05PZT-BT) composite. The coexistence of tetragonal and rhombohedral phases is observed in the studied compound. The high value of dielectric constant and loss is attributed to the contribution of the space charge polarization and of dc conductivity, respectively. The non-Debye kind of relaxation is confirmed in the imaginary part of the electric modulus in the sample. The conductivity data displays metal-semiconductor-like behavior.

3.1 Introduction

Over the past five decades, piezoelectric materials have played a significant role in various commercial applications, such as sensors, actuators, and energy-harvesting devices [1–3]. Their intrinsic characteristic is the ability to convert mechanical energy into electrical signals and vice versa, making them valuable for detecting and measuring physical quantities such as pressure, strain, and acceleration [4]. Lead zirconate titanate (PZT) is widely recognized as a highly advantageous option among different classes of piezoelectric materials due to its exceptional properties and cost-effectiveness, making it the preferred choice for numerous applications across multiple industries [5–8].

Both PZT and La-doped PZT have demonstrated outstanding piezoelectric performance, with higher piezoelectric constants [9], such as d_{33} and d_{31}, which enhance sensitivity and overall device performance. This makes them valuable assets in applications where high sensitivity is crucial, such as ultrasound imaging and vibration sensors [10, 11]. Furthermore, lead-based materials have garnered significant interest in ultrasonic imaging due to their remarkable sensitivity, enabling the reliable detection and measurement of high-frequency sound waves. This characteristic has proven instrumental in improving the overall quality and precision of ultrasound imaging techniques. Its potential impact extends to the field of medical diagnostics, revolutionizing non-invasive imaging of various organs and tissues [12].

It is also worth noting that lead-based materials exhibit exceptional thermal stability, ensuring consistent and dependable performance even under demanding circumstances characterized by fluctuating temperature conditions. The inherent robustness of these systems makes them highly suitable for deployment in a diverse range of industries, including aerospace, automotive, and industrial sectors [13].

The superiority of lead-based piezoelectric materials over their lead-free counterparts in terms of mechanical strength and durability has been well-documented in the literature [14–16]. Thanks to their robustness, they can withstand significant mechanical loads and shocks without compromising performance or structural integrity. This durability is of notable significance, especially in applications that require prolonged and unwavering dependability, as demonstrated in studies involving ultrasonic cleaners and precision positioning systems [15, 16].

Lead-based materials have emerged as a viable and economically advantageous solution due to the abundant availability of lead resources and well-established manufacturing processes [6]. Dedication to research and development efforts has led to significant advancements in optimizing production methodologies, reducing costs, and enhancing commercial possibilities for widespread industrial implementations [17–19]. The widespread use of lead-based materials has laid a solid foundation for their production, understanding, and integration into various commercial devices. The presence of established methods, expertise, and supply chains further enhances the commercial viability of specific endeavors.

The utilization of lead zirconate titanate-barium titanate (PZT-BT) composites has garnered significant attention and interest within the field of lead-based piezoelectric materials. The incorporation of BT into the composition of PZT results in a variety of advantageous characteristics that are employed to enhance its piezoelectric capabilities [20]. BT is well known for its multiple transition features, such as the coexistence of the system's polymorphic phase boundary (PPB). With this structural feature and exceptional electrical properties, it is assumed that BT, acting as a "dopant" within the PZT domain, will induce alterations in its crystal structure and significantly impact its piezoelectric features [21, 22]. Investigating the effect of BT domains on the structural, electrical, and other physical characteristics of PZT-based ceramics is of great importance. It is presumed that an appropriate dopant concentration will enhance the transition temperature and electrical properties of PZT ceramics.

This chapter delves into the structural and electrical properties of 0.05PZT-BT ceramics synthesized through the traditional solid-state synthesis route. The objective is to explore the ferroelectric and dielectric response of 0.05PZT-BT ceramics in comparison to PZT. A comprehensive analysis of the ferroelectric and dielectric properties of PZT is discussed elsewhere [2]. The inherent properties and characteristics of 0.05PZT-BT make it a highly versatile and promising material for a wide range of applications. Its exceptional attributes render it suitable for use in ultrasound imaging, vibration sensing, energy harvesting, and numerous other domains. The present discussion elucidates the ongoing research and development efforts to explore the dielectric, electric modulus, ferroelectric, and conductive behavior and expand the potential capabilities of 0.05PZT-BT materials in emerging technologies.

3.2 Fundamentals of Piezoelectricity and Ferroelectricity

The basic crystal structure of a material that exhibits piezoelectricity is non-centrosymmetric. Quartz is one of the most popular natural crystals exhibiting non-centrosymmetric structure used in many piezoelectric devices. When pressure is applied to such a crystal, an electric charge is developed on the surface of the crystal. It is referred to as the direct piezoelectric effect. The crystal gets deformed when subjected to an electric field. This effect is called the inverse piezoelectric field (Fig. 3.1). The phenomenon of piezoelectricity was commercially used for the first time in ultrasonic submarine detectors developed during World War I after several decades of its discovery, and later scientists synthesized barium titanate ($BaTiO_3$) ceramics as a piezoelectric ceramic material. $BaTiO_3$ (BT), like other piezoelectric ceramics, exhibits a spontaneous polarization as a result of shifting of the center of negative ions non-uniformly relative to that of positive ions in the crystallographic unit cell, giving rise to a non-centrosymmetric structure as shown in Fig. 3.2.

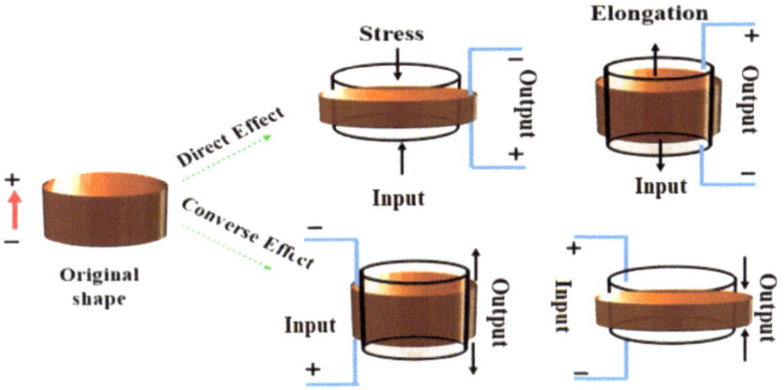

Figure 3.1 Direct and converse piezoelectric effect.

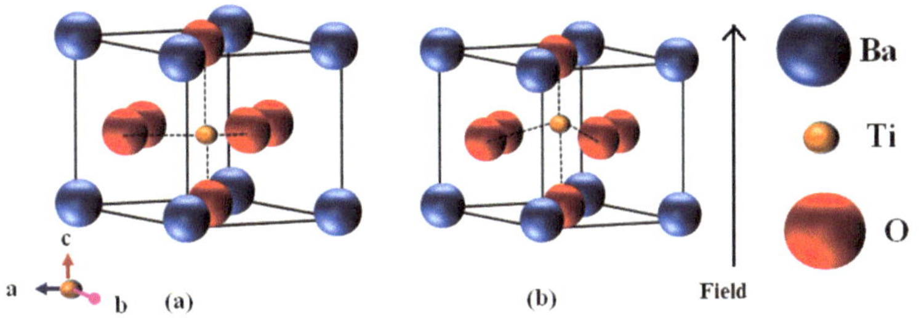

Figure 3.2 (a) Centrosymmetric and (b) non-centrosymmetric representation of the ABO_3 perovskite structure (here $BaTiO_3$ is taken as an example).

Barium titanate belongs to perovskite structure ABO_3 in which A cations (Ba) occupy the A-site, B-cations (Ti) center site, and oxygen anions (O_3) are located at the face centers of the unit cell. BT exhibits a centrosymmetric cubic phase at high temperatures, and when it is cooled down, it undergoes a series of phase transitions (tetragonal, rhombohedral, orthorhombic, or monoclinic). On cooling, where the first non-centrosymmetric structural transition takes place is consistent with Curie temperature. During cooling through the Curie temperature, a crystallographic orientation occurs in different regions (called the domain) of the material from higher symmetry to the lower symmetry of crystal structure. Different regions that separate all domains are called domain walls. Ferroelectricity is the phenomenon in which materials change their directions of spontaneous polarization under the action of the electric field. The electric field needed for the directional change to occur is referred to as the coercive field, which is associated with the motion of ferroelectric domain walls. When polycrystalline ceramics are cooled from the high processing temperatures, they exhibit no net spontaneous polarization (due to cancellation of local area spontaneous polarization) and piezoelectricity at the macroscopic length scale. A polycrystalline material exhibits piezoelectricity at the macroscopic length scale by poling (Fig. 3.3).

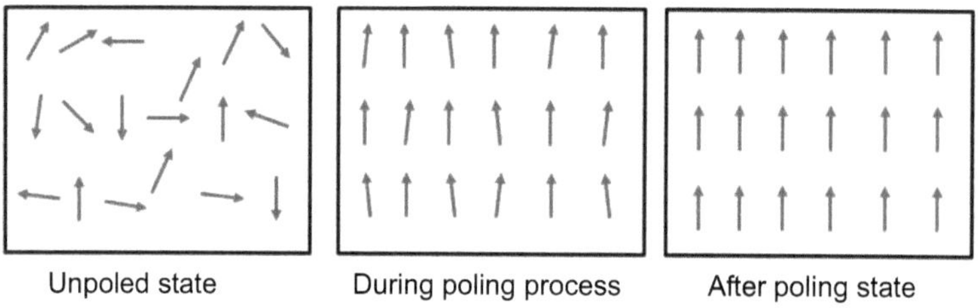

Figure 3.3 Poling process.

The direct piezoelectric effects are found in applications in sensors that can be written as:

$$\boldsymbol{D_i = d_{ij}\sigma_j} \tag{3.1}$$

Here $\boldsymbol{D_i}$ is the dielectric displacement and $\boldsymbol{\sigma_j}$ is the applied stress.

The converse piezoelectric effect is used in actuators. This effect can be expressed by the relation:

$$\boldsymbol{S_i = d_{ij}E_j} \tag{3.2}$$

Here $\boldsymbol{S_i}$ represents the electrostrain, $\boldsymbol{E_j}$ is the applied electric field, and $\boldsymbol{d_{ij}}$ is the piezoelectric coefficient.

In equations (3.1) and (3.2), the coordinate axes are assigned to the 3-direction. When both the applied electric field and measured strain are in the 3-direction, the piezoelectric coefficient of concern is the longitudinal piezoelectric coefficient, d_{33}. Equations (3.1) and (3.2) become

$$D_3 = d_{33}\sigma_3 \tag{3.3}$$

$$S_3 = d_{33}E_3 \tag{3.4}$$

The dielectric constant is one of the most important properties of a piezoelectric material that describes the materials' ability to store electrical potential energy per cycle of the alternate electric field. The dielectric constant is generally found in maximum near-phase transitions and exhibits a temperature-dependent broad peak in diffuse phase transition. Moreover, the electromechanical coupling factor (k_p) is described by the relation:

$$k_p = \frac{\text{Energy converted}}{\text{Input energy}} \tag{3.5}$$

It is the electrical energy output ratio to the total input mechanical energy or vice versa.

Further, PZT in its pure and modified form has been widely used piezoelectric material in sensors and actuators that have dominated the field of $BaTiO_3$. Both donor and acceptor doping PZT produce soft and hard ferroelectric materials for different applications. In this work, we synthesize PZT-BT composite to tune their properties for future application.

3.3 Methods for Synthesis of PZT-Based Piezoelectric Materials

Multiple techniques are currently accessible for synthesizing PZT-based piezoelectric materials. Several frequently employed methods include sol-gel, hydrothermal, chemical vapor deposition (CVD), and solid-state reaction routes. Here, we have adopted the mechanical ball milling method. Figure 3.4 illustrates a ball milling process, a cylindrical device utilized for grinding various materials such as ceramic raw materials and paints. The mechanical process involves using spherical objects, typically made with steel or ceramics, within a rotating cylindrical container to facilitate the mixture and crushing of various substances. During the ball milling process, the mill's rotational motion induces collisions between the grinding media and the solid precursor materials. The collision between particles induces significant mechanical forces, leading to the reduction in particle size and the homogenization of the constituents. The ongoing interaction and friction between the particles and the grinding media facilitate the creation of a uniform blend. The ball milling duration may vary depending on the desired particle size and the material system under consideration. Achieving the desired grinding and mixing stage requires several hours. Cooling systems are employed as a means of preventing the occurrence of excessive heat generation in the milling process. Following the ball milling process, the resultant powder is gathered and may undergo subsequent procedures, including calcination, pressing, and sintering. Including these supplementary procedures produces chemical reactions, increases the density of the substance, and improves its piezoelectric characteristics. Ball milling is frequently employed as a beginning process in conjunction with the solid-state reaction pathway to improve the reactivity and uniformity of the initial substances.

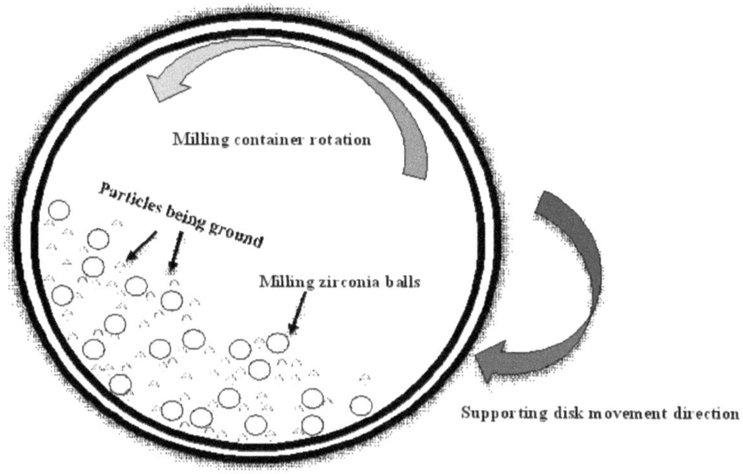

Figure 3.4 A schematic representation of the ball milling synthesis technique.

The solid-state reaction pathway is extensively employed in synthesizing piezoelectric materials containing lead. The process involves the combination and thermal treatment of solid precursor substances to initiate chemical transformations and produce the desired compounds. The initial step involves carefully selecting appropriate materials, commonly metal oxides or carbonates, with the requisite elements for the desired composition. The materials undergo careful weighing and blending to achieve a uniform dispersion of the components. The mixed powders are subjected to grinding to improve their reactivity and reduce their particle size. This step is commonly accomplished through a ball mill or a vibratory mill. Subsequently, the finely crushed mixture undergoes a thermal process known as calcination. The calcination method entails subjecting the powder to high temperatures to induce chemical reactions and facilitate the creation of the intended phases. Following the calcination process, the powder is milled to optimize particle size distribution and increase homogeneity. The milled powder is hydraulically or uniaxial pressed into circular discs. The compact's density and shape are achieved by carefully regulating pressing force. The flow chart for mechanical milling is given in Fig. 3.5.

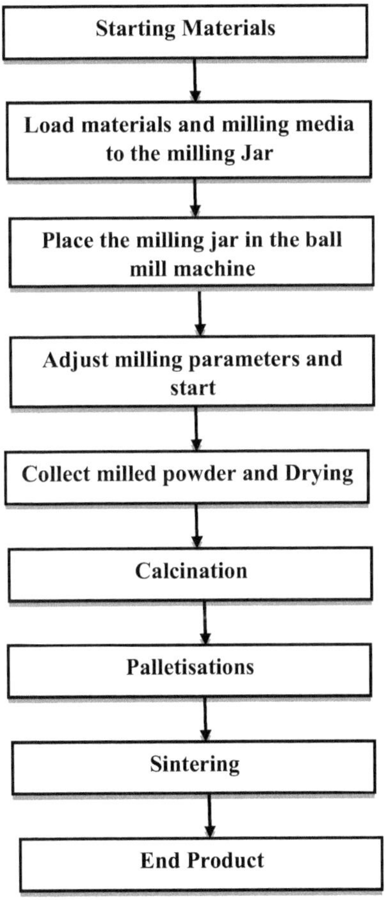

Figure 3.5 Flow chart of mechanical method.

After compression, the specimens are sintered at high temperatures below melting to bind particles and increase density. Sintering increases crystal growth and improves the mechanical and electrical characteristics of the material.

Synthesizing lead-containing piezoelectric materials becomes possible via the solid-state reaction route. After carefully choosing and combining starting components, grinding, calcination, milling, pressing, and sintering follow. The produced materials may be characterized by evaluating their structural and functional characteristics.

Solid-state reaction synthesizes lead-based piezoelectric materials reliably. Grinding, calcination, milling, pressing, and sintering follow properly selected and mixed starting materials. Characterizing materials allows for structural and functional analysis. It can be concluded that the solid-state reaction technique for lead-based piezoelectric materials synthesis is more reliable than others. It involves mixing of precursor materials, calcination, milling, pressing, and sintering to prepare the targeted ceramics for different measurements.

3.4 Characterization Techniques

3.4.1 X-Ray Diffraction

X-ray diffraction (XRD) is a non-destructive technique that may be carried out on powders, thin films, polycrystalline, and bulk materials. The method helps determine crystal structures of metal and alloys, ceramics, minerals, semiconductors, inorganic compounds, polymers, and organic materials. XRD is used to evaluate various physical characteristics, including chemical composition, crystallite size, orientation, lattice strain, state of ordering, and determine phase diagrams. It produces a diffraction pattern, which can be analyzed in many ways, utilizing Bragg's Law ($n\lambda = 2d\sin\theta$) to determine a crystal's phase. While changing the diffraction angle, the intensity and diffraction data are measured. The incident and diffracted beams create the two-times-theta diffraction angle. The diffractometer geometry and material being measured determine the incident and diffracted beam angles. Recently, the XRD method has been extended for extraction of three-dimensional microstructural properties as well as to carry out "in situ" diffraction studies as a function of electric fields, stress, pressure, temperature, etc. The available efficient software enables search and matching of diffraction data of various substances. In situ, combinations of XRD with EXAFS (extended X-ray absorption fine structure), both long-range order and short-range order, information can be collected to study the formation of heterogeneous catalysts. Thus, X-ray powder diffraction is the only method to study micro- and macro-stresses.

3.4.2 SEM

Scanning electron microscope (SEM) is used for micro-scale surface imaging by focusing an electron beam of high-energy electrons on the target surface. The interaction between the target surface and electron beam produces reflected, secondary, and auger electrons, which are utilized for collecting various information. The reflected electron beam is used for surface imaging for particle size, grain boundary, and grain structure information. While secondary, auger, and other kinds of electrons are used for elemental and surface chemical analysis such as backscattered imaging, chemical precipitation information, and generating energy dispersive spectrum.

3.4.3 Impedance Analyzer

Investigations of materials often make use of impedance analysis to get a better understanding of electrical characteristics. Impedance analyzers are used in the measurement of both resistance and reactance. Researchers are able to detect resistance, capacitance, and inductance by making use of a signal consisting of alternating current, voltage, and current.

The electrical devices, sensors, batteries, and electrochemical systems might all stand to benefit from this analytical procedure. They investigate into the factors that contribute to the dissipation, storage, and use of material energy. The frequency and temperature-dependent electrical response of the material or device is analyzed by this spectrum.

Mathematical models and computer software are used in the analysis of impedance data. The research considers a wide range of topics and types of material. Researchers get an improved understanding of the electrical behavior and functioning of the material or device due to this in-depth investigation. In the process of designing, optimizing, and testing electrical devices, material scientists may benefit from doing impedance analysis. This technique finds unwanted capacitance, inductance, impedance differences, and electrical qualities that shift in response to changes in temperature or humidity as well as the passage of time. Researchers and engineers may benefit from doing impedance analysis when it comes to developing electrical systems and devices for a wide range of uses.

3.4.4 d_{33} Meter

The d_{33} meter is a regularly used instrument in material science for measuring the piezoelectric response of different materials. The d_{33} coefficient determines the material's electric charge response to mechanical stress or strain. The d_{33} meter has a measuring cell, electrodes, signal generator, and data gathering system. Electrodes support the sample material in the measuring cell when it is properly prepared. Using a known d_{33} coefficient reference material measures the d_{33} meter. The signal generator applies voltage or force to the sample material, while the electrodes detect the electric charge. Measurements at different frequencies determine a material's piezoelectric response. The d_{33} parameter is calculated using specialized techniques and software. Scientific research, technological development, and piezoelectric material product quality depend on the d_{33} meter. This tool analyzes material properties, improving their performance in sensing mechanisms, actuation systems, and energy harvesting devices.

3.5 Properties of PZT-Based Materials

PZT-based materials possess various important properties useful for various practical uses.

3.5.1 Crystalline Structure

PZT-based ceramic materials' crystalline structure is closely related to the piezoelectric features that PZT ceramics have. PZT is a perovskite-structured ceramic material that generally consists of lead, zirconium, and titanium ions organized in a perovskite arrangement. The piezoelectric behavior of the material may be related to the particular arrangement of the atoms that make up this structure.

3.5.2 Dielectric Constant

For use in capacitors and other energy storage devices, PZT-based materials have a high dielectric constant. The dielectric constant measures the ability of a material to store electrical energy when it is subjected to an electric field. This is a useful parameter of the energy storage devices.

3.5.3 Ferroelectricity and High Curie Temperature

The non-existence of an external electric field does not prevent PZT from retaining its spontaneous electric polarization; this property is known as ferroelectricity. This quality is required for uses like ferroelectric capacitors and non-volatile memory devices. The Curie temperature, at which PZT materials no longer exhibit ferroelectric behavior, appears to be very high. Because of its high Curie temperature, PZT-based materials are well-suited for use in high-temperature environments.

3.5.4 Low Hysteresis and High Piezoelectric Coefficient

The low hysteresis of PZT materials is an advantage. The delayed voltage response to stress is called hysteresis. PZT-based materials are commonly used in control and measurement systems due to their low hysteresis. PZT is a popular material for use in piezoelectric devices because of their high piezoelectric coefficient when put under mechanical stress; PZT-based materials can generate high electric charges. Many different types of sensors, actuators, and transducers depend on this property.

3.5.5 Versatility, Biocompatibility, and Stability

PZT-based materials have the benefit of versatility. Due to the wide variety of PZT forms and compositions, researchers can easily relate their characteristics to an objective. It finds application in a variety of contexts due to its versatility. Ultrasonic transducers and imaging systems benefit from biocompatible PZT formulations. The sensors and actuators used in aerospace and automobiles benefit from PZT materials because of their durability.

3.6 Applications

PZT-based materials are widely used for their usefulness in medical imaging, non-destructive testing, and industrial cleaning utilizing ultrasonic transducers. Automotive pressure sensors and consumer electronics accelerometers use PZT-based sensors. Robotics, adaptive optics, and precision engineering use PZT actuators for positioning and control. Additionally, PZT materials transform mechanical vibrations into electrical energy. The ferroelectric memory devices they support improve memory technology. In gas ranges and grills, PZT ignitors are reliable. Underwater acoustic PZT transducers allow sonar systems to communicate. Nano-positioning stages release ink droplets in microscopy and semiconductor production, while piezoelectric (PZT) actuators release ink in inkjet printers. Furthermore, PZT-based materials are vibration-reduction materials that improve vehicle stability and performance. Based on such a variety of applications of PZT-based materials, we synthesized 0.05PZT-BT material and studied its electrical properties.

3.7 Synthesis of 0.05PZT-BT Ceramic Samples

This chapter introduces a new material (1-x)$PbZr_{0.6}Ti_{0.4}O_3$-x$BaTiO_3$ or (1-x)PZT–xBT. The (1-x)PZT–xBT with x = 0.05 (0.05PZT-BT) was prepared using a traditional method that involves mixing commercially available powders, such as PbO (99%), ZrO_2 (97%), TiO_2 (98%) (Loba Chemie Pvt. Ltd., India), and $BaCO_3$ (Sigma Aldrich, USA). To ensure the formation of desired phases and prevent unwanted pyrochlore phases, separate PZT and BT powders were first formed. Then, the 0.05PZT-BT powder was created by combining the PZT and BT precursors using the same mixed-oxide approach. The powders were mixed for 12 h in an agate mortar and pestle, using acetone as a medium to achieve homogeneity. After drying, the mixed powders were calcined at 1000°C for 3 h in alumina crucibles. The resulting calcined powders were ground, and polyvinyl alcohol (PVA) was added to facilitate pressing with a hydraulic press. The powders were uniaxially compressed into circular pellets of 10 mm diameter and ~1 mm thickness. The pellets were sintered for 2 h at 1100°C. The phases were determined with the use of an XRD. XRD analysis was performed on the Rigaku Ultima IV X-ray powder diffractometer with Cu-K_α radiation (k = 1.5418 Å) and a Ni filter. A thin electrode and a dc electric field (EF) (2.5 kV/cm) poled pellets at 120°C for 30 min before cooling them to ambient temperature. Another sets of pellets were polished and high-temperature silver paste was used to make electrode on both surfaces for electrical measurements.

Dielectric testing was done at 30–160°C using Novocontrol Alpha-A impedance analyzers. The piezoelectric coefficient was calculated using Excel Instruments d_{33} meter. Radiant Technologies' Precision Premier II testers were utilized for the P-E hysteresis loop.

3.8 Structural Analysis

Figure 3.6 shows the XRD patterns of sintered 0.05PZT-BT ceramics. The material 0.05PZT-BT pattern shows a perovskite phase with an impurity peak. The preliminary analysis suggests that this impurity peak belongs to ZrO_2, matching JCPDS file no. 37-1484. The formation of the free ZrO_2 phase in sintered 0.05PZT-BT ceramics is thought to result from compositional changes caused by the evaporation of lead oxide inside surface areas [20, 22]. After being sintered at 1100°C, it has been shown that pure $Pb(Zr_{0.6}Ti_{0.4})O_3$ ceramic has tetragonal and rhombohedral phases that are consistent with JCPDS file nos. 33-0784 and 73-2022 [23]. Here, the peak observed at 2θ ~43–46° (002), shown in the inset of Fig. 3.6, has splitting and broadening, indicating the coexistence of tetragonal and rhombohedral phases in the studied compound. It also suggests that the 0.05PZT-BT may lead to a diffuse phase transition between the tetragonal and rhombohedral phases near the morphotropic phase boundary

(MPB). Low BT content XRD patterns indicated the presence of both tetragonal and rhombohedral phases [24].

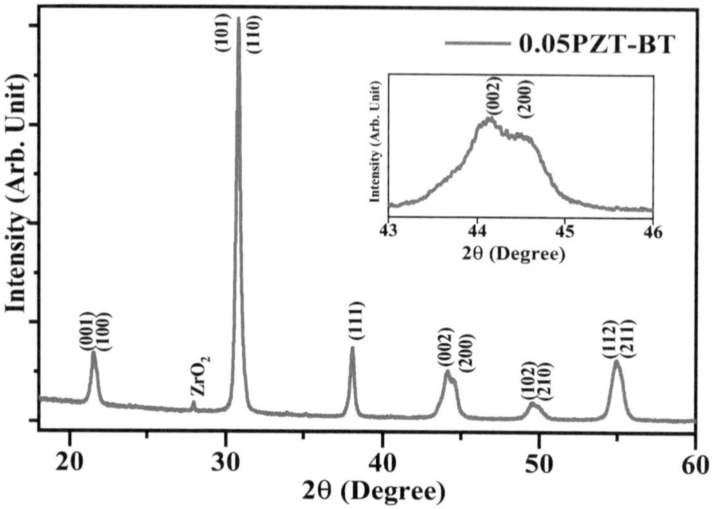

Figure 3.6 XRD graph of 0.05PZT-BT.

3.8.1 Dielectric Analysis

Dielectric characteristics help understand the charge transport nature of the ceramic with respect to the applied frequency and temperature. This analysis includes dielectric permittivity, dielectric loss, polarization properties (electronic, ionic, orientation, space charge, etc.), relaxation processes, etc. The relative dielectric permittivity (ε_r) is the primary parameter used in the dielectric analysis of ceramic materials [25]:

$$\varepsilon_r = \frac{C_p}{C_o} \tag{3.6}$$

Here, "C_p" stands for parallel capacitance, and "C_o" stands for the capacitance of the ceramic in free space. The applied electric field activates various dipoles within the ceramic, which respond differently to the field frequency. These dipoles are the strong function of temperature and frequency. Figure 3.7 shows the real (ε') and imaginary (ε'') parts of dielectric permittivity as a function of frequency measured at different temperatures. In Fig. 3.7a, the dielectric constant (ε') shows the dispersion in higher frequency regions, while the dielectric loss (ε") (Fig. 3.7b) exhibits very clear dielectric dispersion at low-frequency regions. The dielectric dispersion is a common feature of order-disorder ferroelectric materials and may be attributed to the motion of the ferroelectric domain boundary. The higher values of ε' and ε" observed in lower frequency regions are ascribed to the influence of all combined polarization, which gradually decreased with increasing frequency [26]. Among them, the contribution of space charge polarization is more dominant, which may be attributed to increased conductivity in the sample [25]. The dielectric constant (ε') values are 703, 731, 739, 768, 808, 855, and 914, respectively, for the temperatures 40, 60, 80, 100, 120, 140, and 160°C. At grain boundaries, space charge polarization creates a possible barrier. As the charge accumulates at the grain boundary, the real component of permittivity increases [27]. Due to material relaxation, the imaginary part of the dielectric constant (permittivity) decreases with frequency. Charge carriers have less time to react to the electric field at higher frequencies.

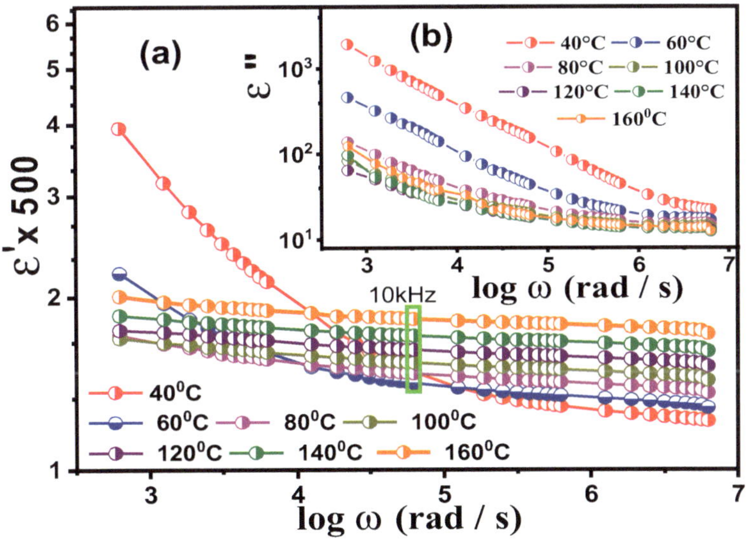

Figure 3.7 (a) Variation of the real part of dielectric constant with frequency at different temperatures. (b) Variation of the imaginary part of dielectric constant with frequency at different temperatures of 0.05PZT-BT.

Polarization and the imaginary part of the dielectric constant decrease since they cannot move and realign with the field [27]. The magnitude of loss tangent, tan δ, decreases with an increase in the frequency of the ac field and increases with an increase in temperature, attributed to the contribution of dc conductivity in 0.05PZT-BT ceramics (Fig. 3.8). No dielectric loss peak is observed in Figs. 3.7b and 3.8 for 0.05PZT-BT ceramics. The tangent loss values at 10 kHz are 0.207, 0.0616, 0.0329, 0.0263, 0.0224, 0.0213, and 0.001 for 40, 60, 80, 100, 120, 140, and 160°C, respectively.

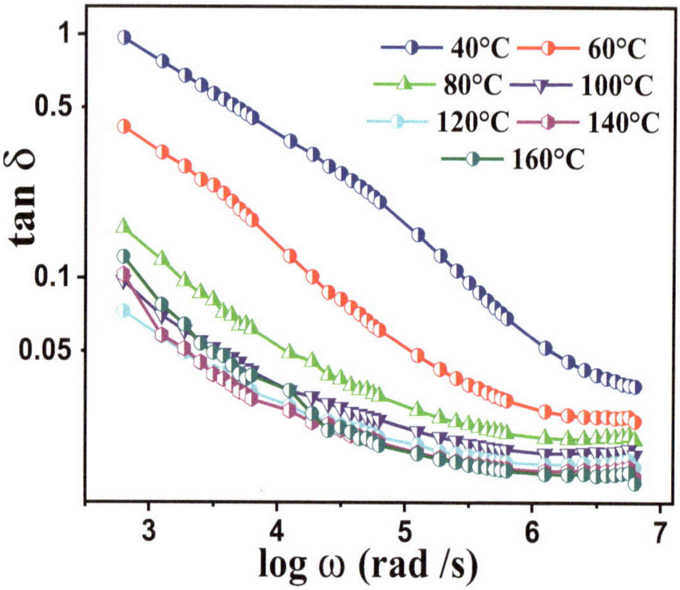

Figure 3.8 Variation of the tanδ with frequency at different temperatures of 0.05PZT-BT.

3.8.2 Modulus Formalism

Electric modulus is an important tool for analyzing polycrystalline materials' relaxation mechanisms. It can distinguish the influence of electrode polarization from the grain boundary involved in the relaxation process. The complex electric modulus can be calculated by following two important relations:

$$M^* = \frac{1}{E^*(\omega)} = M' + jM'' \tag{3.7}$$

$$M' = -\omega C_0 Z'' \quad \text{and} \quad M'' = -\omega C_0 Z' \tag{3.8}$$

Here, Z' and Z'' are the real and imaginary parts of the impedance, respectively. Figure 3.9 shows the variation of the real part of the electric modulus. M' is a function of frequency. It is observed that M' increases with an increase in frequency and decreases with an increase in temperature. However, in the higher frequency regions, M' exhibits a maximum value and dispersion, which may be due to the conductivity relaxation process and the short-range mobility of charge carriers [28]. The frequency dependence of the imaginary part of the electric modulus (M'') at different temperatures is shown in Fig. 3.10. The presence of a broad relaxation peak characterizes the plot of the 40°C graph. The non-Debye kind of relaxation in the sample is supported by observing an irregularity in the widening of the peaks. No relaxation peaks are observed for higher temperature graphs as these peaks are located in lower frequency regions than the measured frequency window. The exponential decrease in the value of M' is observed with increasing frequency for all the given temperatures due to the long-range mobility of charge carriers under the influence of an induced electric field.

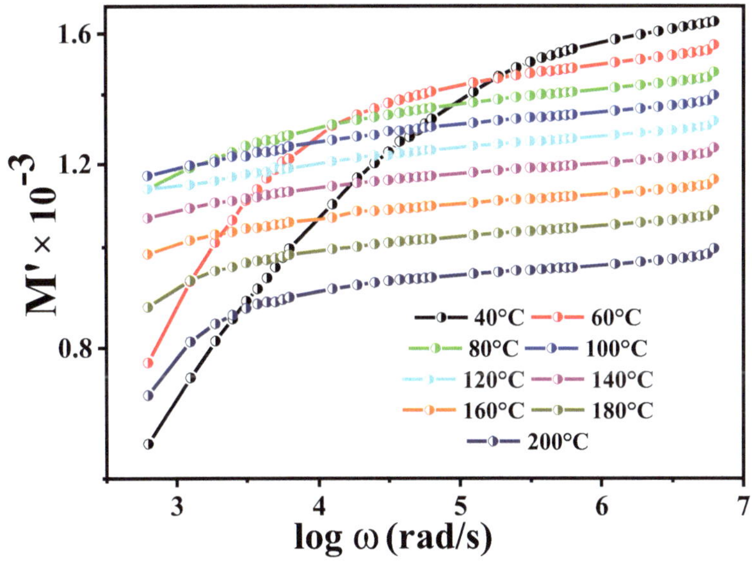

Figure 3.9 Variation of the real part of electric modulus with frequency at different temperatures of 0.05PZT-BT.

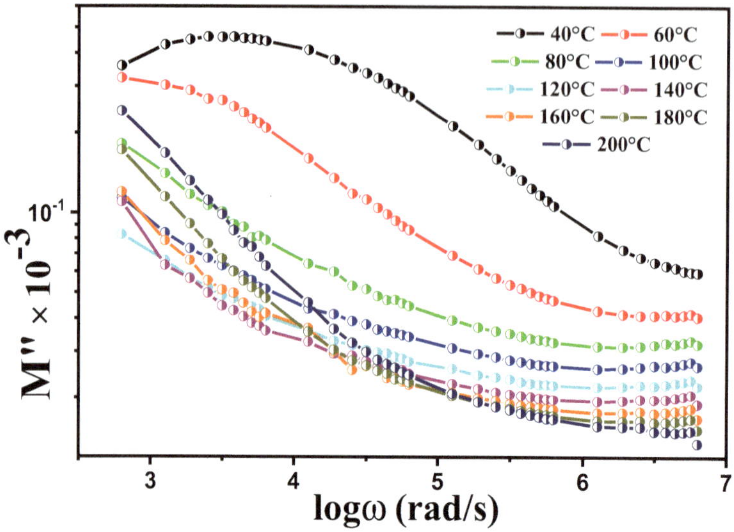

Figure 3.10 Variation of the real part of electric modulus with frequency at different temperatures of 0.05PZT-BT.

3.8.3 Electrical Conductivity Analysis

Figure 3.11 represents the frequency variation of ac conductivity for 0.05PZT-BT at several temperatures. The ac conductivity measurement is an important tool for studying ionic transport features on the materials. The following equation is used to determine this value:

$$\sigma_{ac} = \omega\varepsilon_0\varepsilon'' \tag{3.9}$$

Here, $\omega\,(=2\pi f)$ represents the angular frequency and f represents the applied field frequency. Figure 3.11 illustrates the frequency variation of ac conductivity for 0.05PZT-BT at several temperatures. The conductivity spectrum displays the ionic materials' features and is found to be highly frequency dependent in the high-frequency region, which decreases with decreasing frequency and becomes frequency independent (dc plateau) with conductivity dispersion. This extrapolating in the frequency-independent plateau in the low-frequency region gives dc conductivity. This behavior of the conductivity dispersion is generally explained using Jonscher's universal power law [29]:

$$\sigma(\omega) = \sigma_{dc} + A\omega^s \quad (3.10)$$

Here σ_{dc} is the dc conductivity, A is the temperature-dependent factor, and "s" is the frequency exponent of the charge carriers (ions). The temperature dependence of dc conductivity obtained from the value of the conductivity at 100 Hz is presented in the inset of Fig. 3.11. It is obvious from the figure that the dc conductivity decreases initially with increasing temperature up to 120°C and then starts increasing with a further increase in temperature. The dc conductivity exhibits a thermally activated behavior and obeys the Arrhenius relation:

$$\sigma = \sigma_0 \exp(-E_a / k_B T) \ \ldots \quad (3.11)$$

Here σ_0 is a pre-exponential factor, E_a is the activation energy of this conduction process, k_B is the Boltzmann constant, and T is the absolute temperature. It has been found that there are two linear regions with two different slopes. From the slope of two straight lines of two different temperature regions, the values of E_a have been obtained as: E_a (lower temperature) = 0.469 eV and E_a (higher temperature) = 0.387 eV.

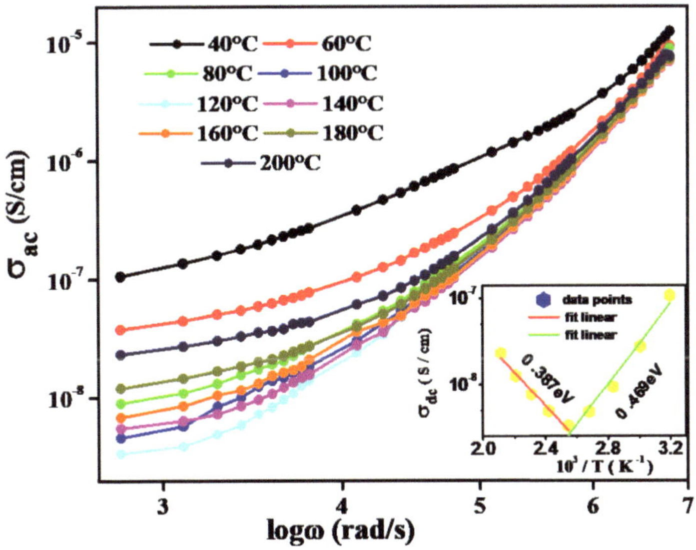

Figure 3.11 Frequency dependence of the ac conductivity (σ_{ac}) at various temperatures in a log-log scale of 0.05PZT-BT. Inset: The variation of σ_{dc} with respect to 10^3/T.

Figure 3.12 shows the temperature-dependent conductivity variation of 0.05PZT-BT at various field frequencies. It is observed that ac conductivity decreases to its minimum value at 120°C with increased temperature, showing metal-like behavior. After 120°C temperature, it starts increasing with temperature exhibiting semiconductor-like behavior. Moreover, the magnitude of ac conductivity increases with an increase in frequency, as shown in the direction of increase by an arrow in Fig. 3.12. A set of graphs exhibit thermally activated linearity in the lower temperature regions according to the relation given in equation (3.11), which has been employed to fit this set of linear regions. From the slopes of straight lines, the value of E_a is calculated in lower temperature regions. The obtained value is: E_a = 0.516 eV.

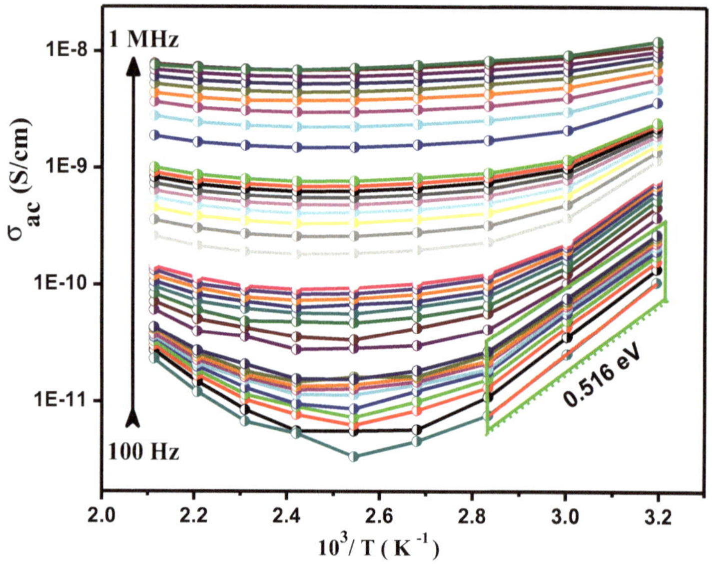

Figure 3.12 The ac conductivity response of 0.05PZT-BT as a function of temperature at various field frequencies. Arrhenius fitting lines showed in green color.

3.8.4 Ferroelectric Analysis

Figure 3.13 shows the ferroelectric response of 0.05PZT-BT ceramic at an electric field (EF) 70 kV, like (a) polarization vs. electric field

(PE) loop, (b) bipolar strain, and (c) unipolar strain. It is observed that the PE loop has a difference in the remanent polarization value for the negative and positive fields, indicating the presence of defect-induced dipoles in the ceramic. These dipoles interact with the EF differently than the system's intrinsic dipoles. Switching them in the direction of the field is hard compared to spontaneously polarized dipoles. Also, they contribute to the strain count; as shown in Fig. 3.13b, there is a considerable difference in the positive and negative strain counts. It should be noted that the PE loop is not saturated at such a high EF (kV), indicating the hard ferroelectric nature of the ceramic. It suggests the studied compound has a high field withstand property, a prime requirement for high field applications such as high voltage transformer sensors.

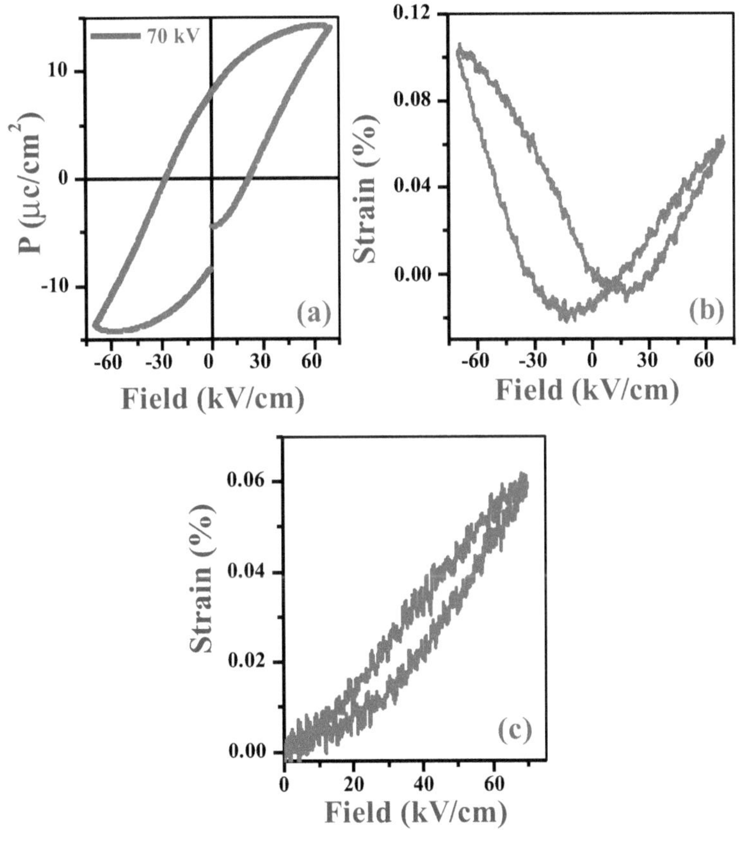

Figure 3.13 Ferroelectric response of the 0.05PZT-BT ceramics at EF 70 kV, (a) PE loop, (b) bipolar strain, and (c) unipolar strain.

The non-electrical breakdown ferroelectric characteristic of 0.05PZT-BT ceramics showed great potential to work at high field compared to PZT and PLZT. However, there is a compromise of strain value in achieving this property. However, one can achieve high strain counts and improved d_{33}* to tune the chemical composition. There are many possibilities to modify this system to enhance its ferroelectric behavior.

3.9 Conclusion and Futuristic Aspects

So far, this chapter discussed a comparative analysis of 0.05PZT-BT ceramics compared to PZT. This chapter also gave a gesture of various synthesis techniques and structural, dielectric, and ferroelectric behavior of ferroelectric compound 0.05PZT-BT. The traditional solid-state reaction method was employed to prepare the composite of low dielectric loss 0.05PZT-BT. XRD analysis showed the coexistence of the tetragonal and rhombohedral phase symmetry; i.e., the system exists under the MPB region. A detailed frequency and temperature-dependent analysis showed the possibilities of ferroelectric diffused phase transition near 100°C, which is absent in the PZT compound. It could be the effect of BT insertion in the PZT domains. However, a sharp structural phase transition is not observed under the dielectric measurement limit (temperature region). The modulus study showed the both short- and long-range mobility of charge carriers. The conductivity analysis exhibited metal-semiconductor-like behavior. The ferroelectric analysis suggested that the synthesized compound has a hard ferroelectric property compared to the PZT. However, the electrostrain value is compromised for this hard ferroelectric characteristic. Also, the PE loop showed the presence of defect dipoles in the system as it has nonequal remanent polarization values. In contrast, the system has high EF withstand properties, which make it suitable for high-energy applications.

However, this system requires a detailed analysis of the background of chemical tunning, temperature, and field-dependent structural transition and their synergistic effect on the physical properties. It presumes that the chemical modifications, which introduce a relaxor nature in the system, such as modification with $SrTiO_3$ and $Bi(Ni_{1-x}Zn_x)O_3$, can improve the electromechanical behavior of the compound. A detailed structural analysis using neutron powder diffraction (NPD) and in-situ XRD helps to understand the phase diagram of the PZT-BT. A frequency-dependent ferroelectric study will enlighten the polarization and strain behavior of the compound in depth.

Acknowledgment

Dr. Dev Kumar Mahato is thankful to the Ministry of Defence, Aeronautics Research and Development Board (ARDB), India for providing financial support under grant no. ARDB/01/2031929/M/I.

References

1. Chen B, Li H, Tian W, Zhou C. PZT based piezoelectric sensor for structural monitoring, *Journal of Electronic Materials* **48**(1), 2916–2923, 2019.
2. Kumar S, Singh D N, James A R, Mahato D K. Probing of the physical characteristics of antiferroelectric $Pb(Zr_{0.6}Ti_{0.4})O_3$; PZT (60/40) ceramics, *Journal of Materials Science: Materials in Electronics* **34**(1), 879(1–12), 2023.
3. Kang M G, Jung W S, Kang C Y, Yoon S J. Recent progress on PZT based piezoelectric energy harvesting technologies, *Actuators* **5**(5), 1–17, 2016.
4. Gao X, Yang J, Wu J, Xin X, Li Z, Yuan X, Dong S. Piezoelectric actuators and motors: Materials, designs, and applications, *Advanced Materials Technologies* **1**(1), 1900716 (1–26), 2019.
5. Rotan M, Zhuk M, Glaum M. Activation of ferroelectric implant ceramics by corona discharge poling, *Journal of the European Ceramic Society* **40**(1), 5402–5409, 2020.
6. Tiwari B, Babu T, Choudhary R N P. Review on different preparation techniques of lead zirconate titanate, *International Research Journal on Advanced Science Hub* **10**(2), 121–126, 2020.
7. James A R, Kumar A. Development of PLZT electroceramics with ultrahigh piezoelectric properties by a novel material engineering approach, *Handbook of Advanced Ceramics and Composites*, Springer, Switzerland, **1**(1), 1–36, 2019.

8. Vacca M, Graziano M, Chiolerio A, Lamberti A. Electric clock for NanoMagnet Logic Circuits, *Field-Coupled Nanocomputing, Paradigms, Progress, and Perspectives*, Springer, Italy, **1**(1), 73–110, 2014.
9. Thakur O P, Prakash C. Structural, dielectric and piezoelectric properties of PLZT (x/60/40) ceramics, *Integrated Ferroelectrics: An International Journal* **122**(1), 100–107, 2010.
10. Sadeghpour S, Kraft M, Puers R. Design and fabrication strategy for an efficient lead zirconate titanate based piezoelectric micromachined ultrasound transducer, *Journal of Micromechanics and Microengineering* **29**(1), 125002 (1–14), 2019.
11. Sathyanarayana C N, Raja S, Ragavendra H M, Procedure to use PZT sensors in vibration and load measurements, *Smart Materials Research* **1**(1), 1–9, 2013.
12. Kamel N A. Bio-piezoelectricity: Fundamentals and applications in tissue engineering and regenerative medicine, *Biophysical Reviews* **14**(1), 717–733, 2022.
13. Soudarthi K, Kaleemullah M, Rao B V S. Impact of piezoelectric materials in electric vehicles, *International Journal of Engineering and Innovative Technology* **10**(2), 28–31, 2020.
14. Panda P K, Sahoo B, Thejas T S. High strain lead-free piezo ceramics for sensor and actuator applications: A review, *Sensors Internationals* **4**(1), 100226 (1–11), 2023.
15. Khan M U, Rehman F, Saleem M, Elahi H, Sung T H, Jabbar H. Optimum driving of ultrasonic cleaner using impedance and FFT analysis with validation of image processing of perforated foils, *Applied Science* **13**(1), 699 (1–18), 2023.
16. Liu C, Guo Y. Modeling and positioning of a PZT precision drive system, *Sensors* **17**(1), 2577 (1–12), 2017.
17. Necira Z, Boutarfaia A, Abba M, Abdessalem N. Synthesis of PZT powder by conventional method at various conditions, *EPJ Web of Conferences* **29**(1), 00038 (1–7), 2012.
18. Shoghi A, Shakeri A, Abdizadeh H, Golobostanfard M R. Synthesis of crack-free PZT thin films by sol-gel processing on glass substrate, *Procedia Materials Science* **11**(1), 386–390, 2015.
19. James A R, Praveen J P, Premkumar M. Mechanically activated PLZT ceramics: Structural and electrical properties, *Materials Research Bulletin* **47**(1), 3459–3462, 2012.
20. Vittayakorn W C, Yimnirun R, Ananta S. Composition, structure and properties of PZT-BT ceramics prepared by two-stage sintering, *Key Engineering Materials* **421**(422), 436–439, 2010.

21. Funsueb N, Limpichaipanit A, Ngamjarurojana A. Electrical properties and microstructure of phase combination in $BaTiO_3$-based ceramics, *Journal of Physics: Conference Series* **1144**(1), 012133–012136, 2018.
22. Miclea C, Miclea C F, Amarande L, Micclea T, Cioangher M. Structural, dielectric and piezoelectric properties of the ferroelectric PZT-BT ceramic compounds, *Journal of Optoelectronics and Advanced Materials* **20**(9), 558–565, 2018.
23. Chaisan W, Yimnirun R, Ananta S, Cann D P. Dielectric properties of solid solutions in the lead zirconate titanate–barium titanate system prepared by a modified mixed-oxide method, *Materials Letters* **59**(1), 3732–3737, 2005.
24. Chaisan W, Ananta S, Tunkasiri T. Synthesis of barium titanate–lead zirconate titanate solid solutions by a modified mixed-oxide synthetic route, *Current Applied Physics* **4**(1), 182–185, 2004.
25. Halder S, Arya B B, Bhuyan S, Choudhary R N P. Analysis of dielectric, impedance and electrical properties of electronic material: $Bi(Ni_{2/3}V_{1/3})O_3$, *Phase Transitions* **93**(7), 709–721, 2020.
26. Xie R J, Akimune Y, Matsuo K, Sugiyama T, Hirosaki N, Sekiya T. Dielectric and ferroelectric properties tetragonal tungsten bronze $Sr_{2-x}Ca_xNaNb_5O_{15}$ (x = 0.05–0.35) ceramics, *Applied Physics Letters* **80**(5), 835–837, 2002.
27. Rayssi C, El.Kossi S, Dhahri J, Khirouni K. Frequency and temperature-dependence of dielectric permittivity and electric modulus studies of the solid solution $Ca_{0.85}Er_{0.1}Ti_{1-x}Co_{4x/3}O_3$ ($0 \leq x \leq 0.1$), *RSC Advances* **8**(1), 17139–17150, 2018.
28. Mahato D K, Dutta A. Sinha T P. Synthesis and electric characterization of rare earth double perovskite Ho_2CdZrO_6 ceramics, *Journal of Electroceramics* **29**(2), 99–105, 2012.
29. Jonscher A K, A new understanding of the dielectric relaxation of solids, *Journal of Materials Science* **16**(1), 2037–2060, 1981.

Chapter 4

Design and Applications of Piezoelectric Biomaterials

Usman Lawal Usman,[a,b] **Sarita Rai,**[c] **and N. B. Singh**[d,e]
[a]*Department of Biology, Umaru Musa Yar'adua University, Katsina, Nigeria*
[b]*Department of Environmental Sciences, Sharda University, Greater Noida, India*
[c]*Department of Chemistry, Dr. Harisingh Gour University, Sagar (MP), India*
[d]*Department of Chemistry and Biochemistry, Sharda University, Greater Noida, India*
[e]*Research Development Cell, Sharda University, Greater Noida, India*
usman.usman@umyu.edu.ng, nbsingh43@gmail.com

Piezoelectric biomaterials have emerged as a promising class of materials with the ability to convert mechanical energy into electrical energy and vice versa. This unique property has led to the development of innovative biomedical devices and therapies, including sensors, actuators, and tissue engineering scaffolds. This chapter discusses the fundamentals of piezoelectricity, including the principles of crystal symmetry, electrical polarization, and mechanical deformation. It then explores different types of piezoelectric biomaterials, including ceramics, polymers, and

Piezoelectric Materials: Design and Applications
Edited by N. B. Singh and Dev Kumar Mahato

ISBN 978-981-5129-40-3 (Hardcover), 978-1-003-59897-8 (eBook)
www.jennystanford.com

composites, and their physical and chemical properties. The chapter also focuses on the application of piezoelectric biomaterials in biomedical engineering, including tissue engineering, biosensors, drug delivery, and neural interfaces. Similarly, a review of the current and potential clinical and biomedical applications of piezoelectric biomaterials, including tissue engineering, regenerative medicine, sensing, and actuation, is highlighted. Furthermore, the chapter highlights the latest developments in the field, including the use of piezoelectric biomaterials for regenerative medicine and the development of implantable devices for monitoring and treating neurological disorders. Characterization techniques such as piezoelectric coefficient measurement, impedance spectroscopy, ferroelectric loop measurements, and scanning probe microscopy are employed to evaluate their piezoelectric properties. X-ray diffraction, mechanical testing, and electrical characterization are also crucial for understanding the crystal structure, mechanical behavior, and electrical properties of these biomaterials. Challenges and opportunities associated with the design and implementation of piezoelectric biomaterials in biomedical applications have also been highlighted. Lastly, the chapter provides an outlook on the future of piezoelectric biomaterials in biomedical engineering, including emerging technologies and potential areas for future research.

4.1 Introduction

Piezoelectric biomaterials represent a fascinating intersection between materials science and biology. These materials possess a unique property known as piezoelectricity, which allows them to generate electric charges when subjected to mechanical stress or vibrations. This remarkable characteristic has attracted significant attention in the field of biomedical engineering due to its potential applications in various areas, including tissue engineering, regenerative medicine, biosensing, and energy harvesting [1]. Piezoelectricity is a phenomenon exhibited by certain crystals, ceramics, and polymers, where an electric charge accumulates in response to applied mechanical stress. In the context of biomaterials, piezoelectric materials have been harnessed for their ability to convert mechanical energy into electrical signals, making them particularly relevant for interfacing with biological systems [2]. In general, piezoelectric biomaterials can be broadly categorized into two main groups: natural and synthetic. Natural piezoelectric biomaterials are derived from biological sources, such as bone, tendon, and certain proteins while synthetic biomaterials with piezoelectric properties are categorized into four groups: naturally occurring piezocrystals, piezoceramics (lead-free, lead-based ceramics and titanates), piezopolymers, and piezocomposites. As single crystals (monocrystalline, anisotropic, defect-free crystal lattice without grain boundaries), polycrystals (random crystallite orientation with grain boundaries or twin boundaries), solid solutions (homogeneous dispersion of minor dispersed phase in the crystal lattice of the matrix phase), and thin films (subnano- to micrometer-thick layer deposited on a solid support), natural crystals and piezoceramics can be found in nature [3]. These materials possess inherent piezoelectric properties, making them well-suited for applications in biomedicine. On the other hand, synthetic piezoelectric biomaterials are engineered through the incorporation of piezoelectric components into biocompatible matrices, such as polymers or ceramics. The piezoelectric effect in biomaterials has opened up a wide range of possibilities in several biomedical applications. For instance, in tissue engineering, piezoelectric scaffolds can provide mechanical stimulation to cells, promoting their growth, differentiation, and tissue regeneration. Additionally, these materials can serve as transducers for energy harvesting within the body, converting mechanical motion into electrical energy to power implantable devices [4]. Another area where piezoelectric biomaterials have shown great promise is biosensing. By integrating these materials into biosensors, researchers can detect subtle mechanical changes or vibrations associated with biological processes. This capability has potential applications in healthcare diagnostics, environmental monitoring, and human-machine interfaces.

4.2 Overview of Piezoelectric Biomaterials

Piezoelectric biomaterials have been explored for their ability to facilitate drug delivery. By applying mechanical stress to drug-loaded matrices, controlled release of therapeutic agents can be achieved. This approach offers precise and localized drug delivery, minimizing systemic side effects and improving treatment outcomes [5]. Despite the immense potential of piezoelectric biomaterials, several challenges need to be addressed for their widespread implementation. One such challenge is the development of biocompatible and stable materials that can withstand the complex and dynamic environment of the human body [6]. Additionally, the integration of piezoelectric materials into existing medical devices and systems requires careful consideration of compatibility, durability, and performance. Research efforts are focused on exploring novel fabrication techniques and material combinations to enhance the piezoelectric properties of biomaterials [7]. For example, researchers are investigating the use of nanomaterials and composites to improve the piezoelectric response and mechanical properties of these materials. By tailoring the composition and structure at the nanoscale, it is possible to optimize the performance and functionality of piezoelectric biomaterials. Moreover, advancements in characterization techniques and modeling approaches are crucial for better understanding the behavior of piezoelectric biomaterials at the molecular and cellular levels. This knowledge can guide the design and optimization of these materials for specific applications as well as facilitate the development of computational models to predict their performance.

In the realm of tissue engineering, piezoelectric biomaterials offer unique advantages. The mechanical stimulation provided by these materials can influence cellular behavior, including cell adhesion, proliferation, and differentiation. By incorporating piezoelectric scaffolds or substrates into tissue engineering constructs, researchers aim to enhance tissue regeneration and functional tissue formation. This approach has shown promise in applications such as bone tissue engineering, where the piezoelectric properties of materials can promote osteogenic differentiation and accelerate bone healing [8]. Piezoelectric biomaterials also have significant potential in biosensing applications. Their ability to convert mechanical signals into electrical signals makes them well-suited for detecting and monitoring various biological processes. For instance, piezoelectric sensors can be utilized to measure forces exerted by cells during cellular processes, such as cell migration, contraction, or interaction with the extracellular matrix [1]. These sensors can provide valuable insights into cell mechanics and behavior, contributing to a better understanding of biological processes and disease mechanisms. In the field of drug delivery, piezoelectric biomaterials offer an innovative approach to the controlled and targeted release of therapeutic agents [9]. Mechanical stress applied to drug-loaded matrices can induce drug release in a controlled manner. By modulating the magnitude and frequency of mechanical stimulation, researchers can achieve precise spatiotemporal control over drug release, potentially improving therapeutic efficacy and minimizing side effects. This approach holds promise for personalized medicine, where drug release can be tailored to individual patient needs. Energy harvesting is another area where piezoelectric biomaterials have garnered interest. Within the human body, mechanical energy is generated through natural movements, such as walking or joint motion. Piezoelectric materials can be used to harvest this mechanical energy and convert it into electrical energy, which can be harnessed to power implantable devices or sensors. This concept has the potential to reduce reliance on external power sources and enhance the sustainability of medical devices [10].

4.3 Importance of Piezoelectric Biomaterials

Piezoelectric biomaterials hold immense importance in the field of biomedical engineering due to their unique properties and potential applications. These materials have garnered significant attention and research interest, as they offer a wide range of possibilities for advancing healthcare and improving patient outcomes [11]. The following are some key areas where piezoelectric biomaterials have shown great promise:

i. Tissue Engineering and Regenerative Medicine: Piezoelectric biomaterials can play a crucial role in tissue engineering and regenerative medicine. The ability of these materials to convert mechanical energy into electrical signals enables them to provide mechanical stimulation to cells, promoting their growth, differentiation, and tissue regeneration. Piezoelectric scaffolds or substrates can be utilized to enhance the regeneration of various tissues, including bone, cartilage, and nerves [12].

ii. Biosensing and Diagnostics: Piezoelectric biomaterials offer excellent opportunities for biosensing and diagnostic applications. By integrating these materials into biosensors, they can detect mechanical changes or vibrations associated with specific biological processes. This capability has potential applications in healthcare diagnostics, monitoring disease progression, detecting biomarkers, and environmental monitoring. Piezoelectric biosensors can provide real-time, sensitive, and label-free detection, making them valuable tools in medical diagnostics [13].

iii. Drug Delivery Systems: Piezoelectric biomaterials have the potential to revolutionize drug delivery systems. By applying mechanical stress to drug-loaded matrices, controlled and targeted release of therapeutic agents can be achieved. This approach allows for precise spatiotemporal control over drug release, leading to improved treatment efficacy, reduced side effects, and personalized medicine. Piezoelectric drug delivery systems have the advantage of being non-invasive and offering localized drug delivery to specific target sites [3].

iv. Energy Harvesting: Piezoelectric biomaterials can harness mechanical energy from the body and convert it into electrical energy. This energy harvesting capability has significant implications for powering implantable medical devices and sensors. By utilizing the natural movements of the body, such as walking or joint motion, piezoelectric biomaterials can generate electricity to sustainably power these devices [8]. This approach eliminates the need for external power sources, thereby improving device portability and longevity.

v. Human-Machine Interfaces: Piezoelectric biomaterials have the potential to enhance human-machine interfaces by enabling touch-sensitive interfaces and prosthetic devices that can detect and respond to mechanical stimuli. These materials can facilitate a more natural and intuitive interaction between humans and machines, leading to advancements in robotics, virtual reality, and haptic feedback systems [14].

vi. Neural Interfaces and Neuromodulation: Piezoelectric biomaterials hold promise in the field of neural interfaces and neuromodulation. These materials can be used to develop interfaces that can detect and stimulate neural activity. By incorporating piezoelectric elements into neural implants or electrodes, researchers aim to improve the sensitivity and selectivity of neural recordings and the precision of neural stimulation [15]. This has applications in areas such as brain-computer interfaces, deep brain stimulation, and neuroprosthetics.

vii. Wearable and Implantable Sensors: The unique properties of piezoelectric biomaterials make them suitable for developing wearable and implantable sensors. These sensors can monitor various physiological parameters, such as pulse, respiration, motion, and biomechanical forces. Piezoelectric sensors can provide real-time data for healthcare monitoring, fitness tracking, rehabilitation, and personalized medicine. Their small size, flexibility, and biocompatibility make them ideal for integration into wearable devices or implantable systems [16].

viii. Biomechanics and Rehabilitation: Piezoelectric biomaterials have the potential to advance the field of biomechanics and rehabilitation. They can be used to measure and analyze mechanical forces and strains in biological tissues, aiding in the understanding of musculoskeletal biomechanics and optimizing rehabilitation protocols [17]. Additionally, these materials can be incorporated into wearable devices or smart textiles to provide real-time feedback and assistive functionalities for individuals with mobility impairments.

ix. Vibrational Energy Harvesting: In addition to energy harvesting from body movements, piezoelectric biomaterials can be used for vibrational energy harvesting from external sources. They can capture and convert ambient mechanical vibrations, such as vibrations from machinery, vehicles, or the environment, into electrical energy. This has potential applications in powering remote sensors, wireless devices, and low-power electronics in various settings, including healthcare, industrial, and environmental monitoring [18]. The importance of piezoelectric biomaterials lies in their versatility and potential to revolutionize multiple aspects of healthcare, diagnostics, and biomedical technology. Their ability to convert mechanical energy into electrical signals, combined with their biocompatibility and adaptability, opens up endless possibilities for improving patient care, enhancing biomedical devices, and advancing our understanding of biological systems. Continued research and innovation in this field are essential to unlock the full potential of piezoelectric biomaterials and translate them into practical and impactful applications.

4.4 Piezoelectric Materials: Properties and Characteristics

Piezoelectric materials are a unique class of materials that exhibit the piezoelectric effect, wherein they generate electric charges when subjected to mechanical stress or vibrations. These materials possess distinct properties and characteristics that make them valuable for various applications in science, engineering, and medicine [19, 20]. Figure 4.1 explores the properties and characteristics of piezoelectric materials.

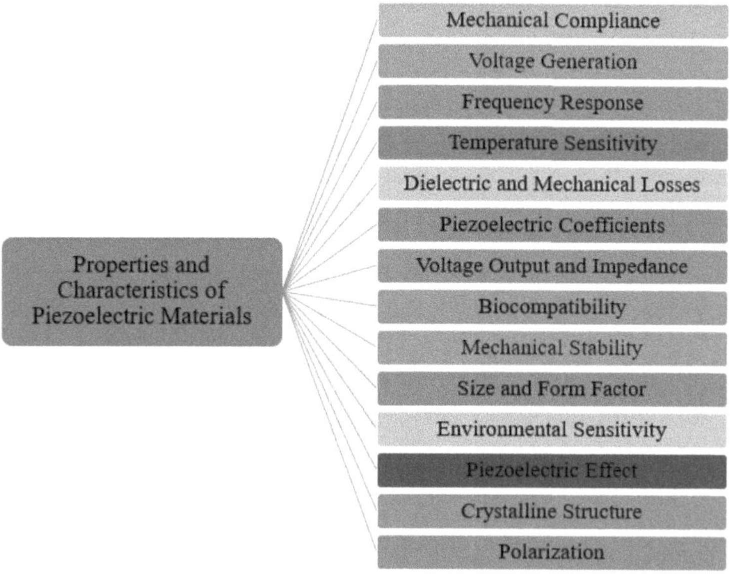

Figure 4.1 An overview of properties and characteristics of piezoelectric materials.

i. Piezoelectric Effect: The piezoelectric effect is the fundamental property of these materials. When mechanical stress or deformation is applied to a piezoelectric material, it generates an electric charge or voltage across its surface. Conversely, when an electric field is applied to the material, it undergoes mechanical deformation. This bidirectional coupling of mechanical and electrical energy makes piezoelectric materials unique and useful for a range of applications [20].

ii. Crystalline Structure: Piezoelectricity is predominantly observed in certain crystalline materials. Crystals such as quartz, tourmaline, and Rochelle salt are naturally piezoelectric, while others like lead zirconate titanate (PZT), barium titanate, and polyvinylidene fluoride (PVDF) can be engineered to exhibit piezoelectric properties. The arrangement of atoms within the crystal lattice gives rise to the asymmetric charge distribution that enables the piezoelectric effect [2].

iii. Polarization: Piezoelectric materials have a permanent electric dipole moment due to the non-centrosymmetric arrangement of their crystal structure. This dipole moment is responsible for the generation of electric charges when the material is mechanically stressed. The direction of the electric dipole moment determines the polarity of the piezoelectric effect, and it can be reversed by applying an electric field [21].

iv. Mechanical Compliance: Piezoelectric materials exhibit mechanical compliance, meaning they can be easily deformed or strained under applied mechanical stress. This property allows them to convert various forms of mechanical energy, such as pressure, vibration, or bending, into electrical signals. The mechanical compliance of piezoelectric materials contributes to their sensitivity and responsiveness to external mechanical stimuli [22].

v. Voltage Generation: The piezoelectric effect in these materials leads to the generation of an electric voltage or charge across their surfaces in response to mechanical stress. The magnitude of the voltage generated is proportional to the applied stress and the material's piezoelectric coefficient. This property makes piezoelectric materials suitable for sensors, actuators, and energy-harvesting devices [22].

vi. Frequency Response: Piezoelectric materials exhibit a broad frequency response, making them capable of converting a wide range of mechanical vibrations or acoustic waves into electrical signals. This property allows for their use in diverse applications, including ultrasound imaging, vibration sensors, and acoustic wave devices [19].

vii. Temperature Sensitivity: The piezoelectric properties of materials can be temperature-dependent. Some piezoelectric materials exhibit changes in their piezoelectric coefficients or loss of piezoelectricity at extreme temperatures. Understanding and accounting for the temperature sensitivity of these materials is crucial for their successful application in various environments [20].

viii. Dielectric and Mechanical Losses: Piezoelectric materials may exhibit some degree of dielectric and mechanical losses, which can affect their performance. Dielectric losses refer to energy dissipation in the material due to electrical resistance, while mechanical losses refer to energy dissipation due to internal friction during mechanical deformation. Minimizing these losses is essential for optimizing the efficiency and sensitivity of piezoelectric devices [23].

ix. Piezoelectric Coefficients: Piezoelectric materials are characterized by their piezoelectric coefficients, which describe the relationship between mechanical stress and generated electric charge or voltage. These coefficients vary among different materials and crystal orientations, and they determine the material's sensitivity and effectiveness in converting mechanical energy into electrical signals. Understanding and optimizing these coefficients are crucial for designing high-performance piezoelectric devices [19].

x. Voltage Output and Impedance: The voltage output of a piezoelectric material depends on the applied mechanical stress and the material's piezoelectric coefficients. The output voltage can range from microvolts to several volts, depending on the specific application and the material used. Additionally, piezoelectric materials have inherent electrical impedance, which affects their electrical response and can be matched with the requirements of the application or electronic circuitry [24].

xi. Biocompatibility: Biocompatibility is an important consideration for piezoelectric biomaterials used in biomedical applications. The material should be non-toxic, non-immunogenic, and capable of interacting safely with biological systems. Biocompatible piezoelectric materials, both natural and synthetic, are being explored for applications such as tissue engineering scaffolds, biosensors, and implantable devices [1].

xii. Mechanical Stability: Piezoelectric materials should exhibit mechanical stability to maintain their piezoelectric properties over time. Mechanical stability ensures that the material can withstand repeated mechanical stress or deformations without undergoing significant changes in its piezoelectric response [20]. This property is essential for the long-term functionality and reliability of piezoelectric devices.

xiii. Size and Form Factor: Piezoelectric materials can be fabricated into various sizes and forms to suit specific applications. They can be engineered as thin films, fibers, and powders or integrated into flexible matrices, depending on the requirements of the device or system. The versatility in size and form factor allows for their integration into a wide range of applications, including miniaturized sensors, wearable devices, and implantable systems [8].

xiv. Environmental Sensitivity: Some piezoelectric materials can exhibit sensitivity to environmental factors such as humidity, temperature, and electrical noise. These factors can influence the performance and stability of the piezoelectric effect. Careful consideration of the operating conditions and environmental factors is necessary to ensure the reliability and accuracy of piezoelectric devices in real-world applications [25]. Table 4.1 shows some examples of piezoelectric materials alongside their properties.

Table 4.1 Piezoelectric properties of commonly used materials

Material	Piezoelectric coefficient (d33, pC/N)	Electro-mechanical coupling factor (k)	Dielectric constant (ε)
Lead zirconate titanate (PZT)	High	High	High
Polyvinylidene fluoride (PVDF)	Moderate	Moderate	Moderate
Barium titanate ($BaTiO_3$)	Moderate	Moderate	High
Zinc oxide (ZnO)	High	Moderate	High
Lithium niobate ($LiNbO_3$)	High	High	High
Quartz	High	Low	Low
AlN (aluminum nitride)	Moderate	Moderate	High
P(VDF-TrFE) (polyvinylidene fluoride-trifluoroethylene)	High	High	Moderate
$BaSrTiO_3$ (barium strontium titanate)	High	High	High
$BiFeO_3$ (bismuth ferrite)	High	High	High

4.5 Different Types of Piezoelectric Materials

Piezoelectric materials encompass a wide range of substances, including both natural and synthetic materials, each with unique properties and applications. Below are some of the commonly used types of piezoelectric materials and their respective applications [11, 26]:

i. Quartz: Quartz is a naturally occurring piezoelectric crystal and one of the most widely used materials in piezoelectric devices. It exhibits excellent piezoelectric properties and high stability over a wide temperature range. Quartz crystals are extensively utilized in oscillators, sensors, filters, and timing devices, including quartz watches, electronic devices, and communication systems.

ii. Lead Zirconate Titanate (PZT): PZT is a synthetic ceramic material composed of lead, zirconium, and titanium oxides. It is one of the most common and versatile piezoelectric materials. PZT ceramics exhibit strong piezoelectric properties, making them suitable for a broad range of applications, such as actuators, sensors, transducers, ultrasonic devices, inkjet printers, and vibration control systems.

iii. Polyvinylidene Fluoride (PVDF): PVDF is a flexible and lightweight piezoelectric polymer. It is biocompatible and suitable for various applications in biomedicine and wearable technology. PVDF films or fibers are used in sensors, energy harvesters, flexible electronics, touch-sensitive devices, and pressure sensors for medical diagnostics, robotics, and human-machine interfaces.

iv. Barium Titanate ($BaTiO_3$): Barium titanate is a ceramic material that exhibits strong piezoelectric properties. It is widely used in ultrasonic transducers, actuators, and sensors. $BaTiO_3$ is also used in capacitors, electro-optic devices, and piezoelectric inkjet printheads.

v. Lead Metaniobate ($PbNb_2O_6$ or PNN): PNN is a ferroelectric material with excellent piezoelectric properties. It is commonly used in high-frequency transducers, ultrasonic imaging devices, and medical imaging systems due to its high sensitivity and ability to generate and receive ultrasound waves.

vi. Lithium Niobate ($LiNbO_3$): Lithium niobate is a single-crystal material with exceptional piezoelectric and electro-optic properties. It finds applications in telecommunications, optical devices, surface acoustic wave (SAW) devices, and frequency conversion in laser systems.

vii. Rochelle Salt (Sodium Potassium Tartrate): Rochelle salt is a naturally occurring crystal that exhibits high piezoelectric sensitivity. It has been used historically in early piezoelectric devices and is still utilized in specific applications such as microphones, hydrophones, and sensitive acoustic devices.

viii. Composite Materials: Composite piezoelectric materials are engineered by combining piezoelectric ceramics or polymers with other materials to enhance specific properties. For example, piezoelectric composites can have improved mechanical flexibility, increased voltage output, or tailored piezoelectric response. These materials find applications in flexible electronics, smart materials, and structural health monitoring systems. These are just a few examples of piezoelectric materials, and there are numerous other materials, both natural and synthetic, being explored for their piezoelectric properties. The choice of material depends on the specific application requirements, such as the desired sensitivity, temperature stability, mechanical properties, and environmental conditions.

4.6 Natural Piezoelectric Biopolymers

Piezoelectricity refers to the property of certain materials to generate an electric charge when subjected to mechanical stress and, conversely, to deform when an electric field is applied to them. While piezoelectricity is often associated with synthetic materials like ceramics and crystals, some natural biopolymers also exhibit this unique property [27]. Table 4.2 shows a few examples of natural biopolymers that possess piezoelectric properties.

Table 4.2 Some natural piezoelectric biopolymers and their properties

Natural biopolymer	Source	Piezoelectric properties	Potential applications	References
Collagen	Animal connective tissues	Exhibits piezoelectric behavior	Tissue engineering, biocompatible sensors	[28]
Silk fibroin	Silkworms and spiders	Possesses piezoelectric properties	Biodegradable electronics, tissue engineering	[29]
DNA	Genetic material in living cells	Demonstrates piezoelectricity	Nanoscale sensors, bioelectronics, DNA-protein interactions	[30]
Chitin	Exoskeletons of insects/fungi	Exhibits piezoelectric behavior	Energy harvesting, biocompatible materials	[31]
Pectin	Plant cell walls	Demonstrates piezoelectricity	Biodegradable sensors	[32]
Alginate	Brown seaweeds	Shows piezoelectric properties	Self-powered biomedical devices, biocompatible sensors	[33]
Cellulose	Plant cell walls	Exhibit piezoelectric properties	Energy harvesting from environmental vibrations and wearable devices	[34]
Gelatin	Animal collagen	Exhibits piezoelectric behavior	Biomedical devices, tissue engineering	[4]
Cotton fibers	Cotton plant	Demonstrates piezoelectricity	Smart textiles, wearable electronics	[35]

Natural biopolymer	Source	Piezoelectric properties	Potential applications	References
Soy protein	Soybeans	Exhibits piezoelectric behavior	Biodegradable sensors	[36]
Keratin	Hair, feathers, claws, and horns of animals	Possess piezoelectric behavior	Biocompatible sensors, biomimetic electronics, and even the design of self-powered biomedical devices	[3]
Nacre [mother of pearl]	Shells of certain mollusks, such as abalones and pearl oysters	Exhibit piezoelectric behavior	Self-powered sensors, structural materials, and even biodegradable energy-harvesting devices	[37]
Chitosan	Chitin found in the exoskeletons of crustaceans and cell walls of certain fungi	Exhibits piezoelectric properties	Wound healing, drug delivery, and tissue engineering	[38]
Fish scales	Fish species, e.g., garfish and black drum	Possess piezoelectric properties	Medical implants and wearable health monitoring devices	[39]
Myosin	Muscle cells	Exhibits piezoelectric behavior	Tissue engineering, muscle-related biomedical devices	[4]
Viral capsids	Engineered viruses	Engineered to possess piezoelectric properties	Nanoscale electronics, self-assembling nanodevices	[40]

(*Continued*)

Table 4.2 *(Continued)*

Natural biopolymer	Source	Piezoelectric properties	Potential applications	References
Chlorophyll	Green pigment in plants	Demonstrates piezoelectricity	Environmentally friendly sensors, energy harvesting	[41]
Phospholipids	Cell membranes	Exhibits piezoelectric behavior	Biomimetic electronic applications, biosensing devices	[42]
Polyhydroxy alkanoates (PHA)	Microbial polyesters	Possess piezoelectric properties	Biodegradable sensors	[43]
Lignin	Plant cell walls	Exhibits piezoelectric behavior	Biodegradable energy-harvesting devices	[44]

The piezoelectricity exhibited by natural biopolymers holds significant potential for various applications due to their biocompatibility, biodegradability, and renewable nature. While the piezoelectric effect in natural biopolymers is generally weaker compared to synthetic piezoelectric materials, their unique properties make them attractive for specific applications, particularly in fields where biocompatibility and sustainability are paramount. The exploration of piezoelectric properties in natural biopolymers continues to uncover a diverse range of materials with unique properties and potential applications [45]. From the tissues of living organisms to the proteins found in nature, these materials offer exciting opportunities for sustainable and biocompatible advancements in various fields.

4.7 Fabrication Methods for Piezoelectric Biomaterials

The fabrication of piezoelectric biomaterials involves techniques that enable the integration of piezoelectric properties into biocompatible matrices. These methods vary depending on the type of biomaterial being used, whether it is a natural or synthetic material. Figure 4.2 depicts some common fabrication methods for piezoelectric biomaterials.

i. Poling of Ferroelectric Materials: Ferroelectric materials, such as PZT or barium titanate ($BaTiO_3$), are typically fabricated as ceramics. The fabrication process involves synthesizing the ceramic material, forming it into the desired shape (e.g., bulk components, thin films, or nanoparticles), and then subjecting it to a poling process. Poling involves the application of a strong electric field to align the spontaneous polarization domains within the material, thereby inducing piezoelectric behavior. Poling can be done using techniques such as electric field poling, corona poling, or domain engineering [5].

ii. Electrospinning: Electrospinning is a technique used to fabricate piezoelectric biomaterials in the form of fibers or non-woven mats. In this process, a polymer solution or melt is electrostatically drawn into a fine fiber using an electric field. By incorporating piezoelectric materials, such as PVDF or its copolymers, into the electrospinning solution, piezoelectric fibers or mats can be produced. Electrospun piezoelectric fibers find applications in tissue engineering scaffolds, wearable sensors, and energy harvesting devices [17].

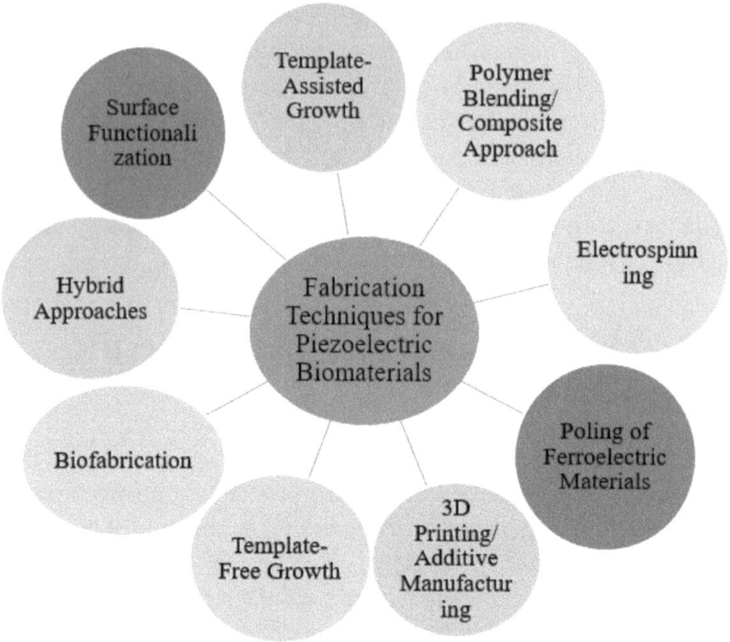

Figure 4.2 An overview of some common fabrication methods for piezoelectric biomaterials.

iii. Sol-Gel Method: The sol-gel method is widely used for fabricating thin films and coatings of piezoelectric biomaterials. It involves the conversion of precursor solutions (sol) into a solid network (gel) through controlled hydrolysis and condensation reactions. The resulting gel is then thermally treated to remove organic components and densify the structure, forming a thin film. By choosing suitable precursors, such as metal alkoxides, piezoelectric materials like PZT or PVDF can be synthesized and deposited onto substrates using techniques like spin coating, dip coating, or spray coating [46].

iv. Polymer Blending/Composite Approach: Piezoelectric biomaterials can also be fabricated through polymer blending or composite approaches. This involves incorporating piezoelectric materials, such as piezoelectric nanoparticles or fibers, into biocompatible polymers or hydrogels. The piezoelectric phase is dispersed within the polymer matrix, providing the desired piezoelectric properties. Various techniques, such as solution casting, melt blending, or 3D printing, can be used to fabricate composite materials. The composite approach allows for the tuning of mechanical properties, piezoelectric response, and biocompatibility to suit specific applications in tissue engineering, biosensing, and drug delivery [47].

v. Template-Assisted Growth: Template-assisted growth methods involve using templates or molds to guide the growth of piezoelectric biomaterials. For example, porous scaffolds with piezoelectric properties can be fabricated by depositing piezoelectric materials onto a pre-existing porous template, followed by template removal. This technique is often employed for creating tissue engineering scaffolds that provide mechanical stimulation to cells and promote tissue regeneration [48].

vi. 3D Printing/Additive Manufacturing: Additive manufacturing techniques, such as 3D printing, have gained popularity for fabricating complex structures and devices with precise control over geometry. Piezoelectric biomaterials can be incorporated into 3D printing processes by either directly printing piezoelectric materials or by integrating piezoelectric components (e.g., sensors or actuators) into the printed structures. This approach allows for the creation of customized and functional piezoelectric devices, such as wearable sensors, tissue engineering constructs, or implantable devices [49].

vii. Template-Free Growth: In some cases, piezoelectric biomaterials can be fabricated through template-free growth methods. These methods involve the direct synthesis or deposition of piezoelectric materials without the need for templates or molds. For example, chemical vapor deposition (CVD) or physical vapor deposition (PVD) techniques can be used to grow thin films or nanoscale structures of piezoelectric materials on substrates. These methods offer precise control over film thickness, composition, and crystal orientation, enabling the fabrication of piezoelectric materials with tailored properties for specific applications [1].

Table 4.3 Comparison of different piezoelectric biomaterial fabrication techniques

Fabrication technique	Description	Advantages	Limitations
Solid-state processing	Involves sintering precursor powders at high temperatures	Produces dense and crystalline materials	Limited control over microstructure and composition
Sol-gel method	Synthesis of materials from precursor sols and subsequent gelation	Homogeneous materials with controlled composition	Complex processing steps and potential for impurity contamination
Electrospinning	Formation of fibers or membranes through electrically charged jets	Ability to create fine and controllable fibrous structures	Difficulty in achieving uniformity and alignment of piezoelectric phases
Chemical vapor deposition	Deposition of thin films from precursor gases	Precise control over film thickness and composition	High equipment and processing complexity
Additive manufacturing [3D printing]	Layer-by-layer deposition of materials using computer-aided design models	Customization and intricate geometries	Limited materials available for 3D printing and potential loss of piezoelectric properties in the process

viii. Biofabrication: Biofabrication techniques involve the use of living cells or tissues to create piezoelectric biomaterials. This approach utilizes biocompatible scaffolds or matrices seeded with cells that have piezoelectric properties or can produce piezoelectric materials. Through cellular activities and extracellular matrix deposition, the cells contribute to the formation of piezoelectric biomaterials with tailored properties for tissue engineering applications [50].

ix. Hybrid Approaches: Hybrid approaches combine multiple fabrication methods to achieve desired properties and functionalities. For instance, a combination of electrospinning and sol-gel techniques can be employed to fabricate composite piezoelectric materials with enhanced mechanical properties and piezoelectric response. Hybrid approaches allow for the incorporation of multiple materials or structures, providing synergistic benefits and expanding the range of achievable properties [51].

x. Surface Functionalization: Surface functionalization techniques can be employed to modify the surface properties of piezoelectric biomaterials. By introducing functional groups or coatings onto the surface, biocompatibility, chemical stability, or interaction with biological entities can be enhanced. Surface modification methods include plasma treatment, chemical grafting, or the deposition of bioactive molecules or coatings. Functionalized surfaces are particularly valuable in applications such as tissue engineering, where optimized interactions with cells and tissues are desired [5]. Table 4.3 depicts some comparative of various piezoelectric biomaterial fabrication techniques.

4.7.1 Design of Piezoelectric Biomaterials

The design of piezoelectric biomaterials involves tailoring the composition, structure, and properties of the materials to achieve specific performance requirements for various biomedical applications. Effective design strategies consider factors such as piezoelectric response, mechanical properties, biocompatibility, and integration with biological systems [7]. Below are key aspects involved in the design of piezoelectric biomaterials:

a. Material Selection: The choice of piezoelectric material is crucial and depends on the desired properties and applications. Natural materials, such as bone, tendon, or proteins, possess inherent piezoelectricity and can be utilized directly or as a source of inspiration for synthetic materials. Synthetic piezoelectric materials, including ceramics, polymers, or composites, can be engineered to exhibit specific piezoelectric properties. The selection should consider factors such as piezoelectric coefficients, mechanical compliance, biocompatibility, stability, and processability [52].

b. Piezoelectric Property Enhancement: Various strategies can be employed to enhance the piezoelectric properties of biomaterials. This includes optimizing the composition, crystal structure, and crystallinity of materials. For example, doping certain ions into piezoelectric ceramics can modify their crystal lattice, leading to improved piezoelectric coefficients. Additionally, nanostructuring techniques can be utilized to enhance piezoelectric performance by introducing size effects and increasing the surface area [20].

c. Structural Design: The structural design of piezoelectric biomaterials is crucial for their functionality and integration with biological systems. It involves considerations such as the shape, size, and architecture of the material. For instance, the design of tissue engineering scaffolds should mimic the native tissue structure to provide mechanical support and enable cell attachment and migration. Micro- and nanostructured surfaces can also be incorporated to enhance cellular interactions and promote tissue regeneration [53].

d. Mechanical Properties: The mechanical properties of piezoelectric biomaterials should be carefully considered to ensure compatibility with the targeted application. The materials should possess suitable elasticity, strength, and flexibility to withstand mechanical forces and deformations without failure. Additionally, matching the mechanical properties of the biomaterials with the surrounding tissues can prevent stress concentration and improve integration [54].

e. Biocompatibility: Biocompatibility is a critical consideration in the design of piezoelectric biomaterials for biomedical applications. Materials should be non-toxic, non-immunogenic, and capable of interacting safely with biological systems. The surface properties, such as surface charge, roughness, and functionalization, can be engineered to promote cell adhesion, proliferation, and tissue integration. Biocompatibility testing, including cytotoxicity assays and in vivo studies, should be conducted to ensure the safety and efficacy of the designed materials [55].

f. Integration and Device Design: Piezoelectric biomaterials are often integrated into devices or systems for specific applications. The design should consider the integration of piezoelectric components, such as sensors or actuators, with electronic circuitry, signal processing, or power systems. The device design should optimize performance, functionality, and usability while considering factors such as size, weight, power requirements, and compatibility with the target application [56].

g. Multifunctionality and Hybrid Designs: Piezoelectric biomaterials can be designed to have multiple functionalities or combined with other materials or components to achieve desired properties. For example, incorporating sensing elements into piezoelectric materials can enable simultaneous sensing and actuation capabilities. Hybrid designs that combine piezoelectric materials with other functional materials, such as conductive polymers or biodegradable matrices, can offer enhanced functionalities for specific applications [3].

h. Computational Modeling and Simulation: Computational modeling and simulation play a crucial role in the design of piezoelectric biomaterials. Finite element analysis (FEA) and molecular dynamics simulations can provide insights into the mechanical behavior, piezoelectric response, and interactions at different length scales. Modeling techniques can aid in optimizing the design parameters, predicting material behavior, and accelerating the development process [57].

i. Stimulus Responsiveness: The design of piezoelectric biomaterials can focus on their stimulus responsiveness, enabling them to generate a specific electrical response to particular mechanical or environmental stimuli. For example, designing materials with a selective response to specific frequencies, pressure levels, or mechanical strains can be advantageous in applications such as biosensing, where targeted detection and measurement are required [3].

j. Biodegradability: In some cases, the design of piezoelectric biomaterials may incorporate biodegradability as a desired property.

Biodegradable piezoelectric materials can be designed to degrade over time, aligning with the natural healing or regeneration process. This feature is particularly important in temporary implants, drug delivery systems, or tissue engineering scaffolds where the biomaterial needs to degrade and be replaced by native tissue over time [54].

4.7.2 Design Considerations for Piezoelectric Biomaterials

The design of piezoelectric biomaterials involves careful consideration of various factors to ensure their optimal performance and suitability for specific applications. Below are key design considerations for piezoelectric biomaterials [7, 58]:

i. Piezoelectric Properties: The primary focus of designing piezoelectric biomaterials is to achieve desired piezoelectric properties, including high piezoelectric coefficients, sensitivity, and stability. The material's piezoelectric response and electromechanical coupling factor [k value] are crucial parameters to consider. Design strategies may involve selecting appropriate piezoelectric materials and optimizing their composition, crystal structure, or nanostructuring to enhance piezoelectric properties.

ii. Mechanical Properties: Piezoelectric biomaterials should possess mechanical properties that align with the intended application. Mechanical considerations include elasticity, strength, flexibility, and fatigue resistance. The biomaterial's mechanical properties should match or closely mimic the target tissue to avoid mechanical mismatch, stress concentration, or mechanical failure. Design strategies may involve selecting appropriate biomaterials, adjusting their composition, or using composite structures to achieve the desired mechanical properties.

iii. Biocompatibility: Biocompatibility is a crucial consideration for piezoelectric biomaterials intended for biomedical applications. Materials should be biologically safe, non-toxic, and compatible with the surrounding biological environment. Design strategies may include selecting biocompatible materials, surface modifications to improve biocompatibility, or incorporating bioactive molecules or coatings to promote cellular adhesion and tissue integration.

iv. Electrical Performance: The electrical performance of piezoelectric biomaterials is critical for their effective integration into devices or systems. Design considerations include electrical impedance, voltage output, electrical noise, and insulation properties. Design strategies may involve optimizing the material composition, surface modifications, or incorporating suitable electrodes to improve electrical performance and minimize signal interference.

v. Durability and Stability: Piezoelectric biomaterials should exhibit long-term durability and stability to maintain their performance over time. Design considerations include resistance to mechanical fatigue, chemical degradation, or environmental factors such as temperature and humidity. Design strategies may involve selecting materials with high chemical stability, optimizing processing techniques, or incorporating protective coatings to enhance durability and stability.

vi. Integration with Biological Systems: The design of piezoelectric biomaterials should consider their integration with biological systems, such as tissues or cells. Factors such as cell adhesion, proliferation, and differentiation should be taken into account. Design strategies may involve surface modifications, scaffold structures, or incorporation of bioactive molecules to promote cellular interactions and tissue regeneration.

vii. Miniaturization and Device Integration: Piezoelectric biomaterials are often integrated into miniaturized devices or systems. Design considerations include size, weight, and compatibility with electronic components or systems. Design strategies may involve microfabrication techniques, additive manufacturing methods, or flexible substrates to enable miniaturization and seamless integration with electronic components.

viii. Stimulus Selectivity and Sensitivity: In certain applications, the design of piezoelectric biomaterials should focus on stimulus selectivity and sensitivity. This involves tailoring the material's response to specific mechanical stimuli, frequencies, or pressure levels. Design strategies may include material composition optimization, structural design, or integrating signal processing techniques to enhance selectivity and sensitivity.

ix. Multifunctionality and Multimodal Sensing: Piezoelectric biomaterials can be designed to have multiple functionalities or enable multimodal sensing. This involves incorporating additional sensing capabilities or combining different sensing mechanisms within the material. Design strategies may involve the integration of additional sensors or utilizing hybrid materials to achieve desired multimodal sensing capabilities.

x. Scalability and Manufacturing: The scalability and manufacturability of piezoelectric biomaterials are crucial for practical applications. Design considerations include compatibility with established manufacturing techniques, cost-effectiveness, and reproducibility. Design strategies may involve selecting materials with scalable synthesis methods, optimizing manufacturing processes, or utilizing additive manufacturing techniques for complex geometries.

xi. Environmental Sustainability: Designing piezoelectric biomaterials with environmental sustainability in mind is becoming increasingly important. Considerations may include the use of eco-friendly materials, reduction of hazardous substances, and energy-efficient manufacturing processes. Design strategies may involve selecting materials with low environmental impact, designing for recyclability or reusability, and minimizing waste generation.

4.8 Characterization Techniques for Evaluating Piezoelectric Properties

Characterizing the piezoelectric properties of biomaterials is essential for understanding their behavior, optimizing their performance, and validating their suitability for specific applications. Several characterization techniques are commonly used to evaluate the piezoelectric properties of materials [59]. Figure 4.3 shows some key techniques employed in the characterization of piezoelectric biomaterials [20, 60].

i. Piezoelectric Coefficient Measurement: Piezoelectric coefficients quantify the relationship between mechanical strain or stress and the resulting electrical charge or voltage. Various methods are used to measure the piezoelectric coefficients, including the d33 meter, Berlincourt method, or interferometric techniques. These measurements involve subjecting the material to controlled mechanical deformation and measuring the corresponding electrical response to determine the piezoelectric coefficients.

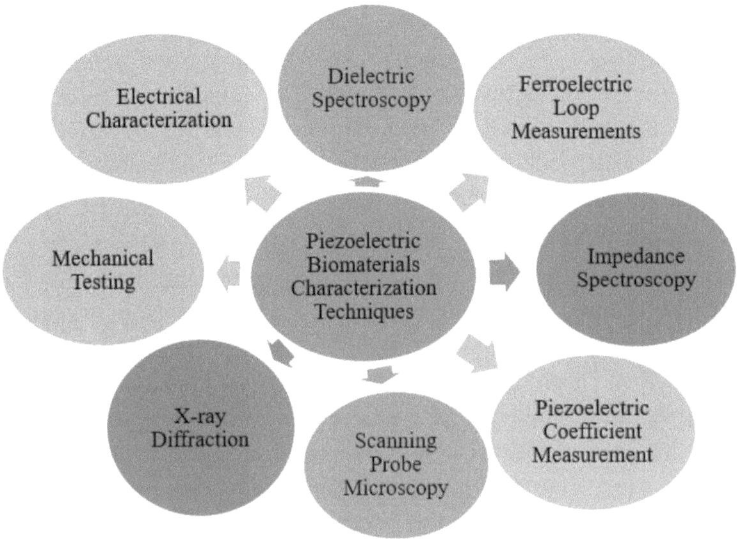

Figure 4.3 An overview of characterization techniques for piezoelectric biomaterials.

ii. *Impedance Spectroscopy:* Impedance spectroscopy is a widely used technique for characterizing the electrical properties of piezoelectric materials. It involves applying a sinusoidal voltage or current signal to the material and measuring the resulting impedance across a range of frequencies. By analyzing the impedance response, parameters such as capacitance, resistance, and resonance frequency can be determined, providing insights into the electrical behavior and performance of the material.

iii. *Ferroelectric Loop Measurements:* Ferroelectric loop measurements are performed to assess the ferroelectric and hysteresis behavior of piezoelectric materials. This technique involves applying an electric field to the material and measuring the resulting polarization as a function of the applied field strength. Ferroelectric loops provide information about the material's remanent polarization, coercive field, and polarization switching characteristics, which are important for understanding the ferroelectric properties and reliability of the material.

iv. Dielectric Spectroscopy: Dielectric spectroscopy is used to investigate the dielectric response and polarization behavior of piezoelectric materials under varying frequencies and temperatures. This technique involves applying an electric field and measuring the resulting dielectric constant and loss tangent across a range of frequencies. Dielectric spectroscopy provides information about the material's electrical polarization, relaxation dynamics, and frequency-dependent behavior, which are crucial for understanding its electrical performance.

v. Scanning Probe Microscopy (SPM): Scanning probe microscopy techniques, such as atomic force microscopy (AFM) and piezoresponse force microscopy (PFM), allow for the direct visualization and mapping of piezoelectric properties at the nanoscale. These techniques utilize a sharp probe to scan the surface of the material, detecting the local piezoelectric response and generating high-resolution images. SPM techniques provide information about the material's piezoelectric domains, domain-switching behavior, and local variations in piezoelectric properties.

vi. X-ray Diffraction (XRD): X-ray diffraction is used to analyze the crystal structure, orientation, and phase composition of piezoelectric materials. By analyzing the diffraction pattern produced when X-rays interact with the material, important information about crystallographic phases, crystal orientation, lattice parameters, and preferred crystallographic orientations can be obtained. XRD is crucial for assessing the crystallographic properties and understanding the structure-property relationship of piezoelectric biomaterials.

vii. Mechanical Testing: Mechanical testing techniques, such as tensile testing, compression testing, or nanoindentation, can provide insights into the mechanical behavior and properties of piezoelectric biomaterials. These tests measure parameters such as Young's modulus, tensile strength, or hardness, which are important for understanding the material's response to mechanical forces and its suitability for specific applications.

viii. Electrical Characterization: Electrical characterization techniques, such as current-voltage (I-V) measurements or leakage current analysis, assess the electrical behavior and performance of piezoelectric materials. These tests provide information about parameters such as electrical conductivity, leakage current, breakdown voltage, and insulation properties. Electrical characterization is essential for evaluating the material's electrical reliability, insulation capability, and performance under applied electric fields.

4.9 Techniques for Optimizing the Piezoelectric Properties of Biomaterials

Optimizing the piezoelectric properties of biomaterials is crucial for enhancing their performance and enabling their successful application in various fields. Several techniques can be employed to optimize the piezoelectric properties of biomaterials [50]. Below are some commonly used techniques:

i. Material Composition Optimization: Adjusting the composition of piezoelectric biomaterials is a key strategy for optimizing their piezoelectric properties. This can involve doping or alloying the base material with specific elements or ions. By carefully selecting the dopant or alloying elements, the piezoelectric coefficients, crystal structure, and domain structure can be modified to enhance the piezoelectric response. For example, in PZT ceramics, substituting certain elements, such as niobium or lanthanum, can improve the piezoelectric properties.

ii. Crystal Orientation Engineering: The piezoelectric response of some materials is highly dependent on the crystal orientation. By carefully controlling the crystal orientation during the fabrication process, the piezoelectric properties can be optimized. This is typically achieved through poling processes, which involve applying a strong electric field to align the crystal domains in a desired orientation. By aligning the crystals in the most favorable orientation, the piezoelectric response can be significantly enhanced.

iii. Nanostructuring: Nanostructuring techniques can be employed to optimize the piezoelectric properties of biomaterials. This involves manipulating the size and shape of the material at the nanoscale. Nanostructuring can lead to enhanced piezoelectric response due to the increased surface area, altered crystal structure, and modified domain configuration. Techniques such as sol-gel processing, CVD, or template-assisted growth can be used to fabricate nanostructured piezoelectric biomaterials.

iv. Strain Engineering: Applying mechanical strain to piezoelectric biomaterials can modify their piezoelectric properties. By subjecting the material to controlled mechanical deformation or strain, the polarization and crystal structure can be manipulated, leading to improved piezoelectric performance. Techniques such as bending, stretching, or applying hydrostatic pressure can be used to induce strain and optimize the piezoelectric response.

v. Poling Optimization: The poling process plays a crucial role in optimizing the piezoelectric properties of materials. The poling parameters, including electric field strength, duration, and temperature, can be optimized to achieve the desired piezoelectric response. Careful control of the poling process ensures that the material's domains are aligned in the most favorable configuration, maximizing the piezoelectric coefficients and electromechanical coupling.

vi. Composites and Hybrid Structures: Incorporating piezoelectric materials into composites or hybrid structures can offer advantages in optimizing piezoelectric properties. Combining different materials with complementary properties can lead to synergistic effects and enhanced overall performance. For example, combining piezoelectric ceramics with polymer matrices can improve mechanical compliance, while maintaining good piezoelectric response. Additionally, using core-shell structures or layered composites can provide precise control over piezoelectric properties.

vii. Surface Functionalization: Surface functionalization techniques can be employed to modify the surface properties of piezoelectric biomaterials, which can affect their piezoelectric performance. Surface modifications can include chemical treatments, coatings, or the deposition of functional molecules. These modifications can improve charge generation, enhance electrical contact, or promote interactions with surrounding biological systems, thereby optimizing the piezoelectric response.

viii. Process Optimization: Optimizing the fabrication processes used to produce piezoelectric biomaterials can significantly impact their piezoelectric properties. Process parameters such as temperature, pressure, and heating/cooling rates can be adjusted to control the material's microstructure, phase purity, and domain configuration. Optimization of the processing steps ensures reproducibility, uniformity, and improved piezoelectric performance.

ix. Computational Modeling and Simulation: Computational modeling and simulation techniques play a crucial role in optimizing the piezoelectric properties of biomaterials. FEA, molecular dynamics simulations, and density functional theory (DFT) calculations can provide insights into the material's behavior at different length scales. Modeling and simulation can aid in understanding the effect of design parameters, material composition, and processing conditions on the piezoelectric properties, facilitating the optimization process.

x. Multiscale Characterization: Comprehensive characterization techniques at different length scales are essential for optimizing piezoelectric properties. This includes macroscopic measurements of piezoelectric coefficients, electromechanical coupling, and mechanical properties, as well as microstructural analysis to investigate crystallographic orientation, domain configuration, and defects. By combining multiple characterization techniques, a holistic understanding of the material's piezoelectric behavior can be obtained, guiding the optimization efforts. Optimizing the piezoelectric properties of biomaterials requires a combination of experimental investigations, computational modeling, and iterative design approaches. By employing these techniques, researchers and engineers can enhance the performance of piezoelectric biomaterials, opening up new possibilities for applications in areas such as healthcare, energy harvesting, sensing, and actuation.

4.10 Challenges in Designing Piezoelectric Biomaterials

Designing piezoelectric biomaterials poses several challenges that need to be addressed to ensure successful development and implementation. These challenges arise from the complex nature of integrating piezoelectric properties into biomaterials while considering their compatibility with biological systems and targeted applications. Below are some key challenges in designing piezoelectric biomaterials [7, 58]:

a. Biocompatibility: One of the primary challenges is achieving biocompatibility while maintaining or enhancing the piezoelectric properties of the biomaterials. Ensuring that the materials are non-toxic, non-immunogenic, and capable of interacting safely with biological systems is crucial. The design process must consider the biocompatibility of the selected piezoelectric materials, surface modifications, and any additional components or coatings to prevent adverse reactions or tissue damage.

b. Mechanical Compatibility: Piezoelectric biomaterials need to possess mechanical properties that match or closely mimic the target tissue or biological system. Mechanical compatibility is crucial to avoid stress concentration, mechanical mismatch, or failure of the biomaterials. Designing materials with suitable elasticity, strength, flexibility, and fatigue resistance that can withstand the mechanical forces and deformations encountered in the application is a challenge.

c. Piezoelectric Coefficients: Achieving high piezoelectric coefficients is important for the effective conversion of mechanical energy into electrical signals. However, optimizing and maximizing the piezoelectric coefficients of biomaterials can be challenging. The selection of appropriate piezoelectric materials, composition optimization, crystal structure control, and domain engineering techniques need to be carefully considered to enhance the piezoelectric response.

d. Stability and Durability: Stability and long-term durability of piezoelectric biomaterials are essential for their reliable performance over time. The challenge lies in maintaining the piezoelectric properties of the materials under various environmental conditions, such as temperature, humidity, and mechanical stress. Design strategies need to address stability issues, including minimizing material degradation, fatigue, or depoling, to ensure the long-term functionality of the piezoelectric biomaterials.

e. Scalability and Manufacturing: Designing piezoelectric biomaterials that are scalable for mass production is a challenge. The fabrication techniques used to produce these materials must be compatible with large-scale manufacturing processes while maintaining their piezoelectric properties and reproducibility.

Developing cost-effective and efficient manufacturing methods is crucial for commercial viability and the widespread adoption of piezoelectric biomaterials.

f. Integration and Device Design: Integrating piezoelectric biomaterials into devices or systems poses challenges in terms of compatibility with electronic components, power requirements, and overall device design. Ensuring seamless integration and efficient electrical contact and addressing potential challenges related to signal processing, noise reduction, and power management are important aspects to consider during the design process.

g. Multifunctionality and Hybrid Designs: Designing piezoelectric biomaterials with multiple functionalities or integrating them with other materials or components can be challenging. The compatibility of different materials, optimizing the interfaces, and ensuring synergistic effects while maintaining the desired properties can be complex. Achieving effective integration and maximizing the performance of multifunctional or hybrid designs requires careful consideration of materials selection, compatibility, and performance optimization.

h. Clinical Translation and Regulatory Considerations: Moving piezoelectric biomaterials from the laboratory to clinical use involves navigating regulatory frameworks and demonstrating safety and efficacy. Meeting regulatory requirements, conducting preclinical and clinical studies, and ensuring the long-term reliability and performance of biomaterials in clinical settings are significant challenges. Collaboration between researchers, clinicians, and regulatory bodies is essential to address these challenges and facilitate clinical translation.

i. Characterization and Validation: Thorough characterization and validation of piezoelectric biomaterials pose challenges due to their complex properties and interactions. Developing comprehensive and standardized characterization techniques for assessing piezoelectric properties, mechanical behavior, biocompatibility, and long-term stability is essential. Validating the designed materials through rigorous testing and validation procedures is crucial to ensure their performance meets the intended application requirements.

4.11 Applications of Piezoelectric Biomaterials

Piezoelectric biomaterials find a wide range of applications in various fields due to their unique ability to convert mechanical energy into electrical signals. These materials offer numerous opportunities for advancements in healthcare, diagnostics, energy harvesting, sensing, and actuation [13, 58]. Figure 4.4 depicts some prominent applications of piezoelectric biomaterials.

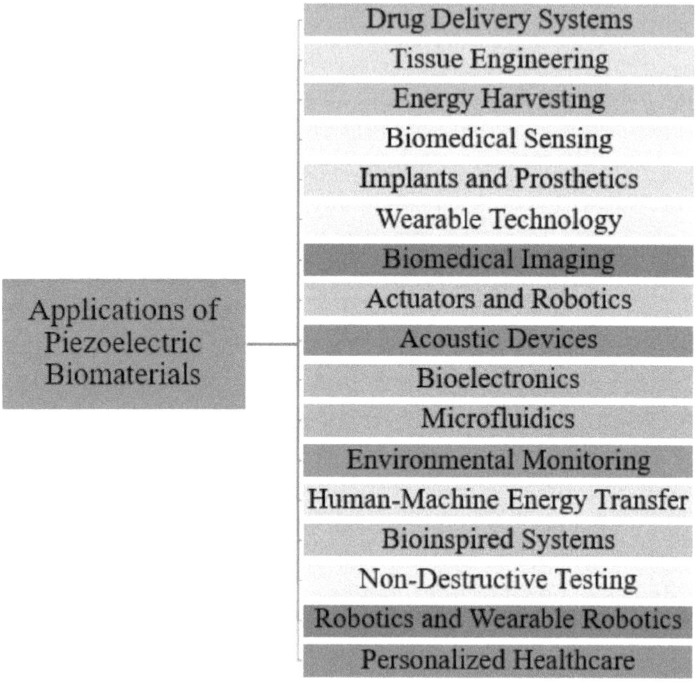

Figure 4.4 An overview of various applications of piezoelectric biomaterials.

i. Biomedical Sensing: Piezoelectric biomaterials are extensively used in biosensing applications. They can be employed to detect and measure biological parameters such as pressure, strain, force, or molecular interactions. Examples include piezoelectric sensors for blood pressure monitoring, tactile sensors for prosthetic limbs, and piezoelectric immunosensors for rapid and sensitive detection of biomarkers in diagnostics.

ii. Energy Harvesting: Piezoelectric biomaterials have great potential for energy harvesting applications, where they can convert mechanical vibrations or movements into electrical energy. These materials can be integrated into wearable devices, implants, or structures to scavenge and store energy from body movements, ambient vibrations, or external mechanical sources. Energy harvested from piezoelectric biomaterials can power various low-power electronic devices or recharge batteries.

iii. Tissue Engineering: Piezoelectric biomaterials have been explored for tissue engineering applications, where they can provide mechanical stimulation and promote tissue regeneration. By incorporating piezoelectric materials into scaffolds or hydrogels, mechanical forces or electrical signals can be applied to cells or tissues, mimicking the native tissue environment and enhancing cellular responses, including cell proliferation, differentiation, and extracellular matrix production.

iv. Drug Delivery Systems: Piezoelectric biomaterials have been investigated for controlled drug delivery applications. By applying mechanical stimuli to piezoelectric materials, they can deform and generate localized electrical fields, facilitating the release of drugs encapsulated within the material. These materials can be used to develop on-demand drug delivery systems with precise spatial and temporal control.

v. Implants and Prosthetics: Piezoelectric biomaterials have the potential to improve the functionality and performance of implants and prosthetic devices. They can be integrated into implants, such as pacemakers or cochlear implants, to harvest energy from the body's movements or vibrations, eliminating the need for battery replacements. Piezoelectric materials can also be utilized in smart prosthetics to enhance sensory feedback or enable active control of limb movements.

vi. Wearable Technology: Piezoelectric biomaterials have applications in wearable technology, enabling the development of self-powered wearable devices and sensors. These materials can be incorporated into textiles, fabrics, or flexible substrates, allowing for the generation of electrical energy from body movements or external mechanical forces. Examples include wearable sensors for healthcare monitoring, gesture recognition devices, and electronic textiles.

vii. Biomedical Imaging: Piezoelectric biomaterials are used in various imaging modalities, such as ultrasound imaging or photoacoustic imaging. Piezoelectric transducers, based on materials like PZT or PVDF, are employed to generate and detect ultrasonic waves, enabling non-invasive imaging of tissues and organs.

viii. Actuators and Robotics: Piezoelectric biomaterials are utilized as actuators in robotic systems and micro-/nano-manipulation devices. Their ability to convert electrical signals into mechanical deformation allows for precise control and manipulation of objects or structures at small scales. Piezoelectric actuators find applications in areas such as microsurgery, lab-on-a-chip devices, micropositioning systems, and haptic interfaces.

ix. Acoustic Devices: Piezoelectric biomaterials are used in the development of acoustic devices such as microphones, hydrophones, and ultrasonic transducers. Their ability to convert acoustic waves into electrical signals and vice versa makes them crucial components in communication systems, underwater sensing, medical imaging, and ultrasonic cleaning devices.

These applications represent just a glimpse of the broad range of possibilities for piezoelectric biomaterials. Ongoing research and advancements in materials engineering, fabrication techniques, and device integration continue to expand the applications of piezoelectric biomaterials, making them versatile tools in numerous fields and contributing to advancements in healthcare, energy, and beyond.

x. Bioelectronics: Piezoelectric biomaterials are utilized in bioelectronic devices for interfacing with biological systems. They can be integrated into electronic circuits or devices to enable direct coupling with biological tissues or cells. Examples include piezoelectric transducers for neural stimulation, implantable sensors for bioelectrical signal monitoring, and bioelectric energy harvesting systems.

xi. Microfluidics: Piezoelectric biomaterials find applications in microfluidic systems, where they can actuate fluid flow or control droplet manipulation. Applying electrical signals to the piezoelectric material can induce fluidic motion, enabling precise control over microscale fluidic processes. This is useful in lab-on-a-chip devices, microreactors, and point-of-care diagnostics.

xii. Environmental Monitoring: Piezoelectric biomaterials have applications in environmental monitoring systems. They can be utilized to develop sensors for detecting and monitoring parameters such as air pollution, water quality, or environmental changes. Piezoelectric sensors can provide real-time, sensitive, and selective measurements, contributing to environmental monitoring and management efforts.

xiii. Human-Machine Energy Transfer: Piezoelectric biomaterials have the potential to enable energy transfer between humans and machines. By incorporating piezoelectric materials into wearable devices or interactive surfaces, mechanical energy generated by human movements or touch can be converted into electrical energy to power electronic devices, reducing the reliance on conventional power sources.

xiv. Bioinspired Systems: Piezoelectric biomaterials can be used to develop bioinspired systems that mimic natural structures and functionalities. For example, by mimicking the piezoelectric properties found in bone or teeth, biomaterials can be designed to enhance the regeneration of bone tissue or facilitate dental applications. Bioinspired systems can also be used for the development of artificial muscles or smart materials that respond to external stimuli.

xv. Non-Destructive Testing: Piezoelectric biomaterials are employed in non-destructive testing techniques, such as ultrasonic testing, for evaluating the integrity and quality of materials or structures. The piezoelectric transducers generate and receive ultrasonic waves, allowing for the detection of flaws, cracks, or material property variations, ensuring the safety and reliability of structures.

xvi. Robotics and Wearable Robotics: Piezoelectric biomaterials have applications in robotics and wearable robotics systems. They can be used to develop actuators, sensors, or energy-harvesting components for robotic devices or exoskeletons. Piezoelectric materials enable precise control, sensing, and power generation, enhancing the performance and functionality of robotic systems.

xvii. Personalized Healthcare: Piezoelectric biomaterials have the potential to contribute to personalized healthcare applications. By integrating them into wearable devices or implants, real-time health monitoring, diagnostics, and personalized treatments can be facilitated. The piezoelectric properties enable the development of wearable sensors, implantable devices, and therapeutic systems tailored to individual patients' needs.

4.11.1 Biomedical Applications of Piezoelectric Biomaterials

Piezoelectric biomaterials offer unique capabilities for various biomedical applications, leveraging their ability to convert mechanical energy into electrical signals. These materials have found significant utility in different areas of healthcare, diagnostics, and therapeutics. Some notable biomedical applications of piezoelectric biomaterials are discussed below [11]:

i. Biomechanical Sensing and Monitoring: Piezoelectric biomaterials are widely used for sensing and monitoring applications in biomechanics. They can be integrated into devices such as pressure sensors, accelerometers, or force sensors to measure biomechanical parameters, including pressure distribution, joint movement, gait analysis, or muscle activity. These sensing technologies find applications in sports performance monitoring, rehabilitation, ergonomic assessment, and biomechanical research.

ii. Implantable Sensors and Devices: Piezoelectric biomaterials are employed in implantable sensors and devices for monitoring various physiological parameters. They can be integrated into implants to measure parameters such as blood pressure, bone strain, or bladder pressure. These sensors provide real-time data for diagnostics, patient monitoring, and therapeutic applications.

iii. Tissue Engineering and Regenerative Medicine: Piezoelectric biomaterials are utilized in tissue engineering and regenerative medicine applications to enhance tissue regeneration and stimulate cellular behavior. By incorporating piezoelectric materials into scaffolds, the mechanical stimuli generated by the materials can promote cell proliferation, differentiation, and tissue growth. These biomaterials find applications in bone tissue engineering, wound healing, cartilage regeneration, and nerve tissue engineering.

iv. Controlled Drug Delivery: Piezoelectric biomaterials are employed in controlled drug delivery systems. By applying mechanical forces or electrical signals to the materials, drug release can be triggered or controlled. These biomaterials enable on-demand drug delivery, precise spatial and temporal control of drug release, and targeted therapies. They find applications in implantable drug delivery devices, transdermal patches, and localized drug release systems.

v. Bioelectronics and Bioelectrodes: Piezoelectric biomaterials are utilized in bioelectronic devices and bioelectrodes for interfacing with biological systems. They can be integrated into electronic circuits or implants to enable direct communication and stimulation of biological tissues or cells. Piezoelectric materials find applications in neural interfaces, bioelectric signal monitoring, cardiac pacing, and biosensing.

vi. Ultrasound Imaging and Therapy: Piezoelectric materials are extensively used in ultrasound imaging and therapy applications. They serve as the transducers that generate and detect ultrasonic waves for medical imaging or therapeutic purposes. Piezoelectric materials enable high-resolution imaging, ultrasound-guided interventions, and targeted therapies such as high-intensity focused ultrasound (HIFU) treatment.

vii. Energy Harvesting for Implants: Piezoelectric biomaterials can harvest mechanical energy from body movements or vibrations and convert it into electrical energy. This energy can be used to power implantable medical devices, eliminating the need for external power sources or battery replacements. Energy harvesting from piezoelectric biomaterials finds applications in pacemakers, neurostimulators, and other implantable devices.

viii. Biomedical Robotics and Prosthetics: Piezoelectric biomaterials contribute to the development of biomedical robotics and prosthetics. They can be used as actuators in robotic systems or prosthetic limbs to provide precise control and natural movement. These biomaterials enable responsive and biomimetic motion in prosthetics and robotic assistive devices.

4.11.2 Potential Applications in Tissue Engineering and Regenerative Medicine

Piezoelectric biomaterials hold great promise for applications in tissue engineering and regenerative medicine. Their ability to generate mechanical stimuli and electrical signals can significantly influence cellular behavior, tissue growth, and regeneration [1]. Below are some potential applications of piezoelectric biomaterials in tissue engineering and regenerative medicine:

a. Bone Tissue Engineering: Piezoelectric biomaterials can be utilized in bone tissue engineering to promote bone growth and regeneration. By incorporating piezoelectric scaffolds or implants into bone defects, mechanical stimulation generated by these materials can stimulate osteogenesis and enhance bone healing. The piezoelectric effect can induce electrical signals that promote cell adhesion, proliferation, and differentiation, leading to accelerated bone formation and improved integration of the implant with the surrounding tissue.

b. Cartilage Tissue Engineering: Piezoelectric biomaterials have the potential to enhance cartilage tissue engineering strategies. By providing mechanical stimulation through piezoelectric scaffolds or implants, these materials can promote chondrogenesis and cartilage regeneration. The mechanical forces generated by the piezoelectric effect can mimic the compressive forces experienced by cartilage tissue, stimulating the production of extracellular matrix components and supporting the development of functional cartilage tissue.

c. Nerve Tissue Engineering: Piezoelectric biomaterials show promise in nerve tissue engineering for promoting nerve regeneration and functional recovery. By incorporating piezoelectric materials into nerve conduits or scaffolds, they can generate electrical signals and mechanical cues that guide nerve cell growth, axonal alignment, and neuronal connectivity. The piezoelectric effect can aid in the regeneration of damaged or severed nerves and facilitate the restoration of nerve function.

d. Wound Healing: Piezoelectric biomaterials can contribute to wound healing by providing mechanical stimulation and electrical cues to enhance tissue regeneration. By incorporating piezoelectric dressings or scaffolds into wound sites, they can generate mechanical forces and electrical signals that promote cellular migration, proliferation, and the synthesis of extracellular matrix components. These materials can accelerate the healing process, improve wound closure, and reduce scar formation.

e. Vascular Tissue Engineering: Piezoelectric biomaterials have the potential to aid in vascular tissue engineering by promoting angiogenesis and vascularization. By applying mechanical stimulation through piezoelectric scaffolds or implants, they can induce endothelial cell proliferation, alignment, and the formation of functional blood vessels. The electrical signals generated by the piezoelectric effect can stimulate the release of angiogenic factors and support the development of vascularized tissue constructs.

f. Muscle Tissue Engineering: Piezoelectric biomaterials offer opportunities for muscle tissue engineering applications. By incorporating piezoelectric scaffolds or constructs, they can provide mechanical stimulation and electrical cues to promote myogenesis, muscle cell alignment, and contractile function. The piezoelectric effect can enhance the differentiation of myoblasts into mature muscle fibers and facilitate the regeneration of functional muscle tissue.

g. Skin Tissue Engineering: Piezoelectric biomaterials can contribute to skin tissue engineering and wound healing applications. By incorporating piezoelectric scaffolds or dressings, they can generate mechanical forces and electrical signals that stimulate cell migration, proliferation, and the production of extracellular matrix components. These materials can facilitate the regeneration of skin tissue and improve the healing of chronic wounds.

h. Organ-on-a-Chip Systems: Piezoelectric biomaterials can be integrated into organ-on-a-chip systems to mimic the mechanical and electrical cues experienced by specific organs. By incorporating piezoelectric materials into microfluidic devices, they can provide dynamic mechanical stimulation and electrical signals that mimic the physiological environment of the target organ. This can enable more accurate modeling of organ function and disease progression, as well as drug screening and toxicity testing.

4.11.3 Use of Piezoelectric Biomaterials in Sensing and Actuation

Piezoelectric biomaterials have gained significant attention for their applications in sensing and actuation. Their ability to convert mechanical energy into electrical signals and vice versa makes them well-suited for these purposes. Here are some notable uses of piezoelectric biomaterials in sensing and actuation [13].

4.11.3.1 Sensing applications

a. Pressure Sensing: Piezoelectric biomaterials can be employed in pressure sensors to detect and measure mechanical pressure variations. When pressure is applied to the material, it generates an electrical charge proportional to the applied pressure. This property enables the development of sensitive pressure sensors for applications such as biomedical sensing, touch sensing, and industrial pressure monitoring.

b. Strain Sensing: Piezoelectric biomaterials can be used as strain sensors to measure mechanical deformation or strain. When subjected to strain, the material generates electrical charges that can be correlated to the applied strain. This property is valuable in applications such as structural health monitoring, robotics, and biomechanical sensing.

c. Acceleration and Vibration Sensing: Piezoelectric biomaterials are employed in accelerometers and vibration sensors. When subjected to acceleration or vibration, the material generates electrical charges that can be measured to determine the magnitude and frequency of the motion. These sensors find applications in automotive systems, wearable devices, and industrial monitoring.

d. Biomedical Sensing: Piezoelectric biomaterials can be utilized for various biomedical sensing applications. They can be integrated into biosensors to detect biomarkers, proteins, or DNA. The biomaterials' ability to generate electrical signals in response to biological interactions enables the development of sensitive and selective sensing platforms for applications in diagnostics, healthcare monitoring, and drug screening.

4.11.3.2 Actuation applications

a. Ultrasound Transducers: Piezoelectric biomaterials are extensively used in ultrasound imaging and therapy as transducers. When an electrical voltage is applied to the material, it undergoes mechanical deformation, generating ultrasound waves that can be used for imaging or therapeutic purposes. These transducers find applications in medical imaging, non-invasive surgery, and targeted drug delivery.

b. Microactuators: Piezoelectric biomaterials are employed as microactuators in microelectromechanical systems (MEMS) and nanoelectromechanical systems (NEMS). By applying electrical signals to the material, it undergoes mechanical deformation, enabling precise and rapid actuation. Piezoelectric microactuators are used in microvalves, microfluidic devices, optical switches, and microsurgical tools.

c. Energy Harvesting: Piezoelectric biomaterials can act as energy harvesters, converting mechanical energy from vibrations or movements into electrical energy. This energy can be used to power low-power electronic devices, eliminating the need for external power sources or batteries. Energy harvesting using piezoelectric biomaterials finds applications in wearable devices, wireless sensors, and self-powered electronics.

d. Haptic Feedback Systems: Piezoelectric biomaterials can be utilized in haptic feedback systems to provide tactile feedback or sensory stimulation. By applying electrical signals to the material, it undergoes controlled mechanical deformation, creating vibrations or forces that can be felt by the user. These systems find applications in virtual reality, robotics, and user interfaces.

e. Biomedical Actuation: Piezoelectric biomaterials are used for actuation in biomedical devices and systems. They can be incorporated into implants or prosthetics to provide controlled movement or stimulation. By applying electrical signals to the biomaterials, precise actuation can be achieved, enabling functions such as drug release, prosthetic limb movement, or neural stimulation.

The use of piezoelectric biomaterials in sensing and actuation is continually advancing, driven by research and technological developments. As these biomaterials continue to evolve, their applications in sensing and actuation are expected to expand further, leading to advancements in various fields, including healthcare, robotics, and consumer electronics.

4.12 Challenges in Piezoelectric Biomaterials

Piezoelectric biomaterials have shown immense potential in various fields, but there are still several important future directions and challenges that need to be addressed to fully realize their capabilities. Below are some key areas of focus and challenges in the development and utilization of piezoelectric biomaterials [17, 61]:

i. Enhanced Performance and Efficiency: Future research efforts aim to improve the performance and efficiency of piezoelectric biomaterials. This involves optimizing the piezoelectric coefficients, electromechanical coupling, and energy conversion efficiency. Enhancing the piezoelectric response and lowering energy losses will expand the range of applications and increase the effectiveness of these biomaterials in energy harvesting, sensing, and actuation.

ii. Biocompatibility and Long-Term Stability: Biocompatibility and long-term stability are crucial considerations for the successful implementation of piezoelectric biomaterials in clinical and biomedical applications. Future research should focus on enhancing the biocompatibility of these materials, minimizing potential cytotoxicity or immunogenicity issues, and ensuring their stability and reliability over extended periods. Long-term studies and in-depth biocompatibility assessments are essential to address potential challenges and ensure patient safety.

iii. Multifunctionality and Integration: The development of piezoelectric biomaterials with multifunctional capabilities and seamless integration with existing systems is a future direction. Incorporating additional functionalities such as sensing, energy storage, or drug delivery within the biomaterials can lead to more versatile and efficient devices. Integration with electronics, wireless communication, and data processing systems will enable smart, self-powered, and interconnected applications.

iv. Miniaturization and Flexible Form Factors: Future efforts will focus on miniaturizing piezoelectric biomaterials and developing flexible, conformable, or stretchable form factors. This will enable their integration into wearable devices, implantable sensors, or biomedical applications where size, shape, and mechanical properties are critical. Advancements in fabrication techniques and materials engineering are required to achieve miniaturization and flexibility while preserving the piezoelectric properties.

v. Biologically Inspired Design: Drawing inspiration from natural systems, future directions in piezoelectric biomaterials involve designing materials and structures that closely mimic biological counterparts. By emulating the unique properties and functionalities found in living organisms, such as hierarchical structures, self-healing capabilities, or adaptive responses, piezoelectric biomaterials can achieve improved performance and better integration with biological systems.

vi. Advanced Fabrication Techniques: Future research will focus on developing advanced fabrication techniques for piezoelectric biomaterials. This includes exploring additive manufacturing (3D printing), nanofabrication, and self-assembly methods to create complex structures with precise control over the piezoelectric properties. Integration of multiple materials and micro-/nanostructures will allow for tailored piezoelectric response and functionality.

vii. Modeling and Simulation: Computational modeling and simulation techniques will play a vital role in the future design and optimization of piezoelectric biomaterials. Advanced modeling approaches, such as multiscale modeling, molecular dynamics simulations, and machine learning algorithms, will enable a better understanding of the materials' behavior, prediction of performance, and optimization of properties. These techniques will guide the design process, accelerate development, and reduce experimental iterations.

viii. Commercialization and Regulatory Considerations: The successful commercialization of piezoelectric biomaterials requires overcoming regulatory challenges and demonstrating their safety, efficacy, and reliability. Future efforts should involve collaboration with regulatory authorities, the establishment of standardized testing protocols, and addressing manufacturing scalability, cost-effectiveness, and quality control. Partnerships between academia, industry, and regulatory bodies are crucial for the translation of piezoelectric biomaterials into commercial products.

ix. Environmental Impact and Sustainability: As with any materials, the environmental impact and sustainability of piezoelectric biomaterials must be considered. Future research should focus on developing eco-friendly synthesis methods, reducing the use of toxic substances, and considering the life cycle assessment of these materials. Incorporating biodegradable or recyclable components can enhance the sustainability of piezoelectric biomaterials, especially in single-use applications.

x. Education and Collaboration: Building interdisciplinary collaborations and fostering education and training programs are vital for advancing the field of piezoelectric biomaterials. Collaborations between materials scientists, biologists, clinicians, engineers, and industry experts will enable the exchange of knowledge and expertise, leading to accelerated progress and innovation.

4.13 Conclusions

Piezoelectric biomaterials have emerged as a promising class of materials with unique properties and capabilities. Their ability to convert mechanical energy into electrical signals, and vice versa, makes them versatile and valuable for a wide range of applications. From sensing and actuation to tissue engineering and regenerative medicine, piezoelectric biomaterials offer exciting possibilities for advancements in healthcare, diagnostics, energy harvesting, and beyond. However, the field of piezoelectric biomaterials is not without its challenges. Biocompatibility, stability, scalability, and integration remain important areas of focus for researchers and developers. Overcoming these challenges will require interdisciplinary collaborations, advancements in fabrication techniques, and regulatory considerations to ensure the safe and effective translation of piezoelectric biomaterials into practical applications. Looking to the future, exciting opportunities lie ahead. Ongoing research efforts aim to enhance the performance and efficiency of piezoelectric biomaterials, improve biocompatibility, and explore new applications. Advancements in fabrication techniques, miniaturization, and flexible form factors will enable the integration of these biomaterials into wearable devices, implantable sensors, and biomedical systems. Moreover, the incorporation of biologically inspired design principles and the development of advanced modeling techniques will further advance the field.

4.14 Future Research Directions

Piezoelectric biomaterials have garnered significant interest and shown immense potential in various applications. To further advance this field and unlock new possibilities, future research directions can focus on the following areas [61, 62]:

i. Novel Piezoelectric Materials: Exploring and developing new piezoelectric materials with enhanced properties and performance is a key research direction. This includes investigating novel compositions, exploring hybrid materials, and exploring advanced manufacturing techniques to fabricate materials with improved piezoelectric coefficients, electromechanical coupling, and stability.

ii. Multiscale Modeling and Simulation: Advanced computational modeling and simulation techniques can provide valuable insights into the behavior and performance of piezoelectric biomaterials. Multiscale modeling approaches that bridge different length scales, from atomistic to macroscopic, can help understand the underlying mechanisms, optimize material properties, and guide the design of new materials and structures.

iii. Biointerface Engineering: Research efforts can focus on enhancing the integration of piezoelectric biomaterials with biological systems. This includes developing surface modifications, bioactive coatings, or functionalization strategies to improve biocompatibility, reduce immune response, and promote tissue integration. Engineering the interface between the biomaterials and the surrounding biological environment is crucial for successful integration and long-term performance.

iv. Bioinspired Design and Tissue-Specific Approaches: Drawing inspiration from natural systems, future research can explore bioinspired designs that mimic the hierarchical structures, composition, and properties found in living tissues. Taking tissue-specific approaches, such as tailoring piezoelectric properties to match the mechanical and electrical requirements of specific tissues, can enhance the biomaterials' effectiveness in tissue engineering and regenerative medicine.

v. Advanced Fabrication Techniques: Further advancements in fabrication techniques can enable the production of complex and functional piezoelectric biomaterials. Research can focus on developing scalable and cost-effective fabrication methods, such as additive manufacturing, electrospinning, or self-assembly, to create biomaterials with controlled microstructures, tailored properties, and precise geometries.

vi. Integration with Bioelectronics: Exploring the integration of piezoelectric biomaterials with bioelectronics can lead to innovative devices and systems. Research can focus on developing seamless integration methods, such as flexible electronics or nanoscale interfaces, to enable direct coupling between piezoelectric materials and electronic components. This integration can open up new opportunities for self-powered sensors, implantable devices, and bioelectronic interfaces.

vii. In Vivo Evaluation and Clinical Translation: Conducting in vivo evaluation and preclinical studies is crucial to assess the safety, efficacy, and long-term performance of piezoelectric biomaterials. Future research should involve comprehensive in vivo testing, including animal models and eventually clinical trials, to validate the performance and functionality of these materials in real-world applications. Collaboration with clinicians and regulatory bodies is essential to navigate the path toward clinical translation.

viii. Smart and Adaptive Piezoelectric Biomaterials: Developing piezoelectric biomaterials that can respond and adapt to external stimuli is an exciting research direction. By incorporating responsive or stimuli-sensitive components, such as shape memory polymers, smart hydrogels, or stimuli-responsive nanoparticles, the properties and functions of piezoelectric biomaterials can be dynamically controlled or modulated. This can lead to new applications in drug delivery, tissue engineering, and personalized healthcare.

ix. Sustainable and Environmentally Friendly Approaches: Considering the environmental impact and sustainability of piezoelectric biomaterials is an important research direction. Future efforts should focus on developing eco-friendly synthesis methods, minimizing the use of toxic substances, and exploring biodegradable or recyclable materials. Sustainable manufacturing processes and life cycle assessments can ensure the ecological footprint of piezoelectric biomaterials is minimized.

x. Industry Collaboration and Commercialization: Collaborations between academia, industry, and regulatory bodies are crucial for the successful commercialization and widespread adoption of piezoelectric biomaterials. Future research can focus on fostering such collaborations, addressing scalability, cost-effectiveness, and regulatory considerations, and facilitating the translation of innovative piezoelectric biomaterials into practical applications that benefit society.

References

1. Kamel NA. Bio-piezoelectricity: Fundamentals and applications in tissue engineering and regenerative medicine. *Biophysical Reviews*, **14**(3), 717–733, 2022. doi: 10.1007/s12551-022-00969-z
2. Covaci C, Gontean A. Piezoelectric energy harvesting solutions: A review. *Sensors*, **20**, 3512, 2020. https://doi.org/10.3390/s20123512
3. Kapat K, Shubhra QTH, Zhou M, Leeuwenburgh S. Piezoelectric nano-biomaterials for biomedicine and tissue regeneration. *Advanced Functional Materials*, **30**, 1909045, 2020. doi: 10.1002/adfm.201909045
4. Lay R, Deijs GS, Malmström J. The intrinsic piezoelectric properties of materials – A review with a focus on biological materials. *RSC Advances*, **11**, 30657–30673, 2021. https://doi.org/10.1039/D1RA03557F
5. Cafarelli A, Marino A, Vannozzi L, Puigmartí-Luis J, Pané S, Ciofani G, et al. Piezoelectric nanomaterials activated by ultrasound: The pathway from discovery to future clinical adoption. *ACS Nano*, **15**(7), 11066–11086, 2021. doi: 10.1021/acsnano.1c03087

6. Jarkov V, Allan SJ, Bowen C, Khanbareh H. Piezoelectric materials and systems for tissue engineering and implantable energy harvesting devices for biomedical applications. *International Materials Reviews*, **67**(7), 683–733, 2022. https://doi.org/10.1080/09506608.2021.1988194
7. Panda S, Hajra S, Mistewicz K, In-na P, Sahu M, Rajaitha PM, et al. Piezoelectric energy harvesting systems for biomedical applications. *Nano Energy*, **100**, 107514, 2022. doi: 10.1016/j.nanoen.2022.107514
8. Liu H, Zhong J, Lee C, Lee SW, Lin L. A comprehensive review on piezoelectric energy harvesting technology: Materials, mechanisms, and applications. *Applied Physics Reviews*, **5**, 041306, 2018. doi: 10.1063/1.5074184
9. Olvera D, Monaghan MG. Electroactive material-based biosensors for detection and drug delivery. *Advanced Drug Delivery Reviews*, **170**, 396–424, 2021. https://doi.org/10.1016/j.addr.2020.09.011
10. Safaei M, Sodano HA, Anton SR. A review of energy harvesting using piezoelectric materials: State-of-the-art a decade later (2008-2018). *Smart Materials and Structures*, **28**, 113001, 2019. doi: 10.1088/1361-665X/ab36e4
11. Chen-Glasser M, Li P, Ryu J, Hong S. Piezoelectric materials for medical applications. In: *Piezoelectricity – Organic and Inorganic Materials and Applications*. 2018, InTech. doi: 10.5772/intechopen.76963
12. Tandon B, Blaker JJ, Cartmell SH. Piezoelectric materials as stimulatory biomedical materials and scaffolds for bone repair. *Acta Biomaterialia*, **73**, 1–20, 2018. doi: 10.1016/j.actbio.2018.04.026
13. Chorsi MT, Curry EJ, Chorsi HT, Das R, Baroody J, Purohit PK, et al. Piezoelectric biomaterials for sensors and actuators. *Advanced Materials*, **31**(1), e1802084, 2019. doi: 10.1002/adma.201802084
14. De Fazio R, Mastronardi VM, Petruzzi M, De Vittorio M, Visconti P. Human–machine interaction through advanced haptic sensors: A piezoelectric sensory glove with edge machine learning for gesture and object recognition. *Future Internet*, **15**(1), 14, 2023. https://doi.org/10.3390/fi15010014
15. Marino A, Genchi GG, Sinibaldi E, Ciofani G. Piezoelectric effects of materials on bio-interfaces. *ACS Applied Materials & Interfaces*, **9**(21), 17663–17680, 2017. doi: 10.1021/acsami.7b04323
16. Wang R, Sui J, Wang X. Natural piezoelectric biomaterials: A biocompatible and sustainable building block for biomedical devices. *ACS Nano*, **16**(11), 17708–17728, 2022. doi: 10.1021/acsnano.2c08164

17. Jiang J, Liu S, Feng L, Zhao D. A review of piezoelectric vibration energy harvesting with magnetic coupling based on different structural characteristics. *Micromachines*, **12**(4), 436, 2021. https://doi.org/10.3390/mi12040436

18. Chandra Sekhar B, Dhanalakshmi B, Srinivasa Rao B, Ramesh S, Venkata Prasad K, Subba Rao PSV, et al. Piezoelectricity and its applications. In: *Multifunctional Ferroelectric Materials*. 2021. doi: 10.5772/intechopen.96154

19. Meng Y, Chen G, Huang M. Piezoelectric materials: Properties, advancements, and design strategies for high-temperature applications. *Nanomaterials*, **12**(7), 1171, 2022. https://doi.org/10.3390/nano12071171

20. Dani SS, Tripathy A, Alluri NR, Balasubramaniam S, Ramadoss A. A critical review: The impact of electrical poling on the longitudinal piezoelectric strain coefficient. *Materials Advances*, **3**, 8886–8921, 2022. https://doi.org/10.1039/D2MA00559J

21. Aabid A, Raheman MA, Ibrahim YE, Anjum A, Hrairi M, Parveez B, et al. A systematic review of piezoelectric materials and energy harvesters for industrial applications. *Sensors*, **21**(12), 4145, 2021. https://doi.org/10.3390/s21124145

22. Mezheritsky AV. Elastic, dielectric, and piezoelectric losses in piezoceramics: How it works all together. *IEEE Transactions on Ultrasonics, Ferroelectrics, and Frequency Control*, **51**(6), 695–707, 2004.

23. Priya S, Yang SC, Maurya D, Yan Y. Recent advances in piezoelectric and magnetoelectric materials phenomena. In: *Composite Magnetoelectrics: Materials, Structures, and Applications*, pp. 103–157. 2015. https://doi.org/10.1016/B978-1-78242-254-9.00006-8

24. Surbhi, Sukesha. Response of piezoelectric materials to the external temperature, electric field and humidity. *Materials Today: Proceedings*, **28**(3), 1951–1954, 2020. https://doi.org/10.1016/j.matpr.2020.05.555

25. Mishra S, Unnikrishnan L, Nayak SK, Mohanty S. Advances in piezoelectric polymer composites for energy harvesting applications: A systematic review. *Macromolecular Materials and Engineering*, **304**(1), 1800463, 2019. https://doi.org/10.1002/mame.201800463

26. Ali M, Bathaei MJ, Istif E, Karimi SNH, Beker L. Biodegradable piezoelectric polymers: Recent advancements in materials and applications. *Advanced Healthcare Materials*, **2300318**, 1–32, 2023. doi: 10.1002/adhm.202300318

27. Kwon J, Cho H. Collagen piezoelectricity in osteogenesis imperfecta and its role in intrafibrillar mineralization. *Communications Biology*, **5**, 1229, 2022. https://doi.org/10.1038/s42003-022-04204-z

28. Niu Q, Wei H, Hsiao BS, Zhang Y. Biodegradable silk fibroin-based bio-piezoelectric/triboelectric nanogenerators as self-powered electronic devices. *Nano Energy*, **96**, 107101, 2022. doi: 10.1016/j.nanoen.2022.107101

29. Pohanka M. Overview of piezoelectric biosensors, immunosensors and DNA sensors and their applications. *Materials*. 2018. PMID: 29562700. doi: 10.3390/ma11030448

30. Kim K, Ha M, Choi B, Joo SH, Kang HS, Park JH, et al. Biodegradable, electro-active chitin nanofiber films for flexible piezoelectric transducers. *Nano Energy*, 2018. doi: 10.1016/J.NANOEN.2018.03.056. Corpus ID: 139733278.

31. Bairagi S, Ghosh S, Ali SW. A fully sustainable, self-poled, bio-waste based piezoelectric nanogenerator: Electricity generation from pomelo fruit membrane. *Scientific Reports*, **10**, 12121, 2020. https://doi.org/10.1038/s41598-020-68751-3

32. Mensur-Alkoy E, Kaya MY, Sarı HA, Pekel O, Olukkent R, Tur YK, et al. Mechanical and electromechanical properties of piezoelectric ceramic fibers drawn by the alginate gelation method. *International Journal of Applied Ceramic Technology*, **17**, 1371–1381, 2020. doi: 10.1111/ijac.13405

33. Song Y, Shi Z, Hu GH, Xiong C, Isogai A, Yang Q. Recent advances in cellulose-based piezoelectric and triboelectric nanogenerators for energy harvesting: A review. *Journal of Materials Chemistry A*, **9**, 1910–1937, 2021. https://doi.org/10.1039/D0TA08642H

34. Zhang M, Gao T, Wang J, Liao J, Qiu Y, Yang Q, et al. A hybrid fibers based wearable fabric piezoelectric nanogenerator for energy harvesting application. *Nano Energy*, **13**, 298–305, 2015. http://dx.doi.org/10.1016%2Fj.nanoen.2015.02.034

35. Omran N, Elnabawy E, Le B, Trabelsi M, Gamal M, Kandas I, et al. Solution blow spun piezoelectric nanofibers membrane for energy harvesting applications. *Reactive and Functional Polymers*, 105365, 2022. https://doi.org/10.1016/j.reactfunctpolym.2022.105365

36. Heinemann F, Launspach M, Gries K, Fritz M. Gastropod nacre: Structure, properties and growth – Biological, chemical and physical basics. *Biophysical Chemistry*, **153**(2–3), 126–153, 2011. https://doi.org/10.1016/j.bpc.2010.11.003

37. Praveen E, Murugan S, Jayakumar K. Investigations on the existence of piezoelectric property of a bio-polymer-chitosan and its application in vibration sensors. *RSC Advances*, **7**, 35490–35495, 2017. https://doi.org/10.1039/C7RA04752E
38. Singh H, Sheetal A, Singh M, Kaur J, Sui T, Loja MAR, et al. Electrical energy generation using fish scale of Rohu fish by harvesting human motion mechanical energy for self powered battery-less devices. *Sensors and Actuators A: Physical*, **349**, 114023, 2023. doi: 10.1016/j.sna.2022.114023
39. Lee BY, Zhang J, Zueger C, Chung WJ, Yoo SY, Wang E, et al. Virus-based piezoelectric energy generation. *Nature Nanotechnology*, **7**(6), 351–356, 2012. doi: 10.1038/nnano.2012.69
40. Mandal R, Dutta G. From photosynthesis to biosensing: Chlorophyll proves to be a versatile molecule. *Sensors International*, **1**(1), 100058, 2020. doi: 10.1016/j.sintl.2020.100058
41. Jákli A, Harden J, Notz C, Bailey C. Piezoelectricity of phospholipids: A possible mechanism for mechanoreception and magnetoreception in biology. *Liquid Crystals*, **35**(4), 395–400, 2008. doi: 10.1080/02678290801905658
42. Pryadko A, Surmeneva MA, Surmenev RA. Review of hybrid materials based on polyhydroxyalkanoates for tissue engineering applications. *Polymers*, **13**(11), 1738, 2021. https://doi.org/10.3390/polym13111738
43. Yadav R, Zabihi O, Fakhrhoseini S, Nazarloo HA, Kiziltas A, Blanchard P, et al. Lignin derived carbon fiber and nanofiber: Manufacturing and applications. *Composites Part B: Engineering*, 110613, 2023. doi: 10.1016/j.compositesb.2023.110613
44. Divya R, Yempally S, Ponnamma D. Piezoelectric properties of biopolymers. In: *Handbook of Biopolymers*. 2023. doi: 10.1007/978-981-16-6603-2_8-1
45. Bhavya P, Verma S, Arya S. In Fabrication Approaches for Piezoelectric Materials, Materials Research Foundations, **131**, 37–60, 2022. doi: https://doi.org/10.21741/9781644902097-2
46. Magnani A, Capaccioli S, Azimi B, Danti S, Labardi M. Local Piezoelectric response of polymer/ceramic nanocomposite fibers. *Polymers*, **14**(24), 5379, 2022. https://doi.org/10.3390/polym14245379
47. Jing Q, Kar-Narayan S. Nanostructured polymer-based piezoelectric and triboelectric materials and devices for energy harvesting applications. *Journal of Physics D: Applied Physics*, **51**, 303001, 2018. doi: 10.1088/1361-6463/aac827

48. Mirzaali MJ, Moosabeiki V, Rajaai SM, Zhou J, Zadpoor AA. Additive manufacturing of biomaterials—Design principles and their implementation. *Materials (Basel)*, **15**(15), 5457, 2022. doi: 10.3390/ma15155457.
49. Najjari A, Mehdinavaz Aghdam R, Ebrahimi SAS, Suresh SK, Krishnan S, Shanthi C, et al. Smart piezoelectric biomaterials for tissue engineering and regenerative medicine: A review. *Biomed Tech (Berl)*, **67**(2), 71–88, 2022. doi: 10.1515/bmt-2021-0265
50. Azimi B, Milazzo M, Lazzeri A, Berrettini S, Uddin MJ, Qin Z, et al. Electrospinning piezoelectric fibers for biocompatible devices. *Advanced Healthcare Materials*, **9**(1), e1901287, 2020. doi: 10.1002/adhm.201901287
51. Yang C, Ji J, Lv Y, Li Z, Luo D. Application of piezoelectric material and devices in bone regeneration. *Nanomaterials*, **12**(24), 4386, 2022. https://doi.org/10.3390/nano12244386
52. Suamte L, Tirkey A, Barman J, Jayasekhar Babu P. Various manufacturing methods and ideal properties of scaffolds for tissue engineering applications. *Smart Materials in Manufacturing*, 100011, 2023. https://doi.org/10.1016/j.smmf.2022.100011
53. Li J, Long Y, Yang F, Wang X. Degradable piezoelectric biomaterials for wearable and implantable bioelectronics. *Current Opinion in Solid State and Materials Science*, 100806, 2020. https://doi.org/10.1016/j.cossms.2020.100806
54. Xu Q, Gao X, Zhao S, Liu Y, Zhang D, Zhou K, et al. Construction of bio-piezoelectric platforms: From structures and synthesis to applications. *Advanced Materials*, 2021. https://doi.org/10.1002/adma.202008452
55. Wu Y, Ma Y, Zheng H, Ramakrishna S. Piezoelectric materials for flexible and wearable electronics: A review. *Materials and Design*, 110164, 2021. https://doi.org/10.1016/j.matdes.2021.110164
56. Nguyen VT, Kumar P, Leong JYC. Finite element modellingand simulations of piezoelectric actuators responses with uncertainty quantification. *Computation*, **6**(4), 60, 2018. https://doi.org/10.3390/computation6040060
57. Yuan H, Lei T, Qin Y, He JH, Yang R. Design and application of piezoelectric biomaterials. *Journal of Physics D: Applied Physics*, **52**, 194002, 2019. doi: 10.1088/1361-6463/ab0532
58. Zhang X, Villafuerte J, Consonni V, Capsal JF, Cottinet PJ, Petit L, et al. Characterizing and optimizing piezoelectric response of ZnO nanowire/PMMA composite-based sensor. *Nanomaterials*, **11**(7), 1712, 2021. https://doi.org/10.3390/nano11071712

59. Arnau A, Soares D. Fundamentals of piezoelectricity. In: *Piezoelectric Transducers and Applications*. 2008. https://doi.org/10.1007/978-3-540-77508-9_1

60. Smith M, Kar-Narayan S. Piezoelectric polymers: theory, challenges and opportunities. *International Materials Reviews*, 2022. https://doi.org/10.1080/09506608.2021.1915935

61. Ji J, Yang C, Shan Y, Sun M, Cui X, Xu L, et al. Research trends of piezoelectric nanomaterials in biomedical engineering. *Advanced NanoBiomed Research*, 2200088, 2023. https://doi.org/10.1002/anbr.202200088

62. Yu Q, Shi W, Li S, Liu H, Zhang J. Emerging advancements in piezoelectric nanomaterials for dynamic tumor therapy. *Molecules*, **28**(7), 3170, 2023. https://doi.org/10.3390/molecules28073170

Chapter 5

Perovskite Structure Materials and Contribution to Technology Advancement

Sanjay Kumar Sinha
Department of Physics, Birla Institute of Technology, Mesra, Patna Campus, Patna-800014, India
sk.sinha@bitmesra.ac.in

Multifunctional smart ceramics mostly have a ternary structure. In the world of ternary ceramics, counting goes to more than one thousand, out of these only half a dozen dominate the world of smart ceramics. Three ternary structures are the most talked, i.e., A_2BO_4 structure, spinal structure, and ABO_3 (i.e., perovskite). Out of this, perovskite has produced many varieties of incredible phases having special properties; many special perovskite structure ceramics are discovered by manipulating the chemical composition and incorporating suitable doping in the perovskite structure.

Piezoelectric Materials: Design and Applications
Edited by N. B. Singh and Dev Kumar Mahato

ISBN 978-981-5129-40-3 (Hardcover), 978-1-003-59897-8 (eBook)
www.jennystanford.com

5.1 Introduction

There is a famous proverb – "My disability has opened my eyes to see my true abilities." This fits very much with perovskite structure materials. Perovskite structure lacks a center of symmetry; it is a kind of defect present naturally in the material. Perovskite structure does not have the same mirror image. This disability in perovskite structure materials makes it very special among the other normal structure materials. Perovski (1792–1856), a Russian mineralogist, first discovered a special type of structure having a lack of center of symmetry in calcium titanate ($CaTiO_3$). Materials similar to calcium titanate having structure ABO_3 are later named perovskite structure. Another important name that contributed most to the field of perovskite is V.M. Goldschmidt (1888–1947), a Norwegian mineralogist, who paved the way for intelligent crystal chemistry involved with the perovskite structure. Goldschmidt formulated a common set of rules that says how elements can grouped and show special characteristics in mineralogy. Based on Gibb's phase rule, he suggested the formulation of special types of materials which later on became the source of discovery of many perovskite materials. Goldschmidt tolerance factor is the basis for stability or distortion of any perovskite structure material [1–4].

Ceramics or processed inorganic compounds are some of the most highly used materials by human beings. There are hundreds of known ceramics. But technologically around half a dozen ceramics, which belong to perovskite or important constituent of perovskite family, have been the most versatile technological applications. For example, quartz is the base for piezo sensors available in rocks and beaches. Mullite, the principal component of pottery and bricks and the basis for advanced high-tech ceramics, belongs to the perovskite family. TiO_2, titanium dioxide, is the main ingredient of electroceramics and these ceramics have abundant application in paint, etc. due to their high volume and many other useful properties. Calcium silicate, the main ingredient of cement and concrete, has the largest tonnage of use in the manufacturing industry. Hard grinding material has applications in all kinds of refractory ceramics such as aluminium oxide.

Multifunctional smart ceramics mostly have a ternary structure. In the world of ternary ceramics, counting goes to more than one thousand, out of these only half a dozen dominate the world of smart ceramics. Three ternary structures are the most talked, i.e., A_2BO_4 structure, spinal structure, and ABO_3 (i.e., perovskite) [5–8]. Out of these, perovskite has produced many varieties of incredible phases having special properties. Many special perovskite structure ceramics are discovered by manipulating the chemical composition and incorporating suitable doping in the perovskite structure.

Recently a team from the University of Sydney discovered a miracle perovskite solar panel that is capable of healing its damage 100% in spacecraft. It is capable of recovering its damaged efficiency (band gap, etc.) in the presence of sun and vacuum [9]. Another recent wonder perovskite structure-related discovery is piezoelectric graphene. Piezoelectric graphene has unparalleled electrical, optical, or mechanical control for applications ranging from touch screens to nanoscale transistors [10–12].

5.2 Basis of Discovery of Perovskite Smart Materials

The contribution of Goldschmidt and Pauling is remarkable as they initially shown the path for the discovery of various perovskite smart materials. The basis for all the new inventions in this field lies in a few facts given below:

- The radius of the constituents' ions of the perovskite structure is the key to the formulation. Goldschmidt proposed the table of radii of elements in his book published in 1911 [13].
- The radius ratio, R_c/R_a, decides the polyhedron formation. It also determines the coordination number of the cation.
- The packing of polyhedral follows a rule which was later codified by Pauling [14].

Goldschmidt introduced tolerance factor (T) in the formation of perovskite structure; tolerance factor decides the structure of interpenetrating octahedra or dodecahedra. It decides how far the ionic radius may be moved from the ideal packing of the perovskite. That is why it is known as the tolerance factor.

Tolerance factor, T, can be represented by the formula

$$T = \frac{(R_a + R_o)}{\sqrt{2}(R_b + R_o)} \tag{5.1}$$

Here,
R_a – the radius of the 'A' cation
R_b – the radius of the 'B' cation
R_o – the radius of the oxygen anion
In an ideal cubic perovskite structure,

$$a = \sqrt{2}(R_a + R_o) = 2(R_b + R_o) \tag{5.2}$$

In above equation's symbols are already described in Equation 5.1.

Let us now take a few examples to illustrate the importance of the tolerance factor in the perovskite structure (Table 5.1) [2].

Table 5.1 Perovskite structure materials have a tolerance factor of up to 0.71

Tolerance factor	Structure	Material	Comments
>1	Tetragonal or hexagonal	$BaTiO_3$ (T = 1.0617)	An ion too big and B ion too small
0.9–1	Cubic	$SrTiO_3$	Both A and B ions have ideal size
0.71–0.9	Rhombohedral/ orthorhombic	$CaTiO_3$	An ion too small and B ion too big
<0.71	Different structure; not perovskite		

5.3 Historical Development of Perovskite Structure Materials

There are many useful inventions in perovskite structure materials from the first discovery of calcium titanate in the year 1839. Table 5.2 gives an account of path changing discovery of various materials having perovskite structures.

Table 5.2 A few important perovskite materials

Perovskite structure material	Year of Invention	Inventor
High dielectric constant material - $BaTiO_3$	1942	Wainer and Salomen (USA)
Piezoelectric material - $PbTiO_3$	1952	Yutaka Takagi, Gen Shirane, and Etsuro Sawaguci (Japan)
Perovskite dielectric for microwave communication	Not known	
Hydrogen storage - $LaCrO_3$	Not Known	
Perovskite catalyst	1952	G. Paravano (USA)

5.4 Interesting Facts about Barium Titanate

During World War II, the demand for dielectric capacitors was high as it required in detonators and many other applications. At that time mica, $KMg_3(AlSi_3)O_{10}(OH_2)$, having dielectric constant 10 was the main source of insulators for making capacitors. The availability of mica was only with the country like India and Brazil. Due to the ongoing Second World War, mica procurement was difficult from India for war allied countries such as England, USA, Russia, or France. Hence, there was a lot of research going on to find an alternative to mica having a dielectric constant in and around 80. The key wartime research goal of the laboratories at that time was to find a suitable alternative to mica. As per the publication of W. R. Buseam, Director of Science Research, Siemens, Germany, during 1941–1942 various earth alkali titanates were invented as alternatives to mica and Germany was exporting those to allied countries. This news put the intelligence agency of the Western country in alert mode and at that time many flights in Germany were shot down suspected of carrying the alternative of mica. During all these stories and pressure finally, Wainer and Soloman, from TAM laboratory, reported that $BaTiO_3$ was the highest dielectric constant material. As per L. E. Cross and Newnham's write-up [15], during the same period of 1945 Russia and Japan also fabricated the $BaTiO_3$.

Barium titanate was initially developed during World War II for the high capacitance capacitor military use. Now, 75 years later, there is tremendous progress observed in the barium titanate-related fields. Currently, barium titanate piezo generators are being used in advanced medical diagnoses such as heartbeat recording and energy harvesting.

5.5 The World of Nano Barium Titanate-Based Piezoelectric Generator

Barium titanate was initially found as one of the good piezoelectric generators. Barium is one of the abundantly available minerals in nature, i.e., baryte or witirite, and hence the progress in this ceramic material has seen tremendous growth. Recently development polymer-barium titanate flexible piezoelectric is ready to be placed in market for day-to-day life application of human beings such as a wearable to generate electricity or to diagnose human health. Commonly this area of development is known as piezoelectric nanogenerators (p-NGs).

Barium titanate NGs have shown high suitability in the application of self-powered devices that generate required energy from ambient sources or attached surroundings. Energy generation by use of p-NGs in pedestalling walk areas, or in exercise machines, or energy generation by sea waves/high wind/horn sound, are becoming quite common nowadays. Barium titanate p-NGs are capable of producing around 11 volts of power at a mechanical force of 10 newtons (N). 10 N is a force that is available in normal sea waves, and the proportionate generation of electricity in full moon day can be imagined where the range of mechanical force even reaches 1000 N. As per a report [16], modified barium titanate p-NG fibres used under shoes can generate an output voltage of 25 volts at a frequency of 0.6 Hz under a force of a human body, i.e., 600 N. The current-voltage generation response of a barium titanate is displayed in Fig. 5.1 [17].

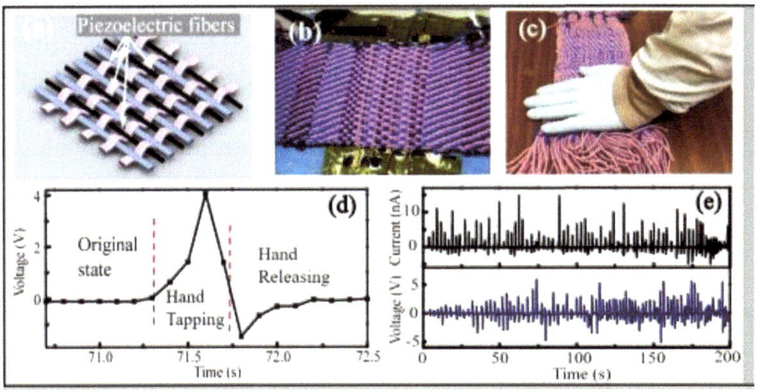

Figure 5.1 Barium titanate p-NG voltage-current response [17].

5.6 Lead Zirconate Titanate (PZT) Ceramics

As mentioned in Table 5.2, the initial starting year of work on PZT was 1952. The interesting high dielectric constant (k) and piezoelectric coefficient (d_{33}) of PZT make it forefront among the other perovskite family materials. Soft bulk PZT exhibits dielectric constant k in the order of 3400, while it exhibits d_{33} having value 593 pC/N [18]. PZT material is widely being used as sensors, transducers, and actuators [18]. Due to higher Curie temperature (T_C, better than barium titanate) and high-temperature stability, PZT is more suitable in applications such as micro electric devices such as surface acoustics wave (SAW) devices, energy harvesting, sensors, etc. Favorable ferroelectric hysteresis makes the PZT most popular for the application of non-volatile memory devices in computer systems.

The property enhancement in thin PZT comparable to bulk one is enormous. Dielectric constant k of PZT increases by more than 10 times once the grain size of the material comes into the nano range [18]. Similarly, there is an enormous improvement in other properties such as piezoelectric coefficient or the actuation level once the PZT material is fabricated to have nano size. There are chemical methods or plasma sputtering techniques to obtain the nano PZT. For high purity and better reproducibility, plasma sputtering technique is considered one of the best among other existing techniques.

The layer deposition of PZT on a substrate is a good example of it. Figure 5.2 shows a schematic diagram of the plasma sputtering technique for growing the PZT thin film. The Zr/Ti ratio in PZT thin film plays a significant role in the enhancement of the relative dielectric constant (μ_r) and d_{33}. The morphotropic phase boundary, i.e., Zr/Ti ratio at which μ_r of PZT becomes optimum is found to be the same in bulk and thin film, and MPB of PZT is 52/48 [18].

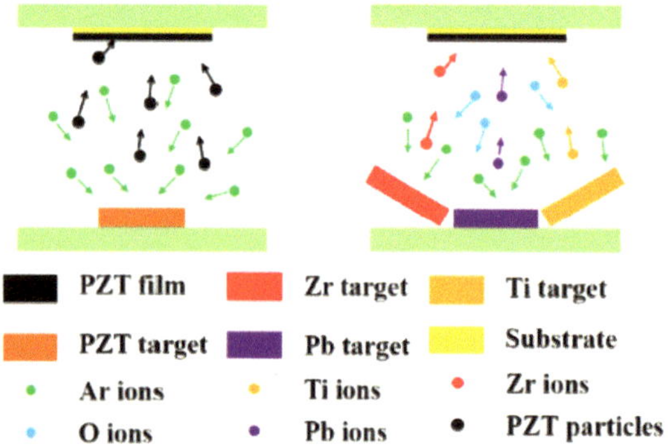

Figure 5.2 Schematic diagram of PZT thin film by the sputtering technique [18].

The film thickness of PZT is also found to have influenced the property of PZT. The thickness size effect could be explained based on the interface of the PZT film and the substrate. Here PZT layer creates residual stress while the substrate layer creates a reduction in μ_r. Therefore, initially with the increase in thickness of PZT, the value of μ_r exhibits decreasing trends due to a decrease in the effect of interface but after a particular thickness there would be saturation in the change of μ_r. It is also reported that there is a change in the orientation of PZT due to increases in thickness, i.e., from (110) to (111) [18].

5.7 Perovskite Dielectrics for Microwave Communication

Microwave dielectrics are different from electroceramics. Electroceramics are ferroelectrics, while microwave dielectrics are paraelectric with a center of symmetry. For microwave tuning, the dielectric mustn't have any defect or strain. Here dielectrics are required to be used in resonator and filter circuits in telecommunication. There are further new emerging applications of microwave dielectrics in the field of telecommunication such as wireless sensors, wireless power transfer, wave absorption, and transparent ceramics with no birefringence [19–21].

If a comparison of the use of perovskite structure materials with the other materials is made, then it can be said that 45% of the use in telecommunication belongs to perovskite structure materials. A comparison chart is displayed in Fig. 5.3 [20].

As mentioned above, microwave dielectrics are used as resonators and filters in mobile communication technology. The miniaturization of the phone system has become possible after the introduction of nano-coated microwave dielectric and other components of microchips. Microwave resonators have well replaced the bulky old cavity resonators and pave the path of manufacturing smaller size of telecommunication system. There are a few technical terminologies that are important to understand the utility of microwave dielectric in the field of telecommunication. These are dielectric constant (ϵ_r), quality factor (Qf), and temperature coefficient of resonant frequency (TCf); these terminologies are discussed in the next section. The high value or low value of ϵ_r, Qf, or TCf in a microwave dielectric decides the application of the material in a communication system. For example, high µr microwave dielectrics are useful in the miniaturization of mobile phones, high ϵ_r and high Qf value are required for the application in the base station, and low ϵ_r dielectrics are needed in millimeter wave range communication [20].

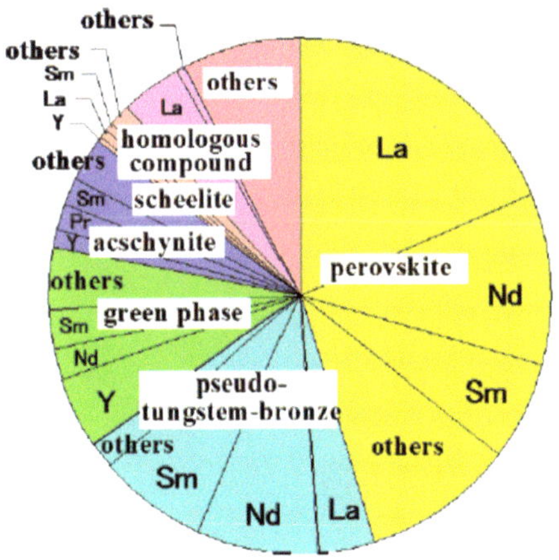

Figure 5.3 Comparison of perovskite microwave dielectrics with other crystal structures [20].

As mentioned above, the three important parameters of a microwave dielectric are dielectric constant (ϵ_r), quality factor (Qf), and temperature coefficient of resonant frequency (TCf).

5.7.1 Dielectric constant (ϵ_r)

The important function of shortening the wave is done by introducing a proper ϵ_r dielectric as per the following formula:

$$\lambda = \frac{\lambda_0}{\epsilon_r} \tag{5.3}$$

Here, λ_0 is the wavelength of the wave in a vacuum.

The above equation illustrates that to shorten the wavelength, i.e., for miniaturization of a mobile set, the ϵ_r is required to be high. For a millimeter wave, there is no need to shorten the wave. In Fig. 5.4, the use of microwave dielectric is demonstrated [20].

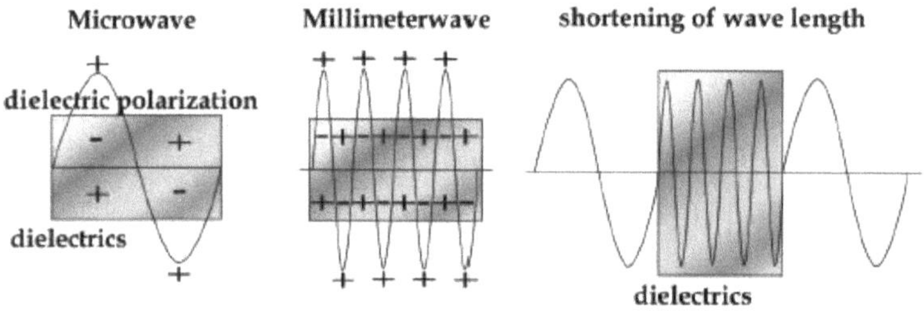

Figure 5.4 Shortening of wavelength by the introduction of microwave dielectrics [20].

The dielectric constant μ_r values also control one other parameter of the wave, i.e., time delay (T) of the communicating wave; it is related to the phase of the signal. The following equation gives the relation of ϵ_r and the T.

$$T = \frac{\sqrt{\epsilon_r}}{c} \tag{5.4}$$

In Equation 5.4, c is the velocity of light.

Time delays need to be adjusted in the communication system for an improvement in the speed of the signal.

5.7.2 Quality Factor (Q_f)

A good microwave dielectric should exhibit good or high response if it is kept in an electromagnetic wave. This measure of response is measured in terms of quality factor (Qf); mathematically it is calculated as the inverse of dielectric loss (1/tan δ).

Therefore,

$$Qf = \frac{1}{\tan\delta} \tag{5.5}$$

The loss factor is related to the dielectric polarization in the presence of the electromagnetic field [22, 23]. In ferroelectric materials, the loss factor is high, and thus they show a low value of Qf. Hence, they are not suitable for the application of communication, but the microwave dielectric materials being paraelectric can meet the requirement of Qf for the application of signal communication.

5.7.3 Temperature Coefficient of Resonant Frequency (TCf)

Propagation of signal is required to be continued without interruption in all the weather and environmental conditions of the globe, and hence a proper tunning of signal is needed from the base communication station. TCf is the tuning of the communicating signal or wave to make it compatible with the environmental condition of the communicating area. Mathematically the TCf is expressed by the following equation:

$$\text{TCf} = -\left(\frac{\alpha + \epsilon_r}{2}\right) \tag{5.6}$$

Here, α is the coefficient of thermal expansion.

Examples of a few important simple perovskite structure-based microwave dielectrics are given in Table 5.3.

Table 5.3 A few examples of microwave dielectric

Material	ϵ_r	Qf (x 103 GHz)	TCf (ppm/0 C)
0.35 $CaTiO_3$ 0.65$LaAlO_3$	37	47	-2
0.7 CaTiO3 0.3NdAlO3	43	47	0
0.64 CaTiO3 0.36LaGaO3	46.5	48	-2.9
0.64 SrTiO3 0.36LaGaO3	26.7	139	-50

5.8 Hydrogen Storage: $LaCrO_3$

In recent years the appreciable capability of storing solid hydrogen has been reported in perovskite structure lanthanum chromite $LaCrO_3$. $LaCrO_3$ has a special property of hydrogen absorption and desorption making it useful in the storage of solid hydrogen. Very high hydrogen storage capacity makes it a basis for producing a promising technology to miniaturize the hydrogen cells required to run vehicles such as trains or buses.

The high hydrogen storage capability in $LaCrO_3$ is due to the presence of a high number of pores in its structure. Figure 5.5 illustrates the capability of hydrogen.

Figure 5.5 The change in the structure of $LaCrO_3$ due to hydrogen absorption (a) before absorption (b) after absorption [24].

5.9 Perovskite Catalyst

Catalyst converter plays a vital role in converting the harmful unburnt hydrocarbon (UHC) emitting gas such as carbon monoxide and nitrogen oxide. The presently available catalyst conversion system converts the UHC into gases such as carbon dioxide and nitrogen. But still, the emitting carbon dioxide and nitrogen gas are causing pollution and zero emissions can be achieved either by introducing a completely new alternative of petrol/diesel or by introducing a more efficient catalyst convertor. Perovskite structure catalyst converters presently being used [25] have a drawback related to decreases in efficiency with the time of use. However, the perovskite catalyst [26] has demonstrated a very high conversion rate of UHC. At present time, it is essential to control the UHC emission to check the global warming.

5.10 Conclusions

Perovskite structure materials are demonstrated as leading technological materials. A recent miracle perovskite solar panel which is capable of healing its damage 100% in spacecraft is a surprising discovery. It is capable of recovering its damaged efficiency (band gap, etc.) in the presence of sun and vacuum. Another recent wonder perovskite structure-related discovery is piezoelectric graphene. Piezoelectric graphene has unparalleled electrical, optical, or mechanical control for applications ranging from touch screens to nanoscale transistors. Barium titanate was initially developed during World War II around 1942, but in around last 75 years, there has been tremendous development in the field of barium titanate. Currently, barium titanate piezo generators are being used in advanced medical diagnoses such as heartbeat recording and energy harvesting. Barium titanate NGs have shown high suitability in the application of self-powered devices that generate required energy from ambient sources or attached surroundings. Due to higher Curie temperature (T_C, better than barium titanate) and high-temperature stability, PZT is more suitable in applications such as micro electric devices such as surface acoustics wave (SAW) devices, energy harvesting, sensors, etc. Favorable ferroelectric hysteresis makes the PZT most popular for the application of non-volatile memory devices in computer systems. Modified barium titanate p-NG fibers used under shoes can generate an output voltage of 25 volts at a frequency of 0.6 Hz under the force of a human body, i.e., 600 N.

References

1. Sato, T., Takagi, S., Deledda, S., Hauback, B. C., & Orimo, S. I. Extending the applicability of the Goldschmidt tolerance factor to arbitrary ionic compounds. *Scientific Reports*, **6**(1), 23592, 2016.
2. Tidrow, S. C. Mapping comparison of Goldschmidt's tolerance factor with perovskite structural conditions. *Ferroelectrics*, **470**(1), 13–27, 2014.

3. Ramos-Terron, S., Jodlowski, A. D., Verdugo-Escamilla, C., Camacho, L., & de Miguel, G. Relaxing the Goldschmidt tolerance factor: Sizable incorporation of the guanidinium cation into a two-dimensional Ruddlesden–Popper perovskite. *Chemistry of Materials*, **32**(9), 4024–4037, 2020.

4. He, J., Chu, Y., Sun, Y., Zhang, R., Li, J., Zhao, L., ... & Li, S. Beyond the limit of Goldschmidt tolerance factor: Crystal surface engineering to boost the α-phase stability of formamidinium-only hybrid inorganic–organic perovskites. *Solar RRL*, **5**(8), 2100188, 2021.

5. Ramadass, N. ABO_3-type oxides—Their structure and properties—A bird's eye view. *Materials Science and Engineering*, **36**(2), 231–239, 1978.

6. Sinha, S. K., Kumari, S., & Chaudhary, R. K. Dielectric and piezoelectric properties of $PbTi_{0.8-x}Se_{0.2}Sm_xO_3$ nanoceramics prepared by high energy ball milling. *Applied Physics A*, **125**(3), 171, 2019.

7. Singh, R. S., & Maiti, K. Manifestation of screening effects and A–O covalency in the core level spectra of A site elements in the ABO_3 structure of $Ca_{1-x}Sr_xRuO_3$. *Physical Review B*, **76**(8), 085102, 2007.

8. Balasubramaniam, K. R., Havelia, S., Salvador, P. A., Zheng, H., & Mitchell, J. F. Epitaxial stabilization and structural properties of $REMnO_3$ (RE = Dy, Gd, Sm) compounds in a layered, hexagonal ABO_3 structure. *Applied Physics Letters*, **91**(23), 1603–1610, 2007.

9. Lele, A., & Lele, A. New materials. In *Disruptive Technologies for the Militaries and Security*, Springer, 79–99, 2019.

10. Yogeswaran, N., Navaraj, W. T., Gupta, S., Liu, F., Vinciguerra, V., Lorenzelli, L., & Dahiya, R. Piezoelectric graphene field effect transistor pressure sensors for tactile sensing. *Applied Physics Letters*, **113**(1), 2018.

11. Kumar, B., Lee, K. Y., Park, H. K., Chae, S. J., Lee, Y. H., & Kim, S. W. Controlled growth of semiconducting nanowire, nanowall, and hybrid nanostructures on graphene for piezoelectric nanogenerators. *ACS Nano*, **5**(5), 4197–4204, 2011.

12. Layek, R. K., Samanta, S., Chatterjee, D. P., & Nandi, A. K. Physical and mechanical properties of poly(methyl methacrylate)-functionalized graphene/poly(vinylidine fluoride) nanocomposites: Piezoelectric β polymorph formation. *Polymer*, **51**(24), 5846–5856, 2010.

13. Goldschmidt, V. M. *Geochemistry*, Clarandon Press, Oxford 1954.

14. Andersen, T. K., Fong, D. D., & Marks, L. D. Pauling's rules for oxide surfaces. *Surface Science Reports*, **73**(5), 213–232, 2018.

15. Newnham, R. E., & Cross, L. E. Ferroelectricity: The foundation of a field from form to function. *MRS bulletin*, **30**(11), 845–848, 2005.

16. Siddiqui, S., Kim, D. I., Roh, E., Trung, T. Q., Nguyen, M. T., & Lee, N. E. A durable and stable piezoelectric nanogenerator with nanocomposite nanofibers embedded in an elastomer under high loading for a self-powered sensor system. *Nano Energy*, **30**, 434–442, 2016.

17. Lu, X., Qu, H., & Skorobogatiy, M. Piezoelectric microstructured fibers via drawing of multimaterial preforms. *Scientific Reports*, **7**(1), 2907, 2017.

18. Ma, Y, Song, J, Wang, X, Liu, Y, & Zhou, J. Synthesis, microstructure and properties of magnetron sputtered lead zirconate titanate (PZT) thin film coatings. *Coatings*, **11**(8), 944, 2021.

19. Sinha, S. K., Kumari, S., & Chaudhary, R. K. Studies of dielectric and piezoelectric properties of $PbTi_{0.8-x}Te_{0.2}Gd_xO_3$ nano ceramics prepared by high energy ball milling. *Journal of Advanced Dielectrics*, **9**(2), 1950017, 2019.

20. Ohsato, H. Microwave dielectrics with perovskite-type structure. In *Perovskite Materials - Synthesis, Characterisation, Properties, and Applications*, 281–330, 2016. DOI: 10.5772/61718281.

21. Sinha, S. K. Role of artificial intelligence and internet of things based medical diagnostics smart health care system for a post-COVID-19 world. In *Implementation of Smart Healthcare Systems Using AI, IoT, and Blockchain* (pp. 131–146). Academic Press, 2023.

22. Kumari, S., Chaudhary, R. K., & Sinha, S. K. Preparation of $PbTi_{0.8-x}Tb_{0.2}Mg_xO_3$ nano ceramics by high energy ball milling and its characterization. *Digest Journal of Nanomaterials and Biostructures*, **13**(4), 1249–1254, 2018.

23. Kumar, M., Sinha, R. K., Yadav, S., Sinha, S. K., & Raole, P. M. Preparation of Ca modified Li_2TiO_3 ceramics pebbles and its activation energy calculation. *International Journal of Materials Engineering Innovation*, **8**(3–4), 273–281, 2017.

24. Lahlou Nabil, M. A., Fenineche, N., Popa, I., & Sunyol, J. J. Morphological, structural and hydrogen storage properties of $LaCrO_3$ perovskite-type oxides. *Energies*, **15**(4), 1463, 2022.

25. Sinha, S., Yadav, S., & Raole, P. M. Sol–gel followed by urea–acetone spherodization for preparation of lithium titanate ceramics pebbles and preliminary characterization. *Fusion Engineering and Design*, **113**, 146–150, 2016.

26. Keav, S., Matam, S. K., Ferri, D., & Weidenkaff, A. Structured perovskite-based catalysts and their application as three-way catalytic converters—A review. *Catalysts*, **4**(3), 226–255, 2014.

Chapter 6

Piezoelectric Materials in Sensors: An Overview

Saroj Kumar Shukla
Department of Polymer Science, Bhaskaracharya College of Applied Sciences, University of Delhi, Delhi-110075, India
sarojshukla2003@yahoo.co.in

The present chapter describes the evolution and integration of piezoelectric properties in different materials for precise sensing of both physical and chemical analytes. The basic principle and mechanistic aspects associated with sensing of different chemicals (i.e. gas, humidity, ammonia, heavy metals, organic pollutants, temperature, pressure, and lights) are presented along with sensing parameters and technical details. Further, the advantageous features of piezoelectric sensors are highlighted along with their superiority over traditional sensors in terms of sensing parameters, self-powered nature, and longer life span. The existing limitations and challenges of these sensors are also highlighted along with the technical details for their progressive commercial applications.

Piezoelectric Materials: Design and Applications
Edited by N. B. Singh and Dev Kumar Mahato

ISBN 978-981-5129-40-3 (Hardcover), 978-1-003-59897-8 (eBook)
www.jennystanford.com

6.1 Introduction

Piezoelectric materials (PMs) are a technically and scientifically important class of energy-producing materials upon applications of mechanical force and stress. Thus produced electrical properties are used to develop the different types of physical and chemical sensors. Initially, the piezoelectricity was reported in limited inorganic materials, but later a wide range of organic and organic-inorganic hybrid materials exhibited the piezoelectric effects and energy [1, 2]. The basic advantageous feature of PMs is the self-generation of current, which adds a self-powered nature in sensors to work without the need for any additional energy source like lithium batteries [3]. Further, the evolution of surface interactive sites in PMs after chemical and physical modifications generates additional features for sensing materials to use as active and passive sensors [4]. The importance of these materials in sensing sciences has been reported very early but currently their applications are exponentially increasing, which is certified by the rise in trends in publication frequency as shown in Fig. 6.1.

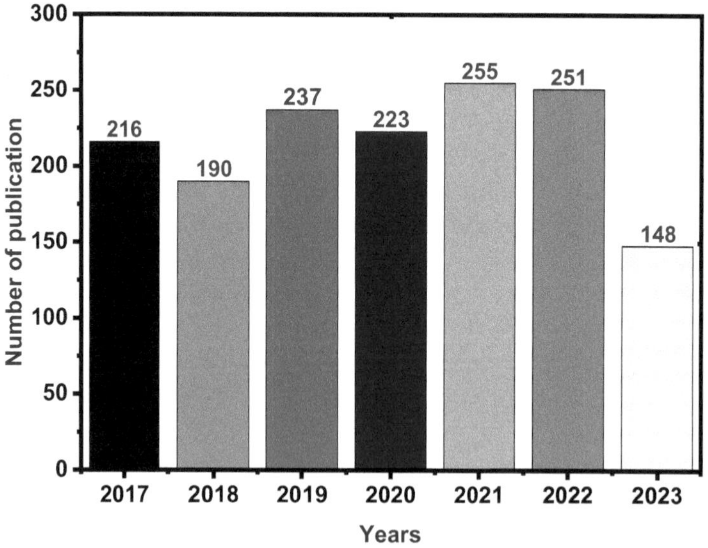

Figure 6.1 Publication frequency of PM sensors (www.scopus.com).

The importance of these materials in sensing sciences has been also compiled in reviews, books, and book chapters by several authors but its updated compilation is required for researchers to update the information [5–8]. In this context, the present book chapters efforts are made to present the holistic pictures of PMs and their applications in the development of different sensors and sensing sciences. The illustration and the scheme are also included to make the topic easy for a common researcher and scientists.

6.2 Overview of PM Sensors

In general, PM sensors work on the principle of piezoelectric effect, which demonstrates the generation of stress-induced energy after evolution of dielectric poles and current after structural rearrangements against a defined center or axis [9]. The extent of piezoelectricity produced in materials is used in different sensing applications after exploring both direct and inverse effects. In the direct principle of piezoelectricity, the applied tensile and compressive stress on a material generates polarization induced voltage, while in the inverse method the observed potential generates the mechanical displacement among the matrix of materials. The stress induced potential change are based on transverse, longitudinal, and shear effects, which may be explored in a wide range of applications including chemical and biosensor sensors. The brief principles of PMs and their use in sensors are shown in Fig. 6.2.

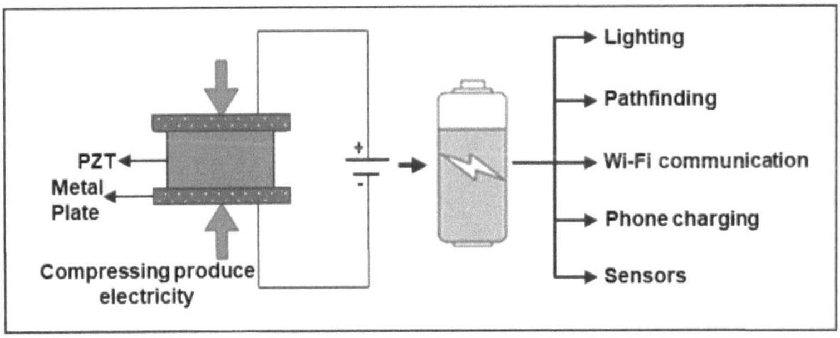

Figure 6.2 Basic principle and applications of piezoelectric materials [10].

The use of the piezoelectric effect in sensing application is based on the quantitative monitoring of change in mass, the induced electric signal produced after observed external pressure by PMs after interactions of physical, chemical, or biological analytes [11]. The inherited advantageous features of piezoelectric sensors are natural produced potential, which supports the self-powered nature, high strength, impact resistance, high signal output, stability, and mechanical strength in sensing materials. The inverse of the piezoelectric effect, i.e., changes in mechanical vibration and molecular frequencies after applying electric field are also explored in sensing of several adsorbed analytes, time, producing signal and berger in automobile vehicles [12]. The inverse piezoelectric effect polarises the sensing elements in the form of mechanical deformation and pressure due to changes in length, volume, thickness, and plane shear for exploration in sensing applications. The basic structure of both direct piezoelectric and inverse piezoelectric sensors is given in Fig. 6.3.

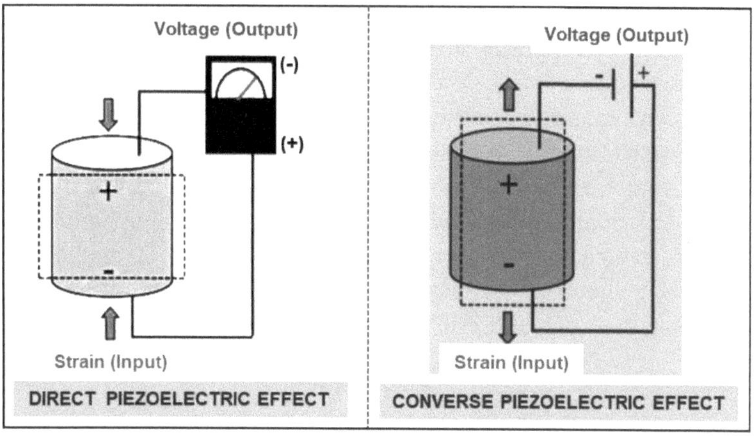

Figure 6.3 Working schematic of both types of piezoelectric sensors.

However, the effectiveness of sensors depends on PMs used in the sensing setup. Initially, piezoelectricity was observed in 1880 by two scientists Pierre Curie and Jacques Curie in the samples of quartz, tourmaline, and Rochelle salt. However, the progress in materials science has demonstrated improved piezoelectricity in nano-confined, hybrid materials and polymer nanocomposites. Temperature is another important parameter for the piezoelectric effect and in past decades several materials have been developed with temperature-dependent piezoelectricity. These materials are also explored for accelerators, hydrophones, acoustic sensors, and piezo drivers [13]. The quartz crystal microbalance is another type of PM sensors, which detect the change in frequency of crystal associated with mass change after adsorption of analytes. Acoustic sensors are the other important class of PM sensors, which are used for sensing of different analytes after monitoring the change in amplitude, velocity, and frequency of propagating waves [14].

The principle of piezoelectric sensor (PS) is based on the monitoring of the electric signal generated and the change properties of propagating waves after applying the external pressure, i.e., acceleration, mechanical stress through external force as well as adsorbed gaseous molecules [15]. The pressure can be applied in the form of external pressure or after the adsorption of external gaseous molecules on PM crystal or sensing elements. Piezoelectric sensing has been used for chemical sensors and biosensors and the schematic of different types of PS is depicted in Fig. 6.4.

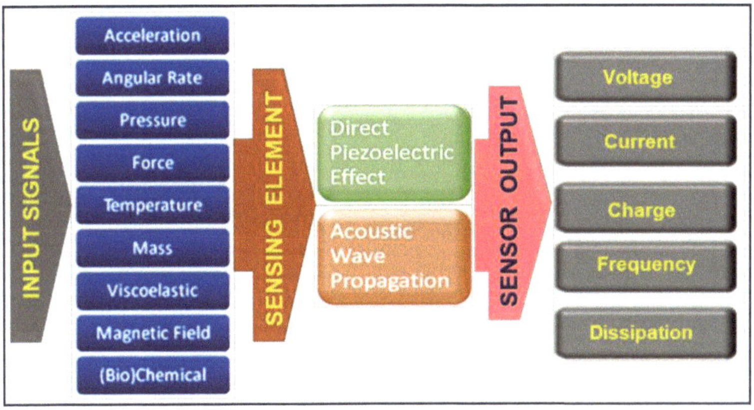

Figure 6.4 Classification of PM-based sensor.

The important analytes of these types of sensors are physical, i.e., light, heat, mechanical vibration, atmospheric and mechanical pressure, and different environmental gaseous molecules. The basic component of the piezoelectric sensor is a thin membrane, which is responsible for transferring the applied force to the PMs to generate electric signals. The schematic of the representative PS sensors is shown in Fig. 6.5a, while the basic configuration of PM sensing elements is given in Fig. 6.5b.

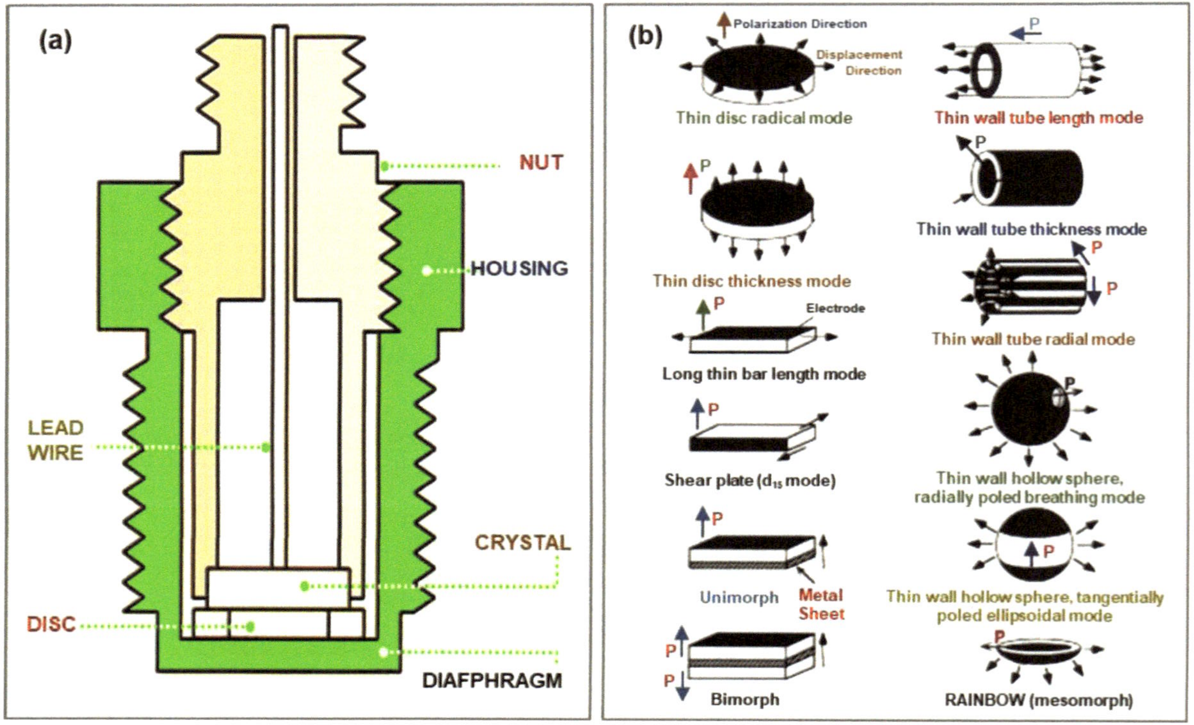

Figure 6.5 (a) Basic components of piezoelectric sensors and (b) configuration of piezoelectric sensor [15].

The advanced version of piezoelectric sensors, i.e., miniaturization, has led to developing the microelectromechanical systems (MEMS) and quartz crystal microbalance (QCM) based sensors for a wider range of applications [16]. The basic approaches of miniaturization to prepare MEMS are as follows:

a. Additive approach: preparation of piezoelectric thin films on silicon substrates containing appropriate insulating and conducting layers.
b. Subtractive approach: bulk micromachining of a PMs, i.e., single crystal, polycrystalline, and piezoceramics.
c. Integrative approach: micromachined structures in silicon bonded onto bulk piezoelectric substrate.

Based on size each approach is used and exhibits specific features are given in Fig. 6.6.

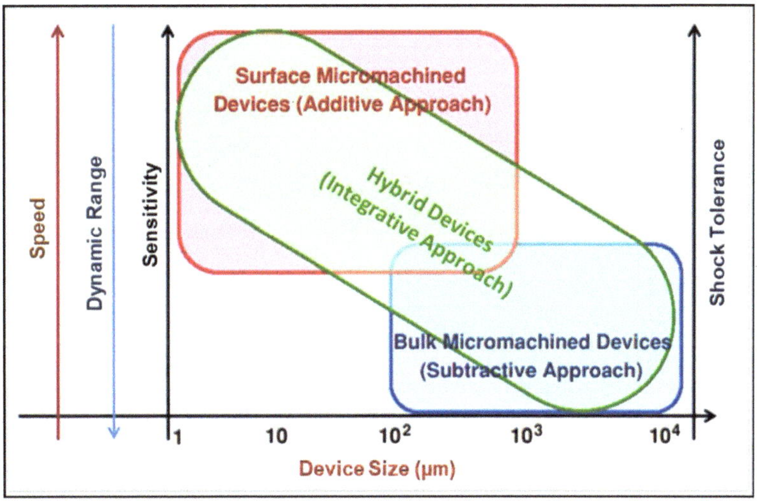

Figure 6.6 Fabrication approach for piezoelectric MEMS.

The development of different materials like nano size and hybrid materials (i.e., polymer composite) has been used for the preparation of PS with better features to sense both physical and chemical parameters.

6.3 Applications of PM Sensors

Self-powered nature of PM sensors is used in sensing a wide range of physical and chemical parameters like composition, intensity, and disturbances. The current advances in synergism between PM sensors and robotics expanded its wide range of applications for different applications. The different applications of PM sensors are shown in Fig. 6.7.

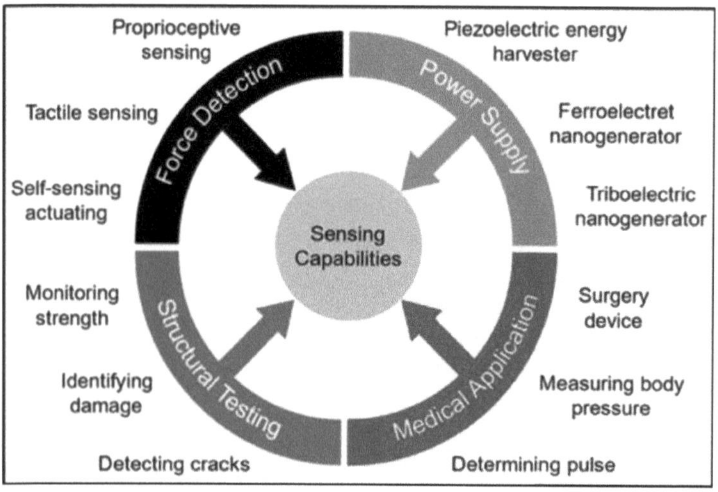

Figure 6.7 Classification of PM sensors [17].

In general, the different sensed analytes are either physical or chemical in nature but are used in structural testing, medical testing, gas sensing, force sensing, and its indicators after using different transducers.

6.4 Physical Sensors

The piezoelectric effect directly reveals the use of PM sensors in monitoring natural and artificial pressure generation during different processes. The different parameters using this class of sensors are pressure, acceleration, and temperature with associated advantageous and disadvantageous features. In a piezoelectric force sensor, the force to be measured is directly applied to the piezoelectric crystal, and the produced current is measured by suitable devices or digitally displayed after due calibration and conversion into digital results. The basic diagram of the piezoelectric force sensor is shown in Fig. 6.8.

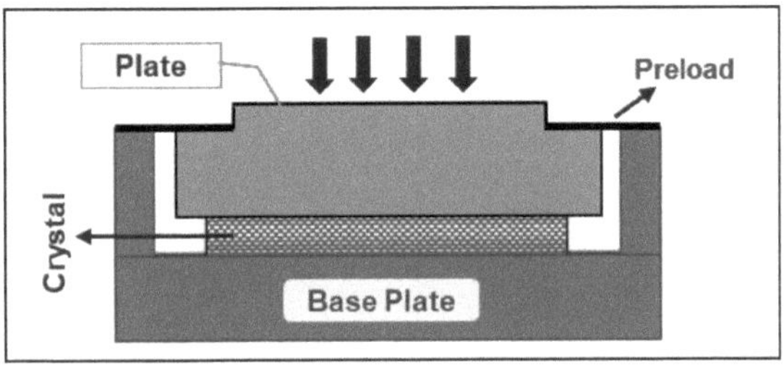

Figure 6.8 Piezoelectric force sensors [18].

The basic need of a physical sensor is PM materials with effective piezoelectric conversion efficiency, which decides the sensitivity of the sensor. The sensitivity of these sensors is equal to the produced open circuit voltage after applied stress, which is equal to the product of g and t, where "g" is the relevant piezoelectric voltage tensor and "t" is the thickness of sensing elements. In general, high sensitivity is the main interest of sensor manufacturers for differentiating the sensing signal and noise.

One of the basic applications of PM-based physical sensors is vibration sensors for different applications, i.e., fluctuating speed, acceleration, and normal vibration. A simple representative vibration sensor suitable for level detection of loose materials under open or pressurized conditions is shown in Fig. 6.9.

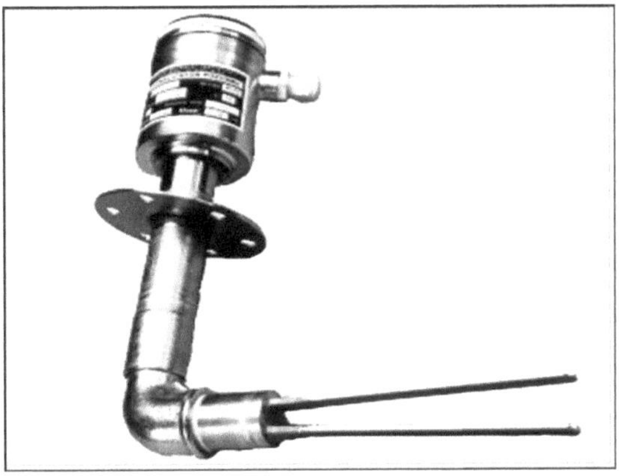

Figure 6.9 Angular vibrational sensor.

Table 6.1 PM-based physical sensors [19–29]

S. No.	Composition	Applications	Properties	Ref.
1	Cellulose nanofibrils film	Disposable piezoelectric sensors for electronics	Sensitivity between 4.7 and 6.4 pC/N	[19]
2	Polyvinylidene fluoride and MoS_2	Mechanical and electronic conformal devices	8.2 folds improved piezoelectric coefficient, i.e., 48.4 pC/N, than PVDF	[20]
3	Carbon fiber and poly(3,4-ethylenedioxythiophene)	Self-powered sensor for different parameters	Sensing response from 0.27 to 3.53 V	[21]
4	Polydimethylsiloxane and zinc oxide	Single-use force and vibration sensors for packaging	Frequency variation range between 50 Hz and 2.5 kHz	[22]
5	Yttrium calcium oxyborate	Prototyped accelerometer	Sensitivity of 5.9 ± 0.06 pC/g	[23]
6	$Ca_3NbGa_3Si_2O_{14}$ crystal	Compressive mode piezoelectric vibration sensor	Excitation frequency is 80–315 Hz and less than 5.2% temperature sensitivity	[24]
7	Bi_2CuO_4 and polydimethylsiloxane	Nanogenerator for harvesting mechanical vibrations	The maximum open circuit voltage is 50 V and 300 nA short circuit current	[25]
8	Aerosol deposited PZT film	Structural health monitoring	Better charge sensitivity of 22.74 pC/g and 4.96 mV/g is voltage sensitivity	[26]
9	Poly(L-lactic acid) fibers	Structural health monitoring	Frequency range up to 500 kHz	[27]
10	Lead zirconium titrate	Blood pressure monitoring	Promises portable wearable devices for continuous monitoring of blood pressure to control hypertension	[28]
11	Implantable piezoelectric soft tissue sensor	Blood pressure monitoring during surgical operation	Better electrical performance of 8.3 V and stability for 6000 cycles	[29]

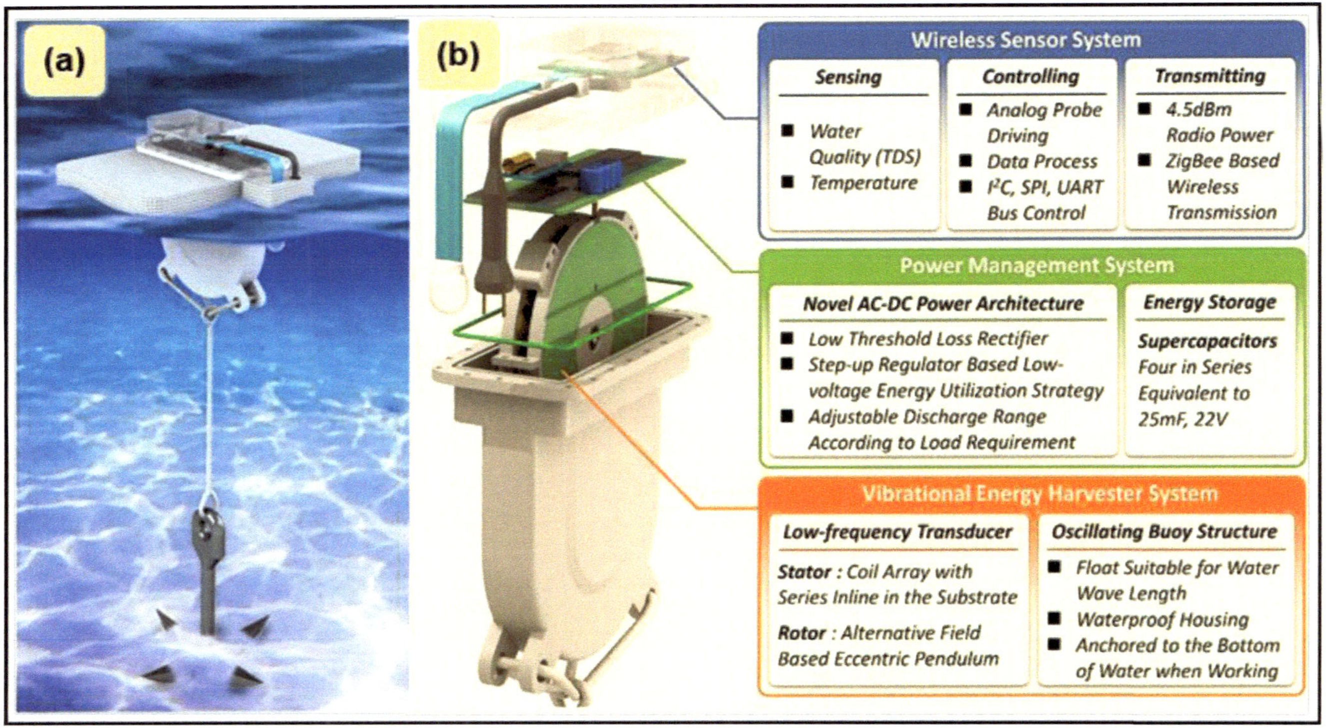

Figure 6.10 Application scenario and composition of the water-wave energy harvester-based self-powered wireless water-quality sensor system [30].

These types of sensors are made up of two pairs of receiving plates and the same supplying plates. Both are connected in a biomorphic system, and the variable voltage fed through supply plates exhibits a change in their thickness directly proportional to the amount of the applied voltage. Further, the value of voltage is transferred to an electronic system for visual display or recordings. Some other important PM based physical sensors are listed in Table 6.1 along with their properties and applications.

The inverse of the piezoelectric effect has been also explored to produce nanogenerators to make self-powered sensors and energy harvesting from mechanical energy for different applications. In this context, Li et al. have developed a self-powered wireless sensor for water monitoring after using a low-frequency electromagnetic-pendulum energy harvester [30]. The proposed sensor is packed in the form of a small float and without the use of an external lithium-ion battery to report data wirelessly about the water wave drives of oceans. The structure and brief details of the proposed sensor are shown in Fig. 6.10.

The wearable flexible sensors have been also reported after dip coating with hydrophobic materials for human movements. The sensor exhibits high sensitivity, i.e., gauge factor of 161, a bending angle of 0–120°, and excellent durability of 100,000 cycles of bending deformation at 30° [31].

6.5 Chemical Sensors

PMs-based crystals are used for sensing organic vapors, gas molecules, metal ions, and biomolecules after coating the crystal with a suitable sensing and selective material [32]. The simple setup of PM-based chemical sensors is given in Fig. 6.11 [33].

QCM is a piezoelectric sensor comprising a thin disc of quartz with electrodes. The electrode acts as a sensing surface and works as label-free advanced sensing of different molecules and biomolecules at nanogram mass. The principle of this technique was proposed by Sauerbrey in 1959, which demonstrates that the frequency of crystal linearly changes against the change of mass due to surface adsorption of analyte, as proposed in Eq. 6.1.

$$\Delta m = -c \times {}^{1}\!/_{n} \times \Delta f \tag{6.1}$$

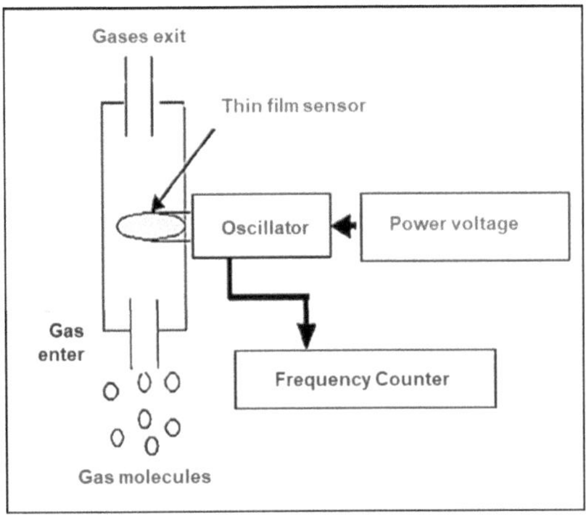

Figure 6.11 Schematic of acoustic sensing setup.

Here, n is the overtone number and C is a constant that depends on the property of the crystal used. The equation supports the growth of QCM sensing technology for different applications. The scheme of the QCM sensor is shown in Fig. 6.12 along with the sensing mechanism and components.

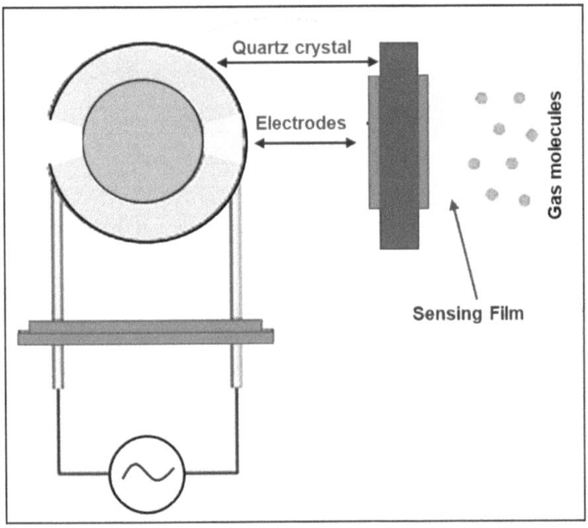

Figure 6.12 Schematic of QCM sensor and mechanism.

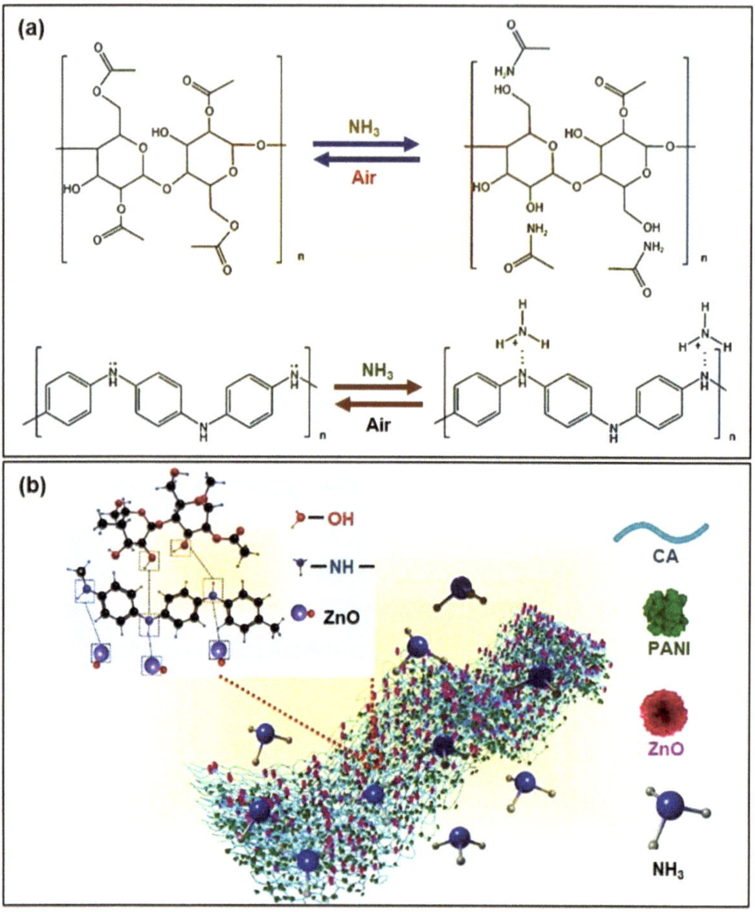

Figure 6.13 (a) Mechanism for ammonia sensing and (b) chemical structure of CA/PANI/ZnO composites [36].

The working efficiency of the QCM sensor depends on the used electrode. In this regard, organic, inorganic, and their composites are used. Current innovated nanomaterials, metal-organic frameworks, and covalent organic frameworks are explored for sensing of different analytes. These materials yield a high sensitivity for effective sensing, while the presence of a metal-organic framework yields catalytic transitions to support the sensing of organic molecules [34, 35]. In this regard, a QCM sensor was reported by Zhang et al. after using zinc oxide nanoflower and cellulose acetate-polyaniline composite (CA/PANI/ZnO) nanofibers for room temperature ammonia sensing.

The molecular surface interaction between composite and ammonia molecules was responsible for the sensing mechanism. The presence of an ester group in cellulose acetate generates corresponding amide molecules. The basic mechanism is shown in Fig. 6.13 [36].

The QCM technology has been also widely used to determine biomolecules, metabolites, and biologically active compounds after using enzymes and non-enzymes [32, 36]. For the surface modification of quartz crystal, different methods and materials are used after optimizing the various parameters. The surface-modified QCM is used for sensing biomolecules in both active and passive principles. The optical photograph of QCM-based sensors is shown in Fig. 6.14 along with the position electrode in both views.

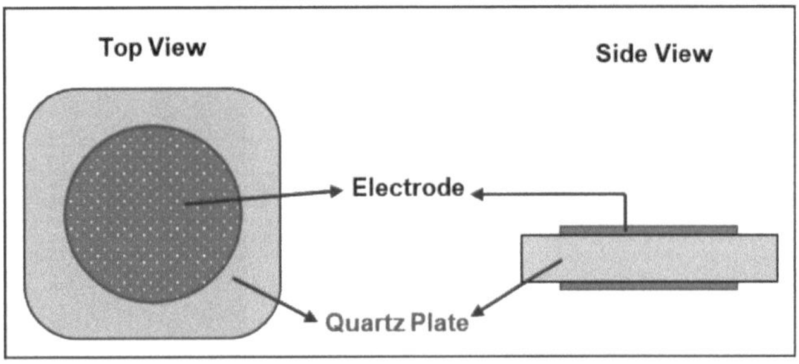

Figure 6.14 Top and side view schematic of QCM biosensors.

The use of nanomaterials, nanocomposite, metal-organic frameworks, and covalent organic frameworks in QCM techniques improves the surface area and effectiveness of sensors. Among the different materials, carbon nanostructures (i.e., graphene and carbon nanotubes) are an important class of 2D nanomaterials with a wide range of applications in the development of electronic, electrical, optical, and sensing devices due to a wide range of surface reactivity [37, 38]. Yola et al. have demonstrated the sensing of zearalenone, i.e., a nonsteroidal estrogenic mycotoxin, on molecularly imprinted QCM with molybdenum disulfide nanoparticle and multiwalled carbon nanotube composite [39]. The proposed sensor exhibits a linear response in the range of 1.0–10.0 ng/L and detection limit of 0.30 ng/L along with the high repeatability, reusability, selectivity, and stability for sensing of zearalenone in rice samples. The other significant QCM-based chemicals and biosensors are listed in Table 6.2, along with details and properties.

Table 6.2 Representative QCM-based biosensors, principle, and properties

S. No.	Composition	Analytes and transducer	Properties	Ref.
1	Molecularly imprinted polydopamine	Proteins	Micromolar range dissociation constant with heterogeneous binding sites	[40]
2	4-*N*-mannosyl benzoic acid and 4-thiophenyl-*N*-mannose	*Escherichia coli*	Detection limit of detection of 3.7 and 6.6 CFU/mL, 2.56×10^3 and 8.99×10^{-5} ng/mL sensitivity in the dynamic range of 10^3 to 10^6 CFU/mL	[41]
3	Molecularly imprinted polymeric membranes	*N*-hexanoyl-L-homoserine lactone	Cost-effective, short time, and better sensitivity in complex environments	[42]
4	Zeolitic imidazolate framework	Acetone	Suitable for breath analysis and diabetic patient	[43]
5	Chitosan-halloysite nanotubes	Humidity sensor	69.1 Hz/% RH sensitivity, less than 2.8% RH hysteresis, better selectivity, stability, and repeatability.	[44]
6	Fe-doped ZnO and graphene oxide	Ethanol	Excellent ethanol sensing response at 25°C and sensitivity up to 10 ppm	[45]
7	MoS_2	CH_4 and CO	Dual sensitivity to CH_4 and CO at ppm level to use in mines	[46]

S. No.	Composition	Analytes and transducer	Properties	Ref.
8	α-$Co(OH)_2$	Humidity	The sensitivity of 23.1 Hz/% RH, 3s response and 4s recovery times along with stability up to 30 days	[47]
9	Enzyme-linked immunosorbent	Necrosis factor	Interference-free detection limit of 1.62 pg/mL	[48]
10	Gold nanoparticles-antibodies conjugate	Procalcitonin	Detection limit and quantification limit of 37.8 ng/L and 104 ng/L, respectively	[49]
11	Au-$BaTiO_3$-FeGa	Sweat and glucose	Optimum sensitivity against the magnetic field of 30 mT	[50]

6.6 Conclusions and Prospects

Piezoelectric effects in sensing applications have been discussed along with the basic principle of piezoelectric sensors. The applications of piezoelectric sensors in the monitoring of physical, chemical, and biosensors have been described with suitable examples and advantageous features. Further, the sensing mechanism and reported parameters are represented with the help of illustrations and figures. Finally, efforts have been made to present the potential of piezoelectric sensors in various fields along with existing limitations and challenges.

References

1. Uchino, K., The development of piezoelectric materials and the new perspective, *Advanced Piezoelectric Materials* (2nd ed.), Woodhead Publishing, 2017.
2. Sekhar, M.C., Veena, E., Kumar, N.S., Naidu, K.C.B., Mallikarjuna, A., and Basha, D.B., A review on piezoelectric materials and their applications, *Crystal Research and Technology*, **58**, 2200130, 2023.
3. Mahapatra S.D., Mohapatra P.C., Aria A.I., Christie G., Mishra Y.K., Hofmann S., and Thakur V.K., Piezoelectric materials for energy harvesting and sensing applications: Roadmap for future smart materials, *Advanced Science*, **8**, 2100864, 2021.
4. Tressler, J.F., Alkoy, S., and Newnham, R.E., Piezoelectric sensors and sensor materials, *Journal of Electroceramics*, **2**, 257–272, 1998.
5. Anton, S. and Sodano, H.A., A review of power harvesting using piezoelectric materials (2003–2006), *Smart Materials and Structures*, **16**(3), R1, 2007.
6. Sawane, M. and Prasad, M., MEMS piezoelectric sensor for self-powered devices: A review, *Materials Science in Semiconductor Processing*, **158**, 107324, 2023.
7. Ju, M., Dou, Z., Li, J.W., Qiu, X., Shen, B., Zhang, D., Yao, F.Z., Gong, W., and Wang, K., Piezoelectric materials and sensors for structural health monitoring: Fundamental aspects, current status, and future perspectives, *Sensors*, **23**, 543, 2023.
8. Mirjalali, S., Mahdavi Varposhti, A., Abrishami, S., Bagherzadeh, R., Asadnia, M., Huang, S., Peng, S., Wang, C.H., and Wu, S., A review on wearable electrospun polymeric piezoelectric sensors and energy harvesters, *Macromolecular Materials and Engineering*, **308**, 2200442, 2023.

9. Heywang, W., Lubitz, K., and Wersing, W. (eds.), *Piezoelectricity: Evolution and Future of a Technology*, Springer Science & Business Media, **114**, 2008.
10. Casini, M., Smart buildings: Advanced materials and nanotechnology to improve energy-efficiency and environmental performance, Woodhead Publishing, 2016.
11. Rupitsch, S.J., *Piezoelectric Sensors and Actuators: Fundamentals and Applications*, Springer, 2019.
12. Yang, H., Zhao, H., Sun, M., Han, Z., and Li, Z., Inverse piezoelectric all-fiber voltage sensor based on tapered fiber, *Optical Fiber Technology*, **78**, 103314, 2023.
13. Sekhar, M.C., Veena, E., Kumar, N.S., Naidu, K.C.B., Mallikarjuna, A., and Basha, D.B., A review on piezoelectric materials and their applications, *Crystal Research and Technology*, **58**, 2200130, 2023.
14. Ali, W.R. and Prasad, M., Piezoelectric MEMS based acoustic sensors: A review, *Sensors and Actuators A: Physical*, **301**, 111756, 2020.
15. Tressler, J.F., Alkoy, S., and Newnham, R.E., Piezoelectric sensors and sensor materials, *Journal of Electroceramics*, **2**, 257–272, 1998.
16. Tadigadapa, S.A.K.M. and Mateti, K., Piezoelectric MEMS sensors: State-of-the-art and perspectives, *Measurement Science and Technology*, **20**, 092001, 2009.
17. Fath, A., Xia, T., and Li, W., Recent advances in the application of piezoelectric materials in microrobotic systems, *Micromachines*, **13**(9), 1422, 2022.
18. Regtien, P.P. and Dertien, E., *Sensors for Mechatronics*, Elsevier, 2018.
19. Rajala, S., Siponkoski, T., Sarlin, E., Mettanen, M., Vuoriluoto, M., Pammo, A., Juuti, J., Rojas, O.J., Franssila, S., and Tuukkanen, S., Cellulose nanofibril film as a piezoelectric sensor material, *ACS Applied Materials & Interfaces*, **8**, 15607–15614, 2016.
20. Islam, M.N., Rupom, R.H., Adhikari, P.R., Demchuk, Z., Popov, I., Sokolov, A.P., Wu, H.F., Advincula, R.C., Dahotre, N., Jiang, Y., and Choi, W., Boosting piezoelectricity by 3D printing PVDF-MoS_2 composite as a conformal and high-sensitivity piezoelectric sensor, *Advanced Functional Materials*, 2302946, 2023.
21. He, Q., Li, X., Zhang, H., and Briscoe, J., Nano-engineered carbon fibre-based piezoelectric smart composites for energy harvesting and self-powered sensing, *Advanced Functional Materials*, 2213918, 2023.
22. Sinar, D. and Knopf, G.K., Disposable piezoelectric vibration sensors with PDMS/ZnO transducers on printed graphene-cellulose electrodes, Sensors and Actuators A: Physical, **302**, 111800, 2020.

23. Kim, K., Zhang, S., Salazar, G. and Jiang, X., Design, fabrication and characterization of high temperature piezoelectric vibration sensor using YCOB crystals, *Sensors and Actuators A: Physical*, **178**, 40–48, 2012.

24. Wang, Y., Liu, X., Bai, L., Cheng, R., Jiang, C., Li, Y., Zhang, J., Chen, H., Li, Y., Yu, F., and Guo, S., Design, fabrication, and characterization of high-temperature piezoelectric vibration sensor based on the Ho: CNGS crystal, *Journal of Alloys and Compounds*, **937**, 168449, 2023.

25. Veeralingam, S., Nanda, O.P., and Badhulika, S., Lead-free Bi_2CuO_4 interspersed into PDMS matrix-based bifunctional piezoelectric nanogenerator for vibrational energy harvesting and visible light photodetection Applications, *Journal of Alloys and Compounds*, 171127, 2023.

26. Gong, X., Kuo, Y.C., Zhou, G., Wu, W.J., and Liao, W.H., An aerosol deposition based MEMS piezoelectric accelerometer for low noise measurement, *Microsystems & Nanoengineering*, **9**, 23, 2023.

27. Yousry, Y.M., Wong, V.K., Ji, R., Chen, Y., Chen, S., Zhang, X., Lim, D.B.K., Shen, L., and Yao, K., Shear mode ultrasonic transducers from flexible piezoelectric PLLA fibers for structural health monitoring, *Advanced Functional Materials*, **33**, 2213582, 2023.

28. Yi, Z., Liu, Z., Li, W., Ruan, T., Chen, X., Liu, J., Yang, B., and Zhang, W., Piezoelectric dynamics of arterial pulse for wearable continuous blood pressure monitoring, *Advanced Materials*, **34**, 2110291, 2022.

29. Wang, C., Hu, Y., Liu, Y., Shan, Y., Qu, X., Xue, J., He, T., Cheng, S., Zhou, H., Liu, W., and Guo, Z.H., Tissue-adhesive piezoelectric soft sensor for in vivo blood pressure monitoring during surgical operation, *Advanced Functional Materials*, 2303696, 2023.

30. Li, M., Zhang, Y., Li, K., Zhang, Y., Xu, K., Liu, X., Zhong, S., and Cao, J., Self-powered wireless sensor system for water monitoring based on low-frequency electromagnetic-pendulum energy harvester, *Energy*, **251**, 123883, 2022.

31. Guan, M., Liu, Y., Du, H., Long, Y., An, X., Liu, H., and Cheng, B., Durable, breathable, sweat-resistant, and degradable flexible sensors for human motion detection, *Chemical Engineering Journal*, **462**, 142151, 2023.

32. Alanazi, N., Almutairi, M., and Alodhayb, A.N., A review of quartz crystal microbalance for chemical and biological sensing applications, *Sensing and Imaging*, **24**, 10, 2023.

33. Mohamed, R., Sallch, M.M., and Yahaya, M., Development of a piezoelectric chemical sensor using metalloporphyrins compounds as a coating material. In 2012 IEEE 8th International Colloquium on Signal Processing and Its Applications (41–43). IEEE, 2012.

34. Torad, N.L., Zhang, S., Amer, W.A., Ayad, M.M., Kim, M., Kim, J., Ding, B., Zhang, X., Kimura, T., and Yamauchi, Y., Advanced nanoporous material–based QCM devices: A new horizon of interfacial mass sensing technology, *Advanced Materials Interfaces*, **6**, 1900849, 2019.

35. Wang, L., Metal-organic frameworks for QCM-based gas sensors: A review, *Sensors and Actuators A: Physical*, **307**, 111984, 2020.

36. Zhou, D., Kang, Z., Liu, X., Yan, W., Cai, H., Xu, J., and Zhang, D., High sensitivity ammonia QCM sensor based on ZnO nanoflower assisted cellulose acetate-polyaniline composite nanofibers, *Sensors and Actuators B: Chemical*, **392**, 134072, 2023.

37. Sengupta, J. and Hussain, C.M., Graphene-based electrochemical nano-biosensors for detection of SARS-CoV-2, *Inorganics*, **11**, 197, 2023.

38. Lin, L.P. and Tan, M.T.T., Biosensors for the detection of lung cancer biomarkers: A review on biomarkers, transducing techniques and recent graphene-based implementations, *Biosensors and Bioelectronics*, **237**, 115492, 2023.

39. Çapar, N., Polat, İ., Yola, B.B., Atar, N., and Yola, M.L. (2023). A novel molecular imprinted QCM sensor based on MoS2NPs-MWCNT nanocomposite for zearalenone determination, *Microchimica Acta*, **190**(7), 262.

40. Lim, H.J., Saha, T., Tey, B.T., Lal, S.K., and Ooi, C.W., Quartz crystal microbalance-based biosensing of proteins using molecularly imprinted polydopamine sensing films: Interplay between protein characteristics and molecular imprinting effect, *Surfaces and Interfaces*, **39**, 102904, 2023.

41. Eshun, G.B., Crapo, H.A., Yazgan, I., Cronmiller, L., and Sadik, O.A., Sugar–lectin interactions for direct and selective detection of *Escherichia coli* bacteria using QCM biosensor, *Biosensors*, **13**, 337, 2023.

42. Acet, Ö. and Odabaşı, M., Detection of N-hexanoyl-L-homoserine lactone via MIP-based QCM sensor: Preparation and characterization, *Polymer Bulletin*, **80**, 6657–6674, 2023.

43. Homayoonnia, S. and Kim, S., ZIF-8/MWCNT-nanocomposite based-resistive sensor for highly selective detection of acetone in parts-per-billion: Potential noninvasive diagnosis of diabetes, *Sensors and Actuators B: Chemical*, **393**, 134197, 2023.

44. Zhang, D., Mao, R., Song, X., Wang, D., Zhang, H., Xia, H., Ma, Y., and Gao, Y., Humidity sensing properties and respiratory behavior detection based on chitosan-halloysite nanotubes film coated QCM sensor combined with support vector machine, *Sensors and Actuators B: Chemical*, **374**, 132824, 2023.

45. Liu, M., Shen, Y., Zhang, Y., Gu, F. and Wang, L., Improved sensing performance of Fe-ZnO/GO mesoporous composites based on QCM, *Journal of Materials Science: Materials in Electronics*, **34**, 79, 2023.

46. Wang, L., Song, J., Ruan, C., Xu, S., and Yu, C., MoS_2 based dual mine gas disaster sensor that operates at room temperature, *Sensors and Actuators A: Physical*, **359**, 114517, 2023.

47. Li, L., Chen, T., He, Z., and Lu, X., α-Co $(OH)_2$-based quartz crystal microbalance humidity sensor for real-time monitoring of respiration, *Physica Status Solidi* (A), 2200707, 2023.

48. Pohanka, M., Piezoelectric biosensor for the determination of tumor necrosis factor alpha, *Talanta*, **178**, 970–973, 2018.

49. Pohanka, M., Quartz crystal microbalance biosensor for the detection of procalcitonin, *Talanta*, **257**, 124325, 2023.

50. Zhang, Q., Xu, M., Liu, Y., Zhang, C., Zhang, R., Fu, Z., Ji, J., Zhang, R., and Sang, S., A magnetostrictive self-powered biosensor based on Au-$BaTiO_3$-FeGa & PDMS, *Journal of Industrial and Engineering Chemistry*, **117**, 247–254, 2023.

Chapter 7

Transition Metal Doped Piezoelectric Materials for Sensor Applications

Suresh Kumar,[a] S. K. Sharma,[b] S. R. Kumar,[c] and K. P. Tiwary[d]

[a]*Department of Physics, Chaibasa Engineering College, Kelende, Jhinkpani, Jharkhand-833215, India*
[b]*Department of Physics, Indian Institute of Technology (Indian School of Mines), Dhanbad, Jharkhand-826004, India*
[c]*Department of Applied Science & Humanities, National Institute of Advanced Manufacturing Technology, Ranchi, Jharkhand-834003, India*
[d]*Department of Physics, Birla Institute of Technology, Mesra, Patna Campus, Bihar-800014, India*
sk93971@gmail.com, kptiwary@bitmesra.ac.in

Just a minority of the piezoelectric (and ferroelectric) materials developed during the past ten years are now utilized for sensor applications. In recent years, synthetic piezoelectric crystals (such as quartz, as opposed to natural quartz), ceramics, polymer sheets, and thin and thick films have been able to be used as transducers, actuators, and sensors. Although light-emitting diodes, solar cells,

Piezoelectric Materials: Design and Applications
Edited by N. B. Singh and Dev Kumar Mahato

ISBN 978-981-5129-40-3 (Hardcover), 978-1-003-59897-8 (eBook)
www.jennystanford.com

and transistors are common semiconductor devices, application requirements frequently necessitate the co-integration of other active materials and component classes. Due to its several intriguing qualities, piezoelectric materials have grown to be extensively employed in a variety of science and technology fields throughout the past several years, including health, electronics, the military, and the oil industry. Lead zirconate titanate (PZT) and barium titanate (BT), which are used to make a variety of electronic and optoelectronic devices, including sensors, capacitors, random access memory, optical modulators, and waveguides, are important classes of piezoelectric and ferroelectric materials with distinguishing characteristics. Because it has been found that the nanostructure of PZTs and BT has more intriguing features, various physical and chemical processes, such as sol-gel, sol-combustion, hydrothermal, solvothermal, precipitation, and solid-state reactions, have been established to create PZT and BT particles that are nanosized. Doping also affects the materials' optical and physical characteristics by stressing their crystal structures. This chapter reviews the creation of piezoelectric materials that have been doped with various transition metals for application in sensors.

7.1 Introduction

History has been divided into several historical periods (such as the Stone, Bronze, and Iron Ages) based on the materials that predominated during these eras, thanks to the considerable impact that materials technology has had on the growth of human society. The so-called Engineering Materials Age, sometimes known as the Synthetic Materials Age, is currently in effect and is characterized by the use of plastic, composites, and other well-designed, man-made engineering materials that outperform conventional materials. These engineering materials and structures, however, have earned the reputation of being "dumb" since they are typically preprocessed and created to respond to a small number of external stimuli. These reactions are usually not the most effective choice in a particular set of conditions, but they are tailored to match the widest possible range of situations in which a structure or substance might be used [1]. The Smart Materials Age will begin at the beginning of the twenty-first century as a result of recent advancements in a number of developing technologies, including biotechnology, biomimetics, nanotechnology, and information technology [2]. Nonliving technologies known as "smart materials systems" integrate the faculties of detecting, movement, reason, and command to respond in a flexible manner to environmental variations.

These responses are practical and typically repeating. Smart materials are useful materials for a variety of technical purposes and are a part of smart systems. They have the capacity for both perception and action. Many of the engineering materials already in use can operate as sensors and actuators if they are built properly. They include, among other things, shape memory alloys, optical fibers, conductive polymers, and piezoelectric ceramics and polymers.

7.2 Piezoelectric Effect

Piezoelectricity is the capability of an object to produce a field of electricity in reaction to exertion or stress. There is an accumulation of electric charge in several solid materials, including crystals, ceramics, and biological elements like bones, proteins, and DNA. When these materials are heated or subjected to mechanical stress, the electrical charge builds up. Greek words "piezein" and "piezo," which both indicate "to press or squeeze something" and "push," respectively, are the roots of the word "piezoelectric." Because of the piezoelectric effect, piezoelectricity is produced. Mechanical values like stress or strain as well as electrical parameters like electric field and electric displacement. are associated in crystals with piezoelectric characteristics. The term "electromechanical coupling" refers to this phenomenon.

A piezoelectric substance produces electricity when it is compressed. It occurs when crystal deformation converts mechanical or kinetic energy into electrical energy. If certain elements are subjected to stress from mechanical contact, they have the ability to produce electricity. A field of electricity is created around a piezoelectric material as a result of the positive (+ve) as well as negative (–ve) charge centers migrating when the material is mechanically expanded. Another possibility is the inverse piezoelectric impact, which can physically distort a piezoelectric substance in reaction to an outside field of electricity.

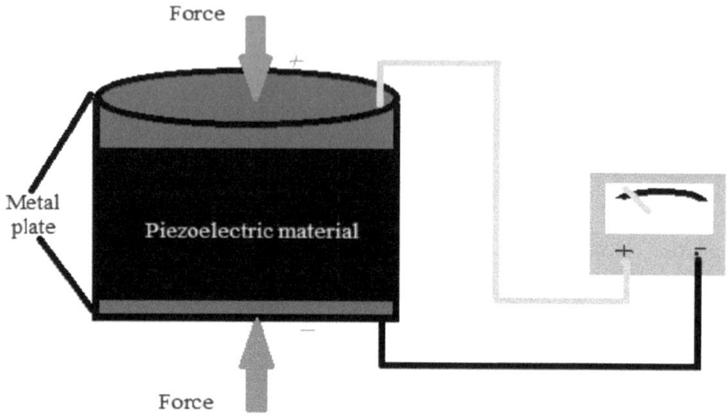

Figure 7.1 Direct piezoelectric effect theory.

Jacques and Pierre Curie discovered in 1880 that certain crystal forms, such as quartz and tourmaline, produce electrical charges under pressure. The "piezoelectric effect" was the name given to this phenomenon. The term "piezo" originates from the Greek language piezein, whose definition is to press/squeeze. Later, they discovered that piezoelectric materials may be deformed by electric fields. "Inverse piezoelectric effect" is the name given to this phenomenon. Following these initial experiments, more research was done to uncover the piezoelectric effect's enormous potential. The sonar gadget, created in the 1900s to find icebergs, was the first real-world use of piezoelectric technology. Intense interest in piezoelectric devices was sparked by this technology's initial application in sonar. New piezoelectric materials and uses were investigated throughout the ensuing decades. Figure 7.1 shows the piezoelectric effect theory.

Piezoelectric technology had its first significant use during World War I. An ultrasonic submarine detector was first being developed in 1917 by P. Langevin and his French colleagues. During the Second World War Period, solitary research groups in the USA, Japan, and USSR focused on developing better capacitor materials. One of these groups discovered that polycrystalline barium titanate ($BaTiO_3$ [BT]), a synthetic heterogeneous anisotropic substance, using heating oxides of metal particles, could be created that had dielectric values 100 times bigger than those of conventionally cut crystals. The material was unable to exhibit a macroscopic piezoelectric effect because hundreds of tiny single crystals were randomly dispersed and macroscopically averaged. But it was not until 1946 [3] that researchers realized that this particular class of polycrystalline substance, also referred to as ceramic ferroelectrics, has been converted to piezoelectricity with the help of an electric field; this procedure can be known as poling. In 1954, Jaffe et al. made the major finding of the lead zirconate titanate (PZT) solid solution that improved piezoelectric performances [4].

7.2.1 Types of Piezoelectric Effect

The piezoelectric effect comes in two different forms.

7.2.1.1 Direct piezoelectric effect

An electric dipole or polarization is produced whenever mechanical stress or friction is applied to these materials, which is caused by the overall migration of the ions, both +ve and −ve, in relation to one another as shown in Fig. 7.2. The crystal becomes a charged material as a result, changing from being a dielectric. The amount of stress or strain put on the crystal determines how much voltage is produced in a direct proportion.

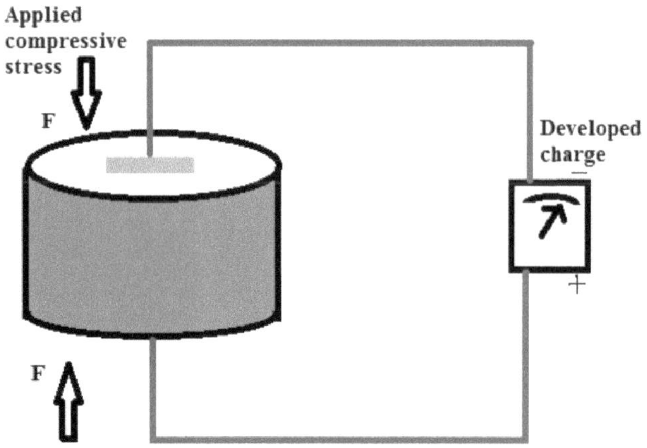

Figure 7.2 Direct piezoelectric effect diagram.

7.2.1.1.1 Use of direct piezo effect

- In Japanese train stations, the "crowd farm" concept has been tested, whereby foot traffic on piezoelectric tiles buried in the pavement can generate electricity.
- In 2008, a nightclub in London built the world's first piezoelectric floor, which generates electricity when people dance on it and can be used to power light bulbs.
- The piezoelectric effect has practical uses in mechanical frequency filters, bulk and surface acoustical wave equipment, etc.
- Hydrophones, ultrasonic imaging, and sound and ultrasound microphones and speakers.
- Pacemaker powering biosensors and piezoelectric guitar pickups.
- Piezoelectric components are used in the generation and detection of sonar waves, as well as single- and dual-axis tilt sensing.

7.2.1.2 Converse piezoelectric effect

Electric dipoles occur when electricity is given to these crystals, causing the dipole movement that leads to crystal deformation and ultimately creating the opposite piezoelectric action as depicted in Fig. 7.3. The opposite piezoelectric effect was later demonstrated in 1881 applying fundamental thermodynamic concepts [5].

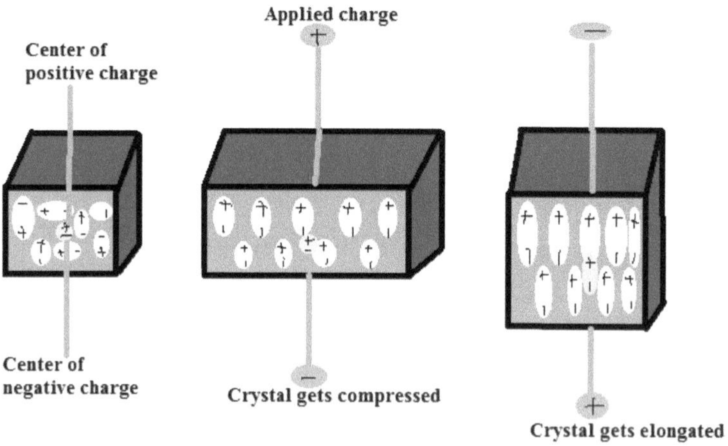

Figure 7.3 Converse piezoelectric effect diagram.

Following are the fundamental equations required for direct as well as converse piezoelectric effect and it can be written as mentioned below [6].

$$\text{Direct effect: } D = dT + eE \qquad \text{(i)}$$

$$\text{Converse effect: } X = sT + dE \qquad \text{(ii)}$$

Here, D is the electrical displacement,
d is the piezoelectric coefficient,
T is stress,
e is permittivity of the material,
E is the electric field,
X is strain, and
s is the mechanical compliance.

7.2.1.2.1 Uses of converse piezoelectric effect

- Motors and actuators.
- The positioning of microscope lenses and their micro-precision alterations.
- Printer needle drivers, miniature motors, and bimorph actuators.
- Actuators with multiple layers for precise placement in optics.
- Fuel valves for autos with injecting mechanisms, etc.

7.2.2 Uses for the Piezoelectric Effect

The following are some examples of piezoelectric effect applications.

- Sonar machinery, pressure detectors, knocking engine detectors, etc. will be a few of the industrial applications for piezoelectric sensors.
- Piezoelectric actuators are used in numerous manufacturing processes, such as piezoelectric drives, stacker controls, striping controls, piezoelectric switches, optical modification, swiftly responding solenoids, ultrasonic extraction, welding with ultrasonic waves together with diesel fuel nozzles.
- Ultrasound visualization, ultrasonic operations, and other applications all make use of piezoelectric transducers within medical applications.

- Devices for consumers such as piezoelectric speakers (cell telephones, listening buds, sound-generating items), musical greeting cards, musical balloons, and piezoelectric printers (a dot-matrix printer, inkjet printer) use piezoelectric actuators. Electric toothbrushes, humidifiers, and piezoelectric buzzers are also offered.
- Instrument pickups and microphones are two examples of musical applications that utilize piezoelectric materials.
- Numerous defense-related technologies, such as Micro Robotics and Course-Changing Bullets, utilize piezoelectricity.
- Other uses for piezoelectricity include making electricity, tennis racquets, MEMS (micro-electromechanical systems), and piezoelectric igniters.

7.3 Piezoelectric Materials

Any substance that contains atoms or molecules that are organized in a highly established sequence based on repetitions of the same fundamental atomic components is referred to as a crystal. Most crystals, including those formed of metals, have symmetric unit cells, however piezoelectric crystals do not. For piezoelectric crystals, electrical neutrality is the norm. The electrical charges are precisely regulated, despite the non-symmetrical arrangement of the atoms: a +ve charge cancels out a nearby –ve charge. However, as a piezoelectric crystal expands or contracts, it deforms, causing some of the atoms to migrate further from one another or more distantly. The removal of the +ve along with –ve polarity results in the formation of net electrical charges. This action, which impacts the entire crystal structure, causes the net +ve together with –ve charges to appear on the crystals' reverse outer sides.

1. The charges in a piezoelectric crystal are often perfectly balanced despite their asymmetrical arrangement. Because the forces (electric dipole moments) of the charges precisely balance out, there is no net charge remaining on the crystal faces.
2. As the crystal expands or contracts, the charges become out of equilibrium.

3. Both +ve and –ve charges can now be seen on the opposing sides of the crystal since the moments of dipole activity have been no lengthy, eliminating one another. A voltage is generated across the opposing sides of a crystal that is piezoelectric when it experiences extreme pressure.

Electricity can be produced by a wide variety of natural and synthetic piezoelectric materials. The first commercially available and most well-known piezoelectric material used in electronic systems is the quartz crystal. Other naturally occurring piezoelectric minerals include tourmaline, topaz, and even bone. The discovery of an entirely novel category of manufactured materials termed ferroelectrics, with piezoelectric properties many times stronger than those of piezoelectric crystals, represented a tremendous advance in research. Only a small portion of the growing variety of piezoelectric materials has been demonstrated to be suitable for transduction components in piezoelectric sensors. Basically, thin films, textures, piezoelectric ceramics, and single crystals made of natural and synthetic materials can be utilized. Barium titanate and lead zirconate titanate, two minerals that adopt piezoelectric characteristics with extremely specific features suitable for particular applications, were developed as a result of scientists' ongoing search for higher-performance materials. With the correct design and material selection, piezoelectric materials are able to identify alterations to force as well as motion at frequencies ranging from below 1 Hz to as high as multiple MHz. Both displacements in the mm range and force variations from mN to kN may be precisely monitored. Industrial piezoelectric ceramic sensors can function in the majority of chemical atmospheres, oil, and dust environments thanks to their sturdy solid-state design. They are resistant to damage from strong stress and/or vibration and work successfully throughout a wide range of temperatures [7]. Figure 7.4 shows the piezoelectric effect and Table 7.1 categorizes various piezoelectric materials with examples.

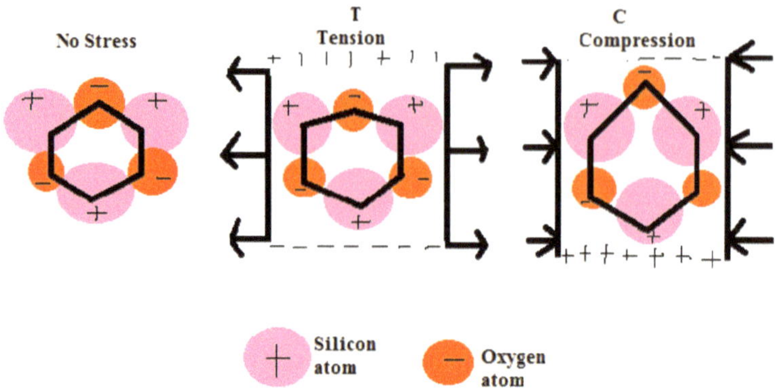

Figure 7.4 (In quartz) Piezoelectric effect.

Table 7.1 Categorization of piezoelectric materials

Piezomaterial types	Examples
Single crystals [non-ferroelectric(1) and ferroelectrics(2)]	Quartz(1), [Zinc blende (sphalerite), Boracite, Tourmaline, Topaz, Sugar, Rochelle salt](2), and Pyrite are examples of natural materials
	Ammonium dihydrogen phosphate (ADP) is a synthetic Ethylene diamine tartrate (EDT) Single crystals of barium titanate ($BaTiO_3$)(2) KDP, or potassium dihydrogen phosphate(2)
Ceramic ferroelectrics	$BaTiO_3$ (ceramic), PZT, PLZT
Ferroelectric polymers	PVDF
Ferroelectric composites	1-3 PZT–polymer composite
Grain-oriented glass ceramics	$2BaO\text{-}3SiO_2\text{-}TiO_2$

7.3.1 Electrostriction

A quadratic component is typically superimposed on the linear behavior in the response of piezoelectric materials. This element is reliant on a quantity known as the electrostrictive coefficient.

In comparison to linear piezoelectric materials, this coefficient is normally lower, but when the electric field is increased, it can become very significant.

7.3.2 Depolarization

A PZT ceramic must be treated carefully moving forward to avoid depolarizing it and losing some or all of its piezoelectric capabilities since once polarized, a PZT ceramic stays indefinitely polarized. The ceramic can get depolarized mechanically/electrically/thermally.

7.3.2.1 Electrical depolarization

Depolarization will take place if the material is exposed to a powerful electric field that has a polarity opposite to the poling field. The grade of the material, the duration of exposure to the depolarizing field, and the poling temperature all have an impact on the field strength required to initiate the depolarization.

7.3.2.2 Mechanical depolarization

When a piezoelectric element absorbs enough mechanical stress to change the orientation of the domains and, as a result, breaks the alignments of the dipoles, this is known as mechanical depolarization. The safety criteria for mechanical stress vary significantly depending on the quality of the materials.

7.3.3 Coupling Factor (k)

It is possible to connect the domains of electricity and mechanics by using piezoelectric materials. As a result, this kind of material can be used to generate electrical energy, absorb mechanical energy, or do both. The efficiency of switching electrical energy to mechanical energy inversely must then be determined using a coefficient.

The coupling factor k_{eff}, which corresponds to this coefficient, is described as follows at frequencies lower than the piezoelectric body's resonance frequency:

$$k_{eff^2} = \frac{\text{Energy converted}}{\text{Imparted input energy to the material}} \quad \text{(i)}$$

The coefficient k for the direct piezoelectric action is given as follows:

$$k_{eff^2} = \frac{\text{Electrical energy}}{\text{Mechanical energy exerted on the material}} \quad \text{(ii)}$$

The concept of k for the opposite piezoelectric effect is:

$$k_{eff^2} = \frac{\text{Mechanical energy generated}}{\text{Electrical energy exerted on the material}} \quad \text{(iii)}$$

7.3.4 Advantages and Limitations of Piezoelectric Materials

7.3.4.1 Advantages

- Piezoelectric materials may function in any temperature range.
- As the best substitute for fossil fuels, they have a low carbon footprint.
- These materials are the greatest at capturing energy because of their properties.
- These materials can be recycled.
- Green energy can be generated using unused energy lost by means of vibrations.

7.3.4.2 Limitations

- When functioning with vibrations, devices like this are capable of picking up unwanted vibrations.
- Resistance and durability limit the usage of systems that harvest energy from paved surfaces and highways.
- The discrepancy in stiffness between pavement and piezoelectric materials.
- These devices' lesser-known characteristics and the study that has been done so far are insufficient to fully utilize them.

7.3.5 Ferroelectricity

Under the context of a realizable field of electricity (=10^6 V/m), ferroelectrics, a subclass of piezoelectric materials, demonstrate spontaneously polarization that can be reorientable. Ferroelectric compositions and the layer structures of perovskite, M_xWO_3, $A_2B_2O_7$, as well as $Bi_2Ti_4O_{11}$ share high dielectric constants and strong electromechanical coupling coefficients. Additionally, they all have corner-linked octahedral Ti^{4+}, Nb^{5+}, or other d^0 ion networks. These transition metals have high permittivity, high piezoelectric constants, and highly polarizable "active" ions that support ferroelectricity, which is necessary for sensing.

There are a total of primarily two categories of charged ions on the periodic chart that are all close to digital "crossover" points, in which hybrid bond production is frequent and different atomic orbitals have similar energies. Among the d^0 ions that are octahedrally coupled to oxygen in the first group are Ti^{4+}, Nb^{5+}, and W^{6+}. The second class of active elements that actively cause polar distortions in ceramic dielectrics are the lone-pair ions, which have two electrons outside of a closed shell in an asymmetric hybrid orbital. The two most conspicuous among the aforementioned lone-pair ions within oxides are lead and bismuth, which are present in several ferroelectrics that exhibit elevated Curie temperatures ($PbTiO_3$, $Bi_4Ti_3O_{12}$, and $PbNb_2O_6$). Lead and bismuth are typically found on pyramidal coordination with oxygen in these compounds, which facilitates spontaneous polarization [8].

The two kinds of contributions to the piezoelectric effect in ferroelectric ceramics that are typically distinguished are the intrinsic contribution, which is brought on by the piezoelectric effect of the lattice, and extrinsic contributions, which are primarily yet not only caused via the rotation in ferroelectric-ferroelastic (non-1800) domain boundaries [9–11]. The everlasting shifting of domain boundaries results in nonlinear as well as hysteretic piezoelectric behavior [12, 13], even at very low applying fields of electricity or pressures. The crystal arrangement and texture of ceramics and the existence of contaminants, additives, imperfections, and specific variations in the composition of the ceramics are a few factors that could alter the domain-wall migration and piezoelectric responsiveness [12, 14]. A domain-wall contribution regarding the piezoelectric behavior of ferroelectric ceramics is now not entirely supported under weak-field circumstances (much beneath the level of the switching field of electricity or stress). The implications of domain-wall migration on the material's characteristics have been thoroughly studied in multiple ferroic systems. It is possible to describe magnetization and magnetic susceptibility in ferromagnetic materials in the low-field area employing the usual Rayleigh law, as shown, for instance, by the treatment of ferromagnetic domain walls moving in a medium where the walls' potential energy varies erratically with their position [15–17]. According to a recent proposal [18, 19], the Rayleigh rule may be expanded to include the piezoelectric effect that occurs on ferroelectric ceramics. By analyzing the answer as a function using the Rayleigh formula, it is feasible to quantify the contribution of the irreparable migration of domain boundary walls to the piezoelectric property of ceramics.

7.3.5.1 Rayleigh relationships

In a variety of ferroelectric ceramics, the dependence of the directly longitudinally piezoelectric d_{33} coefficient on a.c. pressure has recently been empirically proved to be characterized by a linear relationship [18, 19].

$$d_{33A}\,(\sigma_0) = d_{init} + \alpha\sigma_0 \qquad \text{(i)}$$

where σ_0 is the amplitude of the a.c. pressure $\sigma(t)$.

It has been suggested [18, 19] that the Rayleigh-type impact [15–17] to d_{33} from shifting domain walls is what causes the pattern of linear pressure dependency. According to the aforementioned analysis: the unchangeable shifting of domain boundaries is the cause of the expression $\alpha\sigma_0$.

d_{init} is caused by the lattice piezoelectric activity and the reversible migration of domain boundaries. A rise in d_{33} as a percentage of the applied a.c. pressure that results from the irreversible movement of domain walls is known as the Rayleigh coefficient.

Expression $\frac{\alpha\sigma_0}{d_{33}}$ (σ_0) shows the percentage of the entire piezoelectric response that is attributable to the irreversible displacement of domain walls at a certain amplitude of the a.c. pressure.

The piezoelectric charge density's relationship to a.c. pressure is provided by:

$$Q_0 = d_{init}\sigma_0 + \alpha\sigma_0^2 \qquad \text{(ii)}$$

Here, the piezoelectric charge density Q(t) amplitude is Q_0.

A crucial component of the Rayleigh response [15] is the non-analytical expression for the piezoelectric [Q(t) versus σ(t)] hysteresis loop, which is given below:

$$Q = (d_{init} + \alpha\sigma_0)\sigma \pm \frac{\alpha}{2}(\sigma_0^2 + \sigma^2) \quad \text{(iii)}$$

Here, "+" denotes a decrease in air pressure and "–" signifies an increase.

7.3.6 Piezoelectric Materials Are Necessary for the Sensor

The following attributes are ideal in materials for transduction components in sensors:

- Strong piezoelectric sensitivity.
- High stiffness (high modulus of elasticity).
- Electrically good insulation strength and high piezoelectric sensitivity.
- Few hygroscopic properties.
- Direct interaction of mechanical stress and electric polarization.
- Hysteresis is absent.
- All properties are highly stable.
- All properties have little temperature dependency over a wide range of temperatures.
- Mechanical parameters with little anisotropy, such as elastic constants and thermal expansion coefficients.
- Good machinability.
- Cheap manufacture.

7.3.6.1 Quartz

The most significant single crystal still used in piezoelectric sensors as a transducer is quartz. It has the chemical formula SiO_2 and comes in a variety of forms. It is made up of silicon-centered oxygen tetrahedrons. The low-temperature alteration of quartz that is less than 573°C is employed for piezoelectric applications. It is typically referred to as "α-quartz" and is a member of the 32-symmetry original-trapezoid trigonal (original) crystalline technique's category. Quartz has hardness of 7 on the Mohs scale and a density of $2649 \cdot 10^3$ kg/m. It basically cannot dissolve in water and is resistant to the majority of acids and alkalis. At 1710°C, quartz melts.

The mechanical strength of quartz plays a crucial role in its use in piezoelectric sensors. Quartz works well as an electrical insulator. The shift from one structural phase to another occurs at 573°C. The hexagonal point group 622 includes high-temperature quartz alteration, also known as "β-quartz." In the 573–870°C temperature range, it is stable. The high-temperature phase over 573°C is referred to as α-quartz, whereas the low-temperature phase less than 573°C is referred to as β-quartz by some writers.

7.3.6.2 Tourmaline

From a chemical perspective, tourmaline is an aluminum-borosilicate. The formula (Na,Ca) $(Mg,Fe)_3B_3Al_6Si_6$ $(O,OH, F)_{31}$ can be used to represent the highly complex composition of the substance. Many tourmalines contain Li, Mn, Ti, Fe, or Cr, which causes ion charge transfer to provide distinctive coloring. Iron-rich tourmalines are dark in color. Tourmaline, like α-quartz, is a member of the trigonal crystal system and has a ditrigonal-pyramidal structure with a symmetry of 3 m. It should be noted that tourmaline cannot twin and is unable to undergo a phase shift below 900°C, which makes it ideal for use in sensors for an extremely wide temperature range up to about 700°C, particularly in pressure and acceleration sensors. Its main drawbacks are the naturally powerful pyroelectric effect and the challenges of obtaining large quantities of crystals from natural sources.

7.3.6.3 Piezoelectric ceramics

Piezoelectric ceramics are very important for piezoelectric sensors. It is created by compressing a finely crushed powdered combination into the required shape, ideally one made of oxygen-octahedral ferroelectrics, and then sintering it. Piezoelectric ceramics are composed of a large number of ferroelectric grains (crystallites) that each contain domains where the electric dipoles are aligned. When merged, the single domains produce a texture that belongs to the symmetry group ∞/∞/mmm. The single domains have a random orientation at first and commonly have the symmetries mm2, 3m, or 4mm. It is isotropic and lacks piezoelectric properties. Barium titanate ($BaTiO_3$), an initial polycrystalline ferroelectric ceramic with a perovskite structure, was independently found in 1943 by scientists in the USA, Japan, and other countries. It was supplanted as the primary electroceramic material when lead zirconate titanate (PZT) was found in 1954 by Jaffe et al. [20, 21] and its advantages over single crystals were recognized right away [22]. Due to their diverse optical and dielectric properties, transition metals have a number of benefits [23].

7.4 Lead Zirconate Titanate (PZT) and $BaTiO_3$ Piezoelectric Materials

The 1950s saw a tremendous advancement in the field of sensors with the introduction of piezoelectric materials like lead zirconate titanate (PZT), barium titanate ($BaTiO_3$), and its family [24]. These ceramics, which have unusually potent piezoelectric and dielectric features, have been widely employed in sensor and actuator technologies [25–28]. Using these materials has a substantially detrimental effect on the environment because of their high lead content and enhanced toxicity from being volatilized at high temperatures during calcinations as well as sintering. Due to the considerable characteristic acoustic impedances due to those materials' significant density, matching layers must be used. The high relative permittivity of said materials facilitates electrical tuning; besides, it reduces the piezoelectric voltage coefficients.

7.4.1 Lead Zirconate Titanate (PZT)

A large number of piezoelectric ceramic sensor recipes are based on PZT, an acronym for solid solutions with nominal compositions of 52–54 moles percent lead zirconate ($PbZrO_3$) and 46–48 moles percent lead titanate ($PbTiO_3$). The officially licensed trademark of Clevite Corporation is PZT. PZT is an important ferroelectric material that crystallizes as perovskites. Lead atoms make up the unit cell's corners, and oxygen atoms make up its face centers. Ions of Pb and O_2 have radius of roughly 1.4 Å. They come together to create a four-dimensional cubic pattern that is face-centered. Octahedrally related titanium or zirconium ions are located in the middle of the entire cell. The PZT crystal arrangement experiences a displacive step shift with atomic removals around 0.1 Å during cooling from a strong temperature. Near the Curie temperature, the point symmetry of compositions containing titanium switches from cubic m^3m to tetragonal 4 mm. For 0°K, the tetragonal form with its intrinsic polarization along [001] remains. PZT transforms from a cubic perovskite-structured paraelectric state to a ferroelectric phase at the morphotropic phase boundary (MPB) between the tetragonal and rhombohedral states near the Curie point. The mechanical and electric variables interact piezoelectrically very strongly close to this phase boundary. In the beginning, each grain of polycrystalline ferroelectric ceramics (Curie point group 11m) like PZT has domains, which are polarized zones with arbitrarily oriented directions. To reduce the overall elastic potential within a ceramic, these domains form after cooling below the temperature of Curie [29]. Polycrystalline ferroelectrics lack the piezoelectric effect due to this polarization instability. Piezoelectricity may be produced at elevated temperatures below the ferroelectric Curie point, where the domains are easily aligned, through the use of an electric field with static properties that is bigger than the saturation region and lesser than the deformation region. The action is referred to as poling.

Dopants can be added to PZT to substantially change its properties. Cation (metal) vacancies provided by donor dopants in the improvement of domain reorient through crystal structure modification, hence extrinsic contribution to the piezoelectric coefficients. Such electrically "soft" PZTs show significant permittivity, extensive electrical losses, considerable electromechanical interaction, highly significant resistance to electricity, poor mechanical quality parameters, and a weak strong field as a result [30]. Additionally, they have substantial piezoelectric coefficients. On the other hand, O_2 vacancies are because of acceptor dopants in the same direction. Such O_2 vacancies establish the domain boundaries because the defect dipoles correspond with the spontaneous polarization throughout a domain [29]. The PZT thus acquires an electrical "hardness." Less piezoelectric coefficients, little permittivity, minimal losses, small electrical resistivity, excessive Qm, and a strong field of coerciveness are characteristics of a hard PZT [31]. Table 7.2 depicts the various parameters of PZT type ceramic.

Table 7.2 The multicomponent PZT-type ceramics' dielectric and piezoelectric properties (at room temperature) [32]

S. No.	Parameter	Lead zirconate titanate-type ceramics
1	Density (g/cm^3)[a]	7.83
2	Curie temperature (°C)[b]	357
3	ε at T_r[b]	695
4	$\tan\delta$ at T_r[b]	0.030
5	ε at T_C[b]	52,170.00
6	$\tan\delta$ at T_C[b]	0.055
7	Pr ($\mu C/cm^2$)[c]	43.30
8	Ps ($\mu C/cm^2$)[c]	45.11
9	Ec (kV/mm)[c]	1.13
10	a_{rec}	1.077
11	k_p	0.61
12	k_{31}	0.32
13	d_{31} (C/N)	89.92×10^{-12}
14	g_{31} (Vm/N)	16.94×10^{-3}
15	S^E_{11} (m^2/N)	1.23×10^{-11}
16	V_r (m/s)	2065
17	Q_m	51
18	d^*_{33} (pC/N)[d]	185

[a]Mass density (mass/volume); [b]For ν = 1 kHz; [c]For ν = 1 Hz and E = 2.3 kV/mm; [d]For E = 2.0 kV/mm.

A variety of electrical and electronic components, including sensors [33], capacitors [34], random access memory (RAM) [35], optical modulators, and waveguides [36], can be made using lead zirconate titanate (PZT), a significant type of piezoelectric and ferroelectric substances with distinctive features. There are presently numerous physical and chemical techniques that can be used to produce PZT particles at nanoscale sizes, including sol-gel, sol-combustion, hydrothermal, solvothermal, precipitation, and solid-state processes [38–44]. Solid-state procedures have nevertheless been employed to produce iron and niobium doped PZT ceramics.

7.4.1.1 Manganese doped PZT ceramics prepared through co-precipitation method

$Pb(Zr_xTi_{1-x})O_3$ (also referred to as PZT) is the applicable formula for lead zirconate titanate ceramics, well-known ferroelectric materials that are frequently employed in industrial equipment including sensors, actuators, and mechanical filters. It belongs to the family of perovskites, which is called ABO_3 (A = mono, di, or trivalent; B = tri to pentavalent ions). Near the morphotropic phase boundary (MPB), which separates the rhombohedral zirconium-rich stage as well as the tetragonal titanium-rich phase transition, PZT compositions frequently show significant dielectric and piezoelectric activity. Chemically pure PZT ceramics are rarely used; instead, cation dopants are frequently substituted or added to A^{2+} (Pb) or B^{4+} (Zr^{4+}/Ti^{4+}) sites to enhance particular characteristics. The physicochemical properties of the initial raw ingredients have an enormous effect on the way the PZT solid solution develops. It is widely known that the traditional (high-temperature solid-state reaction) method of making PLZT powders results in chemical inhomogeneity and compositional fluctuations. The development of high-quality ceramics has made the powder produced using chemical processes including co-precipitation, spray drying, freeze drying, liquid drying, and sol-gel technique extremely important. These processes are known for their improved purity, homogeneity, and reactivity.

Starting materials for the co-precipitation method of creating ceramics included [$Pb(NO_3)_2$, zirconyl nitrate $ZrO(NO_3)_2{\cdot}2H_2O$, titanium isopropoxide $Ti[(CH_3)_2CHO]_4$, and x weight percent of MnO_2 (where x = 0, 1, 2, and 3). To create a translucent, clear solution of nitrates, stoichiometric amounts of all salts were dissolved in distilled water and swirled on a magnetic stirrer. The aforementioned solution was then treated with titanium isopropoxide, and the precipitation resulted from the addition of some ammonia. Following vacuum filtration, this precipitate was dried for 8 hours at 200°C in an oven. Powders that had been baked were then intermediately ground after being calcined at 800°C. Using a uniaxial hydraulic press, the powder was cold pressed into disks at a pressure of 5.8×10^8 N-m^{-2}. These pellets were then sintered for 2 hours at 1125°C [45].

7.4.1.2 Bi^{3+} and Al^{3+} co-doped lead zirconate titanate ceramics prepared through solid state method

PZT ceramics have been employed as sensors, actuators, transducers, and other kinds of devices for a very long time because of their excellent dielectric and piezoelectric features.

The morphotropic phase boundary (MPB), which separates the $PbTiO_3$ rhombohedral phase from the $PbTiO_3$ tetragonal phase, gives PZT ceramics an ABO_3-type perovskite structure. Cation ion doping at A- or B-sites that is isovalent, supervalent (donor), and subvalent (acceptor) can be broadly classed. Donor doping's potential for creating lead vacancies encourages domain-wall movement. Due of this, PZT ceramics have a greater piezoelectric factor, stronger residual polarization, and a lessened mechanical purity variable, making them appropriate for usage in a variety of sensors and hydrophone applications. In $Pb(Zr,Ti)O_3$ ceramics, Lin Wang et al. investigated the effects of perovskite substitution on increased tetragonality and ferroelectric transition temperature. Bi^{3+} is stereochemically identical to Pb^{2+} due to its lone pair ($6S^2$) and high polarizability. Particular attention has been paid to the impacts of these substitutions since they, respectively, provide PZT ceramics its hard and soft properties (Bi^{3+} as a replacement for Pb^{2+} and Al^{3+} as a replacement for Zr^{4+} and Ti^{4+}). For the solid-state method of manufacturing ceramics, the initial components Pb_3O_4, $SrCO_3$, ZrO_2, TiO_2, Bi_2O_3, and Al_2O_3 were dried at 120°C and weighted stoichiometrically. An additional 0.5 mol% of PbO was influenced to account for PbO volatility throughout calcination as well as sintering. The raw materials were combined using ball milling for 6 hours on deionized water. The resulting powder was subsequently given a perovskite crystalline form by being calcined at 850°C for 4 hours. The powder size was reduced using a second ball milling procedure. After being compressed into a pellet with a diameter of 12 mm, the resulting powder was sintered for 2 hours at temperatures between 1200°C and 1280°C. After the sintered pellets were reduced to a thickness of 0.8 mm, coated with a silver paste, and burned at 720°C, the samples were polarized in a silicon oil bath under a field of approximately 30 kV/cm at 120°C for 30 min.

7.4.1.3 Iron and niobium doped PZT ceramics prepared through solid state reactions

Lead zirconate titanate (PZT) ceramics are a renowned kind of ferroelectric materials that are prominent in many industries, including high-energy capacitors, non-volatile memory (FERAM), ultrasonic sensors, infrared detectors, and electro-optic devices. The conventional formula for them is ABO_3 of the perovskite structure, where site A is occupied by Pb^{2+} (ionic radius = 1.18 Å), and sites B and C are shared, respectively, through Zr^{4+} and Ti^{4+}. Perovskites can have their properties improved by doping different transition metal ions at the A and B sites.

By using the solid-state synthesis technique, lead zirconate titanate ceramic ($(PbZr_{0.68}Fe_{0.14}Nb_{0.14}Ti_{0.04}O_3$) was created. Lead oxide (PbO), zirconium oxide (ZrO_2), and niobium pentoxide (Nb_2O_5) served as the initial reagents for ($PbZr_{0.68}Fe_{0.14}Nb_{0.14}Ti_{0.04}O_3$). In order to make up for the lack of lead concentration and to aid crystallization, 1.5% (molar percent) more lead oxide was added to PZFNT during preparation. The solvent utilized was ethanol. For effective homogenization, raw materials were physically combined in a RETSCH planetary ball mill at 300 rpm for 2 hours. The PZFNT powder was calcined at 900°C for 2 hours after being dried at 150°C to remove the solvent. Having 30 ceramic pellets, PZFNT, cold pressing with an equivalent force of 7 tons produced mm of diameter. These constructions underwent 3 hours of 1200°C sintering heating at a rate of 2°C/min.

7.4.2 Characteristics of Manganese Doped PZT, Bismuth–Aluminum Co-Doped PZT, and Iron–Niobium Co-Doped PZT Ceramics

7.4.2.1 Study of structural properties of manganese doped PZT ceramics

The X-ray diffraction (XRD) is a tool that provides details on a material's crystal structure, atomic space, phase, and other properties. It can be thought of as a quick analytical tool that aids in figuring out whether the substance being studied is homogeneous or has features of bulk composition.

The X-ray diffraction patterns of sintered samples reveal that the existence of a doublet (200) peak has resulted in the tetragonal crystal structure being shared by all samples. Due to the addition of extra PbO during calcination to make up for PbO evaporation at higher temperatures, a little peak of PbO can be seen at around 28°. This peak is of an extremely small amount (insignificant in comparison to the PZT phase) and is responsible for the PbO presence. On Mn doping, it was discovered that tetragonalities, or c/a, of the samples estimated with the X-ray diffraction patterns, decreased. The study reveals a decrease in dielectric constants and an increase in dielectric loss, with an increase in MnO_2 being reported. The Mn doped PZT ceramics and undoped PZT exhibit the hysteresis loop that confirms the ferroelectric behavior [45].

7.4.2.2 Using XRD to study structural characteristics of Bi^{3+} and Al^{3+} co-doped PZT ceramics

Lead zirconate titanate ceramics doped with Al and Bi have been studied in terms of their microstructure, and dielectric, ferroelectric, and piezoelectric features [46]. None of the samples exhibit second phases; instead, they all exhibit a pure perovskite form. It is clear that the samples ($0.00 \leq x \leq 0.04$) show a separation of the (200) peak, proving that the structure is in a tetragonal stage. The composition is found at the morphotropic phase border of the entire system, where tetragonal and rhombohedral stages coexist. The samples ($x = 0.06$), however, display a murky splitting of the (200) peak where a second bulge may be seen, as indicated by the arrow. The phase structure of all samples was pure perovskite, and it was also noticed that when the amount of doping material increased, it gradually transitioned from the tetragonal region to a stage near the MPB. The grain growth might be sped up with the right amount of co-doping of Bi and Al ions, but it could also be slowed down by too much co-doping. The enhancement of piezoelectric and ferroelectric capabilities could result from the correct co-doping of Bi and Al ions into PZT ceramics. This research provided helpful advice for creating PZT-based apps.

7.4.2.3 Using XRD to study structural properties of iron and niobium doped PZT ceramics

When PZFNT ceramics were subjected to XRD analysis, the crystalline structure of the materials was disclosed by crisp, distinct peaks that could only be attributed to the orthorhombic phase as indexed by ASTM pattern 35-0739, as shown in Fig. 7.5. The lattice characteristics can alter with the inclusion of dopants that replace the B-site (Nb^{5+} and Fe^{3+} substitute Zr/Ti locations). The lattice constants of PZNFT ceramic are "a" = 8.2519', "b" = 11.7242', and "c" = 5.8281' after indexing the XRD powder pattern. When the ASTM pattern 35-0739 is compared to the PZT phase diagram, which supports the creation of an orthorhombic phase, we can see a lattice distortion consisting of a lengthened "a" parameter and lowered "b" and "c" parameters [47].

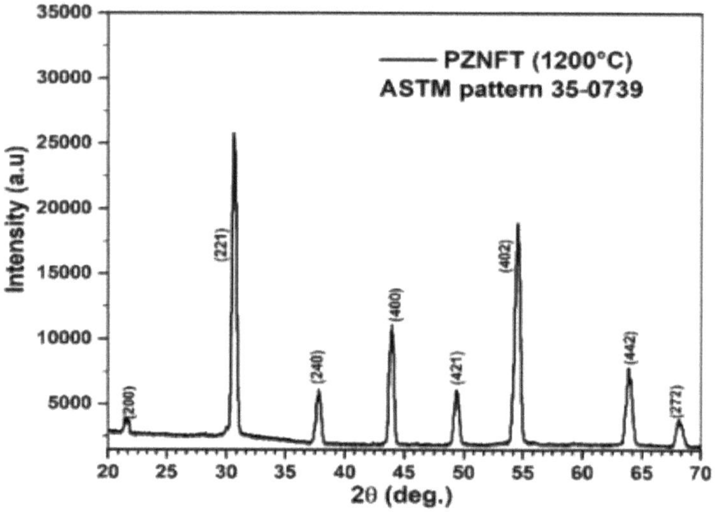

Figure 7.5 Iron and niobium doped PZT ceramics' X-ray diffraction patterns [47].

7.4.2.4 Study of surface morphology of manganese doped PZT ceramics using SEM

SEM micrographs of the sintered samples, as demonstrated in the article [45], showed that Mn doping in the PZT ceramic caused a significant reduction in average grain size. Due to the build-up of extra Mn at grain boundaries during sintering, Mn doping in PZT ceramics inhibited grain development. The Mn doped PZT ceramics and undoped PZT exhibit the hysteresis loop confirming the ferroelectric behavior.

7.4.2.5 Study of surface morphology of Bi^{3+} and Al^{3+} co-doped PZT ceramics

PZT ceramics doped with Al and Bi have been examined in terms of their microstructure, and dielectric, ferroelectric, and piezoelectric characteristics [46]. The largest grain size is observed at $x = 0.02$, and some extremely large pores are present. This is mostly due to the grain expanding too quickly for the pore to be swiftly eradicated. When the doping level reaches 0.04, a large number of small grains develop and the porosity obviously declines. According to the information above, the average grain size increases first and subsequently reduces as the co-doped concentration increases. Although excessive co-doping of Bi^{3+} and Al^{3+} can inhibit growth, effective co-doping of Bi^{3+} and Al^{3+} can significantly speed up the development of grain. This may be the cause of the bulk dissolution limitation for Bi^{3+} and Al^{3+}. It is possible for an excessive amount of Bi^{3+} and Al^{3+} to accumulate within an arrangement close to the boundaries of grains, which would then limit the boundaries' mobility and impede grain growth [46].

7.4.2.6 Study of surface morphology of iron and niobium doped PZT ceramics using SEM

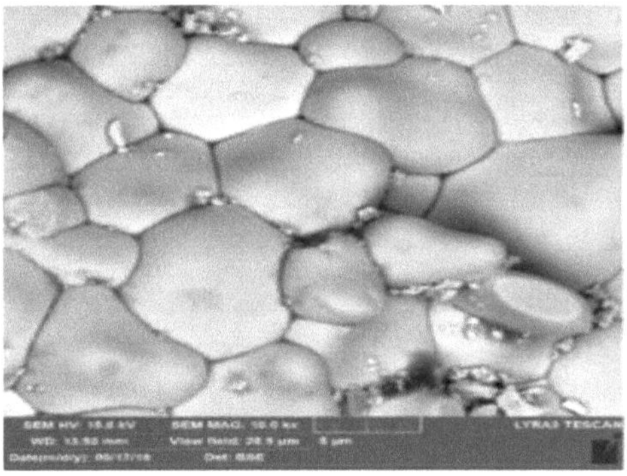

Figure 7.6 SEM micrograph of iron and niobium doped PZT ceramics [47].

The SEM picture presented in Fig. 7.6 illustrates the surface morphology micrograph of iron and niobium doped PZT ceramics made using a solid-state reaction technique. With an average grain size of 6.45 m, the PZFNT morphology exhibits a good compactation structure of ceramics with small pores between grains [47]. Table 7.3 demonstrates the values of K_p and ε_r for PZT with various dopants [48].

Table 7.3 The study of K_p and ε_r in different PZT-based systems [48]

S. No.	Compositions	K_p	ε_r
1	Lead zirconate titanate (52/48)	0.47	707
2	Lead zirconate titanate doped Strontium	0.44	973
3	Lead zirconate titanate doped Calcium	0.49	729
4	Lead zirconate titanate doped Yttrium	0.34	841
5	Lead zirconate titanate doped Lanthanum	0.53	1483
6	Lead zirconate titanate doped Neodymium	0.49	1395
7	Lead zirconate titanate doped (Neodymium + Strontium)	0.49	1776
8	Lead zirconate titanate doped Niobium	0.54	1242
9	Lead zirconate titanate doped (Lanthanum + Niobium)	0.57	1377
10	Lead zirconate titanate doped Tantalum	0.5	1230
11	Lead zirconate titanate doped Iron	0.59	820
12	Lead zirconate titanate doped Aluminum	0.2	1078
13	Lead zirconate titanate doped Magnesium	0.381	983
14	Lead zirconate titanate doped Zirconium	0.446	1396
15	$(Pb_{0.9725}Zr_{0.52}Ti_{0.48})_{0.945}Nb_{0.055}O_3)$	0.591	1529

7.4.3 Barium Titanate ($BaTiO_3$)

Due to its remarkable piezoelectric characteristics, minimal loss, and superior tunability, $BaTiO_3$ (BT)-derived ceramics possess the potential to replace without lead piezoelectric ceramics. Recent developments seem to be surprising [49, 50]. The transition between the phase and electrical characteristics of BT ceramics is affected by Zr, Ca, Sn, or Hf additions [51, 52].

7.4.3.1 Iron doped $BaTiO_3$ made using mechanical milling and conventional solid-state method

Lead-free ceramic, which is among the least well-known materials, $BaTiO_3$, has a variety of uses, including semiconductors, piezoelectric devices, thermositors, and capacitors. The perovskite structure formula for $BaTiO_3$ (BTO) is ABO_3, where Ba is the A site and Ti is the B site. By adding the correct ions to the cation at the A site or B site, the various properties can be enhanced. The ionic radius of a dopant can be used to forecast its site occupancy. The A site of the Ba^{2+} ion is more likely to be occupied by a dopant with a large ionic radius and low valence (rBa^{2+} = 1.35), whereas the B site of the Ti^{4+} ion is more likely to be occupied by a dopant with a small ionic valence (rTi^{4+} = 0.61Å). To enhance their properties, pure and doped BTO materials have been made using a variety of methods, including the sol-gel approach, the autocombustion route, solid-state reactions, and hydrothermal reactions.

The initial components TiO_2, $BaCO_3$, and Fe_2O_3 powders were placed into the mixer containers in the following compositions: $BaTiO_3$ (BT), $BaTi_{0.99}Fe_{0.01}O_3$ (BTF1), $BaTi_{0.98}Fe_{0.02}O_3$ (BTF2), and $BaTi_{0.96}$ $Fe_{0.04}O_3$ (BTF4) for the solid-state method of making ceramics. Three hours were spent milling each composition using various-sized zirconia balls. Each milling stage used a 2:4 ratio of balls, with balls having diameters of 10 and 6 mm, respectively. During the milling process, no control agent was employed. After milling, the powder was calcined at 1100°C for 2 hours, ground in a mortar, compressed under pressure to form cylindrical pellets at 193.11 MPa, and then densified under an atmosphere of air at 1250°C for 2 hours.

7.4.3.2 Ni doped $BaTiO_3$ made using solid state reaction of mixed oxide technique method

$BaTiO_3$ has a conventional ABO_3-type structure, with the 6-fold coordinate B cation in the middle, the 12-fold coordinate A cation in the corner, and the anion, often oxygen, in the center of the face. A and B are cations of different sizes. The A and O ions working together to generate a closely packed array is how the ions are arranged. $BaTiO_3$ is tetragonal at normal temperature and changes to cubic phase over 130°C. Additionally, it can be found in the orthorhombic phase above 0°C and the rhombohedral phase below −90°C. $BaTiO_3$ enters the hexagonal phase above 1460°C. While the fundamental cube is the preached shape, variations in the radii of the A and B cations can change the structure in a variety of ways, among which is tilting.

By employing the necessary quantity of AR grade reagents $BaCO_3$, TiO_2, and NiO as starting materials, the compositions $Ba_{1-x}Ni_xTiO_3$, where x = 0, 0.03, 0.05, 0.07, and 0.10, were synthesized using a traditional process based on solid-state reaction of mixed oxides. In acetone media, the weighted individual reagents were uniformly blended. The well-combined powder was then calcined in an alumina crucible for 2 hours at 1200°C. Using a uniaxial hydraulic press, the calcined powder was compressed into cylindrical pellets that were 8 mm in diameter and 1–2 mm thick at a pressure of ~31.2 × 10^6 Pa. These pellets were subsequently sintered for 5 hours at 1250°C.

7.4.3.3 Using XRD to study structural properties of iron doped $BaTiO_3$ ceramics

It was discovered that pure BT was entirely tetragonal, with no signs of secondary phase development. Figure 7.7b illustrates how the samples BTF1 and BTF2 exhibit minor variations in the symmetry, shape, and location of several diffraction peaks, which may indicate the creation of new phases. The sample (BTF4) with the highest iron content exhibits the emergence of additional peaks, including (103), (104), (201), (202), (203), (204), (205), (213), (214), (207), and (220), which are connected to the production of a hexagonal phase [53].

7.4.3.4 Using XRD to study structural properties of Ni doped $BaTiO_3$ ceramics

The XRD pattern of all the sintered sample powders, as discussed [54], showed that when the samples were doped with Ni ion, diffraction peaks appeared at slightly smaller angles. Since the splitting of the cubic peak at 2θ = 45.235 into the tetragonal peaks (002) at 2θ = 44.853 and (200) at 2θ = 45.339 is not observed, it was initially possible to assume that the XRD patterns of higher concentrations of Ni^{2+} doped BT samples fit well with the peak position of standard BT cubic phase. A closer look at the patterns reveals, however, that the broadening of the peaks in the same diagram does not consistently

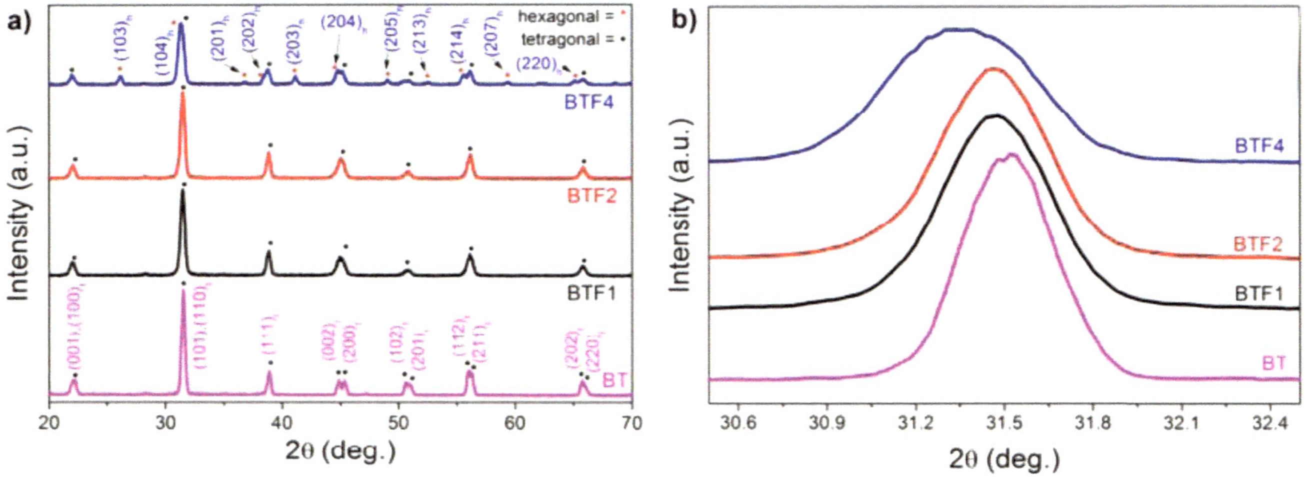

Figure 7.7 The X-ray diffraction patterns of pure and doped samples are displayed in both (a) and (b). Reprinted with permission from the publisher [53].

grow with the Bragg angle. This finding shows that the initially cubic lattices of the objects under study have experienced minimal tetragonal deformation. However, up until $x = 0.10$, the samples do not demonstrate any phase involving the NiO molecule. As a result, it appears that the Ni^{2+} ions at the specified concentration are completely soluble in the BT lattice, with the Ni^{2+} ($r_6{}^{2+} = 0.69$ Å) ions having a potential to replace both the Ba^{2+} ($r_6{}^{2+} = 1.35$ Å) and Ti^{4+} ($r_6{}^{4+} = 0.61$ Å) sites because of their short ionic radius [54].

7.4.3.5 SEM-based study of surface morphology of iron doped $BaTiO_3$ ceramics [53]

The microscopic makeup of the pure and doped BT solid samples is shown in Fig. 7.8. The usual grain size for pure BT is 29.1 μm ± 10.4 in Fig. 7.8a. Figure 7.8b–d illustrates how the subsequent gradual incorporation of Fe^{3+} causes a considerable decrease in the size of the crystals. The microstructure improvement continues as the Fe^{3+} level approaches and eventually a sub-micrometric particle size is acquired. The dopant deposition at the grain borders, which limits grain development during sintering, is responsible for this refinement in grain size. Additionally, the development of a hexagonal phase can be seen in samples BTF2 (Fig. 7.8e) and BTF4 (Fig. 7.8f), which can be seen by the existence of grains with hexagonal flake formations.

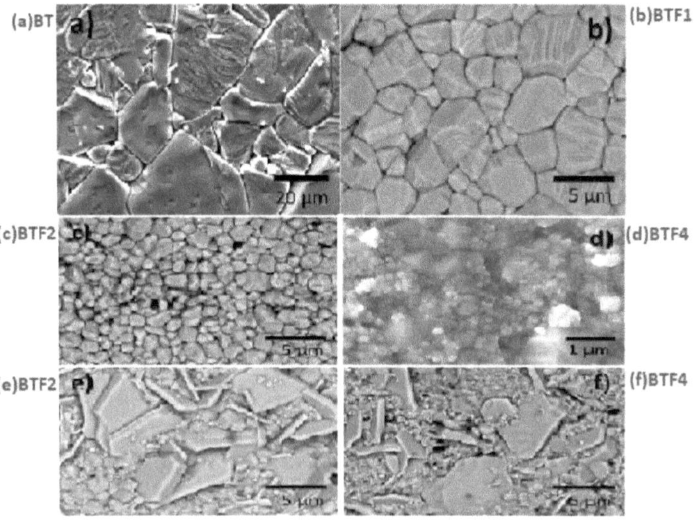

Figure 7.8 SEM images of (a) BT, (b) BTF1, (c and e) BTF2, and (d and f) BTF4. Reprinted with permission from the publisher [53].

7.4.3.6 SEM-based study of surface morphology of Ni doped $BaTiO_3$ ceramics

According to the SEM micrographs described in [54], it is evident that the grain distribution was homogeneous all across the sample's exterior with no chance of porosity. Additionally, the grains' edges are not very crisp, which supports the sample's melting-like nature. Furthermore, Ni doping in BT was observed to intensify this melting appearance.

7.5 Piezoelectric Sensor: Circuit, Specifications, Types, and Applications

A piezoelectric sensor is one that functions according to the piezoelectricity principle. A phenomenon termed piezoelectricity develops when a substance is mechanically stressed, generating electricity. Sensors are tools that are used to find or sense various kinds of physical quantities in the environment. Light, heat, motion, moisture, pressure, vibrations, etc. could all be sources of input. Typically, an electrical signal corresponding to the applied input is the output produced. The output signal may be sent via a network for additional processing or utilized to calibrate the input.

A piezoelectric sensor or transducer is a piece of equipment that makes use of the piezoelectric effect to convert alterations in pressure, acceleration, temperature, strain, or force into an electrical charge. Greek for "push" or "squeeze" is the prefix piezo. The capacity of a piezoelectric material to transform mechanical stress into electrical charge is referred to as the "piezoelectric effect." The amount of pressure applied to materials made of solid piezoelectric crystals determines the amount of piezoelectricity those materials can produce. Piezoelectric ceramics (like PZT ceramic) and single-crystal materials (like quartz) are the two primary sensing components utilized in piezoelectric sensors. Ceramic materials are more sensitive than naturally occurring single-crystal materials, although their great sensitivity gradually fades over time. Figure 7.9 shows the schematic for a piezoelectric sensor.

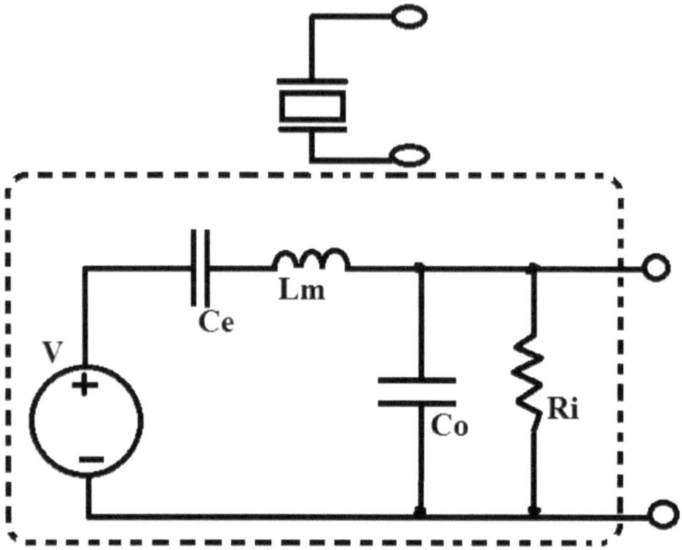

Figure 7.9 Piezoelectric sensor schematic.

The resistance, abbreviated (R_i), is often referred to as insulator resistance. The sensor's inertia is what generates the inductance. The relationship between capacitance (Ce) and the sensor's elasticity is inverse. The sensor must have adequate load and leakage resistance to maintain low frequencies in order for it to correctly react to signals. A sensor is referred to as a pressure transducer in an electrical signal. Sensors are also referred to as primary transducers.

7.5.1 Types of Piezoelectric Sensor

There are many different types of sensors, depending on the input to be measured.

- Temperature is sensed via mercury-based thermometers.
- Cars' emission control devices use oxygen sensors to detect oxygen.
- If there is visible light present, the photo sensor notices it.
- A piezoelectric pressure sensor is a sensor that directly realizes force-to-electricity conversion using piezoelectric components.
- Acceleration piezoelectric sensor.

7.5.2 Piezoelectric Sensor Applications

- Shock detection uses piezoelectric sensors.
- Flow sensors and thickness gauges use active piezoelectric sensors.
- Piezoelectric sensors are also utilized for ultrasonic imaging.
- Passive piezoelectric sensors are employed in microphones, accelerometers, musical pickups, and other devices.
- Sensors are utilized for electro-acoustic, micro-moving measures, optic measurements, and more.
- Used to decrease noise and vibration.
- Control of structural static deformation.
- Used to track structural degradation.
- Employed for technology monitoring processing.
- Used for weighing vehicles.
- A piezoelectric sensor cable is used for this.
- Applicable to navigation and aviation.
- Key fobs, PIN pads, and burglar and intruder alarms.
- Microwaves.
- Ultrasonic apparatus.
- Musical equipment.
- Machine observation.

7.6 Applications of Transition Metal Doped Piezoelectric Materials

- For hydrophone.
- Sensing devices such as pyroelectric detectors for infrared radiation.
- Multilayer ceramic capacitors (MLCs).
- Ultrasonic sensors.
- Infrared detector.
- Electro-optic devices.
- Transducers.
- Memories.
- Micro-electromechanical systems (MEMS).
- Thermistors.

7.7 Conclusions

Lead zirconate titanate and barium titanate, which are used to make a variety of electronic as well as optoelectronic devices, including sensors, capacitors, random access memory, optical modulators, and waveguides, are important groups of piezoelectric and ferroelectric materials with distinguishing characteristics. This chapter describes the development and manufacturing process of pure and Mn, Bi, Al, Fe, and Nb doped PZT-NPs as well as Fe and Ni doped barium titanate piezoelectric materials. Along with the manufacturing process, their structural and morphological characterization have also been studied, as these transition metal doped PZT and BT are useful for various sensor applications.

References

1. Friend, C. Smart materials: The emerging technology. *Mater. World*, **4**(1), 16–18, 1996.
2. Gahdhi, M.V. and Thompson, B.S., *Smart Materials and Structures*. Champman & Hall, London, 1992.
3. Cady, W.G., *Piezoelectricity*, McGraw-Hill, New York, 1946. Dover, New York, 383, 384, 1964.
4. Jaffe, B., Roth, R.S., and Marzullo, S. Piezoelectric properties of lead zirconate-titanate solid-solution ceramics. *J. Appl. Phys.*, **25**(6), 809, 1954.
5. Craig, R.R. and Kurdila, A.J. *Fundamentals of Structural Dynamics*, John Wiley & Sons, Hoboken, New Jersey, Chapter 19, 2011.
6. Haertling, G.H. Ferroelectric ceramics: History and technology. *J. Am. Ceram. Soc.*, **82**(4), 797, 1999.
7. Barrett, R. and Wilcoxon, F. Monitoring vibration with piezoelectric sensors. *Sensors-Peterborough*, **10**, 16, 1993.
8. Newnham, R.E. Molecular mechanisms in smart materials. *MRS Bull.*, **22**, 20, 1997.
9. Zhang, Q.M., Wang, H., Kim, N., and Cross L.E. Direct evaluation of domain-wall and intrinsic contributions to the dielectric and piezoelectric response and their temperature dependence on lead zirconate-titanate ceramics. *J. Appl. Phys.*, **75**, 454, 1994.

10. Herbeit, R., Robels, U., Dederichs, H., and Arlt, G. Domain wall and volume contributions to material properties of PZT ceramics. *Ferroelectrics*, **98**, 107, 1989.

11. Bondarenko, E.I., Topolov, V.Y., and Turik, A.V. The role of 90 domain wall displacements in forming physical properties of perovskite ferroelectric ceramics. *Ferroelectr. Lett.*, **13**(1), 13, 1991.

12. Zhang, Q.M., Pan, W.Y., Jang. S.-J., and Cross, L.E. Domain wall excitations and their contributions to the weak-signal response of doped lead zirconate titanate ceramics. *J. Appl. Phys.*, **64**(11), 6445, 1988.

13. Damjanovic, D., Demartin, M., Shulman, H., Testorf, M., and Setter, N. Instabilities in the piezoelectric properties of ferroelectric ceramics. *Sensors Actuators A*, **53**, 353, 1996.

14. Jaffe, B., Cook, W.R., and Jaffe, H. *Piezoelectric Ceramics.* Academic, New York, p. 135, 1971.

15. Jiles, D. *Introduction to Magnetism and Magnetic Materials.* Chapman and Hall, London, 1991.

16. Kronmuller, H. Statistical theory of Rayleigh's law. *Zeitschrift fur Angewandte Physik,* **30**(1), 9, 1970.

17. N′eel, L. Théorie des lois d'aimantation de Lord Rayleigh et les déplacements d'une paroi isolée. *Cahiers Phys.* **12,** 1, 1942.

18. Damjanovic, D. and Demartin, M. The Rayleigh law in piezoelectric ceramics. *J. Phys. D: Appl. Phys.*, **29**, 2057, 1996.

19. Damjanovic, D. Logarithmic frequency dependence of the piezoelectric effect due to pinning of ferroelectric-ferroelastic domain walls. *Phys. Rev. B*, **55**, 649, 1997.

20. Jaffe, B., Cook, W.R., and Jaffe, H. *Piezoelectric Ceramics.* Academic Press, New York, p. 148, 1971.

21. Ikeda, T. *Fundamentals of Piezoelectricity.* Oxford University Press, New York, p. 209, 1990.

22. Tressler, J.F., Alkoy, S., and Newnham, R.E. Piezoelectric sensors and sensor materials. *J. Electroceram.*, **2**(4), 257, 1998.

23. Mishra, R.K., Kumar, A., Kumar, A., Kumar, D., and Tiwary, K.P. Structural and morphological study of manganese (Mn) doped CdTe nanoparticles. *AIP Conf. Proc.*, **2341**, 1, 2021.

24. Haertling, G.H. Piezoelectric and electro-optic ceramics, in *Ceramic Materials for Electronics*, 2nd edn., edited by RC Buchanan. Marcel Dekker, New York, 1991.

25. Uchino, K. Ceramic actuators: Principles and applications. *Mater. Res. Bull.*, **18**, 42, 1993.

26. Newnham, R.E., Majumdar, S.K., Tressler, R.E., and Miller, E.W. Functional composites for sensors and actuators: smart materials. *The Pennsylvania Academy of Science*, PA, (USA) 1998.

27. Sahoo, B., Jaleel, V.A, and Panda, P.K. Development of PZT powders by wet chemical method and fabrication of multilayered stacks/actuators. *Mater. Sci. Eng. B*, **126**, 80, 2006.

28. Sahoo, B. and Panda, P.K. Dielectric, ferroelectric and piezoelectric properties of (1−x)[Pb0.91La0.09(Zr 0.60 Ti 0.40) O 3]−x [Pb(Mg1/3Nb 2/3)O3], 0≤ x≤ 1. *J. Mater. Sci.*, **42**, 4270, 2007.

29. Damjanovic, D. and Newnham, R.E. Electrostrictive and piezoelectric materials for actuator applications. *J. Intell. Mater. Struct.*, **3**(2), 190, 1992.

30. Jaffe, H. and Berlincourt, D. Piezoelectric transducer materials. *Proc. IEEE*, **53**, 1372, 1965.

31. Haertling, G.H. *Ceramic Materials for Electronics, Processing, Properties, and Applications*, edited by R.C. Buchanan. Marcel Dekkar, New York, p. 165, 1986.

32. Bochenek, D. and Niemiec, P. Microstructure and physical properties of the multicomponent PZT type ceramics doped by calcium, sodium, bismuth and cadmium. *Appl. Phys. A.*, **124**, 775, 2018. https://doi.org/10.1007/s00339-018-2203-3

33. Yang, Y., Hu, Y., and Lu, Y. Sensitivity of PZT impedance sensors for damage detection of concrete structures. *Sensors*, **8**(1), 327, 2008.

34. Jeffrey, S.C., Seung-Hyun, K., Satoshi, W., and Abhijit, C. Characterization of Bi and Fe co-doped PZT capacitors for FeRAM. *Sci. Technol. Adv. Mater.*, **11**, 044402, 2010.

35. Zhang, M.-M., Jia, Z., and Ren, T.-L. Effects of electrodes on the properties of sol–gel PZT based capacitors in FeRAM. *Solid-State Electron.*, **53**(5), 473, 2009.

36. Jiwei, Z., Xi, Y., and Liangying, Z. The optical waveguide characteristics of highly orientated sol–gel derived polycrystalline ferroelectric PZT thin films. *Ceram. Int.*, **27**(5), 585–589, 2001.

37. Nelson, W.G. *Piezoelectric Materials: Structure, Properties and Applications*. Nova Science Publishers, New York, 2010.

38. Khorsand Zak, A. and Abd Majid, W.H. Characterization and X-ray peak broadening analysis in PZT nanoparticles prepared by modified sol–gel method. *Ceram. Int.* **36**(6), 1905, 2010.

39. Liu, C., Zou, B., Rondinone, A.J., and Zhang, Z.J. Sol–gel synthesis of freestanding ferroelectric lead zirconate titanate nanoparticles. *J. Am. Chem. Soc.* **123**(18), 4344, 2001.

40. Ghasemifard, M., Hosseini, S.M., Zak A.K., and Khorrami, G.H. Microstructural and optical characterization of PZT nanopowder prepared at low temperature. *Physica E Low Dimens. Syst. Nanostruct.*, **41**(3), 418, 2009.

41. Deng, Y., Liu, L., Cheng, Y., Nan, C.-W., and Zhao S.-j. Hydrothermal synthesis and characterization of nanocrystalline PZT powders. *Mater. Lett.*, **57**(11), 1675, 2003.

42. Garnweitner, G. and Niederberger, M. Nonaqueous surfactant-free synthesis routes to metal oxide nanoparticles. *J. Am. Ceram. Soc.*, **89**(6), 1801, 2006.

43. Xu, G., Zhao, G., Ren, Z., Shen, G., and Han, G. PVA assisted synthesis of nanosized perovskite PZT powder by a two-stage precipitation route. *Mater. Lett.*, **60**(5), 685, 2006.

44. Ranjan, R., Negi, C.M.S., and Tiwary K.P. Synthesis of Mn^{+2} modified CdS nanoparticles and its application as catalyst in photodegradation of methyl red dye. *Chalcogenide Lett.*, **20**(4), 251, 2023.

45. Yadav, K.L. and Sharma, P. Synthesis and characterization of Mn doped PZT ceramics. *Indian J. Eng. Mater. Sci.*, **15**, 61, 2008.

46. Dan Zhang, Jiangtao Zeng, Liaoying Zheng, Xuezheng Ruan, Xue Shi, and Guorong Li, The microstructure and electric properties of Bi^{3+} and Al^{3+} co-doped PZT ceramics. *Ferroelectrics*, **534**, 212, 2018. https://doi.org/10.1080/00150193.2018.1473692

47. Stancu, V., Amarande, L., Botea, M., Cioangher, M., Tomulescu, A.G., Iuga, A., and Pintilie, L. Structural, dielectric and pyroelectric properties of Nb and Fe doped PZT ceramics. *Digest J. Nanomater. Biostructures*, **14**(1), 225, 2019.

48. Chu, S.Y., Chen, T.Y., Tsai, I.T., and Water, W. Doping effects of Nb additives on the piezoelectric and dielectric properties of PZT ceramics and its application on SAW device, *Sens. Actuators A*, **113**, 198, 2004.

49. Liu, Y., Chang, Y., Li, F., Yang, B., Sun, Y., Wu, J., Zhang, S., Wang, R., and Cao, W. Exceptionally high piezoelectric coefficient and low strain hysteresis in grain-oriented (Ba, Ca)(Ti, Zr)O_3 through integrating crystallographic texture and domain engineering. *ACS Appl. Mater. Inter.* **9,** 29863, 2017.

50. Zhao, C., Wu, H., Li, F., Cai, Y., Zhang, Y., Song, D., Wu, J., Lyu, X., Yin, J., Xiao, D., Zhu, J., and Pennycook, S. J. Practical high piezoelectricity in barium titanate ceramics utilizing multiphase convergence with broad structural flexibility. *J. Am. Chem. Soc.* **140**, 15252, 2018.

51. Kalyani, A.K., Brajesh, K., Senyshyn, A., and Ranjan, R. Orthorhombic-tetragonal phase coexistence and enhanced piezo-response at room temperature in Zr, Sn, and Hf modified $BaTiO_3$. *Appl. Phys. Lett.*, **104,** 252906, 2014.

52. Zhu, X.N., Zhang, W., and Chen, X.M. Enhanced dielectric and ferroelectric characteristics in Ca-modified $BaTiO_3$ ceramics. *AIP. Adv.* **3**, 082125, 2013.

53. Cortés-Vega, F.D.,Montero-Tavera, C., and Yañez-Limón, J.M. Influence of diluted Fe^{3+} doping on the physical properties of $BaTiO_3$. *J. Alloys Compd.*, **847**(1), 156513, 2020. doi: https://doi.org/10.1016/j.jallcom.2020.156513

54. Kumar Yogeswar, Mohiddon Md Ahamad, Srivastava Alok, and Yadav K L, Effect of Ni doping on structural and dielectric properties of $BaTiO_3$. *Indian J. Eng. Mater. Sci.*, **16**, 390, 2009.

Chapter 8

Piezoelectric Materials in the Sound Industry

A. Jaisinghani[a] and A. K. Shukla[b]
[a]*Modern School, Vasant Vihar, New Delhi, 110057, India*
[b]*Department of Physics, Amity Institute of Applied Sciences, Amity University, Noida, G B Nagar, UP, 201313, India*
akshukla@amity.edu

Piezoelectric materials form the basis of several industrial applications utilizing generator and motor actions. The ongoing efforts for miniaturizing the speaker size in the sound industry rely on the motor action of piezoelectric materials. Piezoelectric speakers are currently being developed and used in mobile phones, car audio, flat computers, and public address systems. This chapter reviews the progress in the development of novel piezoelectric materials and their application for sound generation and reception in audible and inaudible ranges.

Piezoelectric Materials: Design and Applications
Edited by N. B. Singh and Dev Kumar Mahato

ISBN 978-981-5129-40-3 (Hardcover), 978-1-003-59897-8 (eBook)
www.jennystanford.com

8.1 Introduction

The word "Piezoelectricity" is derived from the Greek word "Piezo," meaning to press or squeeze, and refers to electricity generated due to mechanical stress [1]. It was discovered in the year 1880 by French physicists Jacques and Pierre Curie. They combined their knowledge of crystal structures along with that of pyroelectricity (a phenomenon in which a material produces electric potential due to a temperature change) to discover the piezoelectric effect, and the two crystals, Quartz and Rochelle, displayed the maximum piezoelectric effect at that time. For the next few decades, piezoelectricity was seen as an interesting laboratory phenomenon without many practical applications.

Finally, the first practical application of piezoelectricity came about in 1917, when Paul Langevin and his co-workers invented an ultrasonic submarine detector. This contained a hydrophone (a device used to record or listen to sound underwater) and a transducer (a device used to convert energy from one form to another) made by gluing quartz crystals in between two steel plates. The transducer was used to emit a high-frequency pulse and the hydrophone was used to hear the echo from the high-frequency pulse after it bounced off an object. By calculating the time between emitting the pulse and hearing the echo, the distance to the object, that rebounded the pulse, can be calculated.

Due to the success of the sonar project, interest in piezoelectricity and its applications increased tremendously. Nowadays, piezoelectricity is used in microphones, speakers, lighters, sensors, buzzers, etc.

8.2 Piezoelectric Phenomenon

Piezoelectricity is the property of a material to convert mechanical energy into electrical energy and vice versa. The word piezoelectricity refers to electricity generated due to mechanical stress. It is observed in materials having a non-centrosymmetric distribution of charge carriers in the lattice [2]. There are two types of piezoelectric effects, (a) direct piezoelectric effect and (b) converse piezoelectric effect, which are detailed in the following sections (Figs. 8.1 and 8.2).

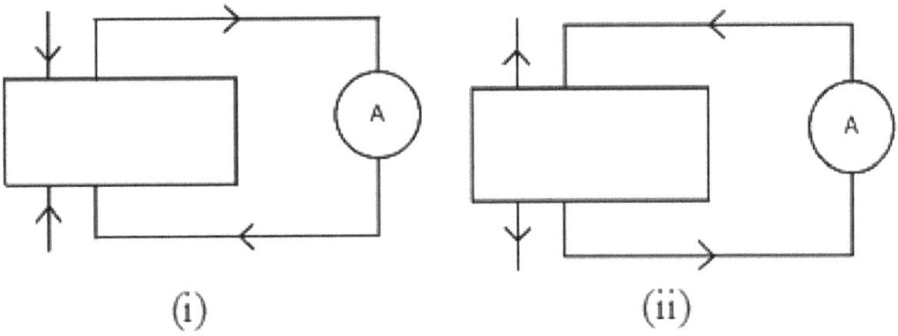

Figure 8.1 The direct piezoelectric effect under (i) compressive force and (ii) extensional force applied to the crystal. The direction of the current changes with the change in the direction of force.

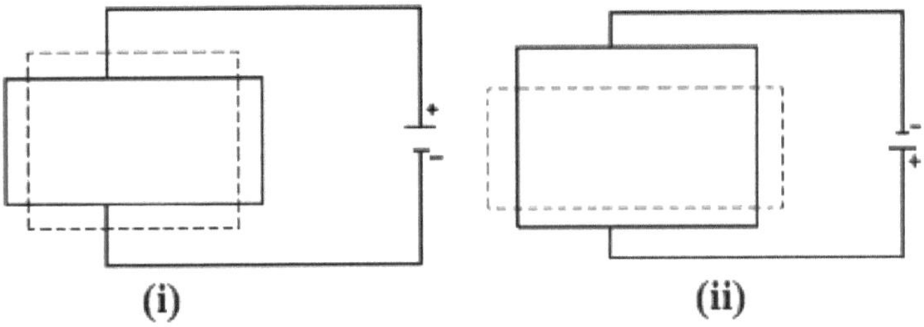

Figure 8.2 The converse piezoelectric effect exhibited by a piezoelectric crystal subjected to an electric field. The longitudinal and transverse strain depends upon the direction of the applied field as depicted in (i) and (ii).

8.2.1 Direct Piezoelectric Effect

The direct piezoelectric effect, also known as the generator effect, refers to the property of a piezoelectric material to produce electric charge due to mechanical stress [2]. When a material is subjected to mechanical stress, the space between the atoms increases or decreases, depending on whether the force is compressing or elongating.

This results in a change in the surface charge density of the material, creating a potential difference across it. This effect is widely used in piezoelectric sensors, mobiles, touchscreen devices, etc.

8.2.2 Converse Effect of Piezoelectricity

The converse, or the motor, effect of piezoelectricity refers to the property of a piezoelectric material to produce mechanical strain due to the application of an electric field [2].

Before applying an electric field, the piezoelectric material is in perfect balance and is absolutely still. The application of an electric field, however, results in the addition or removal of electrons, which shrinks and expands the crystal's structure, thus resulting in a small mechanical force. This effect is used in a variety of applications such as, most importantly, in sonar, ultrasound, and speakers.

8.3 Mechanism of Piezoelectricity

Piezoelectricity depends on the dipole moment of solids. Electric dipole moments can be induced for ions on crystal lattice sites with asymmetrical charge surroundings. The polarization of a crystal, also called dipole density, is the sum of all the dipole moments in a unit volume of the crystal. The piezoelectric effect is caused by the change of polarization, P, due to the application of mechanical stress. Initially, the charges in a piezoelectric crystal are symmetrically arranged in a way such that the electric dipoles present in the crystal cancel out each other and the net dipole moment is observed to be zero. Upon application of mechanical stress, however, the charges are forced out of balance and the dipoles no longer cancel out each other, giving a non-zero polarization in the material. This results in the formation of net positive and negative charges on opposite sides of the crystal and leads to the development of electric potential [3]. The strength and direction of the electric current in a piezoelectric crystal upon application of mechanical stress depends on the following:

1. The amount of mechanical stress applied to the crystal.
2. The orientation of the electric dipoles within the crystal.
3. The internal symmetry of the crystal.

Piezoelectric crystals also show the exact opposite effect, which is called the converse piezoelectric effect, in which the application of an electric field results in mechanical deformation of the crystal.

8.4 Types of Piezoelectric Materials

Piezoelectric materials can be broadly classified into the following different types:

i. Piezoelectric crystals
ii. Piezoelectric ceramics
iii. Piezoelectric polymers

8.4.1 Piezoelectric Crystals

Only certain single crystals that are non-centrosymmetric can exhibit the phenomenon of piezoelectricity [4]. The centrosymmetric crystals do not exhibit the piezoelectric effect. In a centrosymmetric crystal structure, if a line is drawn from the center of symmetry of the crystal to any charge, the line extended in the opposite direction will find the same charge. Under zero mechanical stress, the center of mass of the negative charges at the corners of the crystal coincides with the center of mass of the positive charges. Therefore, there is no net polarization in the crystal, and $P = 0$. Under stress, the center of mass of the negative charges in the crystal still coincides with the center of mass of the positive charges, and hence no piezoelectric effect is observed.

In non-centrosymmetric crystals, however, if a line is extended from the center of symmetry of the crystal to any charge, then the line extended in the opposite direction will find the opposite charge. When the crystal is not under any mechanical stress, the centers of mass of the negative charges and the positive charges coincide. However, when the crystal is under mechanical stress, the positive and negative charges both become displaced inwards and result in a net polarization, P, in the direction of the applied stress. Reversing the direction of the applied stress results in a reversal of the direction of polarization.

Applying an alternating electric field in a non-centrosymmetric crystal results in the ionic charges in the crystal lattice being forced out of balance. This causes a change in the crystal's structure that results in a back-and-forth deformation in the shape of the crystal. It is called the converse piezoelectric effect.

Some examples of piezoelectric crystals are quartz, Rochelle salt, topaz, berlinite, lead titanate, lithium tantalate, lead magnesium niobate, lead magnesium niobate-lead titanate, lead zinc, and niobate-lead titanate.

8.4.2 Piezoelectric Ceramics

The phenomenon of piezoelectricity is also observed in certain ceramics which are found to be better for piezoelectric applications as their shape and electromechanical properties can be tailored, as compared to natural crystals which cannot be modified or engineered to suit a specific design [5]. Hence, piezoelectric ceramics are more widespread than any other piezoelectric material. The name "Ceramics" is derived from the Greek word, "Keramicos", meaning of or for pottery. Ceramics are non-metallic inorganic compounds that are fabricated at high temperatures mostly using the solid-state reaction method. The solid-state reaction method involves weighing ingredients in a stoichiometric ratio, homogenization of ingredients, calcination, crushing, remilling, and granulation of calcined material, shaping the material, and final firing. As grown piezoelectric ceramic material is ferroelectric and has microscopic regions of uniform polarization in it, commonly known as ferroelectric domains. These domains are randomly oriented and thus a non-centrosymmetric condition is not observed in the as-sintered ceramic and the piezoelectric effect is absent. The alignment of the ferroelectric domains in a particular direction in the material is done by the application of a high dc field to the electroded ceramic material normally in an oil bath at elevated temperatures. This method was originally introduced by Gray to make piezoelectric ceramic materials and is known as poling. After the alignment of the domains, the material exhibits anisotropy, and the piezoelectric effect is seen in it. Some examples of piezoelectric ceramics are families of lead zirconate titanate (PZT), barium titanate (BT) and lead titanate (PT), sodium potassium niobate (KNN), and bismuth titanate.

8.4.3 Piezoelectric Polymers

Some polymeric materials show the piezoelectric effect and are classified on the basis of the molecular arrangement and the dipole moment. The piezoelectric polymers are lead-free and offer many advantages over piezoelectric crystals and ceramics. The piezoelectric ceramic and piezoelectric polymer composites have found a variety of applications, particularly in underwater communication systems and medical ultrasound. The charge constant of piezoelectric polymers is low compared to the piezoelectric crystals and ceramics; however, the voltage constant value is superior to them and hence makes the piezo-polymers a better choice for sensor applications [6]. The piezoelectric polymers can withstand high dc electric fields, are flexible and lightweight, and possess low acoustic impedance. A few widely used polymers are polyvinylidene difluoride (PVDF) and its copolymer with tri- and tetrafluoroethylene.

8.5 Applications of Piezoelectric Materials in Sound Generation and Reception

In daily life, we come across different kinds of sounds, out of which only a few are audible. Sound is a mechanical wave and is classified based on the frequency and audibility as follows.

8.5.1 Infrasound

Infrasound refers to the sound wave created due to mechanical vibrations of a frequency lower than 20 Hz. It can be generated naturally by earthquakes, volcanoes, lightning, etc. as well as artificially through sonic booms, explosions, and specialized loudspeakers, especially subwoofers.

The detection of infrasonic waves helps to monitor both seismic and geographic phenomena such as earthquakes, volcanoes, charting rock and petroleum formations below the earth. For example, before a volcanic eruption, the formation of cracks in the tectonic plates due to sub-surface strains and the expansion of gas produce lower frequency sounds that can be detected by the infrasonic sensor [7]. It is also used for medical applications such as studying the mechanics of the heart as well as in defense applications such as infrasonic weapons. Infrasound has a low frequency and long wavelength which enables it to travel far distances [8].

Shuheng and Wenfan have made a high-precision infrasonic sound sensor with a range of 3–30 Hz using polyvinylidene fluoride (PVDF) [8]. Georgiana and Stefan have made use of certain piezoelectric materials such as PZT for military infrasound applications as it is inaudible by humans [9].

8.5.2 Audible Range of Sound

The audible range of sound lies between 20 Hz and 20 KHz and it is the only range of sound humans can hear. It is mainly used in communication, speakers, and entertainment. Piezoelectric materials have a higher frequency response (200 Hz to 100 kHz) and do not reproduce lower frequency sounds with the same quality as required for speakers [10].

Wickham et al. invented the piezoelectric microphone in 1951 [11]. It consisted of a piezoelectric crystal mounted onto a plate which was connected to external wires. Speaking into the microphone would result in small vibrations in the crystal, which would then produce electrical energy that would leave the body through the external wires.

One of the first piezoelectric devices used in this field was a piezoelectric loudspeaker invented by H. W. Kompanek in 1969. This device was made using a number of circular piezoelectric ceramics (PZT) connected to each other in parallel (Fig. 8.3). These were sandwiched between two layers of silver which was then, finally, attached to a thin metal plate as shown in Fig. 8.4 [12].

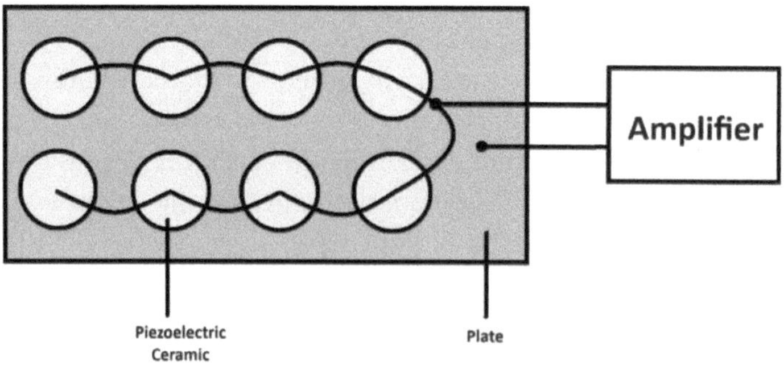

Figure 8.3 The schematic diagram of a piezoelectric utilizing a number of piezoelectric ceramics attached to a common metal plate and amplifier.

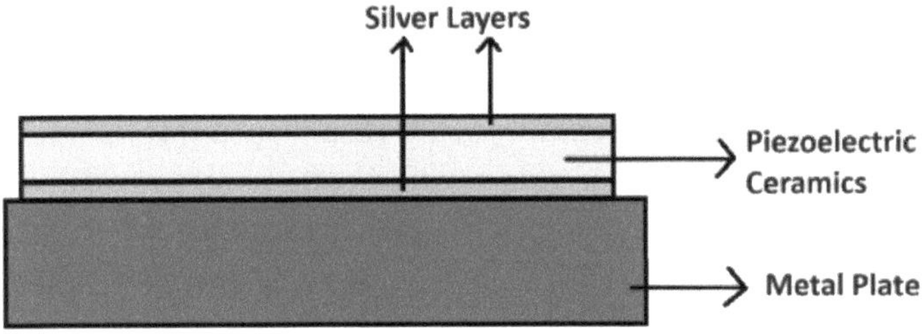

Figure 8.4 Individual piezoelectric ceramic disc connected to the metal plate in a piezoelectric speaker.

By using an amplifier to apply AC voltage across the plate, the piezoelectric ceramics vibrate to produce sound pressure waves that generate sound in the audible frequency range of 200 Hz to 5 kHz. These piezoelectric loudspeakers have been miniaturized and are used, nowadays, in cell phones, buzzers, digital watches, etc. Piezoelectric speakers are useful as they can be easily miniaturized and require relatively less electrical power to produce the same sound pressure levels as compared to driver units and music speakers.

The ability of piezoelectric materials to show higher frequency response and to reproduce similar sound pressure levels as compared to the driver unit with low power consumption formed the basis for the invention of piezoelectric buzzers. G. D. Stephens developed the first piezoelectric buzzer in 1978, which consisted of a piezoelectric crystal mounted on a plate (Fig. 8.5) that was attached to an electrode, a PCB, and an oscillator circuit [13].

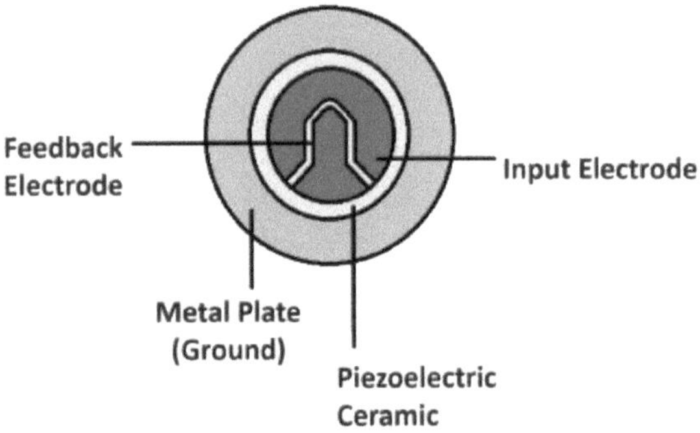

Figure 8.5 Schematic diagram of a piezoelectric buzzer.

A power terminal was used to create a potential in the oscillator circuit on either side of the crystal that resulted in the crystal vibrating with a "buzz" sound. The efficiency of piezoelectric materials over a narrower frequency response in the audible range made them a great material to create cheaper and better buzzers.

Piezoelectrics are found less useful in music speakers due to their inability to reproduce lower frequency ranges of sound and they can produce sound with a frequency of over 1 kHz. The speaker was further improved in the form of a piezoelectric sound generator by Ogawa [14]. The sound generator consists of a monolithic sintered body created by laminating a number of green ceramic sheets which were formed by a plurality of alternating electrodes and cofiring the same. This invention helped overcome the problem caused by the suppression of vibrations, to produce sound, by electrically conducting parts. This would help to accurately produce the desired sound pressure levels. The piezoelectric speaker was further improved on by Ohga in 1981 by increasing its efficiency through the lowering of the fundamental resonance frequency of the diaphragm [1]. In 1990, Josse et al. used piezoelectric bulk acoustic wave resonators as detectors in viscous, conductive fluids [15]. This functions on the principle that the frequency change in a medium is directly proportional to its density and viscosity. By using a piezoelectric to produce a certain frequency, the frequency change through a medium was measured to determine the medium's density and viscosity. Another interesting application of piezoelectrics is in noise suppression. In 1998, Jayachandran et al. used piezoelectrics to reduce the internal noise in aircraft [16]. The small size and low weight of small piezoelectric speakers made them ideal to place inside aircraft. By using a system of piezoelectrics that produce sound waves, that create destructive interference, toward both the walls of the aircraft and the interior of the aircraft itself, the ambient sound would get canceled out, resulting in no net sound in the aircraft.

Recently, there have been several developments regarding the uses and applications of piezoelectrics in the audible range of sound. Lin et al. used piezoelectrics to control vibrations in active machinery in 2016 [17]. This was done by using the piezoelectrics in such a way that their vibrations exhibited destructive interference to cancel out the vibrations in the machinery and result in no net vibration. Liechti et al. developed a high-performance piezoelectric speaker that works using a wafer bonding process and produces a sound pressure level (SPL) of 110 dB in a frequency range of 1–4.5 kHz [18]. Fang et al. developed a method of harnessing energy using piezoelectrics [19]. A function generator is used to amplify any ambient sound in the surroundings. A piezoelectric is used to transduce this sound energy into electrical energy. The energy harvested is minute, however, equalling approximately 33.3 dBμW. Therefore, a combination of a power source and different interface circuits such as a rectifier and a voltage multiplier circuit are used to harvest the ambient sound energy in a meaningful amount. In addition, in 2015, Lang et al. developed a piezoelectric by electrospinning nanofibres [20]. This forms a material that can detect low-frequency sound with a sensitivity of 266 mV/Pa. This sensitivity is five times greater than that of a commercial piezoelectric polymer. The material has also shown a brilliant mechanical to electrical energy conversion ability.

8.5.3 Ultrasound

Ultrasound refers to the sound wave produced due to mechanical vibrations of a frequency >20 kHz [21]. Piezoelectric materials are extensively used for the generation and reception of ultrasonic frequencies (200 Hz to 100 kHz) [22] and find applications in sonar, directional loudspeakers, and medical imaging.

An important application of piezoelectric materials in the ultrasonic spectrum is that of sonar. Sonar is a method of navigation using sound waves and is used in underwater as sound waves travel farther than light or radio waves in that medium. A sonar system consists of an underwater receiver (hydrophone), an underwater speaker (projector), and other electrical devices such as amplifiers and systems for data acquisition and interpretation [23]. In sonar, the project is used to propagate sound waves into the water. If there is an object in the way of the sound wave, the sound wave bounces off the object and returns an "echo" to the hydrophone. The sonar device is also equipped with electric components that can pick up this signal (echo) and use the strength and direction of the signal to locate the object in the water.

Recently, Tressler developed an optimized piezoelectric transducer consisting of a hydrophone, projector, and other electronic components using the material PZT – 5 for the hydrophone and PZT – 8 for the projector [23].

Another interesting application in the field of ultrasound using piezoelectric materials is that of a directional loudspeaker [24] which can be used for private communication (Fig. 8.6). For example, one could stand at a spot in a museum, art gallery, etc. and hear about the object they are looking at without disturbing others in the vicinity. The speaker in question is an air-coupled, single-chip, CMOS-compatible micromachined parametric array loudspeaker based on bimorph AlN piezoelectric ultrasonic transducers [25].

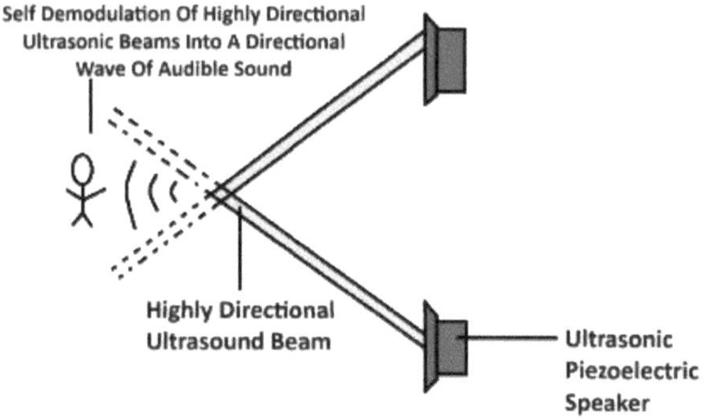

Figure 8.6 Private communication using ultrasound technique.

The array consists of two channels of piezoelectric ultrasonic transducers, with a resonant frequency of around 250 kHz. This array design, using the piezoelectric material molybdenum disulfide, is used to produce two ultrasonic beams in a continuous wave mode. A highly directional audible beam of sound is generated due to the self-demodulation of the ultrasonic beams in the air which can only be heard in a very limited area due to its highly directional nature [26].

8.6 Conclusions

The use of piezoelectric materials in producing and detecting sound in all three regimes, i.e., infrasonic, audible, and ultrasonic range, has been discussed. Piezoelectric materials are being explored for detecting seismic activities and defense applications in the infrasonic range. Suppression of noise in harsh environments using the destructive interference method is a recent area of the application of piezoelectric materials in addition to their use in buzzers and speakers. The capability of generating and receiving ultrasonic waves using piezoelectric materials opened a new area of non-destructive methods for biomedical applications.

References

1. Ohga J. A flat piezoelectric polymer film loudspeaker as a multi-resonance system. *J. Acoust. Soc. Japan* **4**, 113–120, 1983.
2. Shukla A K and Srivastava S L. Ch. 8, Smart materials: Piezoelectrics, In *Synthesis, Characterization and Application of Smart Materials* (Ed. Radheshyam Rai), 155–200, Nova Sci. Pub. Inc. (USA) 2012.
3. Gao X, Cheng Z, Chen Z, Liu Y, Meng X, Zhang X, Wang J, Guo Q, Li B, Sun H, Gu Q, Hao H, Shen Q, Wu J, Liao X, Ringer S P, Liu H, Zhang L, Chen W, Li F, and Zhang S. The mechanism for the enhanced piezoelectricity in multi-elements doped (K,Na)NbO_3 ceramics. *Nat. Commun.* **12**, 1–9, 2021.
4. Trolier-Mckinstry S, Zhang S, Bell A J, and Tan X. High-performance piezoelectric crystals, ceramics, and films. *Annu. Rev. Mater. Res.* **48**, 191–217, 2018.
5. Mercadelli E, Sanson A, and Galassi C. Ch. 6, Porous piezoelectric ceramics, In *Piezoelectric Ceramics* (Ed. Ernesto Suaste-Gomez), 111–128, www.intechopen.com (https://doi.org/10.5772/9942) 2010.
6. Richardson P D. Piezoelectric polymers. *IEEE Eng. Med. Biol. Mag.* **8**, 14–6, 1989.
7. Johnson J B, Anderson J F, Anthony R E, and Sciotto M. Detecting geyser activity with infrasound. *J. Volcanol. Geotherm. Res.* **256**, 105–17, 2013.
8. Shuheng S and Wenfan W. Development of a high precision infrasonic sound sensor. *Int. J. Signal Process. Image Process. Pattern Recognit.* **8**, 211–218, 2015.
9. Golea D G and Ştefan Cozma L. The Use of Piezoelectric Materials for the Purpose of Creating Infrasound and Ultrasound Sources for Terrorist Acts (May 24, 2021). https://dx.doi.org/10.2139/ssrn.3852014
10. Jaffe B, Cook W R, and Jaffe H. *Piezoelectric Ceramics*, Academic Press, 1971; and https://docs.rs-online.com/0f60/0900766b8126cdcc.pdf
11. Wickham P B, Johnston S A, Delavan, and Walworth, Piezoelectric Microphone, US Patent No: 2561084, 1951.
12. Kompanek H W, Loudspeaker With Piezoelectric Wafer Driving Elements, US Patent No: 3423543, 1969.
13. Stephens G. Piezoelectric Buzzer Device, US Patent No: 4122365, 1978.

14. Ogawa T. Piezoelectric Sound Generator, US Patent No: 4742264, 1988.
15. Josse F, Shana Z A, Radtke D E, and Haworth D T. Analysis of piezoelectric bulk-acoustic-wave resonators as detectors in viscous conductive liquids, *IEEE Trans. on Ultrasonics, Ferroelectrics, and Frequency Control*, **37**(5), 359–368, 1990.
16. Jayachandran V, Meyer N E, Westervelt M A, and Sun J Q. Piezoelectrically driven speakers for active aircraft interior noise suppression, *Applied Acoustics*, **56**, 263–277, 1999.
17. Lin R R, Palazzolo A B, Kascak A F, and Montague G. Use of piezoelectric actuators in active vibration control of rotating machinery, *SPIE Proc. Electro-Optical Materials for Switches, Coatings, Sensor Optics, and Detectors*, **1307**(2013), 199–210, 1990.
18. Liechti R, Durand S, Hilt T, Casset F, Dieppedale C. Colin M. High performance piezoelectric MEMS loudspeaker based on an innovative wafer bonding process. *Sensors Actuators A Phys.* **358**, 114413, 2023.
19. Hui L, Idris S, Hassan S, Abd R, and Isa M. Exploring piezoelectric for sound wave as energy harvester. *Energy Procedia* **105**, 459–466, 2017.
20. Fang J, Shao H, Ding X, and Lin T High-sensitivity acoustic sensors from nanofibre webs, *Nat. Commun.*, **7**, 1–7, 2016.
21. Postema M, Attenborough K, Fagan M J, Gilja O H, Hurrell A, Kotopoulis S, Matre K, Wamel A V, and Campbell P A. *Fundamentals of Medical Ultrasonics*, 1st edn, 2011, CRC Press-Taylor & Francis Group, London. https://doi.org/10.1201/9781482266641
22. Manbachi A, Cobbold R S C. Development and application of piezoelectric materials for ultrasound generation and detection. *Ultrasound* **19**, 187–196, 2011.
23. Tressler J F. Piezoelectric transducer designs for sonar applications. In *Piezoelectric and Acoustic Materials for Transducer Applications*, Safari A, Akdoğan E K (eds). Springer, Boston, MA, 2008. https://doi.org/10.1007/978-0-387-76540-2_11
24. Yoneyama M, Fujimoto J, Kawamo Y, and Sasabe S, The audio spotlight: An application of nonlinear interaction of sound waves to a new type of loudspeaker design, *J. Acoust. Soc. Am.*, **73**(5), 1532–1536, 1983.
25. Shao Z, Pala S, Liang Y, Peng Y, and Lin L. A single chip directional loudspeaker based on PMUTS. *Proc. IEEE Int. Conf. Micro Electro Mech. Syst.* 895–898, 2021. doi: 10.1109/MEMS51782.2021.9375191
26. Wygant I O, Kupnik M, Windsor J C, Wright W M, Wochner M S, Yaralioglu G G, Hamilton M F, and Khuri-Yakub B T. 50 kHz capacitive micromachined ultrasonic transducers for generation of highly directional sound with parametric arrays. *IEEE Trans. Ultrason. Ferroelectr. Freq. Control* **56**, 193–203, 2009.

Chapter 9

Piezoelectric Energy Harvesting

Neksumi Musa,[a,b] Cland Makasson Raphael,[a]
Narendra Pratap Singh,[c] and N. B. Singh[d]

[a]*Department of Science Laboratory Technology,*
Adamawa State Polytechnic, Yola, Nigeria
[b]*Department of Environmental Sciences, Sharda University, India*
[c]*Department of Chemistry, UP Autonomous college, Varanasi, India*
[d]*Research and Development Cell, Sharda University, India*
n.b.singh@sharda.ac.in

This chapter offers a comprehen-sive overview of this technology, covering its fundamentals, applications, design aspects, and sustainability considerations. It begins with an introduction, defining piezoelectric energy harvesting and emphasizing its significance, history, and real-world use. The fundamentals of piezoelectricity are discussed, explaining the effect, materials, and transducers like PZT and PVDF. Energy harvesting mechanisms are reviewed, including comparisons with other methods and addressing efficiency and limitations. This chapter explores harvester design and fabrication, discussing components, design factors (resonance, geometry), and fabrication techniques. Mathematical models and simulation tools,

Piezoelectric Materials: Design and Applications
Edited by N. B. Singh and Dev Kumar Mahato

ISBN 978-981-5129-40-3 (Hardcover), 978-1-003-59897-8 (eBook)
www.jennystanford.com

such as finite element analysis (FEA), are highlighted through case studies. Performance optimization techniques like load and impedance matching, along with multimodal and multilayer harvesters, are explained for enhanced efficiency. Real-world applications and case studies illustrate diverse industry uses. The chapter covers current challenges, emerging tech, and prospects. Additionally, it examines environmental aspects, including impact assessment, benefits, and green initiatives' role.

9.1 Introduction

Piezoelectric energy harvesting is a cutting-edge technology that harnesses electrical energy from mechanical vibrations, strain, or pressure applied to certain materials with piezoelectric properties [1]. These materials can uniquely convert mechanical stress or deformation into electrical charge, which can then be stored or used to power electronic devices. In essence, it's a process that allows us to tap into otherwise wasted mechanical energy from various sources in our environment and convert it into a usable electrical form. The significance of piezoelectric energy harvesting lies in its potential to provide a sustainable and environmentally friendly source of power for a wide range of applications. By transforming ambient mechanical energy into electricity, it offers a number of advantages [2]:

i. **Renewable Energy**: Unlike traditional power sources that rely on fossil fuels, piezoelectric energy harvesting is renewable and doesn't deplete natural resources.
ii. **Low Environmental Impact**: It produces minimal environmental pollution and greenhouse gas emissions, making it a greener energy solution.
iii. **Versatility**: Piezoelectric energy harvesting can be applied in various fields and industries, from wearable technology and IoT (Internet of Things) devices to automotive and aerospace systems.
iv. **Energy Efficiency**: It allows for more efficient utilization of available mechanical energy, reducing waste and improving energy conservation.

v. **Remote Power Generation**: In remote or hard-to-reach locations, where traditional power sources are impractical, piezoelectric energy harvesting can provide a reliable source of electricity.

Some notable applications of piezoelectric energy harvesting are given in Fig. 9.1 [3].

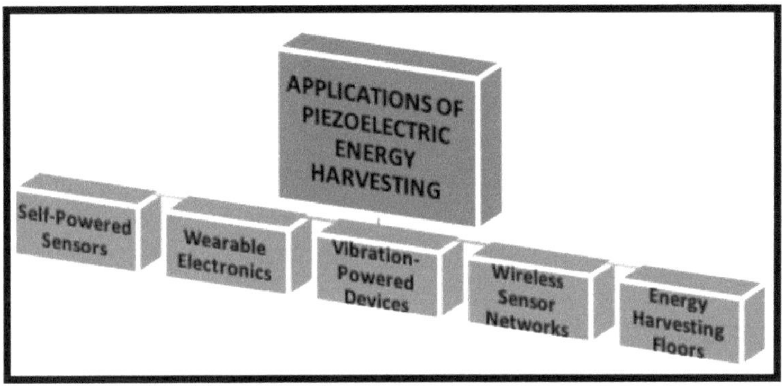

Figure 9.1 Applications of piezoelectric energy harvesting.

The concept of piezoelectricity was first discovered by the Curie brothers, Pierre and Jacques Curie, in the late 19th century [4]. They observed that certain crystals, such as quartz, generated electric charges when subjected to mechanical pressure. This phenomenon laid the foundation for the development of piezoelectric materials and their applications. Over the years, advancements in materials science and engineering have created a wide range of piezoelectric materials, including piezoelectric ceramics, polymers, and composites. These materials exhibit enhanced piezoelectric properties and have expanded the potential for energy harvesting. The field of piezoelectric energy harvesting has seen significant growth in recent decades, driven by the increasing demand for sustainable energy sources and the miniaturization of electronic devices. Researchers and engineers continue to explore new materials, design techniques, and applications to harness mechanical energy efficiently and contribute to the growing field of green energy technology [5].

Piezoelectric energy harvesting is a groundbreaking energy conversion technology that has revolutionized the way we think about energy production. By providing historical context, this introduction effectively highlights the significance of piezoelectric energy harvesting and sets the stage for a deeper understanding of its core principles.

9.2 Fundamentals of Piezoelectricity

The piezoelectric effect is an astonishing and highly useful phenomenon observed in specific materials. It represents a captivating synergy between mechanical stress, electrical charge, and energy conversion. This effect essentially describes the material's ability to effortlessly transition between the realms of mechanical and electrical domains, making it an invaluable property for various applications [6].

9.2.1 Generation of Electric Charge from Mechanical Stress

One facet of the piezoelectric effect is the generation of electric charge or voltage in response to applied mechanical stress or deformation. Imagine, for a moment, a material that exhibits this behavior. When you apply physical pressure, compress, or deform such a material, it responds by producing an electric charge. This might seem almost magical, but it's a well-documented and fundamental principle in materials science. It is worth noting that the piezoelectric effect is reversible, which is equally remarkable. This means that not only can these materials generate electric charge when mechanically manipulated, but they can also undergo mechanical strain or displacement when subjected to an external electric field. In other words, by applying an electric field to these materials, you can make them physically change shape or deform [7].

9.2.2 The Etymology: "Piezoelectric"

The term "piezoelectric" has its roots in ancient Greece, with its components originating from Greek words. "Piezein" translates to "to squeeze or press," and "electricity" is derived from "ēlektron," referring to amber, which was observed to attract lightweight objects when rubbed. This etymology beautifully encapsulates the essence of the piezoelectric effect – the generation of electric charge through mechanical pressure. To further illustrate this concept, consider a common example: quartz crystals in electronic devices like watches. When a quartz crystal experiences mechanical compression or stress (for instance, due to vibrations), it generates a tiny electrical voltage. This voltage, in turn, can be harnessed to keep incredibly accurate time [5].

Conversely, in certain medical devices, an electric field can be applied to a piezoelectric crystal, causing it to deform precisely. This deformation can be utilized in ultrasound machines to generate sound waves for imaging purposes. In essence, the piezoelectric effect embodies the extraordinary interconnectedness of physical phenomena. It bridges the realms of mechanics and electricity, offering a versatile and powerful tool that has found applications across diverse fields, from sensors and energy harvesting to imaging and beyond. Understanding and harnessing this phenomenon has revolutionized technology and continues to drive innovation across various scientific disciplines [8].

9.2.3 Materials and Crystals with Piezoelectric Properties

Not all materials exhibit piezoelectric behavior; it is a property specific to certain crystals, ceramics, and polymers. Here are some materials and crystals known for their piezoelectric properties [9]:

i. **Quartz Crystal**: Quartz is perhaps the most well-known piezoelectric material. It is commonly used in electronic oscillators and watches due to its precise frequency response. When mechanical pressure is applied to a quartz crystal, it generates a voltage across its surface.

ii. **Lead Zirconate Titanate (PZT)**: PZT is a widely used piezoelectric ceramic material. It is highly efficient and can be engineered to have specific piezoelectric properties. PZT finds applications in sensors, actuators, and transducers.

iii. **Polyvinylidene Fluoride (PVDF)**: PVDF is a piezoelectric polymer known for its flexibility and ease of integration into various applications. It is used in wearable technology, sensors, and even musical instruments.

iv. **Rochelle Salt**: This organic crystal exhibits strong piezoelectric properties and was historically used in early electromechanical devices.

v. **Gallium Phosphate ($GaPO_4$)**: $GaPO_4$ is a piezoelectric crystal known for its high-temperature stability. It is utilized in high-temperature sensors and actuators.

vi. **Barium Titanate ($BaTiO_3$)**: $BaTiO_3$ is another piezoelectric ceramic material with applications in sensors and actuators.

9.3 Basic Principles of Piezoelectricity

The fundamental principles underlying piezoelectricity can be understood through the following concepts.

9.3.1 Lattice Symmetry: The Backbone of Piezoelectricity

The remarkable piezoelectric effect owes its existence to the specific crystal structures and lattice symmetries found in certain materials. Lattice symmetry refers to the organized arrangement of atoms or ions within a crystal lattice, and in piezoelectric materials this arrangement is paramount to their unique properties [7]. Central to piezoelectricity is the concept of an electric dipole moment. An electric dipole consists of two equal but opposite electrical charges (one positive and one negative) separated by a small distance. When a material is mechanically deformed, this symmetry-breaking rearrangement of ions within the crystal lattice generates electric dipole moments. These dipole moments are at the heart of the piezoelectric effect [10].

9.3.2 Non-centrosymmetric Crystal Classes: The Piezoelectric Heroes

Piezoelectric materials primarily belong to crystal classes that lack a property known as centrosymmetry. Centrosymmetry implies that a crystal's structure is such that its electrical charges balance out to zero at any given point. In other words, the positive and negative charges are evenly distributed throughout the crystal [11]. However, in non-centrosymmetric crystal classes, this balance is disrupted. When mechanical stress is applied to these materials, it causes a slight shift in the positions of the ions within the lattice. This shift, though tiny, leads to an uneven distribution of charges, resulting in the creation of those electric dipole moments we mentioned earlier.

Some of the most common non-centrosymmetric crystal classes that exhibit piezoelectricity include the following [12]:

i. **Trigonal:** Crystals in the trigonal class have threefold rotational symmetry but lack inversion symmetry. Quartz is a classic example, and its piezoelectric properties have made it a key component in various electronic applications.
ii. **Tetragonal:** In tetragonal crystals, there is a fourfold rotational symmetry but, once again, no inversion symmetry. Lead zirconate titanate (PZT) is a well-known example in this class, widely used in transducers and actuators.
iii. **Hexagonal:** Crystals in the hexagonal class exhibit sixfold rotational symmetry without inversion symmetry. Aluminium nitride is an example, utilized in sensors and resonators.

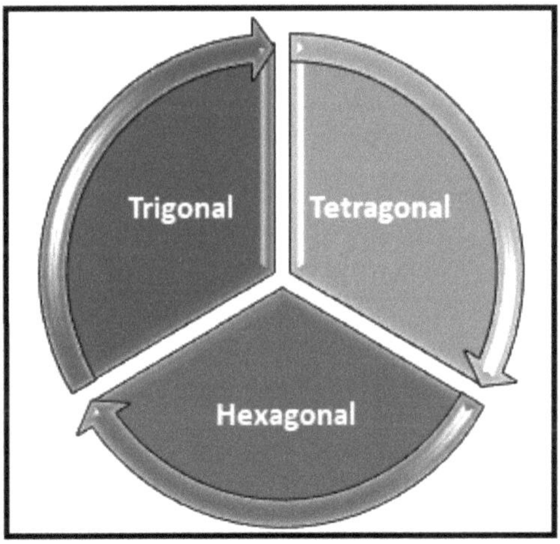

Figure 9.2 Classes of non-centrosymmetric crystal.

These materials owe their piezoelectric prowess to their non-centrosymmetric nature, which allows for the generation of electric dipoles when subjected to mechanical stress. This phenomenon is the cornerstone of piezoelectric technology, enabling applications ranging from sensors that detect pressure or acceleration to energy harvesting devices that convert mechanical vibrations into electrical energy. Understanding the lattice symmetry and crystal structures of these materials is essential for both designing and optimizing their piezoelectric properties for various technological purposes (Fig. 9.2).

9.3.3 Direct Piezoelectric Effect

When mechanical stress or pressure is applied to a piezoelectric material, it leads to the displacement of ions within the crystal lattice. This displacement generates an electric field within the material, resulting in the separation of positive and negative charges and the creation of an electric potential difference or voltage across the material [13].

9.3.4 Converse Piezoelectric Effect

Conversely, when an external electric field is applied to the piezoelectric material, it induces a mechanical deformation or strain in the material. This effect is at the heart of piezoelectric actuators, which can be used for precise positioning or vibration control [13].

9.3.5 Polarization

The piezoelectric effect causes the material to become polarized, meaning it has a net electric dipole moment. This polarization can be temporary and reversible, as in the case of dynamic applications, or it can be permanent and set during the material's manufacturing process [14].

9.3.6 Piezoelectric Coefficient

The piezoelectric coefficient quantifies the relationship between the mechanical stress or strain and the resulting electric charge or voltage. Different materials have varying piezoelectric coefficients, which determine their efficiency in converting mechanical energy to electrical energy or vice versa [15]. Understanding these basic principles of piezoelectricity is essential for harnessing its applications in various fields, from sensors and energy harvesting devices to medical imaging and ultrasonic transducers. The ability to convert mechanical energy into electrical energy and vice versa has made piezoelectric materials invaluable in modern technology and scientific research [16].

9.4 Piezoelectric Transducers: Harvesting the Power of Piezoelectricity

Piezoelectric transducers are extraordinary devices that capitalize on the unique piezoelectric effect to convert mechanical energy into electrical signals or vice versa. They play a crucial role in various fields, from sensing and measurement to energy harvesting and actuation [17]. In this discussion, we'll dive into the types, working mechanisms, properties, and selection criteria of piezoelectric transducers.

Types of Piezoelectric Transducers

Piezoelectric transducers come in various types, each tailored to specific applications. Some of the most common types include the following [18]:

i. **PZT (Lead Zirconate Titanate) Transducers:** PZT is one of the most widely used piezoelectric materials. PZT transducers offer excellent piezoelectric properties, making them suitable for applications such as ultrasonic sensors, actuators, and even musical pickups.

ii. **PVDF (Polyvinylidene Difluoride) Transducers:** PVDF is a flexible and lightweight piezoelectric polymer. PVDF transducers are used in applications where flexibility is essential, such as medical ultrasound probes, hydrophones, and wearable sensors.

iii. **Lithium Niobate ($LiNbO_3$) Transducers:** $LiNbO_3$ is known for its high electromechanical coupling coefficient. It is employed in applications requiring high-frequency performance, including surface acoustic wave (SAW) devices and high-frequency transducers.

iv. **Rochelle Salt Transducers:** Rochelle salt is a naturally occurring piezoelectric material used in early transducers. It is now mostly replaced by synthetic piezoelectric materials due to its moisture sensitivity.

v. **Gallium Phosphate ($GaPO_4$) Transducers:** $GaPO_4$ transducers offer stable piezoelectric properties at elevated temperatures, making them suitable for high-temperature applications like gas flow sensing and aerospace.

9.4.1 Working Mechanism of Piezoelectric Transducers

Piezoelectric transducers operate based on the piezoelectric effect, which we've previously discussed. Here's a step-by-step breakdown of how they work [19]:

i. **Application of Mechanical Stress:** When mechanical stress or deformation is applied to the piezoelectric material (e.g., by pressing, bending, or stretching), it induces a shift in the positions of the charged ions within the crystal lattice.

ii. **Creation of Electric Dipole Moments:** This shift leads to an asymmetrical distribution of positive and negative charges, generating electric dipole moments throughout the material.

iii. **Generation of Voltage or Charge:** As a result of these electric dipoles, a voltage potential develops across the material's surfaces or a charge accumulates.

iv. **Signal Output:** This voltage or charge can be measured and utilized as a signal output. Conversely, if an external electric field is applied to the material, it can induce mechanical deformation or vibrations, allowing the transducer to convert electrical signals into mechanical motion.

9.4.2 Properties and Selection Criteria

Selecting the right piezoelectric transducer involves considering several critical properties and criteria [20]:

i. **Piezoelectric Constants:** These constants, often denoted as d and g, quantify the material's piezoelectric response. A high d indicates strong piezoelectricity for sensing applications, while a high g is crucial for actuator applications.

ii. **Curie Temperature:** The Curie temperature is the point at which a piezoelectric material loses its piezoelectric properties. For high-temperature applications, materials with elevated Curie temperatures are preferred.
iii. **Frequency Response:** Different transducers are designed for different frequency ranges. Ensure your choice aligns with your application's required frequency.
iv. **Sensitivity:** Sensitivity is a measure of the transducer's response to mechanical stress. High sensitivity is desirable for sensors, while lower sensitivity can be suitable for actuators.
v. **Size and Form Factor:** Consider the physical dimensions and shape of the transducer, as these factors can greatly impact its integration into your system.
vi. **Durability and Environmental Factors:** Evaluate the material's resistance to environmental conditions like humidity, temperature, and chemicals, especially if the application is exposed to harsh environments.
vii. **Cost and Availability:** Finally, take into account the cost and availability of the chosen material, as well as its ease of manufacturing into the desired transducer shape.

Piezoelectric transducers are versatile and powerful devices with applications across a wide spectrum of industries. By understanding their types, working mechanisms, properties, and selection criteria, you can make informed decisions when choosing the right transducer for your specific needs.

9.5 Energy Harvesting Mechanisms

Energy harvesting, also known as energy scavenging or power harvesting, is a process that captures and converts ambient energy from the surrounding environment into usable electrical energy. This approach is increasingly important in our quest for sustainable and self-sufficient power sources. Energy harvesting mechanisms tap into various sources of energy, from mechanical vibrations to solar radiation, to generate electricity for a wide range of applications, such as powering remote sensors, wearable devices, and even some low-power electronic gadgets [21].

9.5.1 Overview of Energy Harvesting Techniques

It's essential to understand various energy harvesting techniques, and this document will provide a comprehensive overview of them:

i. **Solar Energy Harvesting**: Solar panels or photovoltaic cells are among the most widely used energy harvesting techniques. They convert sunlight into electricity through the photovoltaic effect. Solar energy harvesting is highly sustainable, particularly in sunny regions, and can be used to power a variety of devices, from solar calculators to remote weather stations [22].

ii. **Thermoelectric Energy Harvesting**: This method relies on the Seebeck effect, where temperature differences between two materials generate an electric voltage. Thermoelectric generators are used to harness waste heat, like that produced by industrial processes or automobile exhaust, to generate electricity. These devices have applications in automotive power generation and waste heat recovery systems.

iii. **Vibration and Mechanical Energy Harvesting**: Vibrational and mechanical energy can be converted into electrical energy using piezoelectric materials or electromagnetic generators. Piezoelectric materials generate voltage when subjected to mechanical stress, while electromagnetic generators use magnetic induction principles to produce electricity from motion. This approach is useful in applications like powering wireless sensors in industrial machinery [22].

iv. **RF (Radio Frequency) Energy Harvesting**: RF energy harvesting captures energy from ambient radio frequency signals, such as Wi-Fi, cellular networks, and RFID readers. Antennas and rectifiers are used to convert RF energy into electricity. These systems are suitable for low-power applications like IoT devices and wireless sensor networks [23].

v. **Hybrid Energy Harvesting**: In some cases, multiple energy harvesting techniques are combined to improve overall efficiency and reliability. For instance, a device might use both solar panels and thermoelectric generators to harvest energy from both sunlight and temperature differences.

vi. **Wind Energy Harvesting**: Wind turbines and windmills have been used for centuries to convert wind energy into mechanical energy, which can then be further converted into electrical energy using generators. This method is particularly useful in off-grid and remote areas [23].

vii. **Harvesting Energy from Vibrations and Motion**: Devices like self-winding watches use kinetic energy from motion to generate electricity. This concept is also being explored for larger scale applications, such as capturing energy from vehicle and pedestrian traffic or ocean waves.

Piezoelectric energy harvesting is a unique and versatile method that distinguishes itself from other techniques in several ways [24]. These are:

i. **Versatility**: Piezoelectric materials can harvest energy from a wide range of mechanical vibrations, including those with low frequencies and small amplitudes. This versatility allows them to be used in various applications where other methods may not be as effective.

ii. **Compactness**: Piezoelectric generators can be small and lightweight, making them suitable for integration into small electronic devices or wearable technology.

iii. **Durability**: Piezoelectric materials are known for their long lifespan and robustness. They can withstand harsh environmental conditions, making them suitable for applications in remote or extreme environments.

iv. **Low Maintenance**: Piezoelectric systems often require minimal maintenance compared to other energy harvesting methods like wind turbines or solar panels, which may need periodic cleaning and upkeep.

However, piezoelectric energy harvesting also has limitations, such as lower efficiency compared to some other methods, particularly in capturing small amounts of energy from low-frequency vibrations [25]. Additionally, the choice of the right piezoelectric material and the design of the harvesting system can be challenging and require careful optimization to achieve the desired performance.

9.5.2 Conversion Efficiency and Limitations

The conversion efficiency of energy harvesting mechanisms is a critical factor that determines their practicality and effectiveness. Efficiency refers to the ratio of the usable electrical energy output to the total energy input from the source. Here are some key factors influencing conversion efficiency and limitations [26]:

i. **Energy Source Variability**: The availability and consistency of the energy source affect efficiency. Solar panels, for instance, are highly dependent on sunlight intensity and duration.
ii. **Material Properties**: The choice of materials, such as the type of piezoelectric material or the semiconductor in solar cells, greatly impacts efficiency. Materials with high conversion coefficients and minimal energy losses are desirable.
iii. **Conversion Mechanism**: The specific mechanism used for energy conversion plays a crucial role. For example, thermoelectric generators are limited by the temperature gradient, while piezoelectric generators are limited by mechanical vibrations' frequency and amplitude.
iv. **Environmental Factors**: Environmental conditions, such as temperature, humidity, and air quality, can affect energy harvesting systems' performance and lifespan.
v. **Energy Storage**: Efficient energy harvesting often requires energy storage solutions, such as batteries or supercapacitors, to store and manage the harvested energy. The efficiency of these storage systems can impact the overall energy harvesting system efficiency.
vi. **Cost**: The cost of materials, manufacturing, and maintenance can be a limitation, especially for large-scale energy harvesting projects.
vii. **Scaling**: Scaling up energy harvesting systems for higher power requirements can pose challenges. For instance, capturing enough energy to power an entire building with piezoelectric generators may not be feasible due to the limitations in available mechanical vibrations.

By harnessing the power of energy harvesting mechanisms, we can pave the way toward a brighter and more sustainable future. These mechanisms provide a promising opportunity to achieve self-sufficiency in power sources while keeping our planet green and healthy. Each method has its advantages and limitations, and the choice of which to use depends on the specific application's requirements and constraints. Piezoelectric energy harvesting, while versatile and durable, is just one of many options available in the growing field of energy harvesting technologies [27]. Researchers and engineers continue to work on improving conversion efficiency and exploring innovative ways to harness energy from the environment, paving the way for a greener and more energy-efficient future.

9.6 Design and Fabrication of Piezoelectric Energy Harvesters

Piezoelectric energy harvesters are devices that convert mechanical vibrations or movements into electrical energy using piezoelectric materials. These energy harvesters have gained significant attention due to their ability to scavenge energy from various sources, such as vibrations from machinery, footsteps, and even environmental vibrations. In this discussion, we will explore the key components, design considerations, and fabrication techniques associated with piezoelectric energy harvesters [28].

9.7 Overview of Components

9.7.1 Piezoelectric Elements

Piezoelectric Materials: These are the core components of piezoelectric energy harvesters. Common materials include lead zirconate titanate (PZT), polyvinylidene fluoride (PVDF), and various composites. These materials exhibit the piezoelectric effect, which generates an electric charge when subjected to mechanical stress or deformation [29].

Piezoelectric Transducers: These are the actual piezoelectric elements used to convert mechanical energy into electrical energy. They come in various shapes and sizes, including plates, beams, and cantilevers.

9.7.2 Electronics

Rectifier Circuit: It converts the alternating current (AC) output generated by the piezoelectric element into direct current (DC) to charge a battery or power an electronic device.

Voltage Conditioning Circuit: It stabilizes and conditions the harvested voltage to ensure a consistent and usable output.

Energy Storage: In many applications, an energy storage component like a supercapacitor or a rechargeable battery is included to store and manage the harvested energy [30].

9.7.3 Mechanical Housing and Mounting

Enclosure: It protects the sensitive components from environmental factors, such as moisture, dust, and physical damage.

Mounting Mechanism: It securely attaches the harvester to the source of mechanical vibrations [31].

9.7.4 Design Considerations

i. **Resonance Frequency:** The resonance frequency of a piezoelectric harvester should match the frequency of the ambient mechanical vibrations. This maximizes energy conversion efficiency. Tuning the resonance frequency can be achieved by adjusting the size, shape, or mass of the piezoelectric element [32].
ii. **Geometry:** The geometry of the piezoelectric element plays a crucial role in energy harvesting efficiency.
iii. **Cantilever Beam**: A simple and commonly used design for harvesting energy from vibrations.
iv. **Bimorph or Multilayered Configuration**: Stacking multiple piezoelectric layers to increase voltage output.
v. **Flexural Mode**: Utilizing bending or flexing of the material to generate strain and harvest energy.
vi. **Load Matching:** Proper load matching ensures that the impedance of the harvesting circuit matches that of the piezoelectric element, maximizing power transfer efficiency.

vii. **Environmental Factors:** Environmental conditions, such as temperature, humidity, and mechanical stress, should be considered in the design to ensure the harvester's durability and reliability.

viii. **Scalability:** Depending on the application, designers may need to consider the scalability of the energy harvester. Some applications may require small, low-power devices, while others may involve larger systems.

9.7.5 Fabrication Techniques and Processes

The fabrication of piezoelectric energy harvesters involves several steps and processes [33]:

a. **Material Selection:** The choice of piezoelectric material depends on the specific application and required performance characteristics. Common materials like PZT and PVDF are commercially available.

b. **Piezoelectric Element Fabrication:** Fabrication techniques include cutting, machining, or depositing the piezoelectric material into the desired shape and size. Techniques like dicing, laser cutting, or photolithography may be used.

c. **Electrical Contacts:** Electrodes are attached to the piezoelectric material to allow for the collection of generated charges. The choice of electrodes and their proper attachment is crucial.

d. **Integration:** The piezoelectric element, electronics, and energy storage components are integrated into a compact housing, taking into account factors like size, weight, and environmental protection.

e. **Testing and Calibration:** The fabricated harvester undergoes testing to determine its resonance frequency, voltage output, and efficiency. Calibration may be necessary to optimize performance.

f. **Mounting:** The harvester is securely mounted to the source of mechanical vibrations. This may involve adhesive mounting, clamping, or other attachment methods.

g. **Performance Optimization:** After installation, the harvester's performance may be further optimized by adjusting parameters such as the load resistance and the mechanical coupling to the vibration source.

To achieve optimal results, it is crucial to approach the design and fabrication of piezoelectric energy harvesters with a multidisciplinary mindset that encompasses materials science, mechanical engineering, electronics, and energy systems [34]. The choice of components, geometry, and design considerations should be carefully evaluated to ensure optimal energy harvesting efficiency for the targeted application. As technology continues to advance, piezoelectric energy harvesters hold promise for powering low-energy electronic devices, and sensors, and even contributing to sustainable energy solutions in various industries.

9.8 Modeling and Simulation

Modeling and simulation are indispensable tools in the field of piezoelectric energy harvesting. They allow researchers and engineers to understand, analyze, and optimize the performance of energy harvesters without the need for costly and time-consuming physical prototypes. In this discussion, we will explore mathematical models for piezoelectric energy harvesting, the use of FEA, and provide case studies and practical applications that highlight the significance of modeling and simulation in this field.

9.8.1 Mathematical Models for Piezoelectric Energy Harvesting

Mathematical models are essential for describing the behavior of piezoelectric energy harvesters and predicting their performance under different conditions. These models typically involve equations that relate mechanical, electrical, and material properties.

9.8.2 Constitutive Equations

The piezoelectric effect equations are fundamental mathematical expressions that describe the underlying physical phenomenon in piezoelectric materials, which is the relationship between mechanical stress and the resulting electric field, or vice versa. Piezoelectric materials are characterized by their unique ability to generate an electric charge when subjected to mechanical deformation (stress) and, conversely, to undergo mechanical deformation when subjected to an applied electric field [35]. These equations play a central role in understanding and quantifying this phenomenon. Let's delve deeper into these equations and the associated coefficients.

9.8.3 Piezoelectric Effect Equations

9.8.3.1 Linear piezoelectric effect (d_{31})

The linear piezoelectric effect relates the induced electric charge (polarization) in a piezoelectric material to an applied mechanical stress. It is described by the equation [36]:

$$P_i = d_{ij} T_j \tag{9.1}$$

The *i*th direction is denoted by P_i, which signifies the polarization or electric charge.

The symbol "d_{ij}" represents the piezoelectric coefficient, which measures the extent of the connection between mechanical stress and electrical charge. For example, d_{31} relates the polarization in the direction perpendicular to the plane of the stress applied in the three-direction.

9.8.3.2 Nonlinear piezoelectric effect (d_{33})

The nonlinear piezoelectric effect takes into account higher order terms and describes how the polarization depends on both linear and quadratic mechanical stress terms. It is represented as [37]:

$$P_i = d + d_{ijkl} T_j T_k \tag{2}$$

In addition to the linear coefficient d_{ij}, this equation includes the quadratic piezoelectric coefficient d_{ijkl}, which describes the nonlinear coupling between mechanical stresses T_j and T_k in the *j*th and *k*th directions, respectively, and the resulting polarization P_i in the *i*th direction.

For example, d_{3333} would represent the nonlinear coupling between stress in the three-direction and the resulting polarization in the three-direction.

9.8.3.3 Significance of piezoelectric coefficients

i. **d_{31} Coefficient**: The linear coefficient d_{31} is particularly important in applications where a single direction of polarization is desired, such as in piezoelectric sensors and actuators. When mechanical stress is applied in a certain direction, this coefficient quantifies the resulting electric charge in a direction perpendicular to the stress [38].

ii. **d_{33} Coefficient**: The nonlinear coefficient d_{33} becomes significant when higher order effects need to be considered. This is particularly relevant in applications where mechanical stress is significant enough to cause nonlinear behavior, such as high-intensity ultrasound transducers and high-power piezoelectric devices.

iii. **Material Characterization**: Accurate determination of these coefficients is essential for characterizing the behavior of piezoelectric materials. Experimental methods are often used to measure these coefficients, enabling engineers to select the most suitable materials for specific applications.

iv. **Device Design**: Engineers use these coefficients to design piezoelectric devices with desired performance characteristics. Depending on the application, they can choose materials and geometries that maximize the relevant piezoelectric coefficients to achieve the desired effect [39].

v. **Energy Harvesting**: In the context of energy harvesting, understanding these coefficients is critical to optimize the design of piezoelectric energy harvesters. By maximizing the relevant coefficients, engineers can enhance the efficiency of energy conversion from mechanical vibrations into electrical energy.

The piezoelectric effect equations, which involve piezoelectric coefficients like d_{31} and d_{33}, are essential for understanding and quantifying the relationship between mechanical stress and electric polarization in piezoelectric materials. These coefficients have wide-ranging applications in various fields, from sensors and actuators to energy harvesting and ultrasound technology. Their precise determination and use are key factors in the successful design and implementation of piezoelectric devices [40].

9.8.3.4 Mechanical equations

Mechanical equations play a critical role in understanding and characterizing the behavior of piezoelectric energy harvesters, especially in the context of FEA and simulations. These equations provide insights into how the harvester's structure responds to external mechanical forces or vibrations. Here, we'll expound on the components and significance of mechanical equations in the context of piezoelectric energy harvesting [41].

9.8.3.5 Displacement equations

Displacement equations describe the movement or deformation of various parts of the energy harvester. They help quantify how the harvester's components respond to mechanical input. Key aspects include the following [42]:

a. **Deformation**: These equations account for the deformation or displacement of the piezoelectric material when subjected to mechanical forces. They are crucial for understanding how the material's shape changes when mechanical vibrations are applied.
b. **Strain**: Strain is a measure of how much a material deforms under mechanical load. Strain equations describe the relationship between deformation and the original size and shape of the material. In piezoelectric materials, strain is directly related to the generation of electrical charge through the piezoelectric effect.

9.8.3.6 Stress equations

Stress equations describe how forces are distributed within the harvester's structure. Stress is a measure of internal resistance to deformation caused by external forces. Key aspects include the following [43]:

a. **Mechanical Stress**: These equations quantify the distribution of mechanical forces within the piezoelectric material or other structural components of the harvester. Understanding stress is crucial for ensuring the material does not exceed its mechanical limits, which can lead to failure or reduced efficiency.

b. **Stress Concentration**: Some parts of the harvester may experience higher stress levels than others. Stress concentration equations help identify these areas and are essential for designing robust and reliable harvesters.

9.8.3.7 Stiffness equations

Stiffness equations describe the resistance of a structure to deformation under mechanical loads. Stiffness is a crucial factor in understanding the harvester's response to vibrations and optimizing its design. Key aspects include the following [44]:

a. **Stiffness Matrix**: In FEA, stiffness matrices are used to represent the relationship between applied forces and resulting displacements. They describe how the harvester's components interact mechanically, allowing for the calculation of deformation and strain throughout the structure.
b. **Eigenvalues and Eigenmodes**: Eigenvalue analysis helps determine the natural frequencies of the harvester, which is vital for ensuring that the harvester operates efficiently at the desired resonant frequency, maximizing energy conversion.

9.8.3.8 Significance of mechanical equations in piezoelectric energy harvesting

Mechanical equations are essential for several reasons [44]:

i. **Design Optimization**: Mechanical equations are used to assess the mechanical integrity of the harvester's components, ensuring they can withstand the expected mechanical stresses without deformation or damage.
ii. **Resonance Analysis**: Understanding the natural frequencies and mode shapes of the harvester through mechanical equations is crucial for optimizing the device's performance. This helps ensure that the harvester operates efficiently at its resonant frequency, where it can harvest the most energy from mechanical vibrations.
iii. **Failure Prediction**: Mechanical equations can help predict areas of high-stress concentration or potential points of failure in the harvester's design. This information is vital for improving the device's durability and reliability.

iv. **Efficiency Improvement**: By analyzing the mechanical behavior of the harvester, designers can make informed decisions about the material choices, geometry, and structural design to maximize energy conversion efficiency.

Mechanical equations are an integral part of the modeling and simulation process for piezoelectric energy harvesters. They provide a comprehensive understanding of how the harvester's structure responds to mechanical inputs, guiding the design and optimization process to create efficient and reliable energy harvesting systems.

9.8.3.9 Circuit equations

Electrical Circuit Equations: These equations model the electrical behavior of the harvester, taking into account electrical components like resistors, capacitors, and inductors.

Voltage Generation Equation: It relates the mechanical input (e.g., force or displacement) to the voltage generated by the piezoelectric element.

Power Generation Equation: It describes the rate at which electrical energy is generated and can be calculated as the product of voltage and current [45].

9.8.3.10 Optimization models

Mathematical optimization techniques, such as linear programming or genetic algorithms, can be employed to maximize energy harvesting efficiency by adjusting various design parameters.

9.8.3.11 Finite element analysis and simulation tools

Finite element analysis (FEA) is a powerful numerical technique widely used for simulating the behavior of piezoelectric energy harvesters. FEA breaks down complex structures into smaller finite elements, allowing for detailed analysis of stress, strain, and electrical behavior. Key aspects of FEA in energy harvesting modeling include the following [46]:

i. **Geometry Modeling:** FEA tools enable designers to create accurate 3D models of the energy harvester's geometry, including the piezoelectric material, mechanical supports, and electrodes.

ii. **Material Properties:** Material properties, such as the piezoelectric coefficients and Young's modulus, are defined in FEA simulations.

iii. **Boundary Conditions:** Boundary conditions, such as fixed or constrained nodes, are set to replicate the physical environment and mechanical loading.

iv. **Piezoelectric Coupling:** FEA software can incorporate piezoelectric constitutive equations to account for the interaction between mechanical and electrical behavior.

v. **Response Analysis:** FEA simulations provide insights into the harvester's response to mechanical vibrations, including stress distribution, strain levels, and voltage generation.

vi. **Parametric Studies:** Design variations can be explored by changing parameters in the simulation, such as the size, shape, or material properties of the harvester.

Common FEA software packages used for piezoelectric energy harvesting simulations include ANSYS, COMSOL Multiphysics, and Abaqus.

9.9 Practical Applications

Modeling and simulation have played a crucial role in the development and optimization of piezoelectric energy harvesters across various applications [47].

1. Structural Health Monitoring

Piezoelectric energy harvesters have been used to power sensors that monitor the structural integrity of bridges, buildings, and aircraft. FEA simulations help optimize harvester placement and performance for specific monitoring systems.

2. Wearable Electronics

Energy harvesters integrated into wearable devices, such as smartwatches and fitness trackers, benefit from simulations to ensure they can generate sufficient power from body movements.

3. Wireless Sensor Networks

Energy harvesters power remote sensors in applications like environmental monitoring and industrial automation. Simulation aids in designing harvester-sensor systems for efficient, long-term operation.

4. **Internet of Things (IoT)**
IoT devices often rely on energy harvesting to eliminate the need for battery replacement. Modeling and simulation help design energy-efficient harvesters for IoT applications.

5. **Automotive Applications**
Piezoelectric energy harvesters in vehicles can scavenge energy from vibrations and convert it into electrical power for sensor systems. Simulation ensures optimal performance under various driving conditions.

6. **Human Motion Energy Harvesting**
Piezoelectric shoe insoles and flooring systems capture energy from footsteps. Simulations assist in optimizing harvester geometries and materials for enhanced energy conversion. In these case studies, modeling and simulation have been instrumental in reducing development costs, shortening design cycles, and improving the overall efficiency of piezoelectric energy harvesting systems. They enable engineers to explore a wide range of design possibilities and assess performance parameters before physical prototypes are built, ultimately accelerating innovation in this sustainable energy technology [48].

9.10 Performance Optimization in Piezoelectric Energy Harvesting

Performance optimization is a crucial aspect of designing piezoelectric energy harvesters to ensure they efficiently convert mechanical vibrations into electrical energy. This process involves various techniques and considerations aimed at maximizing energy harvesting efficiency [49]. Here, we'll delve into techniques for optimizing the performance of piezoelectric energy harvesters, including load matching, impedance matching, and the use of multimodal and multilayer harvesters.

9.10.1 Techniques for Maximizing Energy Harvesting Efficiency

1. Resonance Tuning

Resonance tuning involves adjusting the harvester's natural frequency to match the frequency of the ambient mechanical vibrations. This ensures that the harvester operates at its highest efficiency. Design parameters like the geometry, mass distribution, and stiffness of the harvester can be modified to achieve resonance at the desired frequency [50].

2. Material Selection

Choosing the right piezoelectric material with high coupling coefficients is critical for maximizing energy conversion efficiency. Materials like lead zirconate titanate (PZT) are known for their excellent piezoelectric properties.

3. Geometry Optimization

The geometry of the piezoelectric element plays a significant role in energy harvesting efficiency. Different shapes, such as cantilever beams, bimorphs, or flexural structures, may be optimized to enhance mechanical strain and energy output [51].

4. Electrical Circuit Design

Efficient electrical circuitry is essential for energy harvesting. Techniques like impedance matching and voltage regulation circuits can be employed to ensure that the harvested energy is efficiently transferred and stored.

5. Multimodal Operation

Harvesters can be designed to capture energy from multiple sources or vibration modes, such as bending and torsional vibrations. Multimodal harvesters broaden the range of mechanical inputs they can utilize, increasing overall efficiency.

6. Multilayer Harvesters

Multilayer piezoelectric structures stack multiple layers of piezoelectric material with varying properties, such as thickness or composition. This approach enhances voltage output and mechanical flexibility, allowing for better adaptation to various operating conditions.

7. Vibration Damping
To avoid excessive strain and mechanical stress on the harvester, vibration-damping materials or mechanisms can be integrated to reduce the risk of material fatigue and failure.

9.10.1.1 Load matching and impedance matching

9.10.1.1.1 *Load matching*

Load matching involves selecting an appropriate load resistance or impedance that matches the output impedance of the piezoelectric harvester. This ensures maximum power transfer from the harvester to the load. Mismatched loads can result in energy losses and reduced efficiency.

9.10.1.1.2 *Impedance matching*

Impedance matching is an electrical technique that optimizes the electrical load seen by the piezoelectric element. This is typically achieved by using electrical components such as transformers or matching networks. Impedance matching ensures that the electrical output of the harvester is efficiently transferred to the load, minimizing reflected energy and maximizing power transfer [52].

9.10.2 Multimodal and Multilayer Harvesters

1. Multimodal Harvesters
Multimodal harvesters are designed to capture energy from multiple modes of mechanical vibration. For example, a harvester may be engineered to capture both bending and torsional vibrations simultaneously. This approach broadens the range of mechanical inputs that can be harvested and increases overall energy capture.

2. Multilayer Harvesters
Multilayer harvesters stack multiple piezoelectric layers, often with varying properties such as thickness or composition. This configuration enhances the voltage output of the harvester. The layers can be arranged in series or parallel to optimize energy conversion for specific operating conditions. Additionally, multilayer structures provide mechanical flexibility, allowing them to adapt to a wider range of mechanical vibrations [53].

Performance optimization is a critical aspect of designing efficient piezoelectric energy harvesters. Techniques like resonance tuning, material selection, geometry optimization, load matching, impedance matching, and the use of multimodal and multilayer harvesters all contribute to maximizing energy harvesting efficiency. By carefully considering these factors and tailoring the harvester's design to its intended application, engineers can develop energy harvesting systems that effectively harness mechanical vibrations to generate sustainable electrical power [54].

9.11 Challenges and Future Trends in Piezoelectric Energy Harvesting

Piezoelectric energy harvesting has made significant advancements in recent years, but it still faces several challenges while also offering promising prospects. In this discussion, we will explore the current challenges, emerging technologies, and future trends in the field of piezoelectric energy harvesting.

9.11.1 Current Challenges in Piezoelectric Energy Harvesting

1. Energy Efficiency

One of the primary challenges is improving the energy conversion efficiency of piezoelectric energy harvesters, especially for low-frequency and low-amplitude vibrations. Maximizing power output while maintaining a compact form factor is a complex task [55].

2. Frequency Adaptability

Many piezoelectric harvesters are designed for specific resonant frequencies, making them less adaptable to variable or unpredictable mechanical vibrations. Developing adaptable systems that can operate effectively across a range of frequencies is a challenge.

3. Scalability

Scaling up piezoelectric energy harvesting systems to meet the power demands of larger applications, such as buildings or industrial equipment, remains a challenge. Large-scale piezoelectric harvesters often face issues related to mechanical coupling and structural integrity.

4. Environmental Factors

Harsh environmental conditions, including temperature variations, humidity, and mechanical stress, can negatively impact the long-term performance and durability of piezoelectric harvesters. Robust and reliable designs are needed to overcome these challenges.

5. Energy Storage and Management

Efficient energy storage and management systems are essential for storing and utilizing harvested energy effectively. Optimizing the integration of batteries, supercapacitors, or other energy storage devices with piezoelectric harvesters is an ongoing challenge.

6. Materials and Sustainability

Developing sustainable and environmentally friendly piezoelectric materials is a priority. Many traditional piezoelectric materials contain toxic elements like lead, which poses environmental and health concerns.

7. Cost

Reducing the cost of production and implementation is crucial for making piezoelectric energy harvesting economically viable for a wider range of applications.

9.11.2 Emerging Technologies and Innovations

1. Flexible and Wearable Devices: Advances in flexible and wearable electronics have led to the development of wearable piezoelectric energy harvesters that can harness energy from human motion. These devices have applications in fitness trackers, smart clothing, and medical devices [56].

2. Triboelectric Nanogenerators (TENGs): TENGs are a complementary technology to piezoelectric harvesters. They generate electricity through the triboelectric effect, which occurs when two materials come into contact and generate an electrostatic charge. TENGs can be integrated with piezoelectric systems for enhanced energy harvesting.

3. Smart Materials: Smart materials, such as shape memory alloys and magnetostrictive materials, are being explored alongside piezoelectric materials to create hybrid energy harvesting systems that capture energy from multiple sources, including mechanical, thermal, and magnetic [57].

4. IoT Integration: Piezoelectric energy harvesting is gaining prominence in IoT applications, where low-power, self-sustaining devices are essential. These harvesters can power remote sensors and communication modules, reducing the need for frequent battery replacements [58].

9.12 Prospects and Trends in the Field

1. Energy Harvesting Ecosystems: Future trends involve the development of integrated energy harvesting ecosystems, where multiple energy sources, including solar, thermal, and piezoelectric, work together to provide continuous and sustainable power for various applications.

2. Energy-Aware Design: Energy-aware design principles will become more prevalent in product development. Engineers will focus on designing products with energy harvesting capabilities from the outset, enabling greater energy autonomy.

3. Enhanced Materials: Research into novel piezoelectric and other smart materials will continue, with an emphasis on environmentally friendly, high-performance materials.

4. Artificial Intelligence and Control Systems: AI and advanced control systems will be used to optimize energy harvesting in real-time by adjusting parameters such as load matching, resonance tuning, and energy storage based on environmental conditions and usage patterns.

5. Sustainable Energy Solutions: Piezoelectric energy harvesting will play a role in sustainable energy solutions, including self-powered environmental monitoring, infrastructure health monitoring, and energy-efficient IoT deployments.

Piezoelectric energy harvesting holds great potential for addressing energy sustainability challenges. While current challenges exist, ongoing research and innovation are poised to overcome these obstacles and drive the field toward greater efficiency, adaptability, and widespread adoption in various applications. The future of piezoelectric energy harvesting is likely to be characterized by integration, sustainability, and improved energy autonomy in an increasingly interconnected and energy-aware world.

9.13 Integration and Commercialization of Piezoelectric Energy Harvesting

Piezoelectric energy harvesting technology has been rapidly advancing, with integration into various systems and commercial applications becoming more prevalent. In this discussion, we will explore the integration of piezoelectric energy harvesting in different systems, market trends, commercial viability, and intellectual property considerations.

9.13.1 Integration of Piezoelectric Energy Harvesting in Various Systems

1. **Structural Health Monitoring (SHM)**

Bridges and Buildings: Piezoelectric sensors and energy harvesters are integrated into structures to monitor their health. Vibrations and strains in the structure can be converted into electrical energy and used to power sensors that assess structural integrity.

2. **Wearable Electronics**

Smartwatches and Fitness Trackers: Piezoelectric energy harvesters can capture energy from users' movements, such as wrist motions or footsteps, to extend battery life or even eliminate the need for regular charging.

Implantable Medical Devices: Some medical implants, like pacemakers or glucose monitors, can benefit from piezoelectric energy harvesters, reducing the need for battery replacement surgeries.

3. **Internet of Things (IoT)**

Sensors and Remote Devices: Low-power IoT sensors and devices often employ piezoelectric energy harvesting to operate without the need for battery replacement in remote or inaccessible locations.

4. **Automotive Industry**

Tire Pressure Monitoring: Piezoelectric sensors in tires can monitor pressure and send data wirelessly, powered by energy harvested from tire vibrations.

Shock Absorbers: Piezoelectric devices can capture energy from vehicle vibrations and shocks to supplement the vehicle's electrical system.

5. Footfall Energy Harvesting

Flooring and Pavements: Energy can be harvested from the footsteps of pedestrians and converted into electrical power for lighting, sensors, or other urban infrastructure needs.

9.14 Market Trends and Commercial Viability

Growth of IoT: The increasing deployment of IoT devices in various industries is driving the demand for energy harvesting solutions, including piezoelectric technology, to reduce maintenance costs and extend the operational life of these devices.

Green Energy and Sustainability: Growing awareness of environmental sustainability is pushing industries to adopt energy harvesting technologies, including piezoelectric, as a clean and renewable energy source.

Self-Powered Sensors: The need for autonomous sensors in applications like environmental monitoring, industrial automation, and smart cities is creating significant market opportunities for piezoelectric energy harvesting.

Consumer Electronics: The consumer electronics market, particularly wearable devices, is a promising avenue for piezoelectric energy harvesting. Consumers appreciate the convenience of devices that don't require frequent charging.

Research and Development: Continued research and development efforts are focused on improving energy harvesting efficiency, scalability, and the integration of piezoelectric devices into new applications.

9.15 Intellectual Property and Patent Considerations

As piezoelectric energy harvesting technologies advance, intellectual property and patents play a vital role in protecting innovations and ensuring commercial success. Here are some key considerations:

1. **Patent Protection:** Companies and researchers often seek patent protection for novel piezoelectric energy harvesting technologies, including unique designs, materials, and integration methods.

2. **Patent Landscape:** Understanding the existing patent landscape is crucial to avoiding infringement and identifying areas of opportunity for innovation.

3. **Licensing:** Licensing agreements allow companies to use patented technologies developed by others while respecting intellectual property rights. This can facilitate the commercialization of energy harvesting solutions.

4. **Open Source:** Some initiatives and organizations promote open-source hardware and software for energy harvesting, which can encourage collaboration and innovation while providing access to technology for a wider audience.

5. **International Considerations:** Intellectual property laws and regulations vary by country, so businesses involved in international commercialization should consider patent protection in multiple jurisdictions.

Piezoelectric energy harvesting is a technology that is gaining more and more significance in today's world, with diverse applications in commercial settings. Its potential for generating electricity from mechanical pressure has opened up new avenues for sustainable energy production and has attracted significant attention from industries across the globe. As market demand continues to grow and technological advancements improve efficiency, the integration of piezoelectric energy harvesting into various systems is set to expand. Intellectual property considerations will play a significant role in protecting innovations and ensuring the commercial viability of these energy-harvesting solutions.

9.16 Environmental and Sustainability Aspects of Piezoelectric Energy Harvesting

Piezoelectric energy harvesting is gaining attention not only for its potential to generate electricity from mechanical vibrations but also for its positive environmental and sustainability impacts. In this discussion, we will explore the environmental impact assessment of piezoelectric energy harvesting, the sustainability benefits it offers, and the role of green energy initiatives and regulations.

9.16.1 Environmental Impact Assessment

Environmental Impact Assessment is a process of evaluating the potential consequences of a proposed project or development on the environment. It is conducted to identify any potential negative effects on the environment and to determine ways to mitigate them. The assessment takes into account various factors such as air and water quality, biodiversity, soil health, and noise pollution. The results of the assessment are used to inform decision-making and ensure that any development is done in an environmentally responsible manner [59].

1. **Reduced Carbon Footprint:** Piezoelectric energy harvesting generates electricity without combustion or emissions, contributing to a reduction in greenhouse gas emissions. This is particularly relevant in applications where it replaces batteries or grid-powered sources [60].

2. **Lower Resource Consumption:** Energy harvesting systems, including piezoelectric devices, can extend the operational life of electronic devices and sensors by eliminating or reducing the need for disposable batteries. This reduces the consumption of raw materials used in battery manufacturing and disposal.

3. **Decreased E-Waste:** By enabling longer lasting and self-sustaining electronics, piezoelectric energy harvesting helps reduce electronic waste (e-waste) generated from the disposal of batteries and electronic devices [61].

4. **Sustainable Manufacturing:** The development of environmentally friendly and lead-free piezoelectric materials contributes to more sustainable manufacturing practices in the energy harvesting industry.

5. **Energy Efficiency:** Piezoelectric energy harvesting enhances energy efficiency by capturing wasted energy from mechanical vibrations and converting it into usable electricity.

9.16.2 Sustainability Benefits of Piezoelectric Energy Harvesting

Piezoelectric energy harvesting is a game-changing technology that has the potential to revolutionize the energy industry. With its ability to convert mechanical energy into electrical energy, it offers a clean and reliable source of power. By harnessing this technology, we can achieve sustainable energy solutions that will benefit both the environment and the economy.

1. **Renewable Energy Source:** Piezoelectric energy harvesting is considered a renewable energy source because it relies on mechanical vibrations from sources such as footsteps, machinery, or structural movements, which are continuously available.

2. **Energy Independence:** Piezoelectric harvesters can power remote and off-grid sensors and devices, reducing reliance on traditional energy sources and extending the operational life of equipment in remote locations.

3. **Reduced Environmental Impact:** By reducing the need for battery replacement and disposal, piezoelectric energy harvesting contributes to lower environmental impact and waste management costs.

4. **Enhanced Sustainability:** The sustainable operation of sensors and IoT devices powered by piezoelectric harvesters supports a more sustainable approach to environmental monitoring, industrial automation, and smart cities.

9.17 Contribution to Green Buildings

Integrating piezoelectric energy harvesting into building infrastructure, such as flooring systems, can contribute to green building certifications by reducing energy consumption and supporting energy-efficient designs.

9.17.1 Green Energy Initiatives and Regulations

1. **Renewable Energy Mandates**
Many regions and countries have implemented renewable energy mandates and targets, encouraging the adoption of technologies like piezoelectric energy harvesting to meet sustainability goals.
2. **Energy Efficiency Standards**
Regulations and standards that promote energy efficiency in buildings and devices can drive the adoption of piezoelectric energy harvesting as a sustainable energy source.
3. **Sustainable Development Goals (SDGs)**
Piezoelectric energy harvesting aligns with various United Nations Sustainable Development Goals, including affordable and clean energy (SDG 7), responsible consumption and production (SDG 12), and climate action (SDG 13).
4. **Green Building Certifications**
Green building certification programs, such as LEED (Leadership in Energy and Environmental Design), recognize and reward the integration of energy harvesting technologies into building designs.
5. **Funding and Incentives**
Governments and organizations may offer financial incentives, grants, or tax credits to support the development and deployment of sustainable energy technologies, including piezoelectric energy harvesting.

Generally, piezoelectric energy harvesting presents an environmentally friendly and sustainable approach to generating electricity from mechanical vibrations [62]. Its positive impacts include reduced carbon emissions, extended device lifespans, and decreased e-waste. Green energy initiatives and regulations are driving the adoption of this technology in various applications, contributing to a more sustainable and environmentally conscious future.

9.18 Conclusions

Our exploration of piezoelectric energy harvesting reveals a promising technology with diverse applications and significant sustainability benefits. As we conclude, let's summarize key takeaways, reflect on its role in the energy landscape, and consider prospects.

Key Takeaways

i. Piezoelectric energy harvesting efficiently converts mechanical vibrations into electricity, enabling a wide range of applications.
ii. Material selection, design considerations, and optimization techniques are essential for maximizing performance.
iii. This technology contributes to sustainability by reducing emissions, extending device lifespans, and minimizing electronic waste.
iv. Emerging innovations, such as flexible devices and smart materials, expand their potential.

Role in the Energy Landscape

a. Piezoelectric energy harvesting aligns with global sustainability goals as a clean and renewable energy source.
b. It empowers energy autonomy for connected devices and promotes resource efficiency.
c. Integration into green building designs supports sustainability in urban environments.

Future Outlook

i. Ongoing research aims to enhance efficiency and expand its viability.
ii. Energy harvesting ecosystems will integrate multiple sources for continuous, sustainable power.
iii. Energy-aware designs will prioritize energy autonomy in various applications.
iv. Sustainable materials and regulatory support will drive adoption.
v. Green energy initiatives will further boost its role in cleaning for a more sustainable world.

As we conclude, the concept of piezoelectric energy harvesting offers a promising future for sustainable energy. It aligns with our commitment to a greener world and addresses the increasing energy requirements. The continuous growth and impact of this technology on the energy landscape instill a sense of hope for a cleaner and more sustainable future.

References

1. T. Li and P. S. Lee, "Piezoelectric energy harvesting technology: From materials, structures, to applications," *Small Struct.*, vol. 3, no. 3, 2022, doi: 10.1002/sstr.202100128.
2. N. Sezer and M. Koç, "A comprehensive review on the state-of-the-art of piezoelectric energy harvesting," *Nano Energy*, vol. 80, no. August 2020, p. 105567, 2021, doi: 10.1016/j.nanoen.2020.105567.
3. S. Bairagi, Shahid-ul-Islam, M. Shahadat, D. M. Mulvihill, and W. Ali, "Mechanical energy harvesting and self-powered electronic applications of textile-based piezoelectric nanogenerators: A systematic review," *Nano Energy*, vol. 111, no. April, p. 108414, 2023, doi: 10.1016/j.nanoen.2023.108414.
4. M. Hasani, M. Khazaee, J. E. Huber, L. Rosendahl, and A. Rezania, "Design and analytical evaluation of an impact-based four-point bending configuration for piezoelectric energy harvesting," *Appl. Energy*, vol. 347, no. June, p. 121461, 2023, doi: 10.1016/j.apenergy.2023.121461.
5. S. Chen, W. Chen, R. Yang, and C. Wu, "A piezoelectric wave energy harvester equipped with a sequential-drive rotating mechanism and rotary piezoelectric harvesting component," *Energy Convers. Manag. X*, vol. 20, no. May, p. 100463, 2023, doi: 10.1016/j.ecmx.2023.100463.
6. E. Hassan, S. A. Kouritem, F. Z. Amer, and R. I. Mubarak, "Acoustic energy harvesting using an array of piezoelectric cantilever plates for railways and highways environmental noise," *Ain Shams Eng. J.*, no. August, p. 102461, 2023, doi: 10.1016/j.asej.2023.102461.
7. D. Man et al., "Improving energy harvesting from low-frequency excitations by a hybrid tri-stable piezoelectric energy harvester," *Alexandria Eng. J.*, vol. 76, pp. 153–165, 2023, doi: 10.1016/j.aej.2023.06.031.
8. B. Zhao et al., "A graded metamaterial for broadband and high-capability piezoelectric energy harvesting," *Energy Convers. Manag.*, vol. 269, no. August, p. 116056, 2022, doi: 10.1016/j.enconman.2022.116056.
9. E. Hassan, S. A. Kouritem, F. Z. Amer, and R. I. Mubarak, "Acoustic energy harvesting using an array of piezoelectric cantilever plates for railways and highways environmental noise," *Ain Shams Eng. J.*, no. November 2022, p. 102461, 2023, doi: 10.1016/j.asej.2023.102461.
10. F. Craciun et al., "Flexible composite films with enhanced piezoelectric properties for energy harvesting and wireless ultrasound-powered technology," *Compos. Part B Eng.*, vol. 263, no. May, p. 110835, 2023, doi: 10.1016/j.compositesb.2023.110835.

11. M. Hafizh et al., "Vortex induced vibration energy harvesting using magnetically coupled broadband circular-array piezoelectric patch: Modelling, parametric study, and experiments," *Energy Convers. Manag.*, vol. 276, no. September 2022, p. 116559, 2023, doi: 10.1016/j.enconman.2022.116559.

12. T. Hou, Y. Jiang, J. Wang, and Y. Ren, "Novel optimization strategy of a flow-induced piezoelectric vibration-based energy harvesting structure," *e-Prime - Adv. Electr. Eng. Electron. Energy*, vol. 5, no. March, p. 100227, 2023, doi: 10.1016/j.prime.2023.100227.

13. S. Salem and K. Fraňa, "Harvesting energy of flow-induced vibrations using cylindrical piezoelectric transducers," *Energy Reports*, vol. 9, pp. 279–285, 2023, doi: 10.1016/j.egyr.2022.12.113.

14. G. Selleri et al., "Energy harvesting and storage with ceramic piezoelectric transducers coupled with an ionic liquid-based supercapacitor," *J. Energy Storage*, vol. 60, no. January, p. 106660, 2023, doi: 10.1016/j.est.2023.106660.

15. Y. Wang, M. Hong, J. Venezuela, T. Liu, and M. Dargusch, "Expedient secondary functions of flexible piezoelectrics for biomedical energy harvesting," *Bioact. Mater.*, vol. 22, no. October 2022, pp. 291–311, 2023, doi: 10.1016/j.bioactmat.2022.10.003.

16. Y. Yu, C. Luo, H. Chiba, Y. Shi, and F. Narita, "Energy harvesting and wireless communication by carbon fiber-reinforced polymer-enhanced piezoelectric nanocomposites," *Nano Energy*, vol. 113, no. April, p. 108588, 2023, doi: 10.1016/j.nanoen.2023.108588.

17. N. Turdakyn, Z. Bekezhankyzy, S. Araby, R. Montazami, Z. Bakenov, and G. Kalimuldina, "Investigation of electrospun piezoelectric P(VDF-TrFE) nanofiber-based nanogenerators for energy harvesting," *Energy Reports*, vol. 10, pp. 628–636, 2023, doi: 10.1016/j.egyr.2023.07.010.

18. Y. Yu, Y. Shi, H. Kurita, Y. Jia, Z. Wang, and F. Narita, "Carbon fiber-reinforced piezoelectric nanocomposites: Design, fabrication and evaluation for damage detection and energy harvesting," *Compos. Part A Appl. Sci. Manuf.*, vol. 172, no. January, p. 107587, 2023, doi: 10.1016/j.compositesa.2023.107587.

19. K. Maruyama et al., "Electromechanical characterization and kinetic energy harvesting of piezoelectric nanocomposites reinforced with glass fibers," *Compos. Sci. Technol.*, vol. 223, no. March, p. 109408, 2022, doi: 10.1016/j.compscitech.2022.109408.

20. A. Moayedizadeh and D. Younesian, "Application of the meta-substrates for power amplification in rotary piezoelectric energy harvesting

systems: Design, fabrication and testing," *Energy Reports*, vol. 8, pp. 5653–5667, 2022, doi: 10.1016/j.egyr.2022.04.022.

21. Y. Ma, M. Liu, Y. Feng, H. Zheng, and Y. Wu, "ZnO@Ag modified piezoelectric fibers for higher sensitivity and enhanced energy harvesting," *J. Mater. Res. Technol.*, vol. 20, pp. 2689–2704, 2022, doi: 10.1016/j.jmrt.2022.08.067.
22. A. T. Hazmi, F. B. Ahmad, M. H. Maziati Akmal, A. A. Md Ralib, and F. Binti Ali, "Fungal chitosan for potential application in piezoelectric energy harvesting: Review on experimental procedure of chitosan extraction," *Alexandria Eng. J.*, vol. 67, pp. 105–116, 2023, doi: 10.1016/j.aej.2022.08.020.
23. M. Edla, M. Deguchi, and Y. Y. Lim, "A self-powered H-Bridge joule theory circuit for piezoelectric energy harvesting systems," *Power Electron. Devices Components*, vol. 3, no. March, p. 100015, 2022, doi: 10.1016/j.pedc.2022.100015.
24. H. Liang, G. Hao, and O. Z. Olszewski, "A review on vibration-based piezoelectric energy harvesting from the aspect of compliant mechanisms," *Sensors Actuators A Phys.*, vol. 331, p. 112743, 2021, doi: 10.1016/j.sna.2021.112743.
25. K. J. I. Egbe, A. Matin Nazar, P. Jiao, Y. Yang, X. Ye, and H. Wang, "Vibrational turbine piezoelectric nanogenerators for energy harvesting in multiphase flow fields," *Energy Reports*, vol. 7, pp. 6384–6393, 2021, doi: 10.1016/j.egyr.2021.09.085.
26. Z. Wang, K. Maruyama, and F. Narita, "A novel manufacturing method and structural design of functionally graded piezoelectric composites for energy-harvesting," *Mater. Des.*, vol. 214, p. 110371, 2022, doi: 10.1016/j.matdes.2021.110371.
27. J. X. Wang, J. C. Li, W. Bin Su, X. Zhao, and C. M. Wang, "A multi-folded-beam piezoelectric energy harvester for wideband energy harvesting under ultra-low harmonic acceleration," *Energy Reports*, vol. 8, pp. 6521–6529, 2022, doi: 10.1016/j.egyr.2022.04.077.
28. M. Khalili, S. Ahmed, and A. T. Papagiannakis, "Piezoelectric energy harvesting for powering a novel weigh-in-motion system," *Energy Convers. Manag. X*, vol. 15, no. January, p. 100259, 2022, doi: 10.1016/j.ecmx.2022.100259.
29. F. J. Diaz Sanchez, M. Chung, M. Waqas, V. Koutsos, S. Smith, and N. Radacsi, "Sponge-like piezoelectric micro- and nanofiber structures for mechanical energy harvesting," *Nano Energy*, vol. 98, no. March, p. 107286, 2022, doi: 10.1016/j.nanoen.2022.107286.

30. G. Shan, Y. Kuang, and M. Zhu, "Design, modelling and testing of a compact piezoelectric transducer for railway track vibration energy harvesting," *Sensors Actuators A Phys.*, vol. 347, no. September, p. 113980, 2022, doi: 10.1016/j.sna.2022.113980.

31. Y. Zhang, X. J. Song, Z. X. Zhang, D. W. Fu, and R. G. Xiong, "Piezoelectric energy harvesting based on multiaxial ferroelectrics by precise molecular design," *Matter*, vol. 2, no. 3, pp. 697–710, 2020, doi: 10.1016/j.matt.2019.12.008.

32. Y. Zhang, J. Wang, and C. Lü, "Optimal patching locations and orientations for maximum energy harvesting efficiency of ultrathin flexible piezoelectric devices mounted on heart surface," *Theor. Appl. Mech. Lett.*, vol. 12, no. 3, p. 100341, 2022, doi: 10.1016/j.taml.2022.100341.

33. S. Wang et al., "Core-shell structured silk Fibroin/PVDF piezoelectric nanofibers for energy harvesting and self-powered sensing," *Nano Mater. Sci.*, vol. 4, no. 2, pp. 126–132, 2022, doi: 10.1016/j.nanoms.2021.07.008.

34. H. G. Çetin and B. Sümer, "A flexible piezoelectric energy harvesting system for broadband and low-frequency vibrations," *Procedia Eng.*, vol. 120, pp. 345–348, 2015, doi: 10.1016/j.proeng.2015.08.631.

35. C. Yang et al., "Highly conductive liquid metal electrode based stretchable piezoelectric-enhanced triboelectric nanogenerator for harvesting irregular mechanical energy," *Mater. Des.*, vol. 201, p. 109508, 2021, doi: 10.1016/j.matdes.2021.109508.

36. A. Pasharavesh, R. Moheimani, and H. Dalir, "Nonlinear energy harvesting from vibratory disc-shaped piezoelectric laminates," *Theor. Appl. Mech. Lett.*, vol. 10, no. 4, pp. 253–261, 2020, doi: 10.1016/j.taml.2020.01.032.

37. A. M. Elhalwagy, M. Y. M. Ghoneem, and M. Elhadidi, "Feasibility study for using piezoelectric energy harvesting floor in buildings' interior spaces," *Energy Procedia*, vol. 115, pp. 114–126, 2017, doi: 10.1016/j.egypro.2017.05.012.

38. R. Tavares and M. Ruderman, "Energy harvesting using piezoelectric transducers for suspension systems," *Mechatronics*, vol. 65, no. September 2019, p. 102294, 2020, doi: 10.1016/j.mechatronics.2019.102294.

39. M. Zhu and S. Edkins, "Analytical modelling results of piezoelectric energy harvesting devices for self-power sensors/sensor networks in structural health monitoring," *Procedia Eng.*, vol. 25, pp. 195–198, 2011, doi: 10.1016/j.proeng.2011.12.048.

40. B. Bagchi, N. A. Hoque, N. Janowicz, S. Das, and M. K. Tiwari, "Re-usable self-poled piezoelectric/piezocatalytic films with exceptional energy harvesting and water remediation capability," *Nano Energy*, vol. 78, no. August, p. 105339, 2020, doi: 10.1016/j.nanoen.2020.105339.

41. Y. Amini, M. Heshmati, P. Fatehi, and S. E. Habibi, "Piezoelectric energy harvesting from vibrations of a beam subjected to multi-moving loads," *Appl. Math. Model.*, vol. 49, pp. 1–16, 2017, doi: 10.1016/j.apm.2017.04.043.

42. M. Heshmati and Y. Amini, "A comprehensive study on the functionally graded piezoelectric energy harvesting from vibrations of a graded beam under travelling multi-oscillators," *Appl. Math. Model.*, vol. 66, pp. 344–361, 2019, doi: 10.1016/j.apm.2018.09.002.

43. X. Xu, D. Cao, H. Yang, and M. He, "Application of piezoelectric transducer in energy harvesting in pavement," *Int. J. Pavement Res. Technol.*, vol. 11, no. 4, pp. 388–395, 2018, doi: 10.1016/j.ijprt.2017.09.011.

44. A. Kumar, A. Sharma, R. Kumar, R. Vaish, and V. S. Chauhan, "Finite element analysis of vibration energy harvesting using lead-free piezoelectric materials: A comparative study," *J. Asian Ceram. Soc.*, vol. 2, no. 2, pp. 138–143, 2014, doi: 10.1016/j.jascer.2014.02.001.

45. A. Romani and M. Filippi, "Quasi-synchronous charge extraction for improved energy harvesting from highly coupled piezoelectric transducers," *Procedia Eng.*, vol. 87, pp. 1525–1528, 2014, doi: 10.1016/j.proeng.2014.11.589.

46. L. H. Fang, S. I. S. Hassan, R. A. Rahim, M. Isa, and B. Bin Ismail, "Charaterization of differents dimension piezoelectric transducer for sound wave energy harvesting," *Energy Procedia*, vol. 105, pp. 836–843, 2017, doi: 10.1016/j.egypro.2017.03.398.

47. S. Lee, B. Yeom, Y. Kim, and J. Cho, "Layer-by-layer assembly for ultrathin energy-harvesting films: Piezoelectric and triboelectric nanocomposite films," *Nano Energy*, vol. 56, no. September 2018, pp. 1–15, 2019, doi: 10.1016/j.nanoen.2018.11.024.

48. J. S. Alrowaijeh and M. R. Hajj, "Piezoelectric energy harvesting from flexible delta wings," *Theor. Appl. Mech. Lett.*, vol. 8, no. 4, pp. 267–271, 2018, doi: 10.1016/j.taml.2018.04.005.

49. K. Gkoumas, F. Petrini, and F. Bontempi, "Piezoelectric vibration energy harvesting from airflow in HVAC (Heating Ventilation and Air Conditioning) systems," *Procedia Eng.*, vol. 199, pp. 3444–3449, 2017, doi: 10.1016/j.proeng.2017.09.496.

50. A. Kokkinopoulos, G. Vokas, and P. Papageorgas, "Energy harvesting implementing embedded piezoelectric generators – The potential for the Attiki Odos traffic grid," *Energy Procedia*, vol. 50, pp. 1070–1085, 2014, doi: 10.1016/j.egypro.2014.06.126.

51. Y. S. Shih, D. Vasic, F. Costa, and W. J. Wu, "Magnetic stoppers on single beam piezoelectric energy harvesting," *Phys. Procedia*, vol. 70, pp. 1022–1026, 2015, doi: 10.1016/j.phpro.2015.08.213.

52. F. Laumann, M. M. Sørensen, R. F. Jul Lindemann, T. M. Hansen, and T. Tambo, "Energy harvesting through piezoelectricity – Technology foresight," *Energy Procedia*, vol. 142, pp. 3062–3068, 2017, doi: 10.1016/j.egypro.2017.12.445.

53. H. Liu, C. Quan, C. J. Tay, T. Kobayashi, and C. Lee, "A MEMS-based piezoelectric cantilever patterned with PZT thin film array for harvesting energy from low frequency vibrations," *Phys. Procedia*, vol. 19, pp. 129–133, 2011, doi: 10.1016/j.phpro.2011.06.136.

54. E. Park, N. Bassiri-Gharb, and I. Stern, "Coupling high resonant frequency piezoelectrics to human-scale frequencies for energy harvesting," *Procedia Comput. Sci.*, vol. 109, no. 2016, pp. 771–776, 2017, doi: 10.1016/j.procs.2017.05.318.

55. E. Nilsson, L. Mateu, P. Spies, and B. Hagström, "Energy harvesting from piezoelectric textile fibers," *Procedia Eng.*, vol. 87, pp. 1569–1572, 2014, doi: 10.1016/j.proeng.2014.11.600.

56. D. Alghisi, S. Dalola, M. Ferrari, and V. Ferrari, "Ball-impact piezoelectric converter for multi-degree-of-freedom energy harvesting from broadband low-frequency vibrations in autonomous sensors," *Procedia Eng.*, vol. 87, pp. 1529–1532, 2014, doi: 10.1016/j.proeng.2014.11.590.

57. M. N. Hasan, M. A. Muktadir, and M. Alam, "Comparative study of tapered shape bimorph piezoelectric energy harvester via finite element analysis," *Forces Mech.*, vol. 9, no. September, p. 100131, 2022, doi: 10.1016/j.finmec.2022.100131.

58. S. Shevtsov and M. Flek, "Random vibration energy harvesting by piezoelectric stack charging the battery," *Procedia Eng.*, vol. 144, pp. 645–652, 2016, doi: 10.1016/j.proeng.2016.05.059.

59. N. Musa, B. K. Allam, N. B. Singh, and S. Banerjee, "Investigation on water defluoridation via batch and continuous mode using Ce–Al bimetallic oxide: Adsorption dynamics, electrochemical and LCA analysis," *Environ. Pollut.*, vol. 328, no. December 2022, p. 121639, 2023, doi: 10.1016/j.envpol.2023.121639.

60. N. H. Diyana, A. G. A. Muthalif, M. N. Fakhzan, and A. N. Nordin, "Vibration energy harvesting using single and comb-shaped piezoelectric beam structures: Modeling and simulation," *Procedia Eng.*, vol. 41, pp. 1228–1234, 2012, doi: 10.1016/j.proeng.2012.07.305.

61. P. Cahill, B. Hazra, R. Karoumi, A. Mathewson, and V. Pakrashi, "Data of piezoelectric vibration energy harvesting of a bridge undergoing vibration testing and train passage," *Data Br.*, vol. 17, pp. 261–266, 2018, doi: 10.1016/j.dib.2018.01.009.

62. N. Sezer and M. Koç, "A comprehensive review on the state-of-the-art of piezoelectric energy harvesting," *Nano Energy*, vol. 80, no. October 2020, p. 105567, 2021, doi: 10.1016/j.nanoen.2020.105567.

Chapter 10

Piezoelectric Materials and Their Applications in Electrochemical Processes

Sulekha Sharma,[a] Parveen Dabur,[b] Pooja Ranjan,[c] and Jyoti Tanwar[c]
[a]*Indian Institute of Science and Research, Bhopal (IISERB), MP-462066, India*
[b]*C.M.R.A. Govt Polytechnic Education Society, Sanghi, Rohtak, Haryana-124001, India*
[c]*Department of Chemistry, Hindu Girls College, Sonepat, Haryana-131001, India*
jyot.me1@gmail.com, ranjanpooja011@gmail.com

Energy harvesting is an emerging field that aims to convert ambient energy sources into useful energy that can be used to power various devices. One innovative area of research involves connecting energy-harvesting materials and tools directly to electrochemical methods. This approach allows electrochemical processes to be regulated in the occurrence of piezoelectric materials, which can influence electrochemically occurring reactions through the electric charge

Piezoelectric Materials: Design and Applications
Edited by N. B. Singh and Dev Kumar Mahato

ISBN 978-981-5129-40-3 (Hardcover), 978-1-003-59897-8 (eBook)
www.jennystanford.com

produced by mechanical stress. This chapter provides a summary of the fundamental attributes of piezoelectric materials and compares their electrochemical performance under different conditions. Mechanisms of piezo-electrochemical coupling are discussed in detail and a comparison is made between the different material morphologies employed. This chapter also highlights possible future courses and functions for the advances of piezo-electrochemical fusion systems. Overall, this chapter aims to provide a comprehensive summary of recent studies on how piezoelectric materials can be used to control electrochemical processes, with the goal of inspiring and guiding future research in this emerging field.

10.1 Introduction

Piezoelectricity was first discovered by Pierre Curie and Jacques Curie in 1880. Piezoelectricity is the ability of certain materials to generate an electrically charged system when mechanical stress or vibrations are applied. The word "piezoelectricity" is derived from the Greek words "piezo" meaning "pressure" or "squeeze" and "electricity" referring to the generation of an electric charge. Piezoelectric materials are those that exhibit piezoelectricity. They produce an electric charge when a stress is applied, known as the direct piezoelectric effect [1]. Conversely, a mechanical strain is also developed when an electric field is applied, referred to as the converse piezoelectric effect [2–8]. The beginning of piezoelectricity can be attributed to the distribution of both the positive and negative electrical charges non-centrosymmetrically upon the material within the unit cell. When a piezoelectric material experiences some stress or mechanical vibration, ions are displaced, which brings a change in the unit cell dipole moment. This change creates a net electric charge and generates a potential (piezoelectric) across the material. Piezoelectric materials that belong to nonpolar crystals, such as quartz, do not possess a net electric dipole in a state of zero-stress. However, an electrical dipole when subjected to stress can be generated due to the separating electrically charged centers and the resultant induced piezoelectric potential. On the other hand, piezoelectric materials that belong to polar crystals show spontaneous polarization even in the no electric field state or zero-stress [9, 10]. This is because of the separation between positive and negative charges within the material [11, 12]. Zinc oxide is an example of a polar piezoelectric material. Ferroelectric materials, a subclass of piezoelectric material system, fit in to polar crystals. These materials have a spontaneous polarization and can be permanently altered and moved under exposure to an externally strong electrical field [13]. Barium titanate is an example of a ferroelectric material [14, 15]. The presence of polarization, whether it occurs spontaneously or due to mechanical factors, has a significant impact on the electrical properties of a material. This influence is observed both within the material and at its exterior regions [10, 16]. When polarization is induced, an electric field is produced across the material, leading to the rearrangement of electric charges and disruption of electronic energy states. If the external component of the material mixes with a medium, the redistribution of electric charges results in changing of the electric conductivity. This conductivity alteration is dependent upon the density of electrical charge and the continuity of electronic energy states that can be occupied in between the material and the surrounding medium [16]. This interplay between the material and the medium's electronic properties can have a pronounced effect on electrochemistry. In summary, the electric field induced across a material, as a result of polarization, has a profound influence on its electrical properties both internally and at the material's surface when it interacts with a medium. This interaction can significantly impact electric conductivity, which relies upon the electrical charge density and the electronic energy state continuity between the material and its surroundings. This understanding is particularly relevant in the field of electrochemistry, as electrochemical reactions are typically carried out by an external power source [17–19]. However, there is growing interest in coupling power generation with electrochemical processes, especially in large extent renewable and clean energy approaches [20–22]. These approaches aim to store solar energy along with wind energy and these energies are converted into electrical power for driving electrochemical reactions. On a smaller scale, there is potential to harvest and utilize confined energy sources, such as mechanical energy in the microwatt to milliwatt range [11, 23]. In this regard, systems based on piezoelectric materials have been developed to harness this energy and have found applications in various fields [24–37].

1. It controls the rate of catalysis in the chemical solutions.
2. The etchant solution influences the rate of corrosion of metals.
3. It develops systems having a self-charging potential.
4. Facilitates other electrochemical processes.

In recent decades, the use of piezoelectrically induced electric fields has been explored. This intersection of piezoelectricity and electrochemistry is referred to as piezo-electrochemistry [38–43]. Overall, piezo-electrochemistry represents a promising area of research where the utilization of piezoelectric materials enables the manipulation and control of electrochemical processes using mechanical energy.

10.2 Mechanism of Piezoelectric-Controlled Electrochemistry

Piezo-electrochemical processes deal with the pairing of piezoelectricity and electrochemistry and the chemistry of such reactions is preceded with the help of piezoelectrically generated electric charge and voltage. Also, in electrochemistry the fundamentally quantitative relationships are described by Faraday's laws of electrolysis, published in 1834 by Michael Faraday [44]. Faraday's laws state the relationship between the mass of a substance produced or consumed during an electrochemical reaction (m) and the total electrical charge transferred (Q). These laws can be concise as follows:

a. Faraday's First Law: The mass of a substance produced or consumed during an electrochemical reaction is directly proportional to the total electric charge transferred. Mathematically, this can be expressed as:

$$m = k \times Q$$

where m is the mass of the substance, k is a proportionality constant, and Q is the total charge transferred.

b. Faraday's Second Law: The mass of different substances produced or consumed during an electrochemical reaction is directly proportional to their respective chemical equivalents. The chemical equivalent is the amount of substance that undergoes a

Faraday of charge transfer (96,485 coulombs). Mathematically, this can be expressed as:

$$m = z \times n \times F$$

where m is the mass of the substance, z is the chemical equivalent, n is the number of moles of the substance, and F is Faraday's constant.

These relationships allow for calculating the mass of a substance produced or consumed during an electrochemical reaction based on the total charge transferred. They provide a fundamental basis for understanding and quantifying electrochemical processes, including those influenced by piezoelectric effects. It is important to note that the application of piezoelectricity in electrochemistry introduces an additional factor of the piezoelectrically induced electrical charge or voltage, which can influence the mass of produced material and the overall electrochemical reaction.

The various applications discussed in this chapter are as follows:

1. Dye degradation
2. Water splitting
3. Hydrogen production
4. Self-charging power cell (SCPC)
5. Nanogenerators

Following is a brief discussion of the functions of piezoelectric materials under electrochemical processes.

10.2.1 Dye Degradation

The coupling of piezoelectricity and electrochemistry has significant implications for addressing environmental pollution [45], particularly in the context of harmful organic compounds [46–49]. Industrial and textile dyes are recognized as major contributors to water pollution, with the textile fabricating and dyeing industries alone accountable for a significant portion of water pollution (17–20% according to the World Bank). Most of the hazardous compounds present in wastewater, many of which are difficult to treat, are by-products of textile dyeing processes [50]. In addition to the textile, leather, and paper industries, other sectors such as the food industry, hair colorings, pharmaceuticals, and photoelectrochemical cells also contribute to environmental pollution [50–54]. Addressing water pollution caused by these industries is crucial, and piezo-catalysis offers a potential solution [55]. Piezo-catalysis utilizes natural mechanical energy, an uncontaminated and renewable energy source, to drive catalytic degradation for environmental purification [55, 56]. Piezoelectric nanomaterials are employed in piezo-catalysis, where they go through electrochemical reactions in order to rip water and also degrade organic pollutants [51, 54]. Various piezoelectric materials, including $BaTiO_3$, $PbZ_{rx}Ti_{1-x}O_3$, $NaNbO_3$, MoS_2, $BiFeO_3$, $Ba_{0.8}Sr_{0.2}TiO_3$, $Ba_{1-x}Ca_xTiO_3$, $ZnSnO_3$, $Bi_4Ti_3O_{12}$, and BiOCl, have been studied for their piezo-catalytic breakdown capabilities [54, 57]. External vibrations applied to piezoelectric materials induce electron-hole separation, leading to increased production of reactive oxygen species (ROS) and enhanced pollutant decomposition rates [54, 58]. When combined with light energy above the bandgap, piezo-catalysts with high electron-hole pair generation capabilities exhibit photocatalytic activities. The interaction of oxygen and electrons with each other in the conduction band leads to the generation of superoxide radicals; the oxidized hydroxyl radicals ($^{\cdot}OH$) are generated after water starts to interact with holes in the valence band [59]. The diffusion-controlled kinetics of hydroxyl radicals allow for the oxidation of various compounds in water, making them highly effective in degrading pollutants. Some of the crucial oxidants as ozone, oxygen, and hydrogen peroxide, together with energy sources like ultraviolet (UV) radiation or catalysts, are often employed to generate these radicals [52, 54, 60].

$$\text{(Electrons) } O_2 + e^- \rightarrow [O_2^-]$$

$$\text{(Holes) } OH^- + H^+ \rightarrow [OH^{\cdot}]$$

$$^{\cdot}OH + O_2^- + \text{Dye} \rightarrow \text{Degradation product}$$

The piezo-catalytic dye degradation reaction involves the oxidation of dye molecules by active species such as $^{\cdot}OH$ and O_2^-, resulting in their degradation and discoloration. The fraction of dye degradation can be summed up using the concentration or absorbance values of the dye solution before and after the reaction [54, 61].

$$\text{Dye degradation (\%)} = (1 - C/C_o) \times 100 = (1 - A/A_o) \times 100$$

Reactive oxygen species (ROS) play a crucial role in sterilization, organic transformation, and environmental purification. Piezo-photocatalysis or piezo-catalysis can effectively produce ROS, speeding up associated reactions involved in degrading pollutants [62]. Piezo-catalysis has been demonstrated in various applications. For instance, $BaTiO_3$ piezoelectric nanoparticles [63] have been used as toothpaste abrasives for teeth whitening, while MoS_2 doped with graphdiyne shows excellent catalytic activity in the degradation of tetracycline [64]. Additionally, piezo-photocatalytic systems, such as ZnO nanowires/PDMS platforms and ferroelectric antimony sulfoiodide (SbSI) nanowires, have shown high breaking-down activity in presence of light or ultrasonic vibrations [54]. Different forms of mechanical energies like wind energy, sonic waves, atmospheric pressure, and tidal energy can induce piezo-catalysis and piezo-photocatalysis. The interchange of mechanical energy into electrical energy, described by the piezoelectric strain constant (*d*) and piezoelectric stress constant (*i*), affects the catalytic process [65].

$$d = \left(\frac{\delta e}{\delta E}\right)\eta = \left(\frac{\delta D}{\delta \eta}\right)E \tag{10.1}$$

$$i = \left(-\frac{\delta \eta}{\delta E}\right)e = \left(\frac{\delta D}{\delta e}\right)E \tag{10.2}$$

Equations (10.1) and (10.2) describe the relationship between the strain/stress (e/η) that is forced on by the electrical field (*E*) and electrical displacement (*D*). The piezoelectric effects can be generated by different methods such as vortex-induced shearing forces, physical bending, ultrasonic cavitation, and low-frequency vibrations [66]. Overall, piezo-catalysis and piezo-photocatalysis hold promise for addressing water pollution caused by hazardous organic compounds and facilitating the removal of environmental pollutants.

10.2.2 Water Splitting

In the face of a pressing challenge like pollution, humanity has grown interested in pollution-free, clean, and renewable energy alternatives [67]. Over the past few decades, hydrogen has gained significant popularity. Unlike conventional fossil fuels, hydrogen stands out for its lack of emissions of environmentally detrimental greenhouse gases, which have a harmful impact on our surroundings. As an environmentally conscious energy source, hydrogen is gaining traction [68–70]. It can be generated through various methods such as oxidation, bacterial fermentation, coal gasification, and water splitting. Hydrogen production through water splitting can be achieved using diverse techniques, including electrolysis, photoelectrolysis, pyrolysis, and photobiological water splitting [21, 71–73]. In the field of photocatalytic hydrogen generation, the use of sacrificial materials and noble metals is often necessary. Therefore, design strategies for materials are crucial to achieve the desired band structure and facilitate the separation of the two half-reactions on the material's surface for efficient water splitting. Piezo-catalysts with various morphologies can effectively achieve the separation of reduction and oxidation points of water due to the spatial sharing of charges [54]. The water splitting occurring directly using piezo-catalysis was demonstrated by Hong et al. in 2010 [21], by an application of an electric charge to contained water molecules, and the surface of $BaTiO_3$ microdendrites is used up for piezo-potential generation and after that redox reaction is initiated by ZnO microfibers that produce O_2 and H_2. This process takes advantage of the unique properties of the piezo-catalysts, as shown in Fig. 10.1a. Thereafter, Zhang et al. developed a nanofluidic $BaTiO_3$ suspension that undergoes water splitting when exposed to ultrasound [74]. This suspension, containing a mixture of tetragonal and cubic phases of $BaTiO_3$, exhibited a reduction in particle size from approximately 400 nm to nearly 150 nm during ultrasound transmission. The fine structure of the particles enables the generation of a stable nanofluid. Therefore, the highest percentage of piezoelectrically carried water splitting was obtained with a loading of 5 mg/L of $BaTiO_3$, resulting in a maximum rate of H_2 evolution that is 270 mmol/h/g, as shown in Fig. 10.1b [54, 74].

For successfully achieving water splitting, the potential of the reaction must exceed the water redox potential and be higher than 1.23 V to drive the reactions that produce H_2 and O_2. Zhang et al. developed an indirect piezo-electrochemical method for water splitting, aiming to create a more flexible and cost-effective production approach that utilizes low-frequency vibrations instead of using higher frequency ultrasound [75]. This system utilized water electrolysis and the piezoelectric bimorph cantilever, as depicted in Fig. 10.1c [75]. The electrical output that is generated by the mechanical vibrations was further modified and an electrolyte connection is made for water splitting to produce H_2 and O_2. Overall, the design and utilization of piezo-catalytic materials, such as $BaTiO_3$, in combination with appropriate vibrations or ultrasound, provide avenues for direct or indirect water splitting for efficient hydrogen generation. These approaches offer potential solutions for clean and renewable energy production [54, 75].

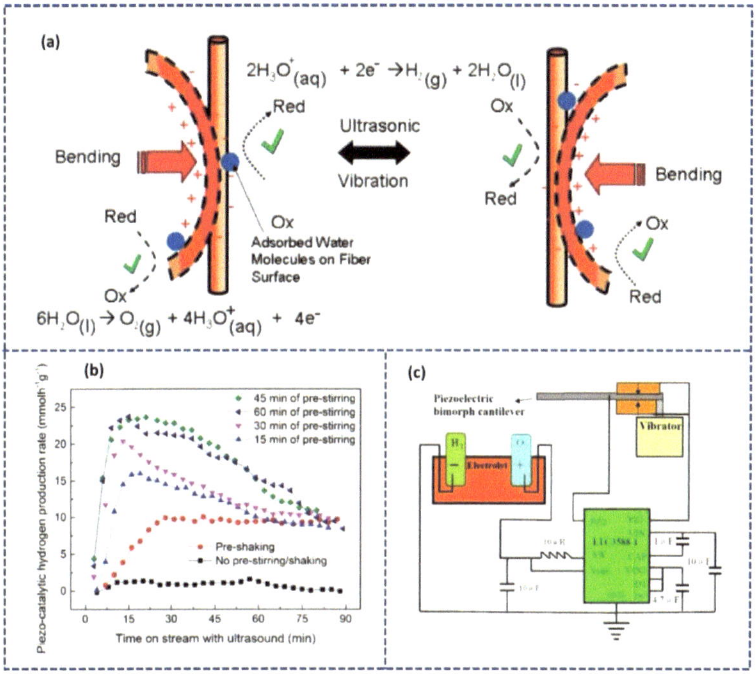

Figure 10.1 (a) Representation of direct water splitting. Reproduced with permission [21]. Copyright 2010, American Chemical Society. (b) Graphical representation of water splitting induced piezoelectrically on loading with $BaTiO_3$. Reproduced with permission [74]. Copyright 2022, Wiley. (c) Water splitting system. Reproduced with permission [75]. Copyright 2013, MDPI.

10.2.3 Hydrogen Production

The challenges of environmental pollution as well as the energy crisis have led to a growing interest in environmentally friendly, clean, and renewable sources of energy [76–78]. Hydrogen fuel, with its abundant chemical energy and lack of greenhouse gas emissions, has emerged as an environmentally friendly energy source [78–81]. Consequently, hydrogen production is a topic of intense research and development and, therefore, various approaches can be employed to harvest hydrogen fuels, including coal gasification, partial oxidation, electrolysis, bacteria fermentation, and water splitting [73, 82]. Piezo-electrochemical water splitting is a recently explored method that combines the piezoelectric properties of materials with electrochemical redox reactions to produce hydrogen [52]. Zhang et al. successfully achieved an indirect piezo-electrochemical hydrogen generation by means of a tool that is made of a piezoelectric bimorph cantilever that is based on PZT-5 ceramics [83]. The rate of hydrogen evolution was comparatively low at approximately 1.21 mmol h^{-1}, and an increase in concentration of ions of electrolyte aqueous solution of $NaHSO_4$ benefited the hydrogen evolution by using piezo-electrochemical water splitting reaction. In another study, approximately 124.1 mmol g^{-1}, which is a very high rate of hydrogen generation, was observed using BFO ($BiFeO_3$) nanosheets under 100 W mechanical vibrations within 1 h [84]. The rates of hydrogen production of BFO (per unit mass) nanosheets varied at different vibration frequencies (28, 45, and 100 kHz) and vibration powers (20.4, 124.1, and 8.3 mmol h^{-1}). The direct water splitting can be utilized for hydrogen and oxygen production by vibrating piezoelectric ZnO microfibers along with BTO ($BaTiO_3$) microdendrites by means of the ultrasonic waves. Rapid hydrogen and oxygen evolution rates were observed, with an initial rate of 3.4×10^{-3} and 1.7×10^{-3} ppm s^{-1} for ZnO microfibers, and a normal hydrogen evolution rate of 1.25×10^{-2} and 9.13×10^{-3} ppm s^{-1} during the first and second piezo-electrochemical reactions, respectively, for BTO microdendrites. The hydrogen production results using piezoelectric materials demonstrated the challenges of driving the necessary potential to prevail over the standard oxidation and reduction potential of water. Therefore, in summary, developing the methods that can remarkably reduce environmental pollution and produce clean energy sources has garnered significant research interest [70]. Piezo-electrochemical processes, including piezo-catalytic hydrogen production, offer potential solutions for hydrogen generation and contribute to environmental sustainability.

10.2.4 Self-Charging Power Cell

Self-charging power systems (SCPSs) encompass the fusion of energy harvesting and energy storage components within power devices [85]. These systems are typically complemented by power management circuits that handle tasks like AC–DC conversion, DC–DC conversion, and power and impedance matching. Over time, endeavors have been made to combine diverse energy-harvesting methodologies with energy storage tools, including thermoelectric generators, solar cells, triboelectric nanogenerators (TENGs), and piezoelectric nanogenerators (PENGs). PENGs and TENGs, typical of these energy-harvesting methods, distinguish themselves for several reasons [86]:

(i) The mechanical energy sources are less influenced by weather and location.
(ii) The uncomplicated components and structures of these nanogenerators enable the realization of versatile SCPSs such as wearables, flexible devices, and stretchable units.
(iii) Human motion not only serves as a power source but also facilitates human–machine interactions in wearable electronics.

As a result, intensive exploration has been committed to crafting diverse categories of SCPSs centered around PENGs and TENGs, emphasizing enhancing their overall performance. Some other examples of the self-charging power cell (SCPC) are also present, which utilizes a piezoelectric polymer, a PVDF separator, to collect electrical energy from mechanical energy and it is stored in battery electrodes with the help of a piezoelectric electrochemical alteration process [86]. The SCPC consists of stainless steel material and is a coin-type cell having an anode, a cathode, and a polarized PVDF film separator [31, 87]. In further research, modifications were made to the electrode and separator sections of the SCPC to enhance its piezo-electrochemical charge/discharge activity, energy conversion efficiency, and stability. In this regard, porous PVDF structures were utilized to create efficient lithium-ion transfer channels, improving performance. To enhance the piezoelectric signals, a PVDF-PZT nanocomposite film, made of lead zirconate titanate (PZT), was used as a piezo separator [88]. The SCPC, with its piezoelectric separator, can be utilized for converting and storing mechanical energy generated by pressing buttons, as demonstrated by a coin-type cell positioned beneath the keys of a calculator. Flexible materials with pores can exhibit improved piezoelectric output under mechanical stress and the absorbent PVDF nanocomposite film, subjected to mechanical stress, enhances the piezoelectric output by producing a higher strain in various directions. The SCPC operates through the piezoelectric potential produced by the PVDF separator. The self-charging process involves three main reactions [9, 54]:

i. The PVDF separator, operating between the anode and cathode in the electrochemically stable region, acts as an insulator before applying the external force (Fig. 10.2a).
ii. In the self-charging region, when a compressing force is encountered, the PVDF separator generates a piezoelectric field, allowing ions to shift from the cathode electrode to the anode electrode through the ion-conducting separator and electrolyte (Fig. 10.2b). This transformation represents the conversion of mechanical energy into electrochemical energy.
iii. In the force release state, when the driving force is disconnected, some ions diffuse backward and reach an equilibrium distribution (Fig. 10.2c).

Besides this, supercapacitors, known for their rapid charge/discharge rate, elevated power density, and longer lifetime, are considered to be promising energy storage technologies [9, 89, 90]. Self-charging supercapacitor power cells (SCSPC) are SCPCs that incorporate supercapacitor electrodes. SCSPCs utilize a piezopolymer separator that is covered with a gel electrolyte that is positioned between supercapacitor electrodes. Siloxene-based SCSPCs, employing siloxene foils as electrodes and electrospun siloxene-integrated PVDF piezoelectric fibers as separators covered with ionogels, have also been developed (Fig. 10.3) [91, 92].

These advancements in SCPCs and SCSPCs demonstrate the potential for harnessing and storing mechanical energy through piezoelectric materials, offering opportunities for self-charging power generation and energy storage systems.

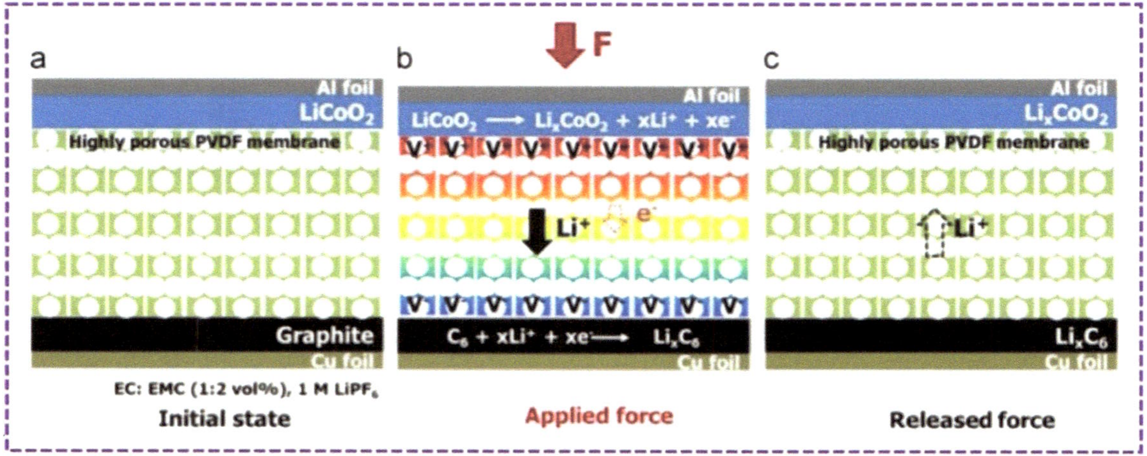

Figure 10.2 Schematic representation of the mechanism of self-charging of piezo-separator PVDF in the Li-ion batteries (a) when in the initial state; here the electrodes are immersed in an electrolyte (liquid). (b) Piezoelectric field generation when the applied external forces compel the migration of Li^+ ions to take place in addition to electron transfer. (c) When the device reaches equilibrium state by removal of the applied force, the diffusion of Li^+ ions reverses as the piezoelectric field disappears. Reproduced with permission [88]. Copyright 2015, Elsevier.

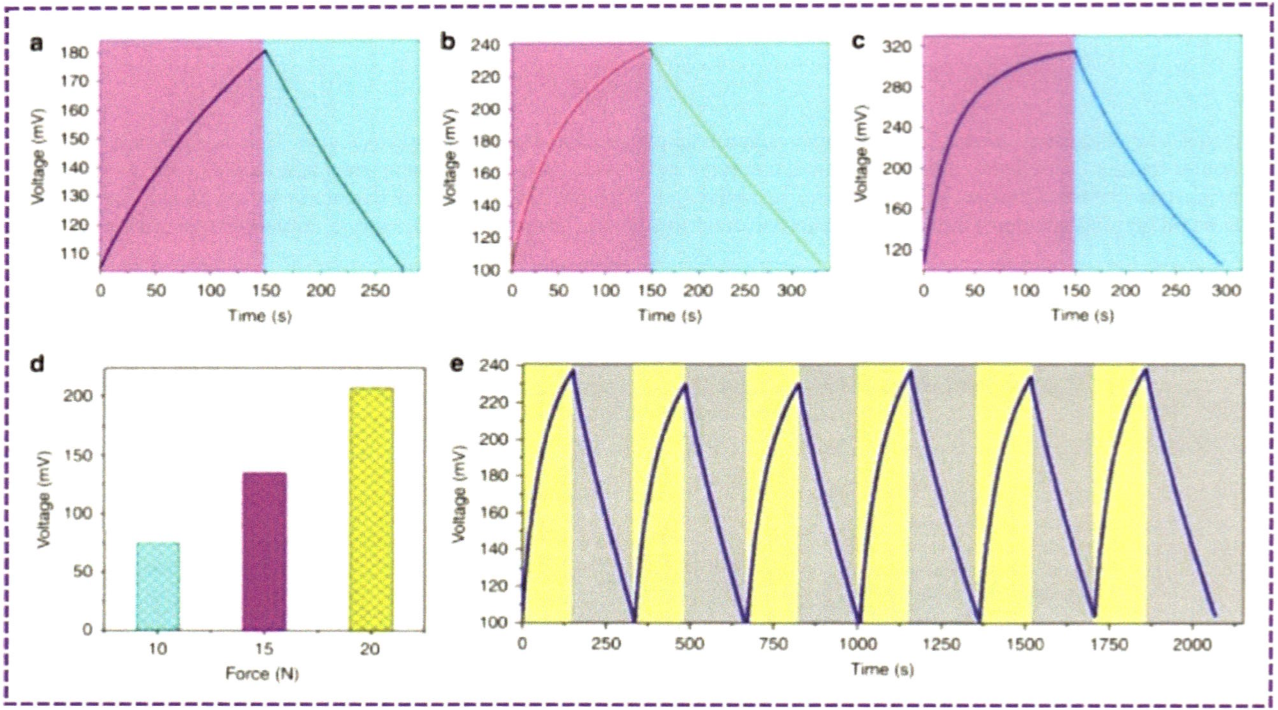

Figure 10.3 (a–c) Graphical representation of the siloxene SCSPCs with self-charging properties, which is further pursued by discharging on applying a steady current, respectively. (d) Demonstration of the siloxene SCSPC's self-charging performance. (e) The six consecutive cycles of siloxene SCSPC's self-charging process. Reproduced with permission [92]. Copyright 2020, Springer Nature.

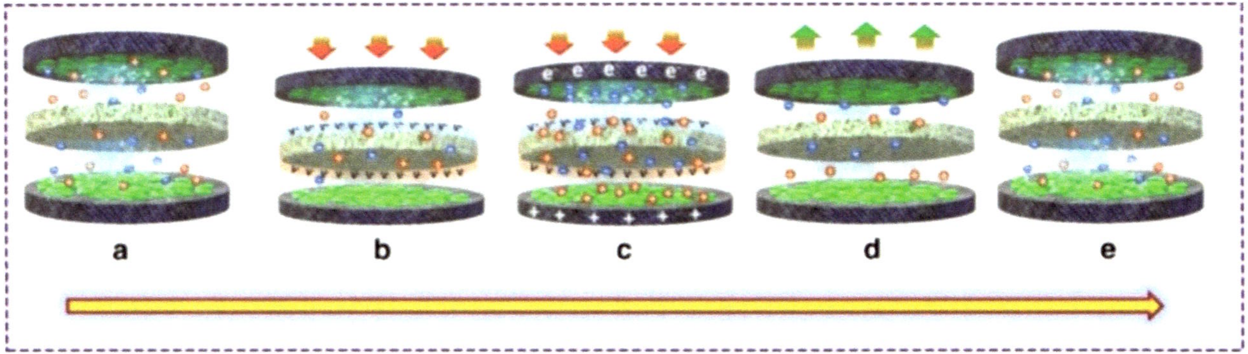

Figure 10.4 (a) The siloxene SCSPC, when no forces are applied, is discharged in its initial state. (b) On compression, the piezoelectric potential is created by the siloxene/PVDF, which moves the ions of the electrolyte in the direction of the electrode surface. (c) The electrochemical reaction and the piezoelectric potential of the siloxene SCSPC in equilibrium state. (d) When compressive force is removed, it results in equilibrium state. (e) A self-charging cycle is shown. Reproduced with permission [92]. Copyright 2020, Springer Nature.

10.2.4.1 Mechanism Followed by Siloxene SCSPC

The siloxene SCSPC operates through an electrochemical process between the siloxene electrodes and the siloxene PVDF piezoelectric fibers. At first, the SCSPC is in an unprompted discharge state. It consists of two siloxene electrodes, which are divided by ionogel-coated siloxene PVDF piezoelectric fibers, with evenly distributed electrolyte ions in the cell. When compression is applied, the siloxene PVDF generates a piezoelectric potential, producing the movement of electrolyte ions in the direction of siloxene electrodes [91, 92]. This leads to the ionic redistribution to the negative and positive electrodes, creating a potential difference and initiating the self-charging process. The electrodes absorb the electrolyte ions during self-charging, increasing the potential difference in opposite directions until reaching a certain voltage value. At this stage, the created piezo potential is further maintained by the distribution of ions of electrolyte, halting the ion transfer and completing the self-charging process. As the compression force is released, the properties of the piezoelectric siloxene-based PVDF piezoelectric fibers diminish, resulting in maintenance of equilibrium by the rearrangement of ions. When the SCSPC is compressed again, the course of self-charging is repeated. This mechanism illustrates the alteration of mechanical energy into electrochemical energy through piezoelectric electrochemical processes. The piezoelectric effect induced by the PVDF piezoelectric fibers drives the movement of electrolyte ions in the direction of the siloxene electrodes, facilitating the self-charging process. When ion migration ceases, the self-charging process also stops. Figure 10.4 illustrates the method of the siloxene-based SCSPC [54, 91, 92].

10.2.5 Nanogenerators

Nanogenerators represent a category of devices meticulously crafted to transmute minute mechanical energy into electrical energy on the nanoscale [93]. Within the realm of energy-harvesting technologies, they specialize in capturing and harnessing even the slightest mechanical forces from diverse origins like vibrations, motion, or pressure, with the aim of generating usable electricity. These innovations frequently leverage concepts such as piezoelectricity, triboelectricity, and other associated principles to realize this energy conversion phenomenon. The piezoelectric nanogenerators are dependent on the principle of the piezoelectric effect, where specific materials generate a potential when they are subjected to mechanical stress or deformation. When a piezoelectric material is mechanically deformed or stressed, it generates a voltage difference across its surfaces. This voltage can be harvested and used as electrical energy. Piezoelectric nanogenerators can be designed using various materials. Nanogenerators, such as those based on MoS_2/PDMS composites, are utilized for cathodic protection to safeguard metal surfaces against corrosion [94]. These nanogenerators can gather mechanical energy from wastewater and serve as active sensors for monitoring water flow and pollution levels. In the case of MoS_2/PDMS nanocomposites, they have an application to metal grids or are fitted in water pipes to harness the mechanical energy of wastewater. For instance, a nanogenerator consisting of a stainless steel grid coated with a 5 wt.% MoS_2/PDMS mixture was developed. Therefore, when they are placed under a kitchen faucet, the nanogenerator converts the mechanical energy of flowing water into electrical energy, with the output voltage comparative to the rate of water flow. The nanogenerator also performs as a self-sufficient energy sensor; it records the water flow rates and thus provides an electrical output accordingly. Another type of nanogenerator, known as the water TENGs, was created using a thin layer of MoS_2/PDMS composite on a flexible substrate. These adaptable water TENGs could be scaled up for industrial applications and attached to polymer tubes. The power generated by the nanogenerator was highest when it was closest to the water inlet, and the output voltage and current density decreased regularly with increasing distance from the inlet. The nanogenerator's performance was affected by some of the factors such as electron transfer speed, distance of flowing water, and impurities present in the water, such as oil content. In order to switch the AC output of the nanogenerator into unidirectional pulses, a full-wave rectifier bridge was added. Thereafter, a capacitor was then charged by means of the rectifier's output, providing a charging curve that varied based on the nanogenerator's position within the tube [94]. Each water TENG served as a self-powered sensor capable of determining impurities in the wastewater, with the power output being influenced by the interaction between nanogenerators and oil in the water [52, 54].

10.3 Conclusions

The book chapter offers an up-to-date exploration of how piezoelectric materials are applied in electrochemical processes. Materials with higher aspect ratios are favored for generating substantial piezoelectric output and optimal deformation. Addressing energy production and environmental concerns is a growing challenge and in this regard piezo-electrochemical applications present fresh perspectives for tackling these issues. The study could lead to techniques for coating materials with specific polarized piezoelectric substances to enhance corrosion resistance. Moreover, piezoelectric material's efficient wastewater cleaning ability might replace photo-cleaning methods. These materials facilitate the change of ambient mechanical vibrations into electrical energy across a broad frequency range for effective energy harvesting. While piezoelectric electrochemical applications have garnered significant attention, evident from the rise in publications, research endeavors, and patents, there remains a need for their actual commercialization.

References

1. Uchino K. The development of piezoelectric materials and the new perspective. In: Uchino K, editor. *Advanced Piezoelectric Materials*. Woodhead Publishing, 2010.
2. Kouritem SA. Array of piezoelectric energy harvesters for broadband natural frequency applications. In: *Proceedings of the ICSV27, Annual Congress of International Institute of Acoustics and Vibration (IIAV), Prague, Czech Republic*. **11**, 2021.
3. Altabey WA, Kouritem SA. The new techniques for piezoelectric energy harvesting: Design, optimization, applications, and analysis. *MDPI*. **15**(18), 6684, 2022.
4. Altabey WA, Kouritem SA. An overview of the topics of the special issue "the new techniques for piezoelectric energy harvesting: Design, optimization, applications, and analysis." *MDPI*. **16**(8), 3357, 2023.
5. Kouritem SA, Bani-Hani MA, Beshir M, Elshabasy MM, Altabey WA. Automatic resonance tuning technique for an ultra-broadband piezoelectric energy harvester. *Energies*. **15**(19), 7271, 2022.

6. Bani-Hani MA, Almomani AM, Aljanaideh KF, Kouritem SA. Mechanical modeling and numerical investigation of earthquake-induced structural vibration self-powered sensing device. *IEEE Sensors Journal.* **22**(20), 19237, 2022.
7. Marino AA, Becker RO. Piezoelectric effect and growth control in bone. *Nature.* **228**(5270), 473, 1970.
8. Katsouras I, Asadi K, Li M, Van Driel TB, Kjaer KS, Zhao D, Lenz T, Gu Y, Blom PW, Damjanovic D. The negative piezoelectric effect of the ferroelectric polymer poly (vinylidene fluoride). *Nature Materials.* **15**(1), 78–84, 2016.
9. Wang ZL, Song J. Piezoelectric nanogenerators based on zinc oxide nanowire arrays. *Science.* **312**(5771), 242, 2006.
10. Masmanidis SC, Karabalin RB, De Vlaminck I, Borghs G, Freeman MR, Roukes ML. Multifunctional nanomechanical systems via tunably coupled piezoelectric actuation. *Science.* **317**(5839), 780, 2007.
11. Zhang Y, Xie M, Adamaki V, Khanbareh H, Bowen CR. Control of electrochemical processes using energy harvesting materials and devices. *Chemical Society Reviews.* **46**(24), 7757, 2017.
12. Sebald G, Guyomar D, Agbossou A. On thermoelectric and pyroelectric energy harvesting. *Smart Materials and Structures.* **18**(12), 125006, 2009.
13. Uchino K. Piezoelectric energy harvesting systems—Essentials to successful developments. *Energy Technology.* **6**(5), 829, 2018.
14. Fei Z, Zhao W, Palomaki TA, Sun B, Miller MK, Zhao Z, Yan J, Xu X, Cobden DH. Ferroelectric switching of a two-dimensional metal. *Nature.* **560**(7718), 336, 2018.
15. Li W, Ji L-J. Perovskite ferroelectrics go metal free. *Science.* **361**(6398), 132, 2018.
16. Starr MB, Wang X. Fundamental analysis of piezocatalysis process on the surfaces of strained piezoelectric materials. *Scientific Reports.* **3**(1), 2160, 2013.
17. Fujishima A, Honda K. Electrochemical photolysis of water at a semiconductor electrode. *Nature.* **238**(5358), 37, 1972.
18. Das D, Veziroğlu TN. Hydrogen production by biological processes: A survey of literature. *International Journal of Hydrogen Energy.* **26**(1), 13, 2001.
19. Grigoriev SA, Porembsky VI, Fateev VN. Pure hydrogen production by PEM electrolysis for hydrogen energy. *International Journal of Hydrogen Energy.* **31**(2), 171, 2006.

20. Mostafaeipour A, Khayyami M, Sedaghat A, Mohammadi K, Shamshirband S, Sehati M-A, Gorakifard E. Evaluating the wind energy potential for hydrogen production: A case study. *International Journal of Hydrogen Energy*. **41**(15), 200, 2016.
21. Hong K-S, Xu H, Konishi H, Li X. Direct water splitting through vibrating microfibers in water. *The Journal of Physical Chemistry Letters*. **1**(6), 997, 2010.
22. Metz WD. Wind energy: Large and small systems competing. *Science*. **197**(4307), 971–973, 1977.
23. Qolipour M, Mostafaeipour A, Tousi OM. Techno-economic feasibility of a photovoltaic-wind power plant construction for electric and hydrogen production: A case study. *Renewable and Sustainable Energy Reviews*. **78**, 113, 2017.
24. Li H, Sang Y, Chang S, Huang X, Zhang Y, Yang R, Jiang H, Liu H, Wang ZL. Enhanced ferroelectric-nanocrystal-based hybrid photocatalysis by ultrasonic-wave-generated piezophototronic effect. *Nano Letters*. **15**(4), 2372, 2015.
25. Qian W, Zhao K, Zhang D, Bowen CR, Wang Y, Yang Y. Piezoelectric material-polymer composite porous foam for efficient dye degradation via the piezo-catalytic effect. *ACS Applied Materials & Interfaces*. **11**(31), 27862, 2019.
26. Zhang Y, Schultz AM, Salvador PA, Rohrer GS. Spatially selective visible light photocatalytic activity of $TiO_2/BiFeO_3$ heterostructures. *Journal of Materials Chemistry*. **21**(12), 4168, 2011.
27. Bermúdez V, Caccavale F, Sada C, Segato F, Diéguez E. Etching effect on periodic domain structures of lithium niobate crystals. *Journal of Crystal Growth*. **191**(3), 589, 1998.
28. Kubel F, Schmid H. Growth, twinning and etch figures of ferroelectric/ferroelastic dendritic $BiFeO_3$ single domain crystals. *Journal of Crystal Growth*. **129**(3), 515, 1993.
29. Nassau K, Levinstein HJ, Loiacono GM. The domain structure and etching of ferroelectric lithium niobate. *Applied Physics Letters*. **6**(11), 228, 2004.
30. Sones CL, Mailis S, Brocklesby WS, Eason RW, Owen JR. Differential etch rates in z-cut $LiNbO_3$ for variable HF/HNO_3 concentrations. *Journal of Materials Chemistry*. **12**(2), 295, 2002.
31. Xue X, Wang S, Guo W, Zhang Y, Wang ZL. Hybridizing energy conversion and storage in a mechanical-to-electrochemical process for self-charging power cell. *Nano Letters*. **12**(9), 5048, 2012.

32. Ramadoss A, Saravanakumar B, Lee SW, Kim Y-S, Kim SJ, Wang ZL. Piezoelectric-driven self-charging supercapacitor power cell. *ACS Nano*. **9**(4), 4337, 2015.

33. Song R, Jin H, Li X, Fei L, Zhao Y, Huang H, Lai-Wa Chan H, Wang Y, Chai Y. A rectification-free piezo-supercapacitor with a polyvinylidene fluoride separator and functionalized carbon cloth electrodes. *Journal of Materials Chemistry A*. **3**(29), 14963, 2015.

34. Wang F, Jiang C, Tang C, Bi S, Wang Q, Du D, Song J. High output nano-energy cell with piezoelectric nanogenerator and porous supercapacitor dual functions – A technique to provide sustaining power by harvesting intermittent mechanical energy from surroundings. *Nano Energy*. **21**, 209, 2016.

35. Parida K, Bhavanasi V, Kumar V, Wang J, Lee PS. Fast charging self-powered electric double layer capacitor. *Journal of Power Sources*. **342**, 70, 2017.

36. Lee B-S, Yoon J, Jung C, Kim DY, Jeon S-Y, Kim K-H, Park J-H, Park H, Lee KH, Kang Y-S, Park J-H, Jung H, Yu W-R, Doo S-G. Silicon/carbon nanotube/$BaTiO_3$ nanocomposite anode: Evidence for enhanced lithium-ion mobility induced by the local piezoelectric potential. *ACS Nano*. **10**(2), 2617, 2016.

37. Kim S, Choi SJ, Zhao K, Yang H, Gobbi G, Zhang S, Li J. Electrochemically driven mechanical energy harvesting. *Nature Communications*. **7**(1), 10146, 2016.

38. Lang SB, Tofail SAM, Kholkin AL, Wojtaś M, Gregor M, Gandhi AA, Wang Y, Bauer S, Krause M, Plecenik A. Ferroelectric polarization in nanocrystalline hydroxyapatite thin films on silicon. *Scientific Reports*. **3**(1), 2215, 2013.

39. Baxter FR, Bowen CR, Turner IG, Dent ACE. Electrically active bioceramics: A review of interfacial responses. *Annals of Biomedical Engineering*. **38**(6), 2079, 2010.

40. Touach N, Ortiz-Martínez VM, Salar-García MJ, Benzaouak A, Hernández-Fernández F, de Ríos AP, El Mahi M, Lotfi EM. On the use of ferroelectric material $LiNbO_3$ as novel photocatalyst in wastewater-fed microbial fuel cells. *Particuology*. **34**, 147, 2017.

41. Jacques E, Kjell MH, Zenkert D, Lindbergh G. Piezo-electrochemical effect in lithium-intercalated carbon fibres. *Electrochemistry Communications*. **35**, 65, 2013.

42. Jacques E, Lindbergh G, Zenkert D, Leijonmarck S, Kjell MH. Piezo-electrochemical energy harvesting with lithium-intercalating carbon fibers. *ACS Applied Materials & Interfaces*. **7**(25), 13898, 2015.

43. Ikeda S, Takata T, Kondo T, Hitoki G, Hara M, Kondo JN, Domen K, Hosono H, Kawazoe H, Tanaka A. Mechano-catalytic overall water splitting. *Chemical Communications.* **20**, 2185, 1998.
44. Bai X, Wei J, Tian B, Liu Y, Reiss T, Guiblin N, Gemeiner P, Dkhil B, Infante IC. Size effect on optical and photocatalytic properties in $BiFeO_3$ nanoparticles. *The Journal of Physical Chemistry C.* **120**(7), 3595, 2016.
45. Lachheb H, Puzenat E, Houas A, Ksibi M, Elaloui E, Guillard C, Herrmann J-M. Photocatalytic degradation of various types of dyes (Alizarin S, Crocein Orange G, Methyl Red, Congo Red, Methylene Blue) in water by UV-irradiated titania. *Applied Catalysis B: Environmental.* **39**(1), 75, 2002.
46. Petrolekas PD, Maggenakis G. Kinetic studies of the liquid-phase adsorption of a reactive dye onto activated lignite. *Industrial & Engineering Chemistry Research.* **46**(4), 1326, 2007.
47. Li Q, Yue Q-Y, Su Y, Gao B-Y, Li J. Two-step kinetic study on the adsorption and desorption of reactive dyes at cationic polymer/bentonite. *Journal of Hazardous Materials.* **165**(1), 1170, 2009.
48. Li J, Cai J, Zhong L, Wang H, Cheng H, Ma Q. Adsorption of reactive dyes onto chitosan/montmorillonite intercalated composite: Multi-response optimization, kinetic, isotherm and thermodynamic study. *Water Science and Technology.* **77**(11), 2598, 2018.
49. Khan HM, Anwar M, Ahmad G. Effect of temperature and light on the response of an aqueous coumarin dosimeter. *Journal of Radioanalytical and Nuclear Chemistry.* **200**(6), 521, 1995.
50. Mondal D, Roy S, Bardhan S, Roy J, Kanungo I, Basu R, Das S. Recent advances in piezocatalytic polymer nanocomposites for wastewater remediation. *Dalton Transactions.* **51**(2), 451, 2022.
51. Xu S, Qian W, Zhang D, Zhao X, Zhang X, Li C, Bowen CR, Yang Y. A coupled photo-piezo-catalytic effect in a BST-PDMS porous foam for enhanced dye wastewater degradation. *Nano Energy.* **77**, 105305, 2020.
52. Qian W, Yang W, Zhang Y, Bowen CR, Yang Y. Piezoelectric materials for controlling electro-chemical processes. *Nano-Micro Letters.* **12**, 1, 2020.
53. Ma J, Ren J, Jia Y, Wu Z, Chen L, Haugen NO, Huang H, Liu Y. High efficiency bi-harvesting light/vibration energy using piezoelectric zinc oxide nanorods for dye decomposition. *Nano Energy.* **62**, 376, 2019.
54. Shaukat H, Ali A, Bibi S, Mehmood S, Altabey WA, Noori M, Kouritem SA. Piezoelectric materials: Advanced applications in electro-chemical processes. *Energy Reports.* **9**, 4306, 2023.

55. Singh G, Sharma M, Vaish R. Flexible Ag@LiNbO$_3$/PVDF composite film for piezocatalytic dye/pharmaceutical degradation and bacterial disinfection. *ACS Applied Materials & Interfaces*. **13**(19), 22914, 2021.

56. Ajmal A, Majeed I, Malik RN, Idriss H, Nadeem MA. Principles and mechanisms of photocatalytic dye degradation on TiO$_2$ based photocatalysts: A comparative overview. *RSC Advances*. **4**(70), 37003, 2014.

57. Fida H, Zhang G, Guo S, Naeem A. Heterogeneous Fenton degradation of organic dyes in batch and fixed bed using La-Fe montmorillonite as catalyst. *Journal of Colloid and Interface Science*. **490**, 859, 2017.

58. Yu D, Liu Z, Zhang J, Li S, Zhao Z, Zhu L, Liu W, Lin Y, Liu H, Zhang Z. Enhanced catalytic performance by multi-field coupling in KnbO$_3$ nanostructures: Piezo-photocatalytic and ferro-photoelectrochemical effects. *Nano Energy*. **58**, 695, 2019.

59. Mushtaq F, Chen X, Hoop M, Torlakcik H, Pellicer E, Sort J, Gattinoni C, Nelson BJ, Pané S. Piezoelectrically enhanced photocatalysis with BiFeO$_3$ nanostructures for efficient water remediation. *iScience*. **4**, 236, 2018.

60. Xue X, Zang W, Deng P, Wang Q, Xing L, Zhang Y, Wang ZL. Piezo-potential enhanced photocatalytic degradation of organic dye using ZnO nanowires. *Nano Energy*. **13**, 414, 2015.

61. Singh S, Khare N. Coupling of piezoelectric, semiconducting and photoexcitation properties in NaNbO$_3$ nanostructures for controlling electrical transport: Realizing an efficient piezo-photoanode and piezo-photocatalyst. *Nano Energy*. **38**, 335, 2017.

62. Li J, Cai L, Shang J, Yu Y, Zhang L. Giant enhancement of internal electric field boosting bulk charge separation for photocatalysis. *Advanced Materials*. **28**(21), 4059, 2016.

63. Lan S, Chen Y, Zeng L, Ji H, Liu W, Zhu M. Piezo-activation of peroxymonosulfate for benzothiazole removal in water. *Journal of Hazardous Materials*. **393**, 122448, 2020.

64. Wu JM, Sun Y-G, Chang W-E, Lee J-T. Piezoelectricity induced water splitting and formation of hydroxyl radical from active edge sites of MoS$_2$ nanoflowers. *Nano Energy*. **46**, 372, 2018.

65. Sartbaeva A, Kuznetsov V, Wells SA, Edwards P. Hydrogen nexus in a sustainable energy future. *Energy & Environmental Science*. **1**(1), 79, 2008.

66. Hu K, Wang R, Maimaitituersun M, Sun H, Liu X, Zhang Y. Synergetic piezo-photocatalytic effect in ultrathin Bi$_2$WO$_6$ nanosheets for

decomposing organic dye. *Journal of Materials Science: Materials in Electronics*. **33**(13), 9845, 2022.

67. Green L. Energy needs versus environmental pollution: A reconciliation? *Science*. **156**(3781), 1448, 1967.
68. Park S, Chang WJ, Lee CW, Park S, Ahn H-Y, Nam KT. Photocatalytic hydrogen generation from hydriodic acid using methylammonium lead iodide in dynamic equilibrium with aqueous solution. *Nature Energy*. **2**(1), 16185, 2016.
69. Zhang P, Guo Y-J, Chen J, Zhao Y-R, Chang J, Junge H, Beller M, Li Y. Streamlined hydrogen production from biomass. *Nature Catalysis*. **1**(5), 332, 2018.
70. Qian W, Yang W, Zhang Y, Bowen CR, Yang Y. Piezoelectric materials for controlling electro-chemical processes. *Nano-Micro Letters*. **12**(1), 149, 2020.
71. Cornejo JA, Sheng H, Edri E, Ajo-Franklin CM, Frei H. Nanoscale membranes that chemically isolate and electronically wire up the abiotic/biotic interface. *Nature Communications*. **9**(1), 2263, 2018.
72. Lin L, Zhou W, Gao R, Yao S, Zhang X, Xu W, Zheng S, Jiang Z, Yu Q, Li Y-W, Shi C, Wen X-D, Ma D. Low-temperature hydrogen production from water and methanol using Pt/α-MoC catalysts. *Nature*. **544**(7648), 80, 2017.
73. Hu Q, Kim DY, Yang W, Yang L, Meng Y, Zhang L, Mao H-K. FeO_2 and FeOOH under deep lower-mantle conditions and Earth's oxygen–hydrogen cycles. *Nature*. **534**(7606), 241, 2016.
74. Zhang Y, Khanbareh H, Dunn S, Bowen CR, Gong H, Duy NPH, Phuong PTT. High efficiency water splitting using ultrasound coupled to a $BaTiO_3$ nanofluid. *Advanced Science*. **9**(9), 2105248, 2022.
75. Zhang J, Wu Z, Jia Y, Kan J, Cheng G. Piezoelectric bimorph cantilever for vibration-producing-hydrogen. *Sensors*. **13**(1), 367, 2012.
76. Tiersten HF. *Linear Piezoelectric Plate Vibrations: Elements of the Linear Theory of Piezoelectricity and the Vibrations Piezoelectric Plates*. Springer, 2013.
77. Wu J, Xu Q, Lin E, Yuan B, Qin N, Thatikonda SK, Bao D. Insights into the role of ferroelectric polarization in piezocatalysis of nanocrystalline $BaTiO_3$. *ACS Applied Materials & Interfaces*. **10**(21), 17842, 2018.
78. Ali A, Pasha RA, Elahi H, Sheeraz MA, Bibi S, Hassan ZU, Eugeni M, Gaudenzi P. Investigation of deformation in bimorph piezoelectric actuator: Analytical, numerical and experimental approach. *Integrated Ferroelectrics*. **201**(1), 94, 2019.

79. Shaukat H, Ali A, Bibi S, Altabey WA, Noori M, Kouritem SA. A Review of the recent advances in piezoelectric materials, energy harvester structures, and their applications in analytical chemistry. *Applied Sciences.* **13**(3), 1300, 2023.

80. Green Jr L. Energy needs versus environmental pollution: A reconciliation? A power generation concept by which pollution of air and water can be reduced is proposed. *Science.* **156**(3781), 1448, 1967.

81. Hickman R. Energy without pollution. *Science.* **170**(3953), 17, 1970.

82. Sigfusson TI. Pathways to hydrogen as an energy carrier. *Philosophical Transactions of the Royal Society A: Mathematical, Physical and Engineering Sciences.* **365**(1853), 1025, 2007.

83. Zhang J, Wu Z, Jia Y, Kan J, Cheng G. Piezoelectric bimorph cantilever for vibration-producing-hydrogen. *Sensors.* **13**(1), 367, 2013.

84. You H, Wu Z, Zhang L, Ying Y, Liu Y, Fei L, Chen X, Jia Y, Wang Y, Wang F, Ju S, Qiao J, Lam C-H, Huang H. Harvesting the vibration energy of $BiFeO_3$ nanosheets for hydrogen evolution. *Angewandte Chemie International Edition.* **58**(34), 11779, 2019.

85. Ikeda S, Takata T, Kondo T, Hitoki G, Hara M, Kondo JN, Domen K, Hosono H, Kawazoe H, Tanaka A. Mechano-catalytic overall water splitting. *Chemical Communications.* **20**, 2185, 1998.

86. Pu X, Wang ZL. Self-charging power system for distributed energy: Beyond the energy storage unit. *Chemical Science.* **12**(1), 34, 2021.

87. Xing L, Nie Y, Xue X, Zhang Y. PVDF mesoporous nanostructures as the piezo-separator for a self-charging power cell. *Nano Energy.* **10**, 44, 2014.

88. Kim Y-S, Xie Y, Wen X, Wang S, Kim SJ, Song H-K, Wang ZL. Highly porous piezoelectric PVDF membrane as effective lithium ion transfer channels for enhanced self-charging power cell. *Nano Energy.* **14**, 77, 2015.

89. Simon P, Gogotsi Y, Dunn B. Where do batteries end and supercapacitors begin? *Science.* **343**(6176), 1210, 2014.

90. Qin T, Peng S, Hao J, Wen Y, Wang Z, Wang X, He D, Zhang J, Hou J, Cao G. Flexible and wearable all-solid-state supercapacitors with ultrahigh energy density based on a carbon fiber fabric electrode. *Advanced Energy Materials.* **7**(20), 1700409, 2017.

91. Maitra A, Karan SK, Paria S, Das AK, Bera R, Halder L, Si SK, Bera A, Khatua BB. Fast charging self-powered wearable and flexible asymmetric supercapacitor power cell with fish swim bladder as an efficient natural bio-piezoelectric separator. *Nano Energy*. **40**, 633, 2017.

92. Krishnamoorthy K, Pazhamalai P, Mariappan VK, Nardekar SS, Sahoo S, Kim S-J. Probing the energy conversion process in piezoelectric-driven electrochemical self-charging supercapacitor power cell using piezoelectrochemical spectroscopy. *Nature Communications*. **11**(1), 2351, 2020.

93. Zhang H, Zhang S, Yao G, Huang Z, Xie Y, Su Y, Yang W, Zheng C, Lin Y. Simultaneously harvesting thermal and mechanical energies based on flexible hybrid nanogenerator for self-powered cathodic protection. *ACS Applied Materials & Interfaces*. **7**(51), 28142, 2015.

94. Lin J-H, Tsao Y-H, Wu M-H, Chou T-M, Lin Z-H, Wu JM. Single-and few-layers MoS_2 nanocomposite as piezo-catalyst in dark and self-powered active sensor. *Nano Energy*. **31**, 575, 2017.

Chapter 11

Application-Oriented Materials Development: High-Power Piezoelectric Materials

Most. Israt Jahan,[a] Mehedi Hasan Jihad,[a] Salahuddin Ammar,[a] and Md. Abu Bin Hasan Susan[b]

[a]*Department of Chemistry, University of Dhaka, Dhaka 1000, Bangladesh*

[b]*Dhaka University Nanotechnology Center, University of Dhaka, Dhaka 1000, Bangladesh*

susan@du.ac.bd

Piezoelectric materials have gained substantial interest owing to their wide-ranging applications across energy harvesting, sensing, actuation, and medical devices. The development of high-power piezoelectric materials (HPPMs) represents a multidisciplinary endeavor that combines materials science, engineering, and physics principles to unlock new frontiers in energy conversion and technological advancement. This chapter provides an overview of the strategies, challenges, and achievements in developing piezoelectric

Piezoelectric Materials: Design and Applications
Edited by N. B. Singh and Dev Kumar Mahato

ISBN 978-981-5129-40-3 (Hardcover), 978-1-003-59897-8 (eBook)
www.jennystanford.com

materials capable of efficient energy conversion and high-power output with enhanced power capabilities. Fundamental principles of piezoelectricity and its significance in various technological domains have been introduced. The subsequent sections highlight the strategies employed in the design and synthesis of HPPMs. These include criteria for material selection, crystallographic considerations, and the role of dopants and defects in enhancing piezoelectric performance. Finally, the potential future directions in the field of HPPMs are also discussed, with a focus on ongoing research efforts and emerging opportunities for innovation.

11.1 Introduction

Piezoelectric energy harvesting devices have gained significant attention as viable energy sources due to their steady operational characteristics and capacity to convert vibrational energy into electrical energy, particularly at relatively low frequencies. Piezoelectric materials (PZMs) in such devices are capable of interchangeably converting energy, for instance, electrical into vibrational or mechanical. They have been found useful as actuators, transducers, and sensors exploiting both the direct and inverse effects. Notably, the use of PZMs capable of withstanding high power and elevated temperatures has witnessed a remarkable upsurge over the past years. Moreover, there has been a growing fascination with PZMs as integral elements in self-powered sensors, actuators, and energy harvesters, marking a contemporary trend in their utilization [1]. Of all these devices, eyes have been pointed on energy harvesters capable of efficiently producing electricity without much limitations. This increased attention is attributable to their potential to effectively transform dispersed mechanical energy into usable electrical power. The advancement of science and technology has led to a notable rise in the usage of electrical and electronic equipment within high-temperature piezoelectric devices, consequently driving up the need for such equipment (Fig. 11.1).

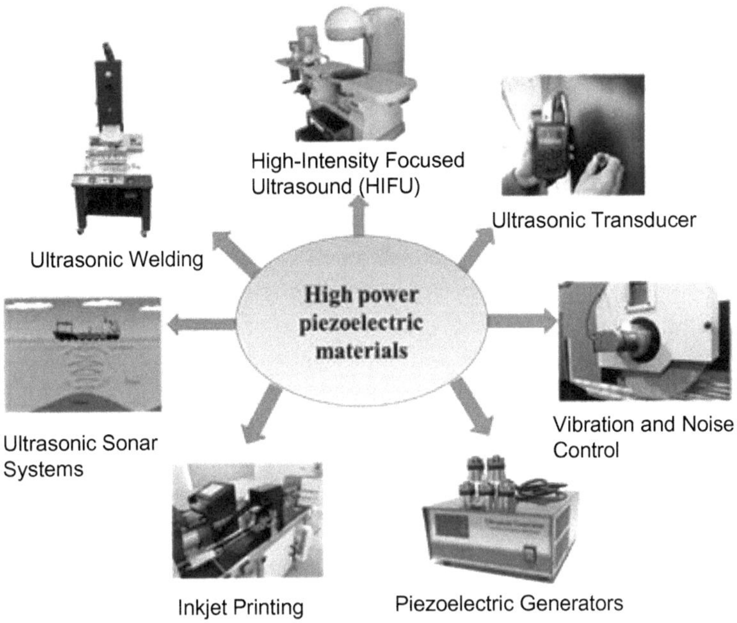

Figure 11.1 Some applications of high-power piezoelectric materials.

Enhancement of the properties of materials at elevated temperatures has emerged as a fascinating domain within the area of functional ceramics [2]. The need to downsize actuators and transducers, while maintaining mechanical energy in a smaller sample without a rise in temperature from inherent losses in high-power piezoelectric materials (HPPMs), has been crucial. Mechanical losses stem from polarization due to rotation and domain wall movement induced by external forces, giving rise to both inherent and external elements [3]. There is a strong demand for PZMs that combine soft and hard attributes, particularly in high-power appliances. Yet, simultaneous attainment of hard characteristics (like a high coercive field (E_c) and high mechanical quality factor (Q_m)) alongside soft traits (such as a high electromechanical coupling coefficient (k) and high piezoelectric constant (d)) in these materials presents a significant challenge. This challenge originates from the approach since controlling the hard behavior of the materials tends to diminish their soft behavior substantially [4]. The generation of substantial mechanical power requires the presence of elevated vibration velocity, often denoted as v_{rms} [5]. Specific materials require a high Q_m and a high d to attain elevated v_{rms} effectively. Additionally, to ensure proficient conversion of electrical to mechanical energy when subjected to elevated electric fields, it becomes imperative to feature high E_c, minimal dielectric/mechanical losses (tanδ), and a substantial k. Consequently, meeting these criteria demands the coexistence of soft and hard attributes within HPPMs. Soft piezoelectrics, recognized for their delicate sensitivity, find extensive use in applications such as medical imaging, sensors, multilayer ceramic capacitors, transducers, and filters [6]. To induce a soft effect in conventional piezoelectric ceramics, typical strategies involve the introduction of ions of high valency [7] or analogous ions [8], which *inter alia* include: Sb^{5+} [9], Nb^{5+} [10], Sr^{2+} [11, 12], La^{3+} [13, 14], and W^{6+} [15] through doping. Furthermore, enhancement of soft characteristics of PZMs involves approaches like the creation of morphotropic phase boundaries (MPB) [16], nano-domains [17], polymorphic phase boundaries (PPB) [18], and polar nano-regions [19]. Hard ceramics constitute an alternative category of PZMs. It is well-known that soft ceramics typically suffer from a noticeable drawback of low Q_m and high tanδ, leading to substantial heat dissipation during operation. In contrast, hard ceramics stand out for their distinguishing attributes of high Q_m and minimal losses. As contemporary electronic ceramic industry technology advances, deployment of PZMs extends far beyond industrial and civilian products, encompassing a diverse array of applications. This spectrum is depicted in Fig. 11.1, spanning military and high-tech domains. HPPMs have extensive applications across diverse industries and technologies, which are crucial for achieving efficient energy conversion and substantial mechanical output. These materials are particularly prevalent in sectors such as ultrasonic cleaning, which serves numerous functions. In ultrasonic cleaning apparatus, high-power piezoelectric transducers are integral components, effectively eliminating contaminants from surfaces. Moreover, HPPMs are pivotal in various other contexts. For instance, the power ultrasonic welding equipment enables the fusion of materials like plastic and metal through high-frequency vibrations. Medical applications benefit from high-intensity focused ultrasound therapy, which leverages these materials to precisely direct ultrasound waves for tasks like tumor ablation and non-invasive surgery. Underwater sonar systems, crucial for navigation, depth sounding, and fish detection, rely on high-power piezoelectric transducers to generate and receive acoustic waves. In printing, piezoelectric actuators in inkjet printers enable high-resolution output by meticulously controlling the ejection of ink droplets onto paper. Beyond these applications, HPPMs contribute to vibration damping and noise control in civil engineering, automotive, and aerospace domains. They also serve as efficient energy harvesters, converting mechanical movements or vibrations into electrical energy, notably for low-power electronic devices and sensors. The versatility and importance of HPPMs are evident in their extensive applications, enhancing performance, efficiency, and precision across various sectors. The development of HPPMs has made significant progress, yet there remains a notable scientific gap that presents an exciting challenge for researchers and engineers. Bridging this gap requires advanced techniques for characterizing and tailoring domain structures on a nano-scale level. One such gap lies in the understanding and manipulation of domain engineering for enhanced power output. Although domain engineering has shown promise in enhancing piezoelectric performance, a comprehensive understanding of how domain boundaries, sizes, and orientations influence power generation remains elusive. High-performance computing can enable researchers to predict how different domain configurations impact power output, guiding experimental efforts. This might involve exploring novel compositions, crystal structures, or engineered defects to promote specific domain patterns favorable for high-power output. By addressing these challenges and advancing our understanding of domain engineering in high-power PZMs, researchers can unlock new avenues for designing and optimizing materials with unprecedented power output capabilities. Over the years, extensive attempts have been made to enhance the power output of piezoelectric devices to meet increasing demands of modern technologies. This chapter presents an in-depth exploration of the multifaceted journey in developing HPPMs, encompassing the principles, challenges, advancements, and potential future directions. It delves into the captivating journey of developing HPPMs, a path defined by intricate material design, ingenious engineering techniques, and a profound understanding of the underlying scientific principles.

11.2 Parameters Related to the Development of HPPMs

Mechanical quality factor: The Q_m measures the energy dissipation or loss when it undergoes mechanical vibrations. A higher Q_m indicates lower energy loss and better efficiency in converting mechanical vibrations to electrical signals or *vice versa* at a specific resonant frequency. Conversely, a lower Q_m value indicates higher energy dissipation and reduced efficiency. In the context of PZMs, Q_m is associated with the sharpness or narrowness of mechanical resonance of a material. Mathematically, Q_m can be described as:

$$Q_m = \frac{\text{Resonant frequency}}{\text{FWHM}}$$

where resonant frequency is the frequency corresponding to the maximum response of the materials to mechanical vibrations, and FWHM refers to full width at half maximum. A higher Q_m is desirable in many applications where efficient energy conversion, sensitive sensors, or accurate actuators are needed.

Electromechanical coupling coefficient: The k is a fundamental parameter that characterizes the efficiency of energy conversion between electrical and mechanical forms in PZMs. It quantifies the strength of interaction between electrical and mechanical behaviors of these materials. In piezoelectric devices mathematically, k is expressed as:

$$k^2 = \frac{\text{Mechanical output}}{\text{Electrical input}}$$

The size and vibration patterns heavily influence the properties of PZMs. As a result, several electromechanical coupling factors exist, such as k_p, k_t, k_{31}, k_{33}, and k_{15}.

Piezoelectric coefficient: The d is a parameter that quantifies the piezoelectric response of a material along a specific direction. It represents the charge produced for every unit of applied mechanical stress or the strain or deformation produced for every unit of the applied electric field. Mathematically,

$$d = \frac{Q}{F}$$

Here, Q is the charge yielded on the surface of the sample, and F is the force applied to the sample. The piezoelectric charge constant d_{ij} is also anisotropic and directions of induced polarization and applied stress are indicated by the subscripts. The axes are denoted as 1, 2, and 3 for the x, y, and z axes, respectively, and 4, 5, and 6 represent the shear about these axes. For instance, d_{33} denotes the induced polarization in the z-axis for every unit stress applied in the z-axis.

Dielectric loss: The tanδ measures the energy dissipation or loss in a dielectric material, including PZMs, when an alternating electric field is applied. It quantifies the ratio of the loss of heat energy to the storage of energy of the material in an electric field during one complete cycle of the field. In PZMs, tanδ is particularly important in applications where alternating electric fields are involved, such as piezoelectric sensors, transducers, and actuators. High tanδ values signify greater energy losses as a result of electrical energy converted to heat, which can be detrimental to the efficiency and performance of devices. Mathematically,

$$\tan\delta = \frac{\text{Imaginary part of complex permittivity}}{\text{Real part of complex permittivity}}$$

Complex permittivity is a complex quantity that describes how the material responds to an alternating electric field. The imaginary part of the complex permittivity represents the loss of energy as heat, and the real part represents the energy stored in the electric field.

Vibration velocity: The v_{rms} in PZMs refers to the speed at which a material undergoes mechanical oscillations or vibrations when applied to an alternating electric field. PZMs exhibit the converse piezoelectric effect, meaning they can deform or vibrate under an electric field, and vice versa – they can generate an electric field when mechanically deformed. Under an alternating electric field, a PZM undergoes rapid mechanical oscillations. The speed at which these oscillations occur is the v_{rms}, which is influenced by several factors, including piezoelectric and mechanical properties of the material, and the frequency and amplitude of the electric field applied.

Dielectric constant: The ε quantifies how much the electric field within a material is diminished in comparison to the electric field in a vacuum. It evaluates how well an electric field can polarize a material, causing the charge to separate within the material. A material with a higher ε can store more electric charge in an electric field, making it more effective at polarizing and redistributing electric charges within the material. Since ε can vary with frequency, measuring it at a specific frequency is common to provide a standardized value relevant to particular applications.

11.3 Strategies for Improving Q_m

11.3.1 Oxygen Vacancy

The impact of vacancies on the Q_m of HPPMs is a topic of great interest and significance. The performance of PZMs is significantly impacted by vacancies, a frequent type of defect. Vacancies, which are imperfections or atoms missing from the crystal lattice of a material, can significantly affect the overall performance of the material, particularly in terms of its Q_m. In perovskite PZMs, two common types of vacancies are A-site and oxygen vacancies. The A-site substitution of high-valent ions with their low-valent counterparts frequently results in oxygen vacancies. Oxygen vacancy has the potential to significantly affect the performance of PZMs. Notably, it has a favorable effect on crucial electromechanical characteristics such as d_{33} and k_p. The Coulomb force influences their interactions with B-site acceptor ions, which frequently result in defect dipoles. To establish charge balance, oxygen vacancies are injected at the B site to replace high-valent acceptor impurity ions.

When T > Curie temperature (T_C), some crystals display a centrosymmetric cubic paraelectric state. In contrast, when $T < T_C$, low-valent ions diverge from the core of the oxygen octahedron. A polar tetragonal ferroelectric state is manifested in Fig. 11.2. It has been thoroughly investigated and described how oxygen vacancies and the generation of P_D interact to increase Q_m, where P_i^O and P_i^V denote conditional probability of finding an O^{2-} ion and O^{2-} vacancy at i site (i = 1, 2, 3, 4), respectively [20]. The distribution of vacancies adds a laborious component to the process of formation and reorientation of defect dipole polarization (P_D). The P_D is a significant phenomenon that can develop when PZMs have defects like vacancies, interstitials, or impurities. The total polarization response, dielectric characteristics, and even piezoelectric performance can all be significantly impacted by the P_D. Due to oxygen vacancy, both the polarization domain alignment and defect dipoles of the material may be impacted. The efficiency of energy conversion may then be impacted by the P_D. Consequently, the P_D is not immediately affected by a change in crystal symmetry. Defect dipoles automatically follow this symmetry-preserving rule. It is generally accepted that the major ways to increase Q_m and reduce losses involve lowering the activity or preventing the reversal of the ferroelectric domain and domain wall motion.

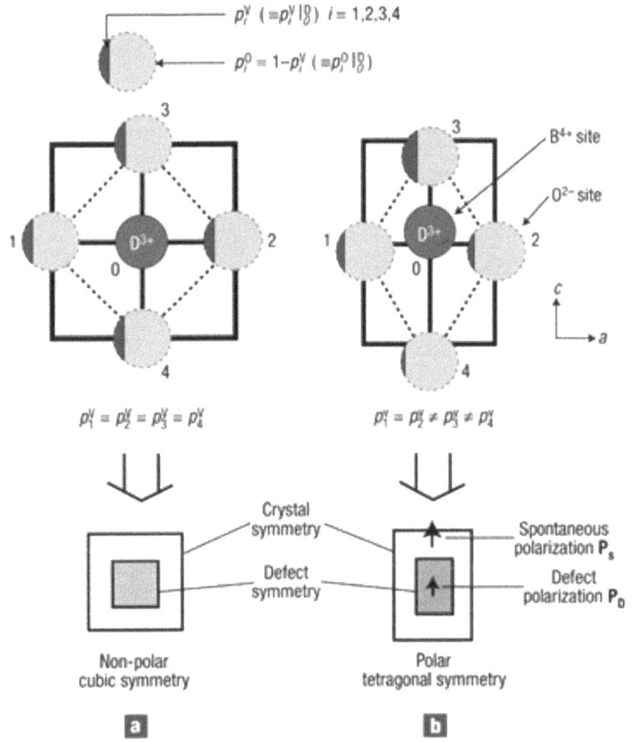

Figure 11.2 Property of point defects conforming symmetry: (a) cubic paraelectric and (b) tetragonal ferroelectric states [20].

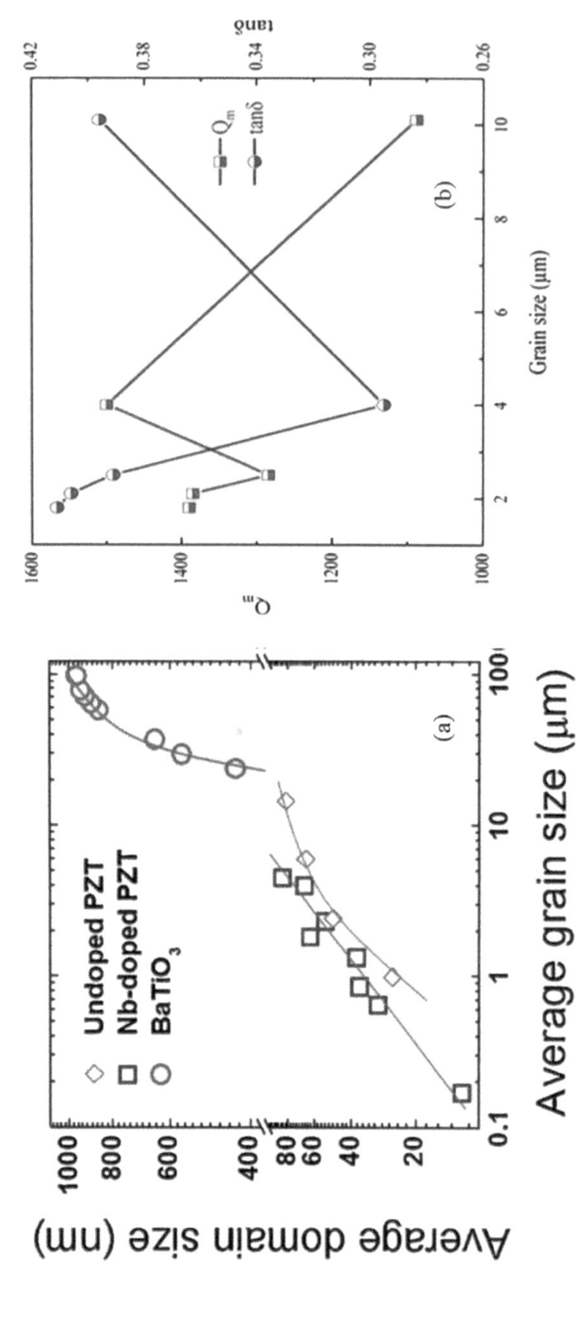

Figure 11.3 An illustration depicting how (a) the size of a domain and (b) the parameter Q_m are interconnected concerning the size of grains [21, 23].

11.3.2 Grain Size

Controlling grain size effectively within reasonable limits improves the electromechanical properties of PZMs. Primary determinants to achieve an optimal Q_m through grain manipulation are domain wall activities and grain boundary effects. These boundaries anchor and stabilize domain walls, limiting domain switching and reducing heat loss caused by domain wall movements. When grain size is reduced, the overall surface area increases, resulting in higher Q_m and lower losses due to improved domain pinning. The grain boundary effect is especially noticeable during this phase. However, as shown in Fig. 11.3a, domain size is not a constant property of a material; rather, it increases as the grain size increases [21, 22]. Smaller domain size tends to induce high activities of domain walls and extrinsic loss of energy, consequently lowering Q_m. The highest Q_m value is attained when the average grain size reaches an optimal point, as illustrated in Fig 11.3b [23]. Notably, this optimum value varies for different piezoelectric systems. Once the grain size surpasses this optimum value, the overall surface area of the system decreases, resulting in a diminished Q_m value. Conversely, if the grain size falls below the optimum value, the lowering in Q_m can be attributed to an increase in activities of the domain wall and extrinsic loss of heat. Thus, the effect of grain size on Q_m originates from the interaction of domain wall activities and grain boundary effects. When the grain boundary effect is greater, smaller grain sizes promote the enhancement of Q_m, whereas larger grain sizes enhance Q_m when domain wall activities are more significant.

11.3.3 Sintering Process

Sintering, which involves increased heating of a material below its melting point to promote particle fusion, is a crucial step in the development of polycrystalline ceramics. This process reshapes the structure of a material and significantly impacts its properties. The process controls a number of factors, including sample quality, element volatilization, composition distribution, density, grain size, and porosity. These factors are particularly vital in dictating the electromechanical attributes of polycrystalline ceramics. Notably, sintering profoundly influences the refinement of PZMs. The sintering of piezoelectric ceramics typically encompasses diverse methods like traditional solid-phase, microwave sintering, spark plasma, and hot-press. Within high-power piezoelectric ceramics fabrication, two primary factors for controlling electromechanical performance and achieving advancements are the sintering temperature and using sintering aids. The chosen sintering temperature profoundly shapes Q_m and electromechanical characteristics during the creation of polycrystalline ceramics. Employing an appropriate sintering temperature and duration can elevate density, diminish defects, and curtail energy losses. Elevated sintering temperatures foster enhanced bonding and reduced porosity, decreasing internal friction and energy dissipation and improving Q_m. Furthermore, sintering temperature impacts grain growth, amplifying mechanical coupling to curtail energy loss and elevate the Q_m. Sintering aids, also called sintering additives or dopants, are minute quantities of substances introduced during the sintering process to enhance the performance of PZMs. These additives optimize electromechanical attributes, decrease energy consumption, and augment material density. Skillful control over sintering temperature, duration, and the incorporation of sintering aids in reducing the mutual constraints among Q_m, d_{33}, and k_p, enabling comprehensive electromechanical performance.

11.4 High Electromechanical Piezoelectric Perovskites

Perovskite-structured PZMs display a more pronounced piezoelectric effect compared to other substances. Perovskite-type PZMs have been interesting due to their distinct crystal arrangement and superior electromechanical performance [24, 25]. The perovskite structure, denoted by the formula ABO_3, as shown in Fig. 11.4, offers notable flexibility regarding chemical composition and overall properties since the atoms are easily interchangeable at both A and B sites [23]. PZMs can be segmented into two primary groups based on their composition: those containing lead and lead-free. Furthermore, Pb-based PZMs can be grouped into lead zirconate titanate (PZT) and lead titanate (PT) based. Conversely, Pb-free PZMs can be divided into three subcategories: those built on the foundation of potassium sodium niobate (KNN), bismuth sodium titanate (BNT), and barium titanate (BT).

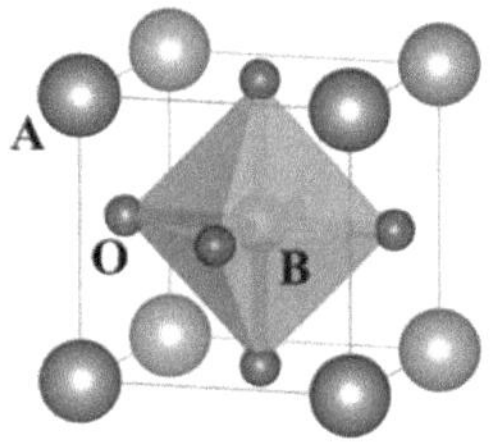

Figure 11.4 Diagram illustrating the optimal ABO_3 configuration in a perovskite structure [23].

11.4.1 Lead-Based Perovskite

Extensive research has been conducted on lead-based perovskite PZMs owing to their exceptional electromechanical capabilities compared to alternative piezoelectric substances. PZT- and PT-based PZMs have notably demonstrated elevated Q_m [26]. The lead-based perovskite PZMs can be further categorized into two distinct subsets: (a) PZT-based perovskites and (b) PT-based perovskites.

11.4.1.1 PZT-based perovskite

PZT materials are desirable for a variety of applications due to their many inherent properties. This stems from their robust k, extensive frequency range, notable temperature stability, elevated T_C, enduring long-term stability, and substantial energy density. Despite these commendable features, certain drawbacks are also noted. PZT materials contain lead, a highly toxic element, and improper handling can give rise to significant environmental and health hazards. PZT materials exhibit hysteresis, indicating that their mechanical and electrical responses are irreversible. On occasion, this phenomenon can result in diminished precision and accuracy. The engagement of high-power applications may result in substantial heat generation within the piezoelectric material due to energy losses. Despite their high T_C, PZT materials possess limitations in withstanding excessive temperatures, potentially jeopardizing their piezoelectric efficacy and leading to thermal malfunction. Conventional commercially available PZT-4 and PZT-8 piezoelectric ceramics boost Q_m up to 500 and 1000, respectively.

However, these values fall short of meeting the demands spurred by steady industrial progress. As a result, numerous researchers have initiated efforts to enhance the electromechanical attributes of PZT materials [27, 28]. Multiple avenues exist for augmenting the overall efficacy of PZT-based PZMs, encompassing techniques like doping, co-doping, sintering, grain size manipulation, aging effects, and enhancement of the T_C. Recent investigations show that the Q_m experiences a reduction with increasing lanthanum concentration. Elevated dopant levels foster lattice defects and disorder, leading to greater energy dissipation and diminished Q_m values. In addition, sintering induces vaporization of PbO and improves agglomeration with the incorporation of lanthanum. This inadvertently promotes enlargement of the grain size, which contributes to greater energy dissipation and reduced Q_m. Furthermore, these studies reveal that ε and electrochemical coupling elevate with augmented lanthanum concentration and sintering temperature [29].

In the realm of doping and co-doping, the attenuation of internal friction through interactions with defects and grain boundaries leads to reduced energy dissipation and an elevation in Q_m values [30]. Moreover, the electrical conductivity of PZT materials experiences an increase due to the presence of oxygen vacancies acting as charge carriers. These vacancies enhance electrical conductivity, rendering the material more conductive. Introducing an electronegative dopant induces oxygen vacancies and consequently influences overall performance. Another intriguing development pertains to the co-doping of F^- ions in PZT material alongside Mn and Mg. This co-doping strategy enhances overall Q_m and d_{33} values as shown in Fig. 11.5, thus promoting high-power conversion. Notably, F^- ions predominantly occupy the O site with a view to optimizing Q_m, differing from A- or B-site considerations [31]. The $Pb(Mn_{1/3}Nb_{2/3})O_3$-$Pb(Zr,Ti)O_3$ (PMN-PZT) system has, at present, been the prime choice for investigating high-electromechanical performance PZMs. The characteristics of PMN-PZT vary depending on the Zr/Ti ratio. For instance, in Fig. 11.6, when the Zr/Ti ratio equals 51/49, a relatively modest Q_m value of 800 is attained near the MPB region. However, with an increase in the Zr content, the Q_m value progressively rises, reaching 1644 at Zr/Ti = 57/43 [32]. A similar observation was reported for $xPb(Yb_{1/2}Nb_{1/2})O_3$-$0.05Pb(Mn_{1/3}Nb_{2/3})O_3$-$(0.95-x)Pb(Zr_{0.48}Ti_{0.52})O_3$+0.25 wt% CeO_2 (PYN-PMN-PZT) ceramics [33]. Near the MPB, the lowest value of Q_m is attained, while the highest values for d_{33} and k_p are noted upon the introduction of Sr as a dopant, followed by controlled incorporation of Mn as a co-dopant. This approach has demonstrated that elevated co-dopant levels amplify Q_m while decreasing ε, tanδ, and other parameters.

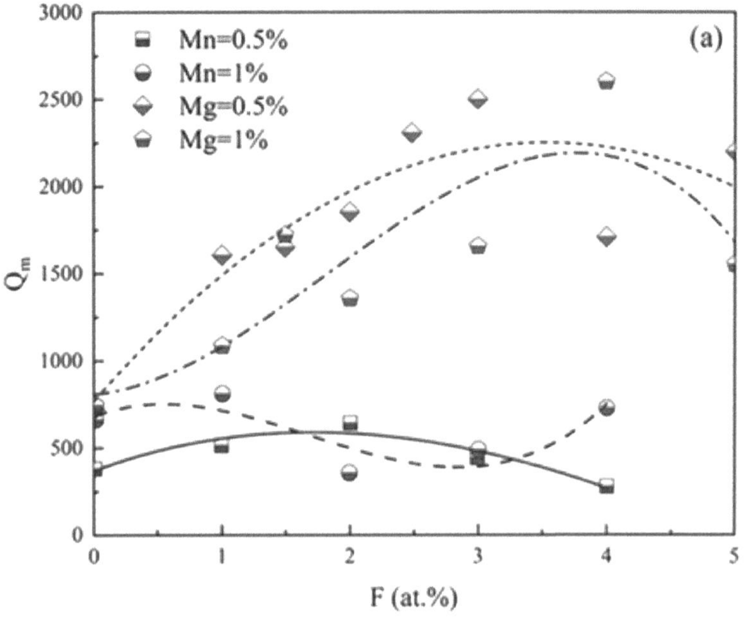

Figure 11.5 Correlation between the Q_m value and the concentration of F in PZT materials with the inclusion of Mn and Mg [31].

Within the single-phase region, the Q_m value surpasses that of the MPB region, and generally a higher Q_m is exhibited by the tetragonal phase compared to the rhombohedral one. Consequently, ceramics near the MPB zone demonstrate a limited ability to enhance Q_m. Alongside intrinsic compositional adjustments incorporating sintering aids is a general practice to enhance electromechanical properties. Table 11.1 highlights important findings addressing the performance of ion-doped piezoelectric ceramics.

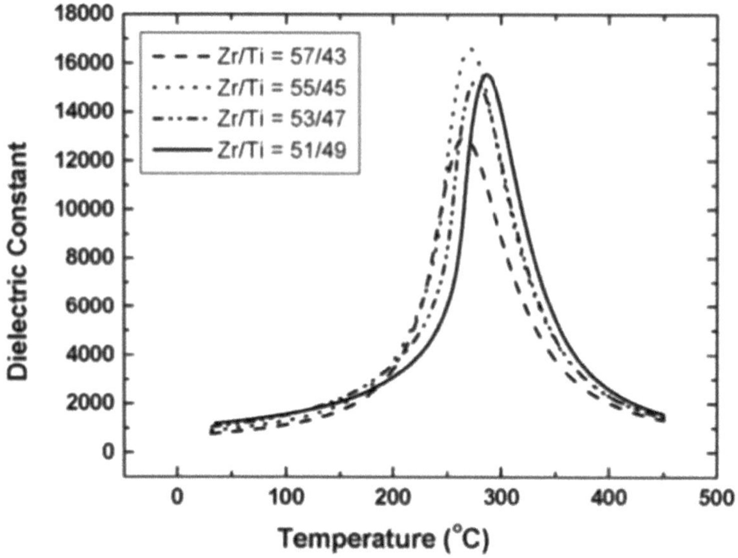

Figure 11.6 Variation of the ε with temperature at distinct Zr/Ti ratio [32].

Table 11.1 Electromechanical performance of PZT ceramics doped with different ions [23]

Ceramics	d_{33} (pC/N)	Q_m	k_p	ε	tanδ
$(Pb_{0.994}La_{0.004})(Zr_{0.95}Ti_{0.05})_{0.98}Nb_{0.02}O_3$	70	304	0.163	310	0.025
PZT-4	289	500	0.58	1300	0.004
$Pb(Zr_{0.52}Ti_{0.48})O_3$	223	500	0.52	730	0.001
$Pb_{0.95}Ba_{0.01}Sr_{0.04}(Zr_{0.53}Ti_{0.47})O_3$ + 1 wt% $LiBiO_2$ + 0.06 wt% Cu + 0.25 wt% $MnCO_3$	255	568	0.58	872	0.0058
PZT + 0.4 wt% MnO_2	200	420	0.51	750	0.008
Nb-$Pb(Zr_{0.52}Ti_{0.48})O_3$ + 1 mol% MnO_2	350	430	0.611	1179	0.009
(Mn,F)-$Pb_{0.89}(BaSr)_{0.11}(Zr_{0.52}Ti_{0.48})O_3$	330	800	–	–	0.11
PZT-8	225	1000	0.51	1000	0.004

Ceramics	d_{33} (pC/N)	Q_m	k_p	ε	tanδ
$Pb(Zr_{0.52}Ti_{0.48})O_3$	–	447	0.24	1101	–
$(Pb_{0.86}Gd_{0.08}La_{0.06})(Zr_{0.52}Ti_{0.48})O_3$	299	564	0.23	412	0.028
$Pb_{0.96}Sr_{0.04}(Zr_{0.53}Ti_{0.47})_{0.9}Mn_{0.1}O_3$	257	756	0.38	974	0.002
PZT + 0.4 wt% MnO_2 + B_2O_3-Bi_2O_3-CdO	250	1000	0.57	1100	0.0036
Sr-$Pb(Zr_{0.52}Ti_{0.48})O_3$ + 1 mol% MnO_2 + 0.9 mol% Nb	207	1000	0.44	928	0.009
(Mg,F)-$Pb_{0.89}(BaSr)_{0.11}(Zr_{0.52}Ti_{0.48})O_3$	160	2600	–	–	0.002

Improved power output is the focus of current research in PZT-based materials, which has generated great attention and so far led to various initiatives. This endeavor has progressed to encompass ternary and quaternary systems employing a methodology akin to binary systems. Table 11.2 presents some selected outcomes showcasing the superior electromechanical capabilities of piezoelectric ceramics based on the PMN-PZT composition.

Table 11.2 Electromechanical performance of PMN-PZT ceramics doped with different ions [23]

Ceramics	d_{33} (pC/N)	Q_m	k_p	ε	tanδ
$[Pb_{0.97}Sr_{0.03}(Mn_{1/3}Nb_{2/3})_{0.07}(Zr_{0.48}Ti_{0.52})_{0.93}O_3]$ + 0.25 wt% CeO_2 + 0.3 wt% Nb_2O_5 + 0.3 wt% CuO + 0.3 wt% Bi_2O_3 + 0.1 wt% Li_2CO_3	–	2414	0.50	1246	–
$0.10Pb(Mn_{1/3}Nb_{2/3})O_3$-$0.91Pb(Zr_{1/2}Ti_{1/2})O_3$ + 0.5 wt% PbO + (Li_2CO_3-CuO-Bi_2O_3)	236	1982	0.584	732	–
$0.9(Pb_{0.94}Sr_{0.06})(Zr_{0.57}Ti_{0.43})O_3$-$0.1Pb(Mn_{1/3}Nb_{2/3})O_3$	154	1644	0.445	501	0.006
$0.07Pb(Mn_{1/3}Nb_{2/3})O_3$-$0.468PbZrO_3$-$0.462PbTiO_3$ + 2CaO-Fe_2O_3	–	2125	0.535	–	0.0025

11.4.1.2 PT-based perovskite

Among the typical piezoelectric substances, PT exhibits a tetragonal structure under normal conditions. PZMs centered around PT are notably less robust than their PZT-based counterparts, owing to the absence of zirconium, which renders the structure less resilient. Consequently, they display restricted mechanical durability and diminished piezoelectric and ferroelectric attributes. Nevertheless, there has been recent progress in developing PT-based PZMs. A thorough electromechanical study on Mn doping on both single crystals and polycrystalline ceramics within the 0.71PMN-0.29PT system shows that the introduction of Mn has led to an increase in the coercive field. This augmentation in the E_c stems from the internal electric field which is generated as Mn is added. Incorporation of an acceptor like Mn into PMN-PT facilitates the formation of oxygen vacancies, which undergo subsequent interaction with the acceptor to give rise to defect dipoles. These, in turn, generate an internal electric field within the crystal lattice, the amplitude of which correlates well with the Q_m. In the context of high Q_m, an internal electric field is thus crucial. Consequently, in Fig. 11.7, the introduction of Mn through doping enhances both E_c and Q_m within the PMN-PT system [34].

Extensive studies have been made on ternary ceramics of the $yPb(In_{½}Nb_{½})O_3$-$(1-x-y)Pb(Mg_{1/3}Nb_{2/3})O_3$-$xPbTiO$ (PIN-PMN-PT) system, both doped with Mn and undoped, focusing on compositions that align with the MPB. A similar outcome is noted in the context of Mn doping within the PIN-PMN-PT system. As the Mn content increases in this system, the coercive field increases, reaching its maximum at a Mn content of 2 mol%. Moreover, at the 2 mol% Mn content level, temperature-dependent ε has been found to decrease. The interplay between ε and E_c signifies a concurrent enhancement of Q_m. Consequently, the overall performance of the PIN-PMN-PT system attains augmentation through the introduction of Mn doping. In this process, Mn, present as Mn^{2+} and Mn^{3+}, acts as a substitution on B sites, creating an oxygen vacancy. This vacancy, in turn, forms defect dipoles, which contribute to attain a notable Q_m as shown in Fig. 11.8 [35].

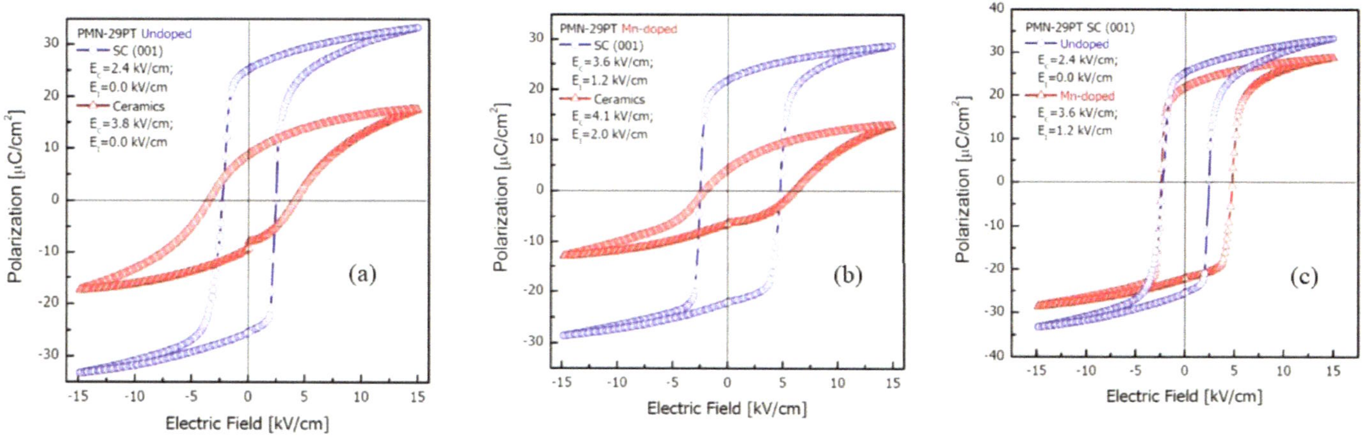

Figure 11.7 Polarization in (a) undoped PMN-PT single crystal and polycrystalline ceramics and (b) doped PMN-PT single crystal and polycrystalline ceramics. (c) Comparison between doped and undoped PMN-PT systems [34].

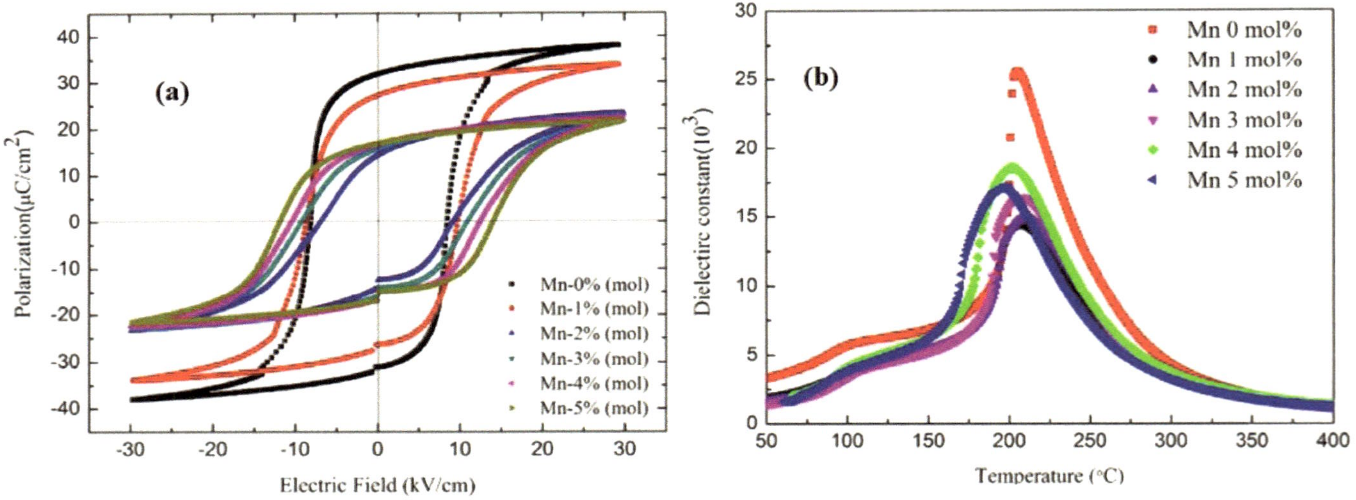

Figure 11.8 (a) Hysteresis curves illustrating the ferroelectric behavior and (b) changes in ε with temperature for PIN-PMN-PT ceramics doped with manganese [35].

Table 11.3 Electromechanical performance of PT-based ceramics doped with different ions [23]

Ceramics	d_{33} (pC/N)	Q_m	k_p	ε	tanδ
$BiScO_3$-$PbTiO_3$-$0.1Pb(Mn_{1/3}Nb_{2/3})O_3$	80	1000	0.2	–	–
$0.36BiScO_3$-$0.64Pb(Ti1_{0.96}Fe_{0.04})O_3$ + 1 mol% Mn	135	350	0.31	782	0.01
$Pb_{0.85}Ba_{0.15}(Mg_{1/3}Nb_{2/3})_{0.4541}(Zn_{1/3}Nb_{2/3})_{0.0982}(Ni_{1/3}Nb_{2/3})_{0.1477}Ti_{0.3}O_3$	7	8284	0.07	7710	0.011
$0.28PbSnO_3$-$0.52PbTiO_3$-$0.2Pb(Mg_{1/3}Nb_{2/3})O_3$	140	570	0.256	1100	0.006
$0.57Pb(Sc_{1/2}Nb_{1/2})O_3$-$0.43PbTiO_3$ + Mn	500	1400	0.64	2500	0.002
$0.24Pb(In_{1/2}Nb_{1/2})O_3$-$0.42Pb(Mg_{1/3}Nb_{2/3})O_3$-$0.34PbTiO_3$ + 1 mol% Mn	430	1296	0.57	1382	0.003
0.71PMgN-0.29PT + Mn	100	500	–	1464	0.005
0.71PMgN-0.29PT	127	200	–	2159	0.025
$0.24Pb(In_{1/2}Nb_{1/2})O_3$-$0.42Pb(Mg_{1/3}Nb_{2/3})O_3$-$0.34PbTiO_3$ + 2 mol% Mn	347	2381	0.57	1214	0.006
$0.58PbSnO_3$-$0.42PbTiO_3$	140	200	0.287	750	0.015
$0.34PbSnO_3$-$0.46PbTiO_3$-$0.2Pb(Mg_{1/3}Nb_{2/3})O_3$	380	690	0.492	2370	0.002
$0.36BiScO_3$-$0.64PbTiO_3$	450	28	0.56	2010	0.04
$0.37BiScO_3$-$0.6PbTiO_3$-$0.03Pb(Mn_{1/3}Nb_{2/3})O_3$ + 3 mol% Fe_2O_3	200	600	0.32	535	0.006

Similar results are observed upon introducing Mn into $Pb(Sc_{1/2}Nb_{1/2})O_3$-PT piezoelectric ceramics. The observed Q_m value exhibits a remarkable amplification by a factor of 28, signifying significant potential for applications demanding high-power output [36]. Ba^{2+} ions within PMN-PZN-PNN-PT piezoelectric ceramics have revealed commendable performance. Upon incorporation of Ba^{2+} ions into the A site, Pb^{2+} ions are displaced. This substitution capitalizes the larger ion radius of Ba^{2+} and comparatively lesser electronegativity than Pb^{2+}. Consequently, addition of Ba^{2+} purifies the quaternary system, a factor contributing to the elevation of Q_m and the reduction of d_{33} within the PMN-PZN-PNN-PT configuration. Notably, the Q_m attains a remarkable value, while d_{33} is modest. However, it is imperative to orchestrate modifications with precision to ensure Q_m enhancement without sacrificing d_{33} [37]. The binary system $BiScO_3$-$PbTiO_3$ (BS-PT) has garnered extensive investigation due to its remarkable piezoelectric attributes and elevated T_C. This system exhibits impressive piezoelectric characteristics, featuring notable d_{33} and 0.56 values of ca. 450 pC/N and 0.56, respectively. However, it is accompanied by a relatively modest Q_m value of 28 and elevated loss. To harness its potential for high-power applications, significant advancements are necessary to enhance the Q_m value [38]. The impact of Fe content on BS-PT piezoelectric ceramics has undergone comprehensive scrutiny, revealing that introducing Fe^{2+} leads to a remarkable augmentation of both Q_m and d_{33}, attaining concurrent values of 600 and 250 pC/N [39]. Table 11.3 represents a few results to demonstrate the improved electromechanical properties of PT-based piezoelectric ceramics.

11.4.2 Lead-Free Perovskite

Due to environmental concerns related to the toxicity of lead, there has been a significant effort to develop lead-free alternatives [40, 41]. Lead-free perovskite PZMs offer the potential for environmentally friendly and sustainable piezoelectric devices. Some of the PZMs of this kind explored are BT, BNT, and KNN.

11.4.2.1 KNN-based perovskite

A mixed piezoelectric system, KNN comprises sodium niobate ($NaNbO_3$) (NN) and potassium niobate (KN), both of which are orthorhombic. This system displays a boundary called the MPB at approximately equal parts of potassium (K) and sodium (Na), around 50% each [42]. The phenomenon of MPB is closely connected with the polymorphic phase transition (PPT), associated with how the boundary undergoes alterations in response to temperature variations. The d_{33} value of KNN obtained by the solid-state reaction is around 80 pC/N. Since alkalis evaporate at high temperatures [43] and the sintering temperature range is narrow [44], achieving dense KNN and excellent characteristics is challenging. Various approaches have so far been suggested and tested to address these challenges to produce KNN ceramics with impressive piezoelectric properties. These include ion substitutions, sintering aids, preparation techniques, and designing the phase boundary near room temperature. By incorporating diverse doping elements, scientists have successfully developed KNN-based ceramics that exhibit properties comparable to PZT. Over the years, there have been notable advancements in KNN-based ceramics, indicating their potential to replace ceramics based on PZT. In advancing KNN-based ceramics, incorporating ions and compounds through doping proves to be an effective strategy for augmenting their piezoelectric attributes. This includes elements like Zr^{4+}, Ag^{+}, Li^{+}, Sb^{5+}, and Ta^{5+} and compounds like $BaTiO_3$, $AgSbO_3$, and $CaTiO_3$, among others. For instance, the introduction of Ag^{+} or $CaTiO_3$ has been observed to enhance T_C and d_{33}. The use of Li^{+} and titanates (such as $CaTiO_3$, $SrTiO_3$, $MgTiO_3$, and $BaTiO_3$) also emerges as a means to enhance the electric properties of KNN-based ceramics while improving temperature stability with a shift in the temperature corresponding to orthorhombic-tetragonal ($T_{O\text{-}T}$) transition below ambient temperature.

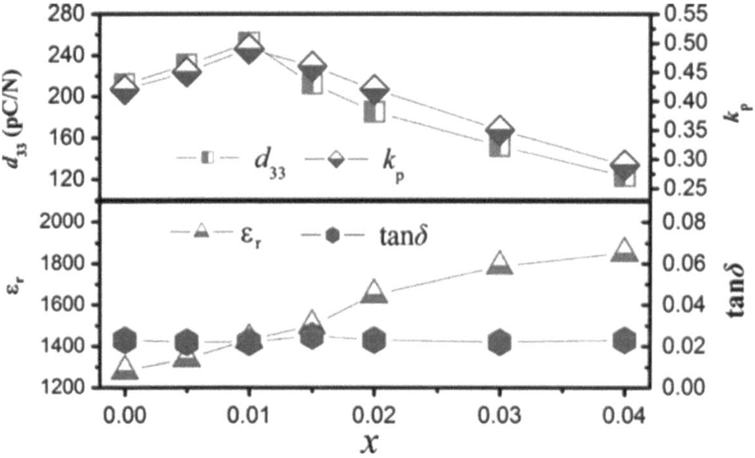

Figure 11.9 The piezoelectric and dielectric characteristics of ceramics containing (1-x)KNLNSxCT with varying values of x [45].

An illustrative case is $(1\text{-}x)[(K_{0.5}Na_{0.5})_{0.94}Li_{0.06}](Nb_{0.94}Sb_{0.06})O_3\text{-}xCaTiO_3$, which reveals tetragonal symmetry in a single-phase structure for the ceramics modified with $CaTiO_3$. With the increase in the x, the T_C and the PPT decrease. As depicted in Fig. 11.9, the d_{33} and k_p values exhibit an initial ascending trend as the $CaTiO_3$ content increases, reaching their maximum values (252 pC/N and 49%) at x = 0.01, followed by a decline with further increase in x [45]. Since K and Na are volatile, the attainment of dense KNN-based ceramics using the conventional method of sintering in the solid state is somewhat hindered. Nonetheless, the addition of sintering aids can enable low-temperature sintering, effectively mitigating the alkali metal volatilization issue. Sintering aids also enhance the electrical properties and density. For instance, $0.98K_{0.5}Na_{0.5}NbO_3$-$0.02BiScO_3$ ceramics doped with 0.8 mol% MnO_2 led to favorable electrical attributes: a d_{33} value of 288 pC/N, ε of 1591, k_p of 0.46, and T_C of 328°C [46]. Furthermore, incorporation of CuO in the $0.96(K_{0.46}Na_{0.54})$-$(Nb_{0.95}Sb_{0.05})O_3$-$0.04Bi_{0.5}(Na_{0.82}K_{0.18})_{0.5}ZrO_3$ system using traditional fabrication exhibited intriguing outcomes [47]. Once CuO content is increased from 0 to 2.5 wt%, sintering temperature decreased from 1095°C to 950°C while still achieving a noteworthy d_{33} value of 310 pC/N. CuO can also serve as a sintering aid to enhance the Q_m. In the (1-x)$(K_{0.5}Na_{0.5})NbO_3$-x$MgTiO_3$ ($0.01 \leq x \leq 0.1$) ceramics doped with 1% CuO, the Q_m reached 1204. In the (1-x)$K_{0.5}Na_{0.5}NbO_3$-x$BiScO_3$ ceramics with CuO, superior electrical properties (d_{33} of 224 pC/N, k_p of 40%, and T_C of 331°C) accompanied by a high Q_m of 288 have been achieved. A comprehensive study involving the doping of 13 different additives in $0.96(K_{0.48}Na_{0.52})(Nb_{0.95}Sb_{0.05})O_3$-$0.04Bi_{0.5}(K_{0.18}Na_{0.82})_{0.5}ZrO_3$ ceramics has been conducted. By optimizing the choice and the amount of additives, remarkable piezoelectric properties, high T_C, and significant converse d_{33} have been successfully achieved [48]. In a heating cycle, a KNN undergoes successive phase transitions: rhombohedral to orthorhombic (R-O) at around 123°C, then orthorhombic to tetragonal (O-T) at approximately 200°C, and finally progressing to a cubic phase at about 410°C. A significant breakthrough was noted in developing KNN ceramics co-modified with lithium (Li^+), tantalum (Ta^{5+}), and antimony (Sb^{5+}) using an innovative reactive template grain method [49]. This specialized system, featuring an O-T phase boundary, showcases an exceptional d_{33} of 416 pC/N achieved through strategic phase boundary construction. The theory of PPT suggests that the transition temperature from R-O ($T_{R\text{-}O}$) and the

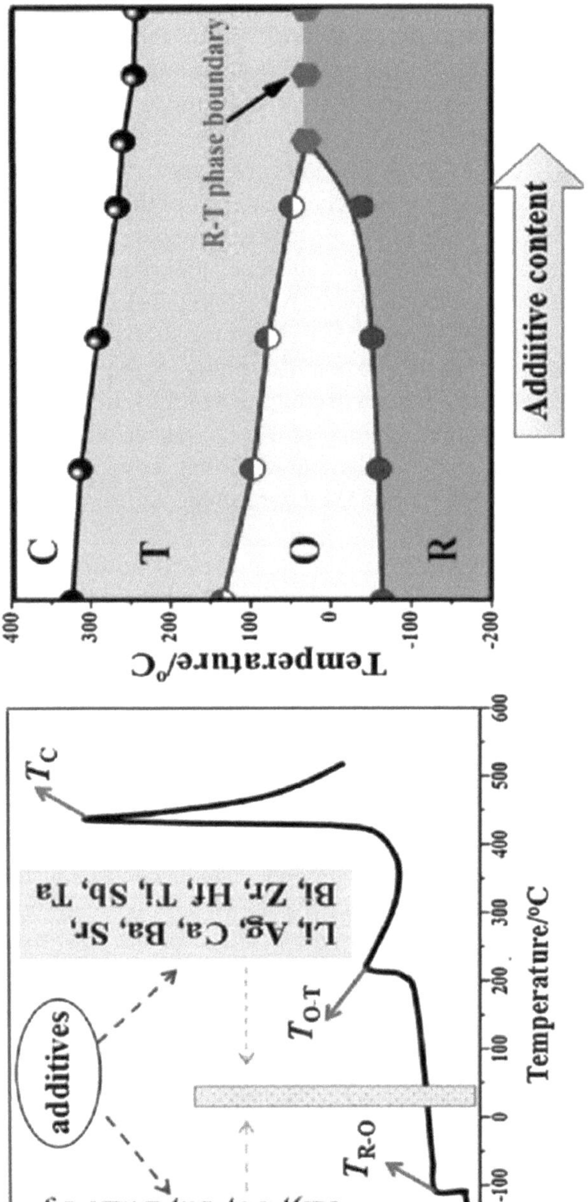

Figure 11.10 Conceptual proposal for novel phase interfaces within KNN materials [50].

T_{O-T} have been adjusted to be in the vicinity of room temperature as a consequence of dopants. As a result, properties are enhanced due to the availability of a more significant number of potential polarization states. After extensive investigation, a novel concept for phase boundary design has been introduced as illustrated in Fig. 11.10 [50]. This involves incorporating specific additives into the KNN matrix to achieve a simultaneous shift of the T_{O-T} and T_{R-O} to room temperature to form a fresh phase boundary. Remarkably, ceramics featuring this newly established R-T phase boundary exhibit exceptionally high piezoelectricity [51–54], marking an exciting advancement. Thus, there is substantial potential within KNN-based ceramics, particularly in attaining high-performance electromechanical capabilities reflected through ultrahigh Q_m, thus enabling their application in high-power scenarios. Beyond the notable challenge of alkaline element volatility, other drawbacks for KNN-based ceramics include the hygroscopic nature of alkaline elements and low ε.

Consequently, our focus should shift toward addressing the issues of high hygroscopicity and volatility in KNN-based ceramics, while also finding ways to enhance d_{33} and ε without compromising the elevated Q_m.

11.4.2.2 BT-based perovskite

BT is the first discovered perovskite PZM. It was one of the earliest materials to exhibit piezoelectric properties and has played a significant role in developing piezoelectric technology. Substantial uses of BT are reported in many electromechanical systems, including transducers, sensors, actuators, and other high-power applications. However, upon subsequent discovery of PZT, its use diminished gradually. BT prepared using traditional solid-phase synthesis and sintering resulted in a d_{33} value of 180 pC/N, which has been insufficient for high-power applications. Various hydrothermal preparation and sintering processes have been developed to make it applicable for high-power applications. The maximum achievable d_{33} values of BT ceramics by different fabrication methods are depicted in Fig. 11.11. While a few methods have been successful in increasing d_{33}, they don't significantly improve Q_m. To enhance Q_m, one can manipulate the grain size. Various fabrication techniques result in BT materials featuring distinct grain sizes. Increasing the grain size causes a proportional increase in Q_m until the grain size reaches a material-specific threshold value.

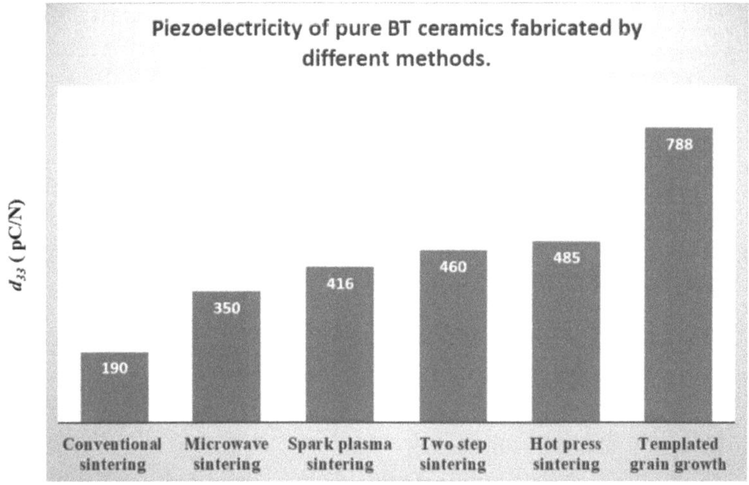

Figure 11.11 Variation in piezoelectric properties of BT ceramics produced through diverse fabrication techniques [55–60].

Table 11.4 shows the relationship between grain size and Q_m. The development of BT faces two major challenges: the need for high sintering temperatures to achieve dense BT and the relatively low T_C. Reduced sintering temperatures can be achieved by incorporating various sintering aids to improve electromechanical performance. For example, when BT is doped with 1 mol% Li_2O, significant improvements are observed, with Q_m reaching 720 and d_{33} measuring 185 pC/N. It is now possible to fine-tune the phase transition temperature of BT by introducing substitutions of various elements such as Zr, Hf, Sn, and Ca, leading to further improvements in its electromechanical properties. An example of this is evident in $(Ba_{0.99}Ca_{0.01})(Ti_{0.99}Co_{0.005}Nb_{0.005})O_3$ doped with 1 mol% Li_2O, where an even more impressive performance is achieved [61]. One of the limitations of BT-based ceramics is the lower T_C, which remains below 100°C. Efforts to raise the T_C have led to successful outcomes, as demonstrated by doping 2 mol% MnO_2 in $(Ba_{0.97}Ca_{0.03})(Ti_{0.96}Sn_{0.035}Hf_{0.035})O_3$, yielding a T_C value of 123°C alongside a moderate Q_m of 340 [62]. Furthermore, the use of $MnCO_3$-doped $(Ca_{0.1}Ba_{0.9})TiO_3$ piezoelectric ceramics has resulted in exceptional results, showcasing ultrahigh Q_m values of 2400 and a significantly elevated T_C of 133°C [63]. The future challenge for BT lies in increasing T_C to make it applicable for high-power applications.

Table 11.4 The relationship between the mechanical Q_m and grain size

Sintering method	Grain size (μm)	Q_m
Two-step sintering	2.5	51
Spark plasma sintering	7.5	530
Traditional high-temperature sintering	26	936

11.4.2.3 BNT-based perovskite

BNT exhibits rhombohedral symmetry at room temperature and contributes to its piezoelectricity. BNT also displays a relatively high T_C of ca. 320°C. Pure BNT piezoelectric ceramics show a moderate Q_m of roughly 366 and a low ε of around 345 at room temperature [64]. Numerous efforts have been made to enhance the electromechanical properties of BNT-based piezoelectric ceramics. Notables are doping modifications, optimized sintering techniques, and compositional adjustments. MnO and MnO_2 doping, particularly, have drawn considerable interest due to their beneficial impact on enhancing BNT ceramics. The incorporation of Mn through doping contributes to a reduction in mechanical, dielectric, and piezoelectric losses. However, to optimize the mechanical property, the appropriate Mn concentration is crucial. In the context of the $Bi_{0.5}(Na_{0.88}K_{0.08}Li_{0.04})_{0.5}(Ti_{1-x}Mn_x)O_3$ system, the optimal Mn content is found to be 1.5 mol%, resulting in a remarkable enhancement of Q_m from 402 to 974 [65]. B-site acceptor doping is another frequent approach for improving performance of BNT. Interestingly, bismuth deficit $x \leq 0.02$ can produce oxygen vacancies in $0.88Bi_{0.50-x}Na_{0.50}TiO_3$-$0.08Bi_{0.50-x}K_{0.50}TiO_3$-$0.04BaTiO_3$ (BNKBT88-xBi) ceramics [66]. BNKBT88 piezoelectric ceramics have a low Q_m of around 150 without a bismuth deficiency. On the other hand, ceramics with Bi deficiency exhibit a low tanδ and a high Q_m of 1200, which is attributable to the pinned domain walls caused by intrinsic flaws generated by Bi non-stoichiometry. Although BNT ceramics have many advantages, a significant barrier arises from their inherent deficiencies in critical parameters such as k_p and d_{33}, which are critical in high-power scenarios. These intrinsic limitations limit their ability to effectively transform and transmit large amounts of energy, limiting their suitability for demanding applications requiring high-power output.

11.5 Non-perovskite Piezoelectric Materials

The phrase "non-perovskite PZMs" describes a class of substances that exhibit piezoelectric action without the perovskite crystal structure. These non-perovskite materials have unique crystal structures and chemical makeups in contrast to perovskite-based PZMs. Due to several factors, such as the demand for new applications, increased performance, and environmental considerations, researchers are constantly investigating and developing these alternative PZMs. Some examples of non-perovskite PZMs include tungsten bronze (TB) structured materials, bismuth layered structured ferroelectric (BLSF) ceramics, niobate structured materials, etc.

11.5.1 TB Structured Materials

TB structured materials belong to a category demonstrating the piezoelectric effect and possessing crystal structures similar to those of tungsten bronze. In a singular tetragonal TB cell, there exist ten distorted BO_6 octahedra sharing a corner and generating A1, A2, and C interstices quadrilateral, pentagonal, and triangular, respectively, as depicted in Fig. 11.12. The structural representation uses the formula $(A1)_2(A2)_4(C)_4(B1)_2(B2)_8O_{30}$. Different cations can occupy interstices and the central position of the oxygen octahedron to bring about changes in the physical properties of the compounds with the TB structure [67].

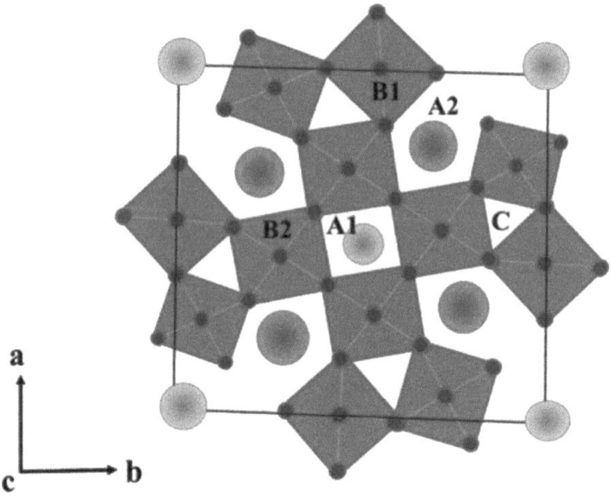

Figure 11.12 Unit cell arrangement of the crystal structure of tetragonal TB as viewed along the c-axis.

Among the non-perovskite configurations are ferroelectric TB ceramics: both lead-free and lead-containing. TB structured compounds exhibit noteworthy traits such as high T_C, substantial spontaneous polarization (P_s), and significant pyroelectricity. Consequently, this group of ceramics has gained recent prominence as an integral part of the piezoelectric ceramic system [68–70]. However, their practical application is hampered by challenges including limited sintering density, instability at low temperatures, and complications with polarization. To address the restricted domain states accessible to symmetric bronze-based structures, ceramics with strong texturing and grain orientation are often manufactured. $PbNb_2O_6$ (PN) is a notable functional ceramic within the lead-based tungsten bronze piezoceramics class. Exhibiting features like low Q_m, pronounced anisotropy, and high T_C, it finds application in ultrasonic transducers as an alternative to PZT when operating beyond 250°C [71]. In the case of $Pb_{0.95}Sr_{0.05}Nb_2O_6$ samples, a novel strategy involving the controlled generation of vacancies via an A-site deficit was successfully employed to finely adjust both structure and performance, yielding impressive outcomes. This approach led to decreased oxygen vacancies, deformation of NbO_6 octahedra, and erratic motion of A-site cations. Consequently, a visibly smaller domain emerged in piezoresponse force microscopic images, contributing to enhanced piezoelectric and dielectric capabilities. These suggest a practical means of enhancing dielectric and piezoelectric properties within the PN-based ceramics system while maintaining a high T_C [72]. The temperature-dependent ε_r and $\tan\delta$ curves for both sets of samples are shown in Fig 11.13a. The enhanced domain structure distinctly fosters improved dielectric and piezoelectric capabilities. Remarkably, both ε_r and d_{33} values increased while retaining a high T_C. This underscores the beneficial influence of A-site element deficit in composition design, leading to a substantial rise in ε_r and d_{33} while nearly maintaining T_C in PN-based ceramics. However, the extensively used lead-free, non-perovskite TB ferroelectric ceramics, designed for tunable devices, notably differ from perovskites. In applications necessitating lower-frequency actuators and sensors, polycrystalline ceramic systems offer a cost-effective alternative. Elevating T_C without introducing hazardous lead compounds could provide advantages for such structures.

Among them, strontium barium niobate (SBN or $Sr_{1-x}Ba_xNb_2O_6$) is particularly significant, boasting a structure akin to tetragonal TB and playing a pivotal role in various technological fields, including pyroelectric detectors, spatial light modulators, electro-optic devices, and surface acoustic wave devices [73–75].

Recent advances include the successful application of the Czochralski method to generate calcium barium niobate ($Ca_xBa_{1-x}Nb_2O_6$), known as CBN100x in the form of large single crystals with a specific composition of $x = 0.28$ [76, 77]. CBN28, a member of the tetragonal TB structure like SBN, demonstrates optical properties akin to SBN. Notably, CBN28 distinguishes itself with a significantly higher ferroelectric phase transition temperature compared to SBN61, which is only 79°C. Utilizing a two-step sintering process, partially filled tungsten bronze ceramics of $Ca_xBa_{1-x}Nb_2O_6$ were manufactured and analyzed for their dielectric properties and phase transition behavior. Ceramics like CBN22, CBN30, and CBN38 exhibit a diffuse to an incommensurate lock-in phase transition at temperatures above ambient temperature. The final sintering temperature significantly influences the characteristics of CBN ceramics [78]. Furthermore, an evaluation of high-power characteristics involved subjecting a $(Sr,Ca)_2NaNb_5O_{15}$ (SCNN) plate to intense vibrations, assessing it through an analysis of electrical transient response. Intriguingly, compared to $(Bi,Na,Ba)(Ti,Mn)O_3$ disks and randomly oriented SCNN, the Q_m of the SCNN plate with a crystal orientation increased nearly fivefold [79]. Figure 11.14 shows the temperature-induced fluctuations in the Q_m for SCNN and hard-PZT specimens, as well as the relationship between temperature increase and vibration velocity.

Another recent study investigated the increase in temperature in a rectangular SCNN column when subjected to increasing v and it was compared with that of hard-PZT. Interestingly, SCNN exhibited an increase in temperature, which was lower than that of PZT, with the rise showing a linear relationship with v. This phenomenon was associated with the distinct characteristics of radial vibration mode. The high-power characteristics of SCNN were studied using a rectangular column with modifying electrode designs for electrical contact in a longitudinal mode. The electrical transient responses of burst voltage with changing the sample temperature using a continuous driving sample were systematically measured and compared with hard-PZT.

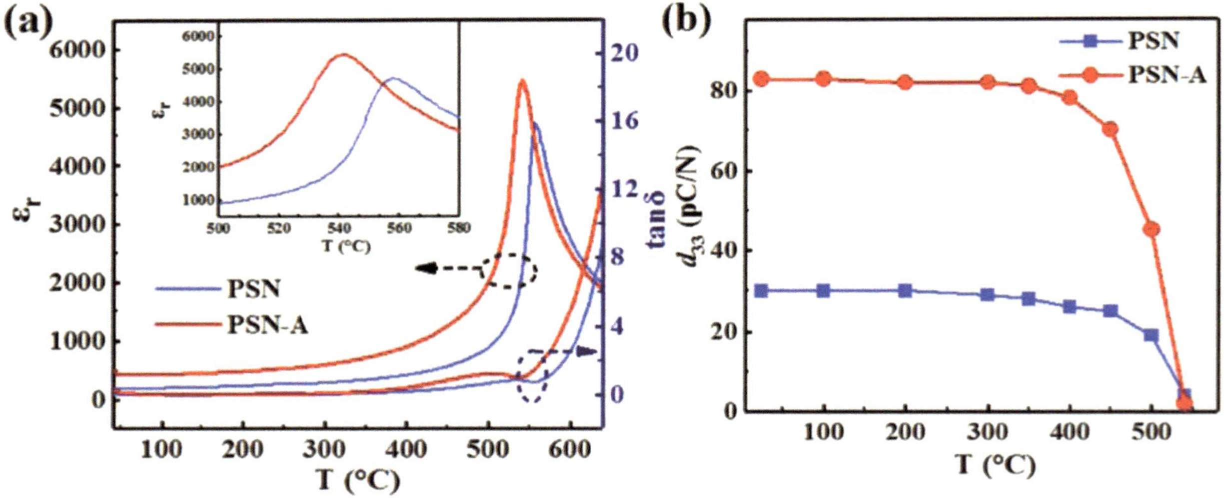

Figure 11.13 (a) The temperature-dependent behavior of relative ε and tanδ at a frequency of 1 kHz. The inset provides an amplified view of the curves specifically within 500–600°C and (b) the influence of annealing temperature on the d_{33} values for both constituents with a fixed annealing duration of 2 hours [72].

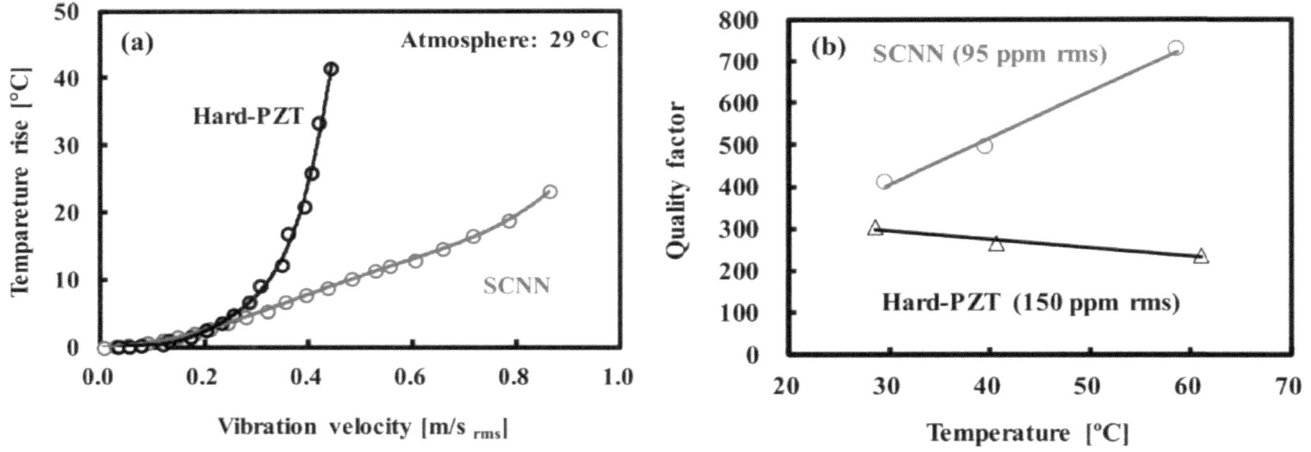

Figure 11.14 (a) The relationship between temperature elevation and v_{rms} and (b) temperature-related changes in the Q_m for SCNN and hard-PZT specimens [79].

11.5.2 BLSF Materials

The distinctive structure of BLSF, as shown in Fig. 11.15, features alternating layers along the c-axis direction, which resemble both bismuth oxide layers in the form of fluorite-like $(Bi_2O_2)^{2+}$ and perovskite $(A_{m-1}B_mO_{3m+1})^{2-}$ layers. BLSF has the general formula of $(Bi_2O_2)^{2+}(A_{m-1}B_mO_{3m+1})^{2-}$, with "m" representing the number of BO_6 octahedra layers in the perovskite-like block [80, 81]. BLSF stands out from perovskite systems due to several distinct electromechanical advantages, including high T_C, robust breakdown field strength, elevated Q_m, and exceptional stability in resonance frequency. These show significant prospects of BLSF materials for applications in areas like high frequency, high temperature, high power, and information storage.

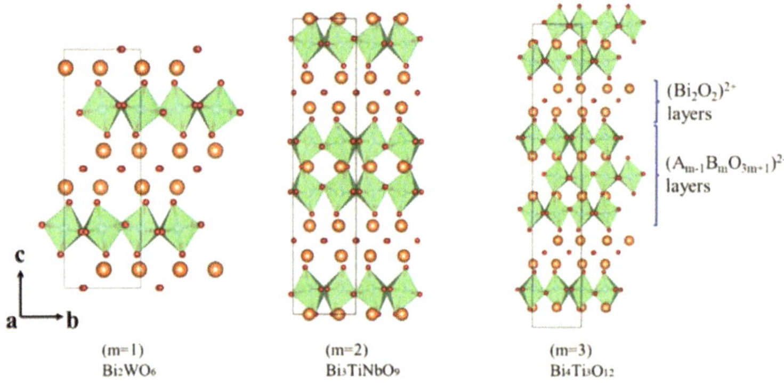

Figure 11.15 The crystal lattice of certain materials belonging to the BLSF category [23].

However, despite these strengths, BLSF has its limitations, including reduced piezoelectric activity, relatively small residual polarization, pronounced tanδ and higher E at elevated temperatures. In specific cases, measurements of d_{33} can even be notably lower by an order of magnitude. The lead-free piezoelectric system $Na_{0.5}Bi_{4.5}Ti_4O_{15}$ (NBT) belongs to the group of composite solutions that conduct oxide ions, specifically the Aurivillius subgroup of BLSFs with "m" set at 4 [82–86]. NBT, known for its high T_C and impressive piezoelectric and electromechanical attributes, was chosen due to its lead-free nature. Creating NBT-based ceramics with A-site vacancies was accomplished using conventional solid-state fabrication methods. The introduction of (Nd, Ce) alteration in NBT-based ceramics led to a noticeable enhancement in piezoelectric activity [87]. Furthermore, polycrystalline ceramics of sodium lanthanum bismuth titanate, $Na_{0.5}La_{0.5}Bi_4Ti_4O_{15}$ (NLBT) were synthesized with cobalt modifications through standard solid-state manufacturing techniques. The comprehensive analysis covered piezoelectric and dielectric characteristics. Incorporating cobalt led to improved piezoelectric properties in NLBT ceramics, with a slightly elevated T_C compared to pure NLBT ceramics. The observed substantial dielectric irregularities and losses in NLBT ceramics were significantly reduced by cobalt treatment.

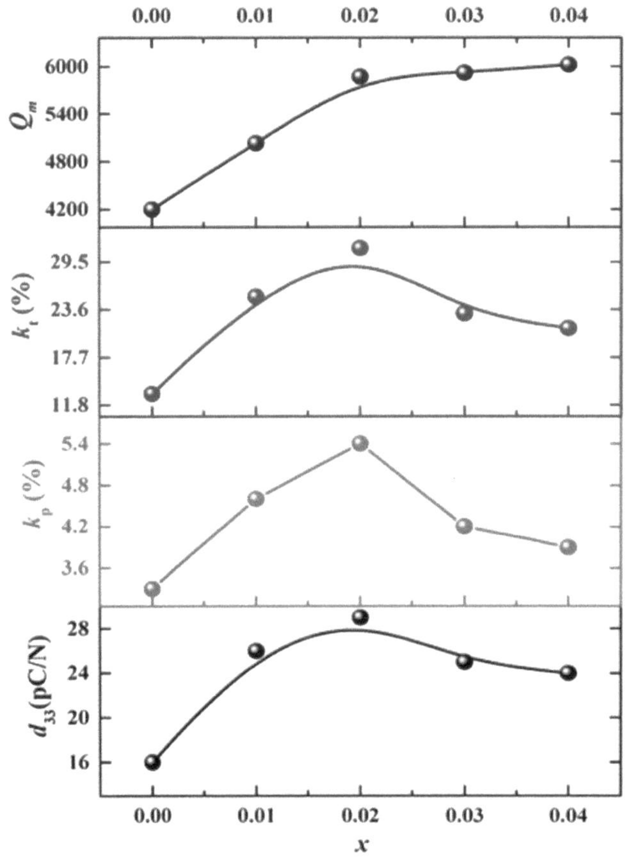

Figure 11.16 The electrical characteristics of the NBT ceramics co-doped with Mn and Nb at room temperature [89].

Additionally, cobalt-modified NLBT ceramics consistently demonstrated favorable piezoelectric properties based on thermal annealing tests [88]. To enhance the modest piezoelectric performance of high-temperature piezoelectric ceramics NBT, a co-doping approach involving Mn/Nb was applied. Traditional solid-state reaction methods were used to investigate Mn/Nb co-doped $Na_{0.5}Bi_{4.5}Ti_{4-x}(Mn_{1/3}Nb_{2/3})_xO_{15}$ (NBT-100x) ceramics in detail. With increasing doping level "x," the mechanical Q_m displayed linear growth as shown in Fig 11.16. This phenomenon is attributed to the substitution of Mn^{2+} and Nb^{5+} for B-site Ti^{4+} due to their closely matched ionic radii [89]. Mn^{2+} compounds influenced domain wall pinning, affecting the piezoelectric properties [90, 91]. The incorporation of more Mn/Nb compound ions constrained domain wall motion in NBT ceramics, resulting in an elevated mechanical Q_m. Figure 11.17 [92] compares ceramics composed of Bi_3TiNbO_9 (BTN) with varying compositions, highlighting fluctuations in both d_{33} and T_C. This pattern of reduced T_C alongside improved d_{33} is commonly seen in various ions-doped BLSFs during enhancements [93]. Particularly in this study, the ceramics with $x = 0.2$ exhibited high d_{33} and T_C concurrently, attributed to the synergistic impact of Gd and Cr co-doping on BTN. Due to its extraordinary thermal stability, this has been an outstanding candidate for use in ultrahigh-temperature piezoelectric sensors, which operate at temperatures higher than 500°C.

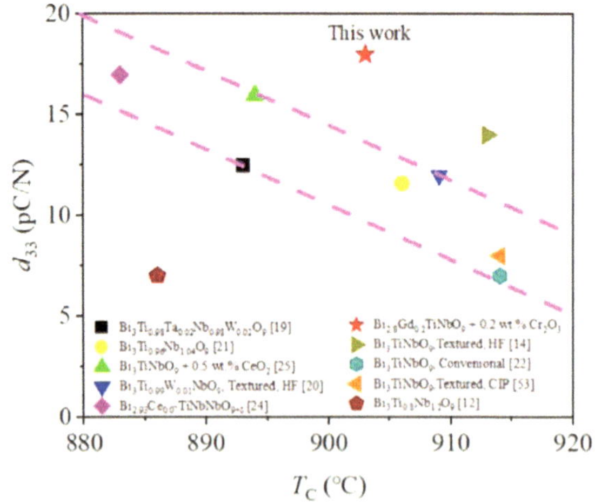

Figure 11.17 Comparison of the d_{33} and T_C of ceramics based on different compositions of BTN [92].

11.5.3 Niobate Structured Materials

$LiNbO_3$ (LN) single crystals present a promising choice for advanced linear actuators and find usage in optical applications and surface acoustic wave devices. These actuators effectively overcome the challenges associated with ceramic PZT actuators, particularly for their non-linear behavior and hysteresis properties. Noteworthy is the substantial T_C of 1210°C of LN crystals, ranking it as a lead-free alternative well-suited for high-temperature actuation. This attribute enhances its environmental compatibility compared to PZTs [94]. Figure 11.18 analyzes and visualizes the temperature-dependent patterns of the d (d_{ij}) and the k (k_{ij}) in LN crystals. While the effective coupling factors k_{31} and k_{33} rise with increasing temperature, k_{15} and k_{21} display marginal decreases. At standard temperatures, the d (d_{15}) demonstrates the highest value. In the temperature range under investigation, the d_{21} consistently demonstrates a limited variation of less than 12.5%. The alteration in relative dielectric permittivity percentage is heavily influenced by the fluctuation in the piezoelectric d_{33}, which experiences fluctuations exceeding 80% [95]. This phenomenon is exemplified by the case of $Li_xNa_{1-x}NbO_3$ (LNN) mixed-perovskite piezoelectric ceramics. Over time, comprehensive research has been dedicated to comprehend the polymorphic and structural changes that arise during the phase transition of LNNs. Several methods have been developed to enhance the piezoelectric properties of LNNs, involving considerations of stoichiometric ratios, chemical modifications, and process optimization. Among these methods, heat treatment has emerged as the most efficient approach to elevate the Q_m of LNN ceramics.

LNN ceramics exhibit superior Q_m values compared to their lead-free counterparts, especially in terms of enhanced stability with increasing v. This distinctive trait positions them favorably for high-power applications. The quest for electromechanical properties, including the k_p and Q_m, establish LNN piezoelectric ceramics as suitable contenders for high-power applications [96–98]. The traditional linear relationship between strain and the driving electric field becomes inapplicable due to the resonance frequency utilization of piezoelectric vibrators. In scenarios necessitating high power, the evaluation of operational performance shifts from strain to v. The remarkable high-power performance of LNN materials is substantiated by the fluctuation in Q_m values observed across different piezoelectric ceramics, attributed to variations in v [99].

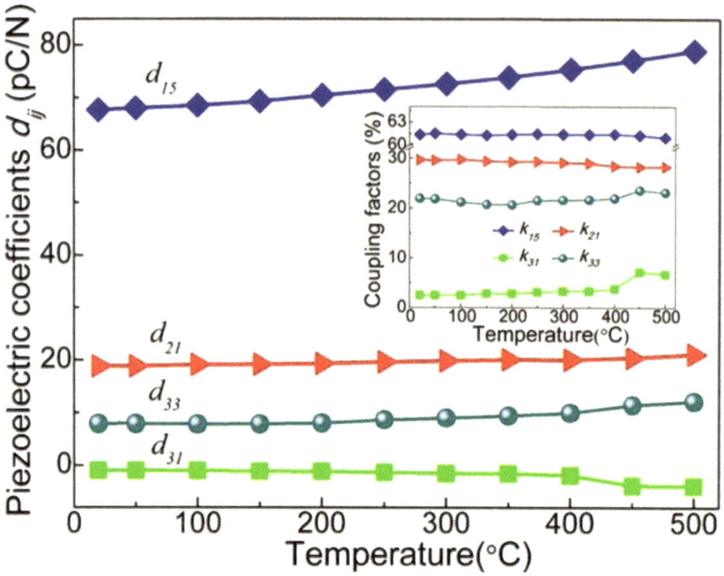

Figure 11.18 Changes in the temperature-dependent behaviors of the k_{ij} (inset) and d (d_{ij}) [95].

Piezoelectric characteristics of LNN in a binary system are notably affected by the proportions of alkali elements. Investigating the Na-rich region, various stoichiometric ratios within the LNN system have been explored [100]. The introduction of LN to NN leads to a reduced lattice volume, promoting domain wall motion owing to the smaller radius of Li^+ compared to Na^+. The nature of the phase transition is correlated with an increase in remnant polarization (P_r) observed after Li doping, as depicted in Fig. 11.19a. The pronounced polarization in regions enriched with rubidium (Rb) can be attributed to the significant displacement of Na by the substantial off-centering of Li^+, as seen in the same figure. A comparison of Q_m and d_{33} values for LNN with different x values is presented in Fig. 11.19b. Notably, the d_{33} value falls in the range of 40 pC/N for $0.04 \leq x \leq 0.12$, surpassing the values of NN or LN [101–103]. The Q_m value also displays considerable variation based on composition, with LNN2, LNN4, and LNN6 reaching values exceeding 600. Moreover, T_C in the Li-rich range surpasses 800°C, rendering LNN appealing for applications requiring high temperatures [91].

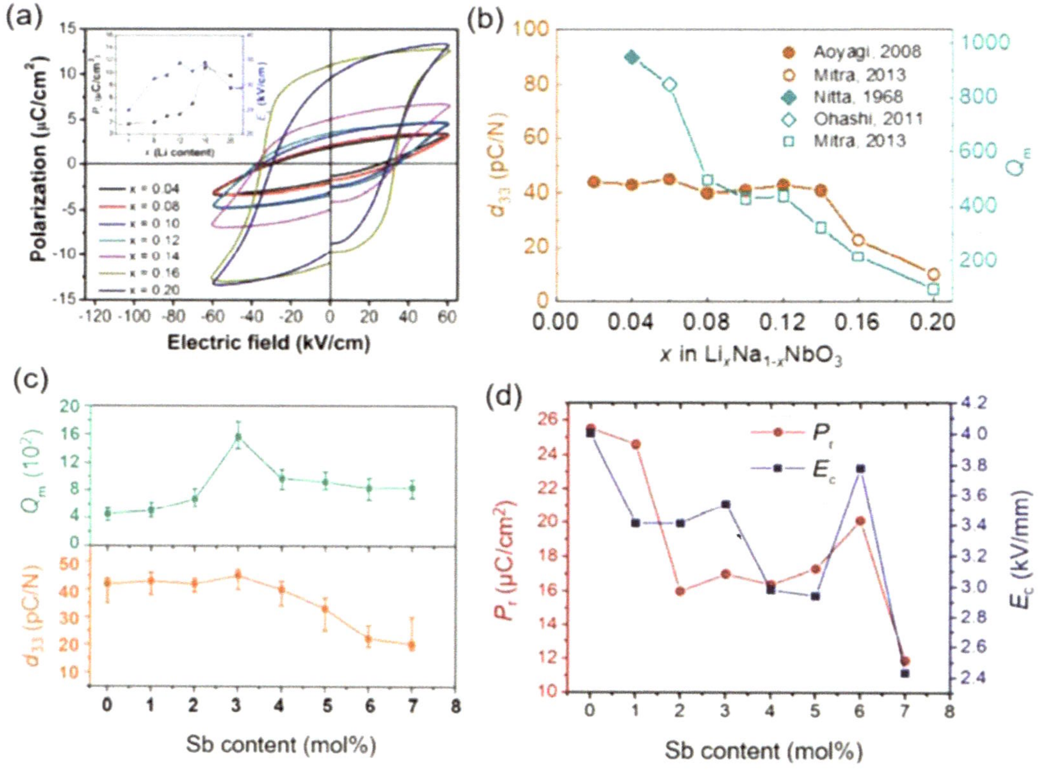

Figure 11.19 (a) Polarization-electric field hysteresis loops and (b) Q_m and d_{33} as a function of in LNN ceramics. (c) Q_m and d_{33} and (d) P_r and E_c as a function of Sb^{5+} content in $LNNSb_x$ [99].

11.6 Future Scopes for Efficient HPPMs

HPPMs have been extensively studied and several advancements have been made in their efficiency and power capabilities. However, researchers might explore numerous future scopes and directions to achieve even more efficient HPPMs. New PZMs with specialized compositions and crystal structures are available for investigation and synthesis. This could entail combining different components and dopants to improve piezoelectric characteristics and power conversion efficiency. Piezoelectric performance may be enhanced by investigating nanostructured materials. To increase the piezoelectric capabilities, alignment of domain and orientation can be more precisely controlled by nano-engineering the crystal structure at the nano-scale. Finding potential material candidates with desirable piezoelectric properties can be achieved much more quickly by using cutting-edge computational methods like machine learning and quantum simulations. Researchers can boost electromechanical capabilities and power conversion efficiency of PZMs by comprehending and modifying their domain structures. Developing PZMs with multiple functionalities, such as piezoelectric and ferroelectric properties combined with other desirable characteristics, can lead to more efficient and versatile applications. It is possible to produce HPPMs with dependable and consistent properties on a wide scale by developing new manufacturing processes and techniques. To lessen the negative effects of the production and disposal of PZMs on the environment, researchers might investigate eco-friendly and sustainable PZMs. Integration of PZMs with other cutting-edge technologies, such as nano-generators, may present new opportunities for high-power applications. To advance the discipline and create more effective HPPMs for manifold applications, such as harvesting energy, actuation, sensing, and more, requires innovative materials synthesis techniques, and rigorous testing methodologies.

11.7 Conclusions

The evolution of HPPMs within the realm of materials science is a testament to remarkable innovation and advancement. These materials, which were once mere curiosities, have now emerged as essential components across diverse industries. This journey of development has been characterized by a delicate equilibrium between conflicting factors – optimizing Q_m while preserving vital d and k_p. This pursuit has yielded groundbreaking insights into manipulating grain structures, capitalizing on domain interactions, and refining sintering techniques. As progress continues, challenges related to stability, scalability, and material compatibility persist. However, the potential impact of HPPMs is undeniably significant. The evolution of HPPMs exemplifies the power of human ingenuity to transform scientific understanding into practical applications. From energy harvesting to medical diagnostics, these materials are poised to reshape industries and elevate technological capabilities. As we look ahead, the ongoing research and innovation in this field promise to unlock even greater possibilities, ensuring that the journey of HPPMs remains one of continuous exploration and advancement.

Acknowledgment

The authors acknowledge the Ministry of Science and Technology, Government of Bangladesh for a research grant under Special Allocation for Science and Technology. The authors also acknowledge support from the DU-UGC, Bangladesh research grant for the themed research.

References

1. Yang Z, Zhou S, Zu J, Inman D. High-performance piezoelectric energy harvesters and their applications. *Joule*. **2**(4), 642, 2018.
2. Karim S, Tan J, Kareem K, Sultan S, Murtaza A, khan MT, Abbas N, Fareed Z, Shaikh MN, Hussain S, Sadiq M, Bakhsh K. Preparation, structure and dielectric/piezoelectric properties of $BiSc0_3$-$PbTi0_3$-Pb $(Mn_{1/3}Nb_{2/3})$ 0_3 high temperature piezoelectric ceramics. 44th International Conference on "Aerospace Science and Engineering (ICASE 2015)," Institute of Space Technology, Islamabad, Pakistan. **275–281**, 2015.
3. Uchino K. Piezoelectric ultrasonic motors: Overview. *Smart Materials and Structures*. **7**(3), 273, 1998.
4. Leng H. Fundamental Understanding of Crystallographic Texturing and Defect Engineering for High Power Piezoelectric Ceramics. Dissertation, Materials Science and Engineering, *The Pennsylvania State University*, 2022.

5. Zhang J, Fang Z, Shu C, Zhang J, Zhang Q, Li C. A rotational piezoelectric energy harvester for efficient wind energy harvesting. *Sensors and Actuators A: Physical.* **262**, 123, 2017.
6. Saito H, Chazono H, Kishi HK, Yamaoka NY. X7R multilayer ceramic capacitors with nickel electrodes. *Japanese Journal of Applied Physics.* **30**(9S), 2307, 1991.
7. Moure C, Villegas M, Fernandez J, Duran P. Microstructural and piezoelectric properties of fine grained PZT ceramics doped with donor and/or acceptor cations. *Ferroelectrics.* **127**(1), 113, 1992.
8. Ramam K, Chandramouli K. Dielectric and piezoelectric properties of combinatory effect of A-site isovalent and B-site acceptor doped PLZT ceramics. *Ceramics-Silikaty.* **53**(3), 189, 2009
9. Zhang J, Hao W, Gao Y, Qin Y, Tan Y, Wang C. Large decrease of characteristic frequency of dielectric relaxation associated with domain-wall motion in Sb^{5+}-modified (K, Na) NbO_3-based ceramics. *Applied Physics Letters.* **101**(25), 252905, 2012.
10. Promjun T, Funsueb N, Ngamjarurojana A. Effect of Nb, Ta and Sb addition on structure and electrical properties of PZT ceramics. *Materials Today: Proceedings.* **17**, 1602, 2019.
11. Zheng H, Reaney IM, Lee WE, Jones N, Thomas H. Effects of strontium substitution in Nb-doped PZT ceramics. *Journal of the European Ceramic Society.* **21**(10–11), 1371, 2001.
12. Kongphimai J, Photankham W, Phewphong S, Namthaisong A, Wattanasarn H. Dielectric and ferroelectric properties of $Pb(Zr_{0.53}Ti_{0.47})O_3$ ceramics modified with Sr. *Integrated Ferroelectrics.* **187**(1), 14, 2018.
13. Pdungsap L, Udomkan N, Boonyuen S, Winotai P. Optimized conditions for fabrication of La-dopant in PZT ceramics. *Sensors and Actuators A: Physical.* **122**(2), 250, 2005.
14. Prabu M, Banu IS, Gobalakrishnan S, Chavali M. Electrical and ferroelectric properties of undoped and La-doped PZT (52/48) electroceramics synthesized by sol–gel method. *Journal of Alloys and Compounds.* **551**, 200–207, 2013.
15. Cao Z, Li G, Or SW, Zeng J, Zheng L, Yin Q. Preparation and the temperature dependence of electromechanical properties of Ca^{2+}–W^{6+} co-doped Pb $(Zr,Ti)O_3$ ceramics. *Journal of Alloys and Compounds.* **496**(1–2), 13, 2010.
16. Ahart M, Somayazulu M, Cohen RE, Ganesh P, Dera P, Mao HK, Hemley RJ, Ren Y, Liermann P, Wu Z. Origin of morphotropic phase boundaries in ferroelectrics. *Nature.* **451**(7178), 545, 2008.

17. Lv X, Zhang XX, Wu J. Nano-domains in lead-free piezoceramics: a review. *Journal of Materials Chemistry A*. **8**(20), 10026, 2020.

18. Iamsasri T, Tutuncu G, Uthaisar C, Wongsaenmai S, Pojprapai S, Jones JL. Electric field-induced phase transitions in Li-modified $Na_{0.5}K_{0.5}NbO_3$ at the polymorphic phase boundary. *Journal of Applied Physics*. **117**(2), 024101, 2015.

19. Xu G, Zhong Z, Bing Y, Ye ZG, Shirane G. Electric-field-induced redistribution of polar nano-regions in a relaxor ferroelectric. *Nature Materials*. **5**(2), 134, 2006.

20. Ren X. Large electric-field-induced strain in ferroelectric crystals by point-defect-mediated reversible domain switching. *Nature Materials*. **3**(2), 91, 2004.

21. Liu G, Zhang S, Jiang W, Cao W. Losses in ferroelectric materials. *Materials Science and Engineering R: Reports*. **89**, 1, 2015.

22. Cao W, Randall CA. Grain size and domain size relations in bulk ceramic ferroelectric materials. *Journal of Physics and Chemistry of Solids*. **57**(10), 1499, 1996.

23. Chen L, Liu H, Qi H, Chen J. High-electromechanical performance for high-power piezoelectric applications: Fundamental, progress, and perspective. *Progress in Materials Science*. **127**, 100944, 2022.

24. Hao J, Li W, Zhai J, Chen H. Progress in high-strain perovskite piezoelectric ceramics. *Materials Science and Engineering R: Reports*. **135**, 1–57, 2019.

25. Fan L, Chen J, Ren Y, Pan Z, Zhang L, Xing X. Unique piezoelectric properties of the monoclinic phase in Pb $(Zr,Ti)O_3$ ceramics: Large lattice strain and negligible domain switching. *Physical Review Letters*. **116**(2), 027601, 2016.

26. Dursun S, Mensur-Alkoy E, Unver MU, Alkoy S. Enhancement of electrical properties in the ternary PMN–PT–PZ through compositional variation, crystallographic texture, and quenching. *Journal of the American Ceramic Society*. **103**(4), 2499, 2020.

27. Jaffe H. Piezoelectric ceramics. *Journal of the American Ceramic Society*. **41**(11), 494, 1958.

28. Zhang S, Xia R, Lebrun L, Anderson D, Shrout TR. Piezoelectric materials for high power, high temperature applications. *Materials Letters*. **59**(27), 3471, 2005.

29. Pdungsap L, Udomkan N, Boonyuen S, Winotai P. Optimized conditions for fabrication of La-dopant in PZT ceramics. *Sensors and Actuators A: Physical*. **122**(2), 250, 2005.

30. Aleem MA, Nawaz H, Shuaib M, Qaisar S, Akbar MS. Piezoelectric and pyroelectric properties of Sr-doped PZT (PSZT) with minor manganese additions. *Journal of Physics: Conference Series.* **439**(1), 012025, 2013.
31. Boucher E, Guyomar D, Lebrun L, Guiffard B, Grange G. Effect of (Mn, F) and (Mg, F) co-doping on dielectric and piezoelectric properties of lead zirconate titanate ceramics. *Journal of Applied Physics.* **92**(9), 5437, 2002.
32. Priya S, Kim HW, Uchino K. Low temperature coefficient of resonance frequency composition in the system $Pb(Zr, Ti)O_3$-$Pb(Mn_{1/3}Nb_{2/3})O_3$. *Journal of the American Ceramic Society.* **87**(10), 1907, 2004.
33. Chen H, Fan C. Fabrication and properties of PYN–PMN–PZT quaternary piezoelectric ceramics for high-power, high-temperature applications. *Materials Letters.* **64**(6), 654, 2010.
34. Oh HT, Joo HJ, Kim MC, Lee HY, Oh HT, Joo HJ, Kim MC, Lee HY. Effect of Mn on dielectric and piezoelectric properties of 71PMN-29PT [$71Pb(Mg_{1/3}Nb_{2/3})O_3$-$29PbTiO_3$] single crystals and polycrystalline ceramics. *Journal of the Korean Ceramic Society.* **55**(2), 166, 2018.
35. Qi X, Sun E, Wang J, Zhang R, Yang B, Cao W. Electromechanical properties of Mn-doped $Pb(In_{1/2}Nb_{1/2})O_3$-$Pb(Mg_{1/3}Nb_{2/3})O_3$-$PbTiO_3$ piezoelectric ceramics. *Ceramics International.* **42**(14), 15332, 2016.
36. Zhang S, Xia R, Lebrun L, Anderson D, Shrout TR. Piezoelectric materials for high power, high temperature applications. *Materials Letters.* **59**(27), 3471, 2005.
37. Talanov MV, Shilkina LA, Verbenko IA, Reznichenko LA. Impact of Ba^{2+} on structure and piezoelectric properties of PMN–PZN–PNN–PT ceramics near the morphotropic phase boundary. *Journal of the American Ceramic Society.* **98**(3), 838, 2015.
38. Eitel RE, Randall CA, Shrout TR, Rehrig PW, Hackenberger W, Park SE. New high temperature morphotropic phase boundary piezoelectrics based on $Bi(Me)O_3$–$PbTiO_3$ ceramics. *Japanese Journal of Applied Physics.* **40**(10R), 5999, 2001.
39. Jia H, Hu X, Chen J. Temperature-dependent piezoelectric strain and resonance performance of Fe_2O_3-modified $BiScO_3$-$PbTiO_3$-$Pb(Nb_{1/3}Mn_{2/3})O_3$ ceramics. *Journal of the European Ceramic Society.* **39**(7), 2348, 2019.
40. Rödel J, Jo W, Seifert KT, Anton EM, Granzow T, Damjanovic D. Perspective on the development of lead-free piezoceramics. *Journal of the American Ceramic Society.* **92**(6), 1153, 2009.
41. Bell AJ, Deubzer O. Lead-free piezoelectrics—The environmental and regulatory issues. *MRS Bulletin.* **43**(8), 581, 2018.

42. Singh R, Kulkarni AR, Harendranath CS. Effect of sintering temperature on composition, microstructure and electrical properties of $K_{0.5}Na_{0.5}NbO_3$ ceramics. *Physica B: Condensed Matter.* **434**, 139, 2014.

43. Zhou J, Ma Q, Wang P, Cheng L, Liu S. Influence of rare-earth Nd, Dy, and Ho doping on structural and electrical properties of $(Na_{0.53}K_{0.47})_{0.942}Li_{0.058}NbO_3$ based lead-free piezoceramics. *Ceramics International.* **40**(1), 2451, 2014.

44. Yin Q, Sun Z, Shi S, Jiang X, Tian C, Bao M. Structure and electrical properties of $K_{0.5}Na_{0.5}$ NbO_3–$SrTiO_3$ lead-free piezoelectric ceramics with $LiSbO_3$ doping. *Journal of Materials Science: Materials in Electronics.* **24**, 4258, 2013.

45. Wu J, Xiao D, Wang Y, Wu W, Zhang B, Li J, Zhu J. $CaTiO_3$-modified $[(K_{0.5}Na_{0.5})_{0.94}Li_{0.06}](Nb_{0.94}Sb_{0.06})O_3$ lead-free piezoelectric ceramics with improved temperature stability. *Scripta Materialia.* **59**(7), 750, 2008.

46. Li X, Jiang M, Liu J, Zhu J, Zhu X, Zhu J, Xiao D. Enhanced piezoelectric properties in Mn-doped $0.98K_{0.5}Na_{0.5}NbO_3$–$0.02BiScO_3$ lead-free ceramics. *Journal of the American Ceramic Society.* **92**(7), 1625, 2009.

47. Wu J. High piezoelectricity in low-temperature sintering potassium–sodium niobate-based lead-free ceramics. *RSC Advances.* **4**(96), 53490, 2014.

48. Wang X, Zheng T, Wu J, Xiao D, Zhu J, Wang H, Wang X, Lou X, Gu Y. Characteristics of giant piezoelectricity around the rhombohedral-tetragonal phase boundary in (K, Na)NbO_3-based ceramics with different additives. *Journal of Materials Chemistry A.* **3**(31), 15951, 2015.

49. Saito Y, Takao H, Tani T, Nonoyama T, Takatori K, Homma T, Nagaya T, Nakamura M. Lead-free piezoceramics. *Nature.* **432**(7013), 84, 2004.

50. Zheng T, Wu J, Xiao D, Zhu J. Recent development in lead-free perovskite piezoelectric bulk materials. *Progress in Materials Science.* **98**, 552, 2018.

51. 51.Wang X, Wu J, Xiao D, Zhu J, Cheng X, Zheng T, Zhang B, Lou X, Wang X. Giant piezoelectricity in potassium–sodium niobate lead-free ceramics. *Journal of the American Chemical Society.* **136**(7), 2905, 2014.

52. Wu B, Wu H, Wu J, Xiao D, Zhu J, Pennycook SJ. Giant piezoelectricity and high Curie temperature in nanostructured alkali niobate lead-free piezoceramics through phase coexistence. *Journal of the American Chemical Society.* **138**(47), 15459, 2016.

53. Xu K, Li J, Lv X, Wu J, Zhang X, Xiao D, Zhu J. Superior piezoelectric properties in potassium–sodium niobate lead-free ceramics. *Advanced Materials*. **28**(38), 8519, 2016.

54. Zheng T, Wu H, Yuan Y, Lv X, Li Q, Men T, Zhao C, Xiao D, Wu J, Wang K, Li JF. The structural origin of enhanced piezoelectric performance and stability in lead free ceramics. *Energy & Environmental Science*. **10**(2), 528, 2017.

55. Bechmann R. Elastic, piezoelectric, and dielectric constants of polarized barium titanate ceramics and some applications of the piezoelectric equations. *The Journal of the Acoustical Society of America*. **28**(3), 347, 1956.

56. Takahashi H, Numamoto Y, Tani J, Tsurekawa S. Piezoelectric properties of $BaTiO_3$ ceramics with high performance fabricated by microwave sintering. *Japanese Journal of Applied Physics*. **45**(9S), 7405, 2006.

57. Shen ZY, Li JF. Enhancement of piezoelectric constant d_{33} in $BaTiO_3$ ceramics due to nano-domain structure. *Journal of the Ceramic Society of Japan*. **118**(1382), 940, 2010.

58. Karaki T, Yan K, Adachi M. Barium titanate piezoelectric ceramics manufactured by two-step sintering. *Japanese Journal of Applied Physics*. **46**(10S), 7035, 2007.

59. Zhang JL, Ji PF, Wu YQ, Zhao X, Tan YQ, Wang CL. Strong piezoelectricity exhibited by large-grained $BaTiO_3$ ceramics. *Applied Physics Letters*. **104**(22), 222909, 2014.

60. Wada S, Takeda K, Muraishi T, Kakemoto H, Tsurumi T, Kimura T. Preparation of [110] grain oriented barium titanate ceramics by templated grain growth method and their piezoelectric properties. *Japanese Journal of Applied Physics*. **46**(10S), 7039, 2007.

61. Ul R, Marchet P, Pham-Thi M, Tran-Huu-Hue LP. Improved properties of doped $BaTiO_3$ piezoelectric ceramics. *Physica Status Solidi (a)*. **216**(22), 1900413, 2019.

62. Tsai CC, Chao WH, Chu SY, Hong CS, Weng CM, Su HH. Enhanced temperature stability and quality factor with Hf substitution for Sn and MnO_2 doping of $(Ba_{0.97}Ca_{0.03})(Ti_{0.96}Sn_{0.04})O_3$ lead-free piezoelectric ceramics with high Curie temperature. *AIP Advances*. **6**(12), 125024, 2016.

63. Tanaka D, Yamazaki J, Furukawa M, Tsukada T. High power characteristics of $(Ca,Ba)TiO_3$ piezoelectric ceramics with high mechanical quality factor. *Japanese Journal of Applied Physics*. **49**(9S), 09MD03, 2010.

64. Someno S, Nagata H, Takenaka T. High-temperature and high-power piezoelectric characteristics of $(Bi_{0.5}Na_{0.5})TiO_3$-based lead-free piezoelectric ceramics. *Journal of the Ceramic Society of Japan.* **122**(1426), 406, 2014.

65. Hejazi M, Taghaddos E, Gurdal E, Uchino K, Safari A. High power performance of manganese-doped BNT-based Pb-free piezoelectric ceramics. *Journal of the American Ceramic Society.* **97**(10), 3192, 2014.

66. Yesner G, Safari A. Improved resistivity in bismuth deficient morphotropic phase boundary 0.88BNT-0.08BKT-0.04 BT ceramics. *In 2017 Joint IEEE International Symposium on the Applications of Ferroelectric (ISAF)/International Workshop on Acoustic Transduction Materials and Devices (IWATMD)/Piezoresponse Force Microscopy (PFM).* 110, 2017, IEEE.

67. Wu S, Sun C, Yang X, Hu C, Liu L, Fang L. Effect of strontium substitution on the structure and dielectric properties of unfilled tungsten bronze $Ba_{4-x}Sr_xSmFe_{0.5}Nb_{9.5}O_{30}$ ceramics. *Ceramics International.* **46**(7), 9240, 2020.

68. Gao TT, Chen W, Zhu XN, Zhu XL, Chen XM. Dielectric and ferroelectric characteristics of $Ba_4Pr_2Fe_2Nb_8O_{30}$ tungsten bronze ceramics. *Materials Chemistry and Physics.* **181**, 47, 2016.

69. Huo Z, Wang H, Xu J, Rao G, Wang J, Zhou C, Yuan C. Antiferroelectric behavior and giant strain in BNKT ceramics complex $Cs_2Nb_4O_{11}$ tungsten bronze phase. *Ceramics International.* **46**(8), 10067, 2020.

70. Tiwari S, Vitta S. Magnetoelectric and magnetodielectric coupling and microwave resonator characteristics of $Ba_{0.5}Sr_{0.5}Nb_2O_6/CoCr_{0.4}Fe_{1.6}O_4$ multiferroic composite. *Scientific Reports.* **8**(1), 11619, 2018.

71. Soejima, J., Nagata, K. $PbNb_2O_6$ ceramics with tungsten bronze structure for low Q_m piezoelectric material. *Japanese Journal of Applied Physics,* **40**(9S), 5747, 2001.

72. Fang R, Zhou Z, Liang R, Dong X. Significantly improved dielectric and piezoelectric properties by defects in $PbNb_2O_6$-based piezoceramics. *Ceramics International.* **47**(19), 26942, 2021.

73. Lenzo PV, Spencer EG, Ballman AA. Electro-optic coefficients of ferroelectric strontium barium niobate. *Applied Physics Letters.* **11**(1), 23, 1967.

74. Bhalla AS, Guo R, Cross LE, Burns G, Dacol FH, Neurgaonkar RR. Measurements of strain and the optical indices in the ferroelectric $Ba_{0.4}Sr_{0.6}Nb_2O_6$: Polarization effects. *Physical Review B.* **36**(4), 2030, 1987.

75. Ke S, Huang H, Fan H, Chan HL, Zhou LM. Dielectric and ageing behaviour of strontium barium niobate with barium strontium titanate additives. *Journal of Physics D: Applied Physics.* **40**(21), 6797, 2007.

76. Eßer M, Burianek M, Klimm D, Mühlberg M. Single crystal growth of the tetragonal tungsten bronze $Ca_xBa_{1-x}Nb_2O_6$ (x = 0.28; CBN-28). *Journal of Crystal Growth.* **240**(1-2), 1, 2002.

77. Song H, Zhang H, Xu X, Hu X, Cheng X, Wang J, Jiang M. Growth and properties of Ca0. $28Ba_{0.72}Nb_2O_6$ single crystals. *Materials Research Bulletin.* **40**(4), 643, 2005.

78. Ke S, Fan H, Huang H, Chan HL, Yu S. Dielectric, ferroelectric properties, and grain growth of $Ca_xBa_{1-x}Nb_2O_6$ ceramics with tungsten-bronzes structure. *Journal of Applied Physics.* **104**(2), 024101, 2008.

79. Doshida Y, Tamura H, Tanaka S, Harada T, Shimizu H. High-power properties of (Sr, Ca) $2NaNb_5O_{15}$ piezoelectric ceramics in a longitudinal mode. *Japanese Journal of Applied Physics.* **59**(SK), SKKA07, 2020.

80. Zhang F, Li Y. Recent progress on bismuth layer-structured ferroelectrics. *Wuji Cailiao Xuebao/Journal of Inorganic Materials.* **29**(5), 449, 2014.

81. Wang CM, Wang JF, Zheng LM, Zhao ML, Wang CL. Enhancement of the piezoelectric properties of sodium lanthanum bismuth titanate ($Na_{0.5}La_{0.5}Bi_4Ti_4O_{15}$) through modification with cobalt. *Materials Science and Engineering: B.* **171**(1–3), 79, 2010.

82. Aurivillius B. Mixed bismuth oxides with layer lattices I. The structure type of $CaNb_2$ Bi_2O_9. *Arkiv kemi.* **1**(54), 463, 1949.

83. Subbarao EC. Crystal chemistry of mixed bismuth oxides with layer-type structure. *Journal of the American Ceramic Society.* **45**(4), 166, 1962.

84. Subbarao EC. A family of ferroelectric bismuth compounds. *Journal of Physics and Chemistry of Solids.* **23**(6), 665, 1962.

85. Du H, Shi X. Dielectric and piezoelectric properties of barium-modified Aurivillius-type $Na_{0.5}Bi_{4.5}Ti_4O_{15}$. *Journal of Physics and Chemistry of Solids.* **72**(11), 1279, 2011.

86. Jannet DB, El Maaoui M, Mercurio JP. Ferroelectric versus relaxor behaviour in $Na_{0.5}Bi_{4.5}Ti_4O_{15}$–$BaBi_4Ti_4O_{15}$ solid solutions. *Journal of Electroceramics.* **11**, 101, 2003.

87. Jiang XP, Fu XL, Chen C, Tu N, Xu MZ, Li XH, Shao H, Chen YJ. High performance Aurivillius type $Na_{0.5}Bi_{4.5}Ti_4O_{15}$ piezoelectric ceramics with neodymium and cerium modification. *Journal of Advanced Ceramics.* **4**, 54, 2015.

88. Wang CM, Wang JF, Zheng LM, Zhao ML, Wang CL. Enhancement of the piezoelectric properties of sodium lanthanum bismuth titanate ($Na_{0.5}La_{0.5}Bi_4Ti_4O_{15}$) through modification with cobalt. *Materials Science and Engineering: B.* **171**(1–3), 79, 2010.

89. Zhao T, Shi K, Fei C, Sun X, Quan Y, Liu W, Zhang J, Dai X. Structure, electrical properties, and thermal stability of the Mn/Nb co-doped Aurivillius-type $Na_{0.5}Bi_{4.5}Ti_4O_{15}$ high temperature piezoelectric ceramics. *Crystals.* **13**(3), 433, 2023.

90. Tang Y, Shen ZY, Du Q, Zhao X, Wang F, Qin X, Wang T, Shi W, Sun D, Zhou Z, Zhang S. Enhanced pyroelectric and piezoelectric responses in W/Mn-codoped $Bi_4Ti_3O_{12}$ Aurivillius ceramics. *Journal of the European Ceramic Society.* **38**(16), 5348, 2018.

91. Zhang X, Jiang G, Guo F, Liu D, Zhang S, Yang B, Cao W. Mn doping effects on electric properties of 0.93($Bi_{0.5}Na_{0.5}$)TiO_3-0.07Ba ($Ti_{0.945}Zr_{0.055}$)O_3 ceramics. *Journal of the American Ceramic Society.* **101**(7), 2996, 2018.

92. Chen Y, Zhou H, Wang Q, Zhu J. Doping level effects in Gd/Cr co-doped Bi_3TiNbO_9 Aurivillius-type ceramics with improved electrical properties. *Journal of Materiomics.* 8(4), **906**, 2022.

93. Slabki M. High-power properties of lead-based and lead-free ferroelectric ceramics. Doctoral dissertation, Darmstadt, Technische Universität Darmstadt, 2022.

94. Kawamata A, Hosaka H, Morita T. Non-hysteresis and perfect linear piezoelectric performance of a multilayered lithium niobate actuator. *Sensors and Actuators A: Physical.* **135**(2), 782, 2007.

95. Chen F, Kong L, Song W, Jiang C, Tian S, Yu F, Qin L, Wang C, Zhao X. The electromechanical features of $LiNbO_3$ crystal for potential high temperature piezoelectric applications. *Journal of Materiomics.* **5**(1), 73, 2019.

96. Yuzyuk YI, Gagarina E, Simon P, Reznitchenko LA, Hennet L, Thiaudiere D. Synchrotron X-ray diffraction and Raman scattering investigations of (Li_xNa_{1-x}) NbO_3 solid solutions: Evidence of the rhombohedral phase. *Physical Review B.* **69**(14), 144105, 2004.

97. Nitta T. Properties of sodium-lithium niobate solid solution ceramics with small lithium concentrations. *Journal of the American Ceramic Society.* **51**(11), 623, 1968.

98. Pozdnyakova I, Navrotsky A, Shilkina L, Reznitchenko L. Thermodynamic and structural properties of sodium lithium niobate solid solutions. *Journal of the American Ceramic Society.* **85**(2), 379, 2002.

99. Li Z, Wang J, Liu YX, Lu JT, Xu Z, Soon PS, Bi K, Chen C, Wang K. Sodium lithium niobate lead-free piezoceramics for high-power applications: Fundamental, progress, and perspective. *Journal of Advanced Ceramics.* **12**(1), 1, 2023.

100. Mitra S, Kulkarni AR, Prakash O. Diffuse phase transition and electrical properties of lead-free piezoelectric $(Li_xNa_{1-x})NbO_3$ ($0.04 \leq x \leq 0.20$) ceramics near morphotropic phase boundary. *Journal of Applied Physics.* **114**(6), 064106, 2013.

101. Aoyagi R, Iwata M, Maeda M. Piezoelectric properties and depolarization temperature of $NaNbO_3$-$LiNbO_3$ lead-free piezoelectric ceramics. *Key Engineering Materials.* **388**, 233, 2009.

102. Reznichenko LA, Shilkina LA, Razumovskaya ON, Dudkina SI, Gagarina ES, Borodin AV. Dielectric and piezoelectric properties of $NaNbO_3$-based solid solutions. *Inorganic Materials.* **39**, 139, 2003.

103. Pardo L, Duran P, Millar CE, Wolny WW, Jimenez B. High temperature electromechanical behaviour of sodium substituted lithium niobate ceramics. *Ferroelectrics.* **186**(1), 281, 1996.

Chapter 12

Future Uses of the Piezoelectric Effect for Energy Production

Babita Tripathi, Sunil Chauhan, and R. C. Singh
Centre for Solar Cell and Renewable Energy, Department of Physics, School of Basic Sciences and Research, Sharda University, Uttar Pradesh 201310, India
babita.tripathi@sharda.ac.in

The worldwide energy problem and environmental pollution triggered mostly due to overexploitation of conventional energy spurred scholars to investigate substitute energy applications that can capture various energies present in an ecosystem. Mechanical energy is popularly known among available energies capable of being absorbed and turned into usable electrical power. Due to higher electromechanical coupling reason and piezoelectric coefficient than its electrostatic, electromagnetic, and triboelectric transductions, piezoelectric transduction gets considered as a general mechanical energy harvesting mechanism. Micro- and nanoscale material and

Piezoelectric Materials: Design and Applications
Edited by N. B. Singh and Dev Kumar Mahato

ISBN 978-981-5129-40-3 (Hardcover), 978-1-003-59897-8 (eBook)
www.jennystanford.com

their manufacturing process advancements support piezoelectric generators' fabrication with beneficial characteristics like improved electromechanical coupling reason, piezoelectric coefficient, flexibility, stretch-ability, and integrate-ability for multiple systems. In addition, as technology advances in the electronics sector, tiny devices consuming low power are being made available in the ecosystem. As a result, it is expected that piezoelectric generators will power many devices in the near future. This chapter focuses on an in-depth evaluation of the current situations in piezoelectric energy harvesting. Recent studies about advancements in inorganic, organic, composite, and biocompatible natural piezoelectric substances have also been discussed. Piezoelectric energy harvesting usages on nanoscale, microscale, and mesoscale in many sectors have been discussed. The targeted sectors such as transport, structures, aerial systems, water systems, smart applications, microfluidics, biomedicals, wearable and implantable electronics, and tissue regeneration are taken into consideration. The breakthroughs, restraints, and possible advancements of the materials as well as uses of piezoelectric energy harvesting technologies are evaluated. As a summary, the highlighted analysis covers a large series of piezoelectric substances used to support clean power sources to wireless electronics in a variety of sectors.

12.1 Introduction

In 1880, Nobel prize winners Pierre Curie and Jacques Curie identified the piezoelectricity phenomenon (mechanical to electrical energy conversion) in natural crystals, which were quartz, tourmaline, Rochelle salt, and others. Lippman hypothesized the converse piezoelectricity based on thermodynamic principles. Curies discovered some relationships between piezoelectricity and crystal structure, but Voigt thoroughly confirmed them in 1894 [1]. Natural piezoelectric crystals were used for sonar applications during World War I, but their hygroscopic nature was the principal problem. The discovery of piezoelectricity in barium titanate ($BaTiO_3$, or BT) during WWII was regarded as an important achievement in the history of piezoelectric materials [2]. However, lead zirconate titanate (PZT), discovered in 1952, is regarded among the most promising piezoelectric elements that has been widely studied and frequently employed due to its superior piezoelectric capabilities as compared to other materials. Unfortunately, PZT is very hazardous due to the presence of lead oxide, and its toxicity is exacerbated by its ability to vaporize at high temperatures, notably at the time of calcination and sintering, resulting in the pollution of the environment [1]. During the discovery of PZT, the environmental issue was not regarded as a major concern. Thus, despite the fact that lead-less piezo substances were invented parallelly to the invention of PZT, researchers' attention focused on PZT-based substances because of the lower piezoelectric capabilities of lead-free applications. However, as time passed, great changes occurred. As a result, there is a lot of interest in developing biocompatible and ecologically benign lead-free piezoelectric materials to replace lead-based PZT materials. As a result of the widespread application of piezoelectric elements in multiple industries, research on lead-free materials, which had practically stalled in the 1990s, has regained impetus [2]. Piezoelectric substances are extensively employed for generators, sensors, actuators, and three-transducer systems as shown in Fig. 12.1 [3]. The generators and sensors avail a straightforward piezoelectric effect, which denotes the transition of mechanical energy to an electrical energy with the help of a dielectric displacement. Electrical energy is transitioned into mechanical energy with the help of actuators using the reverse piezoelectric effect. Finally, both forward and backward effects are applied in transducers, within the single device. However, the development of a piezoelectric nanogenerator (NG) in 2006 may be considered a big step in the history of piezoelectric substances in the energy harvesting area [4]. Several piezoelectric-material-based nanogenerators have been researched thus far, and making the device flexible is regarded as a critical difficulty for its wearable application. Unlike inorganic piezoelectric materials, piezoelectric polymers are noted for their flexibility, durability, lightweight, and low acoustic impedance. However, polymers have lesser piezoelectric characteristics than their inorganic counterparts [5]. Thus, improving the piezoelectricity of polymers becomes a difficult issue, and composite structures are advocated as the best method of producing high-performance flexible piezoelectric materials.

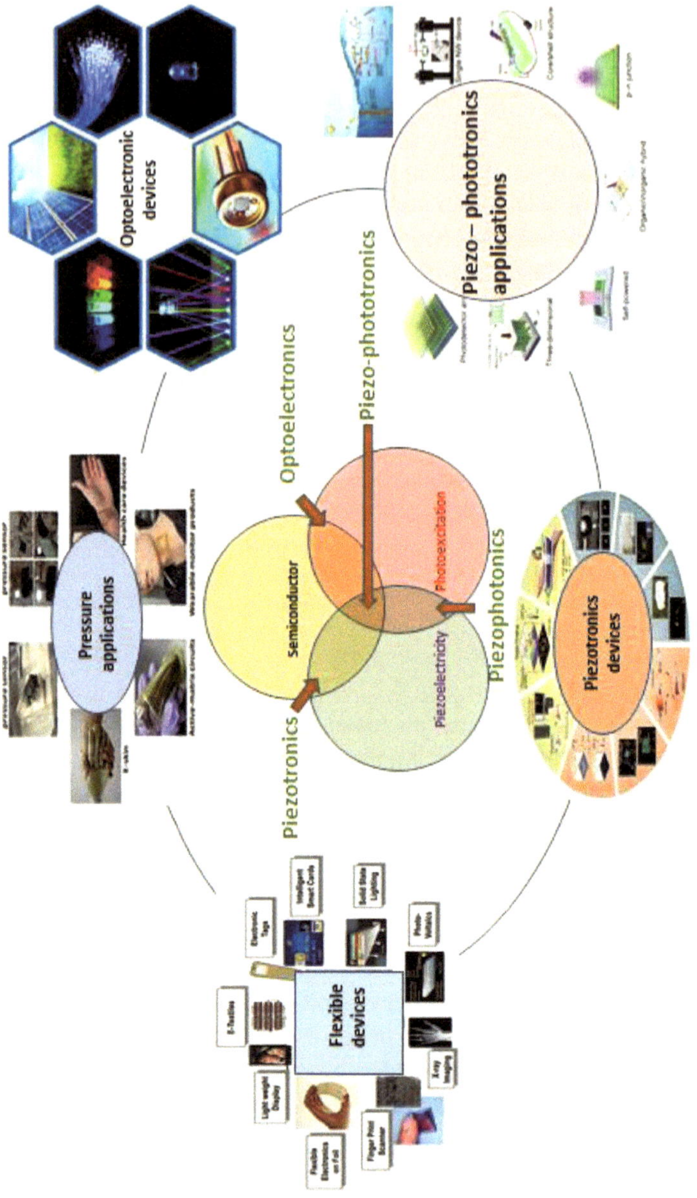

Figure 12.1 Piezoelectric effect for energy production [2–4].

12.2 Types of Piezoelectric Materials

The piezoelectric substances are classified into three parts crystalline, ceramics, and polymers. In the area of ceramic piezoelectric materials, wide band gap materials are used, for example, zinc oxide, lead zirconate titanate, and gallium nitride. These inorganic ceramics have crystal phases and can be easily fabricated in different shapes and directions [6]. However, polymer-based piezoelectric substances show less young modulus and high stress constants; that is why polymer-based piezoelectric substances are applied in sensors and actuators, for example, polyvinylidene fluoride (PVDF), PVDF-HFP, etc. [7]. On the grounds of environmental safety, the piezoelectric energy harvesting materials are classified into three parts: lead-equipped piezoelectric substances, lead-less piezoelectric substances, and hybrid piezoelectric substances [8].

12.2.1 Lead-Based Piezoelectric Materials

Lead zirconate titanate (PZT) is extremely appealing as well as a completely researched lead-based piezo system. It is a solution containing ferroelectric lead titanate ($PbTiO_3$) and anti-ferroelectric lead zirconate ($PbZrO_3$) [7]. This multicomponent solid solution application may be built up into a variety of compositions by differing the Zr:Ti ratio. The structure also varies with the Zr:Ti ratio, with "Zr" controlling structure (Zr:Ti = 54:46) being rhombohedral and "Ti" controlling structure ($Zr:Ti_4$ = 8:52) being tetragonal. By morphotropic phase boundary (MPB), both rhombohedral and tetragonal phases are present together in intermediate compositions. PZT has a formula $Pb(ZrxTi_{1-x})O_3$, which relates to the perovskite group having a general formula ABO_3. Here "A" refers to a divalent cation (Pb^{2+}) that captures the cube corners, "B" refers to a tetravalent cation (Zr^{4+} or Ti^{4+}) that captures the structure-centered place, whereas "O" refers to oxygen atoms at the face-centered places of the cube. By bringing electric field to the PZT unit cell, the Zr^{4+} or Ti^{4+} ion shifts to a different place (upside/downside) in the same direction as the applied field. Furthermore, substituting "A-site" with bigger rare-earth ions (alkali metal ions) and B-site with smaller transition metal ions is a broad method of generating a wide variety of perovskite structures with improved piezoelectric capabilities [8,

9]. Besides PZT, many lead-containing compositions and complex systems like $Pb_2Nb_2O_6$, (Pb $(Zn_{1/3}Nb_{2/3})O_3$)-$PbTiO_3$ (PZN-PT), $Pb(Mg_{1/3},Nb_{1/3})O_3$-$PbTiO_3$ (PMN-PT), $Pb(Ni_{1/3}Nb_{1/3})O_3$-Pb(Zr,Ti) O_3 (PNN-PZT), and $Pb(Sc_{1/2}Nb_{1/2})O_3$-$PbTiO_3$ are also investigated because of their interesting performances [10, 11].

12.2.2 Lead-Free Piezoelectric Materials

Piezoelectric ceramics that are lead-free are those that do not contain PbO in their composition. Over the last two decades, intensive research has been conducted around the world to discover alternates to lead-free piezoelectric compounds that can gradually substitute the ones with lead. In many applications, the usage of PZT is restricted because of its high density, and a low-density lead-free piezoelectric material would be a significant benefit in these applications [10]. However, researchers should keep their attention on improving the Curie temperature of the applications for practical use. Lead-free piezoelectric substances are split into groups named as: perovskite ceramics, wurtzite semiconductors, and polymers [11]. Because of their comparatively large piezoelectric constant, perovskite-type ferroelectrics like $BaTiO_3$ (BT), $(Bi_{0.5}Na_{0.5})TiO_3$ (BNT or bismuth sodium titanate), $KNbO_3$, $NaTaO_3$, and others are the most investigated lead-free piezoelectric compounds [12, 13]. Among them, barium titanate (BT) is the most widely used and extensively researched substance. Like PZT, the piezoelectricity in $BaTiO_3$ is caused by Ti displacement along the polarization direction, which also causes spontaneous polarization and ferroelectric characteristics. Other early lead-less piezo applications were dependent on alkali metal-based bismuth sodium titanate (BNT), bismuth potassium titanate (BKT), potassium sodium niobate (KNN), bismuth layered structures (BLSF), and so on [12]. The prime disadvantages these systems possess include: (i) a limited sintering temperature range, (ii) lower piezoelectric characteristics, (iii) immense corrosiveness because alkalis exist, and (iv) volatility, making stoichiometry tough to maintain [5]. Wurtzite semiconductors are another form of lead-free piezoelectric material that is not ferroelectric, like ZnO, ZnS, AlN, GaN, and CdS. Among these, ZnO has received the greatest attention [13]. It should be noted that most wurtzite semiconductors have poor piezoelectric performance when compared to perovskite-type ferroelectrics. However, their stability and suitability in terms of integration into electrical systems make them valuable piezoelectric materials in electronic devices [2]. There are also natural (cellulose, collagen, gelatin, PLLA) and synthetic polymers (PVDF and respective copolymers, polyamides, parylene-C, nylon-11, and so on) that can be termed lead-free piezoelectric materials [14]. There are three types of polymers: bulk polymer, voided charged polymer, and composite polymer. Ramadan et al. provide a comprehensive background of piezoelectric polymers, their theory, and piezoelectric properties [15]. The piezoelectric mechanism exists in bulk polymers in the form of solid polymer films due to their molecular structure and organization. The voided charged polymer is a polymer film with gas voids and surfaces charged in such a manner that internal dipoles occur. These dipoles' polarization varies with piezoelectric polymers, piezopolymers in bulk, amorphous semicrystalline, piezoelectric polymer composites, and polymers with voids in their charge. The polymer matrix filler is made of the piezoelectric material [16]. The applied stress on the polymer sheet that results in piezoelectric response is referred to as electric poling effect. The piezoelectric composite polymer is the other group. These are polymer constructions having piezoelectric fillers integrated. The fundamental purpose of this composite construction is to take advantage of the mechanical elasticity that polymers contain and a great electromechanical coupling of piezoelectric fillers (usually ceramic nanoparticles) [17].

12.2.3 Polymers as Piezoelectric Materials

It has already been discovered that piezoelectric substances tend to be found in either form as ceramic or polymeric in nature. Electroactive polymers provide distinct benefits as compared to ceramics in particular systems due to their affluence of production at lower temperatures, lower density, lower toughness, elasticity, and mechanical robustness like stiffness and high stresses on chances to fail. They look beneficial too when it comes to their biocompatibility of harvesters and sensors to be implanted [18].

Polymer-equipped piezoelectrics have been considered for several uses to support smart and multifunctional devices, including sensors, transducers, actuators, energy harvesting and storage devices, and fibers, foams, thin films, textiles, and coatings. As they possess a high degree of polarization and the presence of net molecule dipole incidents in their macromolecular build, ferroelectric polymers look quite suitable to support energy harvesting as well as storage. Because of the piezoelectric reaction of the polymers, molecular dipoles in these elements are capable to be instantly polarized as well as orientated when an electric field is applied, owing to mechanical stretching and interactions with nanoparticles [19]. Polarized and electroactive polymers, such as vinylidene fluoride (VDF)-based fluoropolymers, odd-numbered nylon, poly-L-lactide, polyurethane, and liquid crystal elastomers, are particularly fascinating. Their electroactive action is strongly governed by macromolecular structure, and it involves elements of the respective crystalline build, chain conformation, dipole orientation in crystalline areas in the direction of applied poling field, processing settings, and post-treatment procedures [20]. Electroactive polymers possess lower permittivity and piezoelectric coefficient than ferroelectric ceramics. Because of their smaller piezoelectric activity, considerable attempts have been expended for increasing the polymers' piezoelectric coefficients and adjusting the permittivity, and the hypothesis adopted to develop the electrical characteristics consists of changing the molecular configuration, creating composites, governing processing settings and the ways of post-treatment such as mechanical stretching and electrical poling, and so on. Due to their greater piezoelectric coefficient and simplicity of production, PVDF and its copolymers look to be favored for energy harvesting applications [21]. Various PVDF copolymers are devised having the goal of enhancing PVDF characteristics and making it more compatible with increasing technological demands. The molecular configuration of normally examined monomers and copolymers is depicted. Due to the presence of a steric hindrance effect, PVDF chains often support a realization of the ferroelectric phase when a massy co-monomer is added to it. When the TrFE content exceeds mol%, the genesis of a third fluoride atom in the TrFE [–CHF–CF_2] co-monomer drives polymer chains to line up in an expanded planar zigzag all-trans conformation [22]. However, when the TrFE level exceeds 20%, the nucleation of the phase is unaffected by processing ambience or electrical poling. The existence of large –CF_3 groups with PVDF chains provides broader room for dipole reorientation. For a solvent cast film with an HFP concentration of 5 mol%, an upper value of remnant polarization (80 mC P(VDF-TrFE) P(VDF-HFP) 16 m^2) was attained and this resulted in an upper piezoelectric coefficient of 30 pC N1. As a result, P(VDF-HFP) is a potential material for several piezoelectric and ferroelectric applications, such as configuration of magneto-electric sensors as well as actuators [23]. In the event of the P(VDF-CTFE) copolymer exhibited, with less than 16 mol%, 26 CTFE concentration resulted in a higher piezoelectric coefficient, up to 140 pC N1. Due to this, a favorability of PVDF copolymers against PVDF is described as the phase is often available irrespective of the ways of their application, and this is ascribed to the steric restriction that a co-monomer possesses, which elaborates the inter-chain distance and decreases the activation energy for the phase transition. When CTFE is added to PVDF-TrFE at varying percentages, P(VDF-TrFE-CTFE) allows for the production of substances with lesser Curie temperatures as compared to PVDF-TrFE and an equivalent amount of electro-calorific effect [24]. Similar to high-energy electron irradiation, inclusion of chemical monomers such as CTFE brings imperfections to the P(VDF-TrFE) ferroelectric build. With such an event, transformation of polymers is observed, from a normal ferroelectric to a relaxer ferroelectric, making them multifunctional for their unmatchable characteristics compared to P(VDF-CTFE) ferroelectric polymers, like high electro-strictive strain and high dielectric constant [25].

12.2.4 Hybrid Piezoelectric Materials

Nanomaterials, nanocomposites, and hybrid materials have seen rapid progress in recent years. Nanomaterials are materials having grain sizes less than 100 nm or particles with at least one dimension less than 100 nm. At the nanoscale, matter constituents perform functions that differ from those performed by bulk materials and individual atoms. Nanomaterials have a unique electronic nature; they follow quantum mechanical principles, and this fact offers nanomaterials an essential place in semiconductor research [26]. Single-phase materials, on the other hand, frequently fail to exhibit suitable characteristics for certain applications. Making a composite, or merging two or more classes of materials in a single substance, is one technique to overcome such issues. Since the 1980s, piezoelectric composites have been conceptualized, prototyped, produced, and used in so many applications ranging from medical imaging to military missions. In such composites, each component is responsible for delivering a set of equivalent desirable traits (rather than different properties). The piezocomposites are classed based on their "connectivity" between different components. A common two-phase piezoelectric composite can be categorized into numerous varieties based on its physical connectivity [27].

12.3 Importance of Piezoelectric Substances for Energy Production

Energy harvesting from ambient settings is a new technology that has a wide range of uses, from remote-based devices to eco-friendly energy. Many power harvesting applications have used electroactive materials such as piezoelectric and electrostriction to support transformation of mechanical energy to electrical energy. In piezoelectric energy harvesting, piezoelectric action turns mechanical strain into electric current or voltage. This strain might originate via a variety of origins. Human movements, less frequent earthquake tremors, and auditory sounds to name a few [28]. Because piezoelectric energy harvesting has only been studied since the late 1990s, it is yet a new application. Furthermore, nanoscale piezoelectricity is a growing academic topic. Harvesting surrounding mechanical energy at the nanoscale has enormous promise to support some electronic items and succeed for self-powered electronic items [29]. A capacity to self-power electronic system packaging allows for the omission of massy energy storage systems and the omission of heavy battery devices. Nanogenerators (NGs) are such gadgets that could give a new approach to power devices without applying batteries. Polyvi, for example, is a piezoelectric polymer. Piezoelectric polymers such as polyvinylidene fluoride (PVDF) and its copolymers, such as P(VDF-TrFE) and P(VDF-HFP), create electrical polarization in reaction to mechanical stimuli or electric fields (or vice versa). Wang et al. have proposed a different study topic on nanopiezoelectronics, which promises that piezoelectric materials can be an alternative solution for energy harvesting applications. They exhibited the outstanding concept of zinc oxide (ZnO) nanorods or nanowire array-based NGs, which are one-dimensional (1D) nanomaterial having diameters lower than 100 nm and lengths beyond 1 m. PVDF and related copolymers P(VDF-TrFE), P(VDF-HFP), and P(VDF-CTFE), among others, are excellent alternatives for similar applications [26, 32]. Because of its superior ferroelectric and piezoelectric capabilities over PVDF homopolymer, the comparatively low-cost P(VDF-HFP) has piqued the scientific and technological attention of the copolymer [30]. Furthermore, the improvement of piezoelectricity with the inclusion of different inorganic and organic fillers into P(VDF-HFP) has prompted us to examine its piezoelectric effect for mechanical energy harvesting and other multifunctional applications [31]. Furthermore, P(VDF-HFP) offers various advantages, including easy processing conditions, cost-effectiveness, acceptable flexibility, low weight, the capacity to be molded into complicated geometries, and the potential to be functionalized to increase device performance [32].

12.4 Models of Piezoelectric Materials

The word piezoelectricity hails from a Grecian word "piezein," or "to press," i.e., stress-based electricity. If mechanical stress is applied over piezoelectric materials, electrical polarization changes, thus creating the direct piezoelectric effect. The piezoelectricity produces electrical current in an external circuit, which works as an electromechanical generator for energy harvesting. This piezoelectric effect describes a characteristic of substances to produce dispersed electrical charges of opposite sign as and when an external pressure produced by mechanical deformations is applied. Electrical charges collect at two ends of the material body, allowing creation of an electric dipole and piezo-potential generation. When exposed to an electric field, piezoelectric materials experience mechanical strain; the so-caused reverse action is known as the converse piezoelectric effect [33]. The opposite piezoelectric effect is widely used in actuator, vibration damping, and acoustic emitter applications [34]. By allowing a crystalline solid and respective ion distribution within a unique unit cell, an optimal piezoelectricity mechanism is offered. In absence of applied pressure, centers of the positive and negative charges become non-centrosymmetric inside the unit cell planes, which is the most common example of piezoelectricity. This occurs in a few ceramics, including barium titanate ($BaTiO_3$), lead zirconate titanate (PZT), aluminum nitride (AlN), and zinc oxide (ZnO). There are 32 crystallographic classes based on the geometry and symmetry of the unit cell, 21 of which are non-centrosymmetric in nature and all except one (owing to additional symmetry elements) display piezoelectricity. The absence of symmetry in the ion distributions in these crystalline materials causes the presence of an electrical dipole, which causes the piezoelectric response. Because of the absence of symmetry, piezoelectric materials for energy harvesting have a well-defined polar axis, and the application of stress relative to this polar axis is critical for energy harvesting applications; the polar axis is sometimes referred to as the three-direction or z-direction, for example, the crystalline structure of lead zirconate titanate (PZT) [35]. There are three main operation modes of piezoelectric energy harvesting: continuous mode, burst mode, and resonance mode [36].

12.4.1 Continuous Mode

In this mode, the piezoelectric material is subjected to continuous vibrations, and the energy is continuously harvested and stored. This mode is suitable for applications where there is a constant source of vibrations, such as in rotating machinery or vehicles [37].

12.4.2 Burst Mode

In this mode, the piezoelectric material is exposed to intermittent or sporadic vibrations. The energy is harvested and stored during the periods of vibrations and can be used during the periods of no or low vibrations. This mode is suitable for applications where the vibrations occur in bursts, such as in human motion or wind gusts [38].

12.4.3 Resonance Mode

In this mode, the piezoelectric material is tuned to resonate at a specific frequency, which maximizes the energy harvesting efficiency. When the external vibrations match the resonance frequency, the energy harvesting is most effective. This mode is suitable for applications where the vibrations have a specific frequency, such as in certain types of machinery or structures. Due to specific crystallographic structure, the properties of piezoelectric materials differ based on the crystal's orientation [39]. Figure 12.2 shows three axes and three rotations used to describe these orientations. Typically, axis 3 is aligned with the sample's thickness, and axis 1 is aligned with its longest side. All mechanical and electrical parameters are tensors of varying orders. They are used for separating the piezoelectric and elastic coefficients measured along different axes [40].

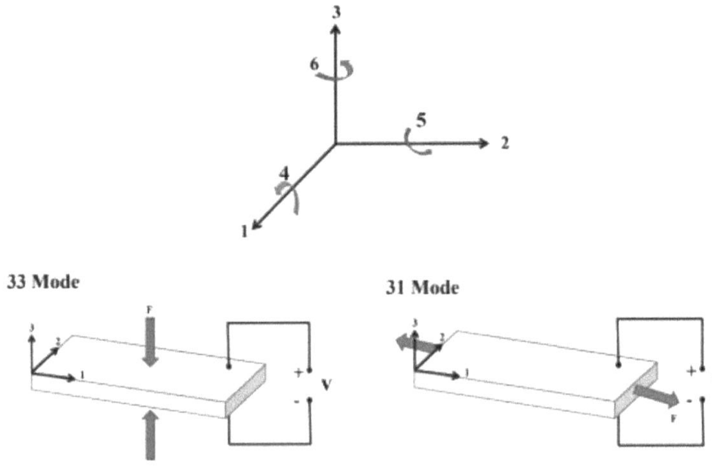

Figure 12.2 Models of piezoelectric materials.

They are denoted in matrix form and are shown as the direction of general voltage and applied stress. To fully characterize the piezoelectric behavior, 36 flexibility coefficients, 18 piezoelectric coefficients, and 9 permittivity coefficients exist. To fully characterize the piezoelectric behavior of a material, 63 coefficients must be known, which is a huge amount. Though piezoelectric materials are not centrosymmetric, they contain adequate symmetry to diminish the coefficients [41].

12.5 Effect of Piezoelectricity on Mechanical Energy Harvesting

The applied force produces an electrical charge that is capable to be used in energy harvesting in direct piezoelectric effect. The piezoelectric effect illustrates the capacity of various substances to create spatially dispersed electrical charges of opposite sign in reaction to an outside force that causes mechanical deformation [42]. An accumulation of electrical charges at two ends of the material body supports the production of an electric dipole, which causes a formation of piezo-potential. Connecting the material to an external load, and in the absence of short-circuited contacts, free electrons flow via the external circuit to balance the polarization charges and produce a new equilibrium state. The application of a dynamic strain does result in the development of a pulsed continuous stream of current flowing through the external circuits, which can be recorded and stored [43].

12.6 Polymer-Based Devices for Mechanical Energy Harvesting and Versatile Applications

Piezoelectric polymer compounds carry numerous applications, including actuation, wireless sensing, and energy harvesting devices [44]. Because of related multifunctionality, higher dependability, lower prices, and higher precision, these systems have revolutionized the trend of PEH devices. One of the notable application areas is the biomedical sector, where polymers are investigated to achieve a viable material for artificial muscle actuators, as intrusive medical robots for diagnosis and microsurgery, and as actuator implants for encouraging bone and tissue formation [45]. Polymeric substances are appropriate for likewise applications as they have biocompatibility and improved impedance matching to human tissues and body fluids. However, numerous energy harvesting applications of piezoelectric polymeric composites have been reported, ranging from portable biomedical devices (wearable or implantable) to biomedicines [46]; wireless sensor networks

[47]; robotics, optics, environmental monitoring, ultrasonics, and acoustic devices; underwater and electro-acoustic transducers; and piezoelectric roads [48]. Furthermore, the creation of PEH devices has been emphasized based on human body movement concurrently. Trusting this technology, piezoelectric shoes [49], electronic skin and health monitoring patches [50, 51], e-textiles [52], piezoelectric dancing floor, and other devices that use vibrations produced by human activities such as walking, jogging, breathing in and out, and dancing are developed [53]. The construction of piezoelectric devices from lightweight polymeric systems has been crucial, and commercial interest is growing. The piezo-based power sensors are also used for environmental monitoring and industrial automation. Such types of power sensors collect data without any other external source. Piezo-based technology is used in wearable devices like smart-watches, fitness trackers, and medical monitoring devices [54]. The generated energy resolves the battery-changing problems. In the area of remote monitoring applications, wireless piezoelectric energy can be used in remote locations to collect data without any external power source. In the area of automotive and transportation, piezo-based energy systems can be integrated into vehicles to change mechanical vibrations into electrical energy. This energy is capable of being used to support road light systems and recharge the vehicle's battery. The various human powered piezoelectric devices can be used as body stickers, chips, electronic devices, rechargeable batteries, and backpacks to generate electricity due to body movement. For infrastructure monitoring of structure of buildings, bridges, towers, and power plants, piezoelectric-energy harvesting technology can be used [55]. The proliferation of Internet of things (IoT) needs low cost and renewable energy. The piezoelectric energy harvesting can be useful for development of efficient energy and wireless devices [56].

12.6.1 Aerial Applications

The piezoelectric energy harvesting processes are used for aerial applications like unmanned aerial vehicles (UAVs), drones, energy harvesting wings, and location and environment monitoring systems [57].

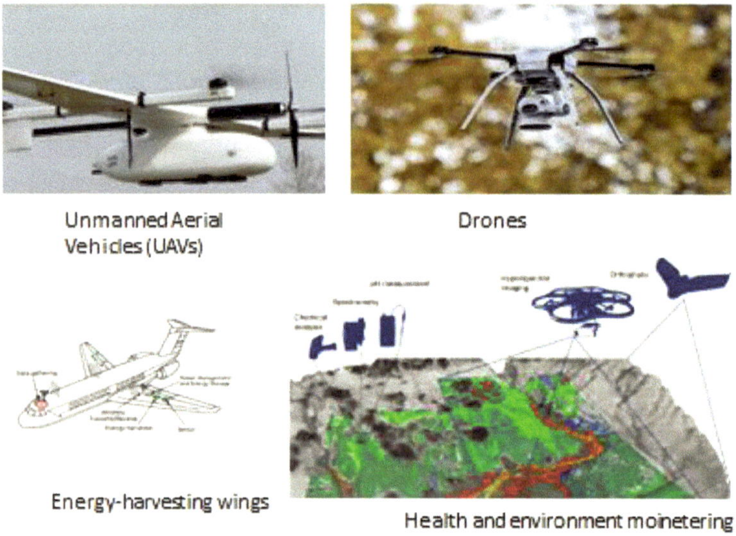

Figure 12.3 Aerial applications of piezoelectric energy harvesting technology.

Piezo-based UAVs are used to capture and convert the vibrations and movements of the aircraft into electrical energy. The generated energy is used as supporting power supply in UAV communication systems and recharging batteries [58]. Drones are used for their power supply systems that restrict their flight time limits. The electrical energy produced by mechanical vibrations increases their flight time and reduces their battery scarcity. When the piezoelectric materials are integrated into the wings of an aeroplane, the airflow during flight hour easily converts their air-based mechanical vibration into electrical energy. The energy harvested by wings is used in sensors and communication devices without any external power source [59]. These are just a few examples of how piezoelectric energy harvesting can be applied in aerial applications. The technology carries a capability for enhancing the efficiency, endurance, and sustainability of various aerial systems by harnessing otherwise wasted mechanical energy [60].

12.6.2 Underwater Applications

Piezoelectric energy harvesting technology has various underwater applications like wave energy harvesting, tidal energy harvesting, underwater monitoring system, communication system, and navigation and tracking systems [61]. The ocean waves continuously generate mechanical vibrations; the piezoelectric devices capture this mechanical energy and convert it to electrical energy. The tides in oceans are regular and predictable natural movements of water caused by the gravity of the moon and the sun. The mechanical energy generated during tidal movements is transformed to energy by using piezoelectric harvesting technology. These generated energies are used for the autonomous monitoring sensors. Since wireless communication is not easy under water, due to the reduction of electromagnetic signals [62], the piezo-based energy harvesting solves the power solution for signal transmission. The underwater robotics or vehicles can be used to harvest energy without any other power supply [63]. Underwater infrastructure, pipelines, and damages can be easily detected by piezoelectric sensors that prevent the accidental situation. Overall, the underwater applications of piezoelectric energy production give a sustainable and better solution for power-devices and sensors by reducing maintenance budget and providing eco-friendly impact on water bodies [64].

12.6.3 Piezoelectric Energy Harvesting by Using Footsteps, Walking, and Finger Tapping

Piezoelectric energy harvesting is a process that transforms mechanical energy into electrical energy by using piezoelectric substances. Piezoelectric substances exhibit a property of generating an electric charge when exposed to mechanical stress. One potential application of piezoelectric energy harvesting is using footsteps [65]. This can be achieved by embedding piezoelectric elements into the flooring where foot movement is very high, such as shopping complexes, airports, or rail stations. When people walk over these piezoelectric elements, the mechanical stress from their footsteps will generate an electric charge, which can be harnessed and stored for later use. Another application of piezoelectric energy harvesting is through walking [66]. This can be achieved by incorporating piezoelectric elements into the soles of shoes. As a person walks or runs, the mechanical stress from their movement will generate an electric charge, and it has the capacity to power small electronic applications, such as fitness trackers or smartphones. Additionally, piezoelectric energy harvesting can be used through finger tapping [67]. By incorporating piezoelectric elements into touch-sensitive surfaces, such as smartphone screens or computer keyboards, the mechanical stress from finger tapping can be transformed into electrical energy [68]. This energy has the capacity to power these devices or charge their batteries. Overall, piezoelectric energy harvesting through footsteps, walking, and finger tapping offers a potential solution for harvesting wasted mechanical energy from our everyday activities and converting it into electrical energy for various applications [69]. Since it can power autonomous wearable devices, piezoelectric energy harvesting from human movement has recently gained significant attention. In order to transform mechanical energy from human body movements into electrical energy, a wearable piezoelectric generator based on boron nitride nanosheets is used; under a periodic mechanical push force of 80 kg, the generator generated a maximum output voltage and output power of 22 V and 40 W, respectively. Different areas of the human body had the generator attached to their skin, which produced output voltages of 2.5 V on the foot, 1.98 V on the elbow, 0.48 V on the neck, 0.75 V on the wrist, and 1.05 V on the knee under varying conditions, in order to capture low-frequency biomechanical energy from body movement for powering wearable electronics. Through a PVDF-based curved piezoelectric generator, 476 LED (light-emitting diode) lamps could be powered by the system's output power density of 3.9 mW/cm^2. At 35 Hz, it was reported that the average output power was 45 V and the average output current was 225 mA. The created technology was also included in a watch strap and a shoe footbed. In comparison to the watch generator, which produced an average output voltage of 22 V and an average output current of 50 A, the insole generator produced an average output voltage of 14 V and 18 A [70].

A piezoelectric generator with a P(VDF-TrFE)/BT nanocomposite can be used as self powered sensor to harvest energy by using foot steps [71]. The three energy sources from walking activity that can be captured to provide electrical power are acceleration pulses from the heel strike, acceleration from leg swing, and the compressive force acting on the shoe owing to a person's weight. The output voltage and power grew from 1.06 to 3.55 V and 0.03 to 0.35 mW, respectively, as the walking speed increased from 2 to 8 km/h [71]. When a PZT-5H piezoelectric ceramic device with a steel frame was fitted into a 90 kg human shoe, with an output rate of 1.43 mW, the system converted 0.4% of the applied force into electrical power. A piezoelectric tile was used for harvesting energy from foot movements. To prevent the piezoelectric module from shattering, a spring-loaded tile with a tip mass was used. The result of a freely dropped steel ball on the designed system gave a maximum output power of 55 mW [72].

12.6.4 Biomedical Applications

Piezoelectric energy harvesting has a wide range of biomedical applications, which include ultrasonic imaging, transducers for therapeutic ultrasound, biomedical sensors, biomechanical energy instruments, bone-tissue engineering, and periprosthetic applications [73]. In ultrasound imaging devices, when an electrical voltage is passed to a piezoelectric material, it produces ultrasonic waves that can go through tissues. The reflected ultrasound waves create brief images of body organs. The therapeutic ultrasound waves can be used for tissue ablation, targeted drug delivery, and tissue repairing process [74]. The transducers are used to deliver ultrasound energy precisely to the targeted part of the tissue. Piezoelectric biomedical sensors are used to measure force, pressure, and acceleration in medical devices like pacemakers and orthopedic implantations. These biomedical sensors are capable of monitoring physiological situations of the body like blood pressure, heart rate, and pulse rate [75].

Energy harvesting: Piezoelectric materials are capable of being used to harvest biomechanical energy generated through the body's movement or organ functions. This energy can then be converted to electrical energy and used to power medical applications, like implantable sensors or pacemakers, without a requisition of replaceable batteries [76].

Bone and tissue engineering: Piezoelectric materials can be incorporated into scaffolds used in tissue engineering to stimulate cell growth and tissue regeneration. The mechanical stimulation provided by the piezoelectric materials can promote cell adhesion, proliferation, and differentiation, causing enhanced tissue healing and reproduction [77].

Periprosthetic: Piezoelectric materials can be used in periprosthetic devices, such as cochlear implants and deep brain stimulation electrodes. In cochlear implants, piezoelectric materials transform sound vibrations as electrical indicators that can accelerate the auditory nerve. In deep brain stimulation, the piezoelectric materials produce electrical pulses that can modulate neural activity and treat conditions like Parkinson's disease [78].

Overall, piezoelectric materials offer significant potential for various biomedical applications, ranging from diagnostic imaging and therapeutic devices to biomedical sensors and tissue engineering scaffolds. Continued research and development (R&D) in this field can support further stimulations in biomedicine sphere.

12.7 Conclusion and Future Prospects of Piezoelectric Energy Harvesting

In conclusion, piezoelectric energy harvesting offers a promising solution for converting mechanical vibrations and movements into electrical energy. It has been successfully applied in numerous applications such as powering small electronic systems, sensors, and wireless communication systems. The development of flexible and wearable piezoelectric materials has expanded the potential for energy harvesting in new environments and industries. These materials can be integrated into clothing, shoes, and even infrastructure such as roads and buildings to harvest energy from human movements or vibrations caused by vehicles. One of the main advantages of piezoelectric energy harvesting is that it can provide a continuous and sustainable power source in situations where traditional batteries may not be practical or accessible. This is particularly valuable in remote or hard-to-reach locations where frequent battery replacements are not feasible. Furthermore, the increasing interest in renewable energy sources and the need for energy-efficient technologies have further fueled the research and development of piezoelectric energy harvesting. The technology has great potential for integration into smart homes, the Internet of Things (IoT), and autonomous systems. However, there are still several challenges that need to be addressed for piezoelectric energy harvesting to reach its full potential. These include improving the efficiency of energy conversion, optimizing the design of piezoelectric materials, and developing scalable manufacturing techniques. Future prospects for piezoelectric energy harvesting involve advancements in nanotechnology and the use of new materials with superior piezoelectric properties. Researchers are exploring the use of nanowires, thin films, and composite materials to enhance the efficiency and performance of piezoelectric energy harvesters. Moreover, the integration of piezoelectric materials with other energy harvesting technologies such as solar or thermoelectric can create hybrid systems that can generate power from multiple sources, further enhancing the energy harvesting potential. Overall, the future of piezoelectric energy harvesting looks promising as advancements in materials and technologies continue to drive innovation and improve efficiency. With further research and development, it has the potential to become a widely adopted method for sustainable and eco-friendly energy generation.

References

1. Ali F., Raza W., Li X., Gul H., Kim K.-H., "Piezoelectric energy harvesters for biomedical applications," *Nano Energy*, **57**, 879–902, 2019, doi: 10.1016/j.nanoen.2019.01.012.
2. Bowen C. R., Kim, H. A., Weaver P. M., Dunn S., "Piezoelectric and ferroelectric materials and structures for energy harvesting applications," *Energy Env. Sci*, **7**(1), 25–44, 2014, doi: 10.1039/C3EE42454E.
3. Beeby S. P., Tudor M. J., White N. M., "Energy harvesting vibration sources for microsystems applications," *Meas. Sci. Technol.*, **17**(12), R175–R195, 2006, doi: 10.1088/0957-0233/17/12/R01.
4. Sun Y., Chen J., Li X., Lu Y., Zhang S., Cheng Z., "Flexible piezoelectric energy harvester/sensor with high voltage output over wide temperature range," *Nano Energy*, **61**, 337–345, 2019, doi: 10.1016/j.nanoen.2019.04.055.
5. Hu D., Yao M., Fan Y., Ma C., Fan M., Liu M., "Strategies to achieve high performance piezoelectric nanogenerators," *Nano Energy*, **55**, 288–304, 2019, doi: 10.1016/j.nanoen.2018.10.053.

6. Yan J., Liu M., Jeong Y. G., Kang W., Li L., Zhao Y., Deng N., Cheng B., Yang G., "Performance enhancements in poly(vinylidene fluoride)-based piezoelectric nanogenerators for efficient energy harvesting," *Nano Energy*, **56**, 662–692, 2019, doi: 10.1016/j.nanoen.2018.12.010.
7. Wang D. W., Mo J. L., Wang X. F., Ouyang H., Zhou Z. R., "Experimental and numerical investigations of the piezoelectric energy harvesting via friction-induced vibration," *Energy Convers. Manag.*, **171**, 1134–1149, 2018, doi: 10.1016/j.enconman.2018.06.052.
8. Xie X. D., Wang Q., "Energy harvesting from a vehicle suspension system," *Energy*, **86**, 385–392, 2015, doi: 10.1016/j.energy.2015.04.009.
9. Chen J., Oh S. K., Nabulsi N., Johnson H., Wang W., Ryou J.-H., "Biocompatible and sustainable power supply for self-powered wearable and implantable electronics using III-nitride thin-film-based flexible piezoelectric generator," *Nano Energy*, **57**, 670–679, 2019, doi: 10.1016/j.nanoen.2018.12.080.
10. Takahashi H., Numamoto Y., Tani J., Matsuta K., Qiu J., Tsurekawa S., "Lead-free barium titanate ceramics with large piezoelectric constant fabricated by microwave sintering," *Jpn. J. Appl. Phys.*, **45**(1L), L30, 2005, doi: 10.1143/JJAP.45.L30.
11. Assi M. A., Hezmee M. N. M., Haron A. W., Sabri M. Y., Rajion M. A., "The detrimental effects of lead on human and animal health," *Vet. World*, **9**(6), 660–671, 2016, doi: 10.14202/vetworld.2016.660-671.
12. Fan Y., Zhou Z., Liang R., Dong X., "Designing novel lead-free $NaNbO_3$-based ceramic with superior comprehensive energy storage and discharge properties for dielectric capacitor applications via relaxor strategy," *J. Eur. Ceram. Soc.*, **39**(15), 4770–4777, 2019, doi: 10.1016/j.jeurceramsoc.2019.07.021.
13. Fei C., Liu X., Zhu B., Li D., Yang X., Yang Y., Zhou Q., "AlN piezoelectric thin films for energy harvesting and acoustic devices," *Nano Energy*, **51**, 146–161, 2018, doi: 10.1016/j.nanoen.2018.06.062.
14. Guan X., Xu B., Gong, J., "Hierarchically architected polydopamine modified $BaTiO_3$@P(VDF-TrFE) nanocomposite fiber mats for flexible piezoelectric nanogenerators and self-powered sensors," *Nano Energy*, **70**, 104516, 2020, doi: 10.1016/j.nanoen.2020.104516.
15. Shi K., Sun B., Huang X., Jiang P., "Synergistic effect of graphene nanosheet and $BaTiO_3$ nanoparticles on performance enhancement of electrospun PVDF nanofiber mat for flexible piezoelectric nanogenerators," *Nano Energy*, **52**, 153–162, 2018, doi: 10.1016/j.nanoen.2018.07.053.

16. Kim H. S., Kim J.-H., Kim J., "A review of piezoelectric energy harvesting based on vibration," *Int. J. Precis. Eng. Manuf.*, **12**(6), 1129–1141, 2011, doi: 10.1007/s12541-011-0151-3.

17. Wu N., Wang Q., Xie X., "Ocean wave energy harvesting with a piezoelectric coupled buoy structure," *Appl. Ocean Res.*, **50**, 110–118, 2015, doi: 10.1016/j.apor.2015.01.004.

18. Siang J., Lim M. H., Salman Leong M., "Review of vibration-based energy harvesting technology: Mechanism and architectural approach," *Int. J. Energy Res.*, **42**(5), 1866–1893 2018, doi: 10.1002/er.3986.

19. Jella V., Ippili S., Eom J.-H., Pammi S. V. N., Jung J.-S., Tran V.-D., Nguyen V. H., Kirakosyan A., Yun S., Kim D., Sihn M. R., Choi J., Kim Y.-J., Kim H.-J., Yoon S.-G., "A comprehensive review of flexible piezoelectric generators based on organic-inorganic metal halide perovskites," *Nano Energy*, **57**, 74–93, 2019, doi: 10.1016/j.nanoen.2018.12.038.

20. Priya S., Song H.-C., Zhou Y., Varghese R., Chopra A., Kim S.-G., Kanno I., Wu L., Ha D. S., Ryu J., Polcawich R. G., "A review on piezoelectric energy harvesting: Materials, methods, and circuits," *Energy Harvest. Syst.*, **4**(1), 3–39, 2019, doi: 10.1515/ehs-2016-0028.

21. Sarker M. R., Julai S., Sabri M. F. M., Said S. M., Islam Md. M., Tahir M., "Review of piezoelectric energy harvesting system and application of optimization techniques to enhance the performance of the harvesting system," *Sens. Actuators Phys.*, **300**, 111634, 2019, doi: 10.1016/j.sna.2019.111634.

22. Pakhira M., Chatterjee D. P., Mallick D., Ghosh R., Nandi A. K., "Reversible stimuli-dependent aggregation-induced emission from a 'nonfluorescent' amphiphilic PVDF graft copolymer," *Langmuir*, **37**(16), 4953–4963, 2021, doi: 10.1021/acs.langmuir.1c00310.

23. Kim K.-S., Park S.-Y., Choi S., Lee H., "Ionic liquid–polymer gel electrolytes based on morpholinium salt and PVdF(HFP) copolymer," *J. Power Sources*, **155**(2), 385–390, 2006, doi: 10.1016/j.jpowsour.2005.05.018.

24. Adhikary P., Garain S., Mandal D., "The co-operative performance of a hydrated salt assisted sponge like P(VDF-HFP) piezoelectric generator: An effective piezoelectric based energy harvester," *Phys. Chem. Chem. Phys.*, **17**(11), 7275–7281, 2015, doi: 10.1039/C4CP05513F.

25. Chen Y., Chen X., Zhou D., Shen Q.-D., Hu W., "Low-temperature crystallization of P(VDF-TrFE-CFE) studied by Flash DSC," Polymer, **84**, 319–327, 2016, doi: 10.1016/j.polymer.2016.01.003.

26. Alluri N. R., Saravanakumar B., Kim S.-J., "Flexible, hybrid piezoelectric film ($BaTi_{(1-x)}Zr_xO_3$)/PVDF nanogenerator as a self-powered fluid

velocity sensor," *ACS Appl. Mater. Interfaces*, **7**(18), 9831–9840, 2015, doi: 10.1021/acsami.5b01760.

27. Banerjee S., Bairagi S., Wazed Ali S., "A critical review on lead-free hybrid materials for next generation piezoelectric energy harvesting and conversion," *Ceram. Int.*, **47**(12), 16402–16421, 2021, doi: 10.1016/j.ceramint.2021.03.054.
28. Sun C., Shi J., Bayerl D. J., Wang X., "PVDF microbelts for harvesting energy from respiration," *Energy Environ. Sci.*, **4**(11), 4508, 2011, doi: 10.1039/c1ee02241e.
29. Wang Z. L., Song J., "Piezoelectric nanogenerators based on zinc oxide nanowire arrays," *Science*, **312**(5771), 242–246, 2006, doi: 10.1126/science.1124005.
30. He X., Yao K., Gan, B. K., "Phase transition and properties of a ferroelectric poly (vinylidene fluoride-hexafluoropropylene) copolymer," *J. Appl. Phys.*, **97**(8), 084101, 2005, doi: 10.1063/1.1862323.
31. Huan Y., Liu Y., Yang Y., "Simultaneous stretching and static electric field poling of poly(vinylidene fluoride-hexafluoropropylene) copolymer films," *Polym. Eng. Sci.*, **47**(10), 1630–1633, 2007, doi: 10.1002/pen.20843.
32. Senthilkumar N., Babu K. J., Gnana Kumar G., Kim A. R., Yoo D. J., "Flexible electrospun PVdF-HFP/Ni/Co membranes for efficient and highly selective enzyme free glucose detection," *Ind. Eng. Chem. Res.*, **53**(25), 10347–10357, 2014, doi: 10.1021/ie500755m.
33. Wang Z., He L., Gu X., Yang S., Wang S., Wang P., Cheng G., "Rotational energy harvesting systems using piezoelectric materials: A review," *Rev. Sci. Instrum*, **92**(4), 041501, 2021, doi: 10.1063/5.0039730.
34. Jin H., Gao X., Ren K., Liu J., Qiao L., Liu M., Chen W., He Y., Dong S., Xu Z., Li F., "Review on piezoelectric actuators based on high-performance piezoelectric materials," *IEEE Trans. Ultrason. Ferroelectr. Freq. Control*, **69**(11), 3057–3069, 2022, doi: 10.1109/TUFFC.2022.3175853.
35. Habib M., Lantgios I., Hornbostel K., "A review of ceramic, polymer and composite piezoelectric materials," *J. Phys. Appl. Phys.*, **55**(42), 423002, 2022, doi: 10.1088/1361-6463/ac8687.
36. Zhang T.-Y., Hack J. E., "Mode-III cracks in piezoelectric materials," *J. Appl. Phys.*, **71**(12), 5865–5870, 1992, doi: 10.1063/1.350483.
37. Li X.-F., Fan T.-Y., Wu X.-F., "A moving mode-III crack at the interface between two dissimilar piezoelectric materials," *Int. J. Eng. Sci.*, **38**(11), 1219–1234, 2000, doi: 10.1016/S0020-7225(99)00072-5.

38. Blackburn J. F., Cain M. G., "Non-linear piezoelectric resonance analysis using burst mode: a rigorous solution," *J. Phys. Appl. Phys.*, **40**(1), 227–233, 2007, doi: 10.1088/0022-3727/40/1/019.

39. Yang W., Sun S., Hu J., Tang L., Qin L., Li Z., Luo W., "Deep learning model as an inversion tool for resonant ultrasound spectroscopy of piezoelectric materials," *Appl. Phys. Lett.*, **120**(18), 184101, 2022, doi: 10.1063/5.0086238.

40. Tian Y., Zhang W., Tan Z., Cho C., "Chiral edge states for phononic crystals based on shunted piezoelectric materials," *Extreme Mech. Lett.*, **50**, 101568, 2022, doi: 10.1016/j.eml.2021.101568.

41. Shaukat H., Ali A., Bibi S., Altabey W. A., Noori M., Kouritem S. A., "A review of the recent advances in piezoelectric materials, energy harvester structures, and their applications in analytical chemistry," *Appl. Sci.*, **13**(3), 1300, 2023, doi: 10.3390/app13031300.

42. Ma B., Cheng L., Bai S., Jia X., Ma J., Zhao J., Wang L., Qin Y., "3D spirally coiled piezoelectric nanogenerator for large impact energy harvesting," *Nano Energy*, **111**, 108412 2023, doi: 10.1016/j.nanoen.2023.108412.

43. Sasmal A., Senthilnathan J., Arockiarajan A., Yoshimura M., "Two-dimensional metal-organic framework incorporated highly polar PVDF for dielectric energy storage and mechanical energy harvesting," *Nanomaterials*, **13**(6), 1098, 2023, doi: 10.3390/nano13061098.

44. Soin N., Shah T. H., Anand S. C., Geng J., Pornwannachai W., Mandal P., Reid D., Sharma S., Hadimani R. L., Bayramol D. V., Siores E., "Novel '3-D spacer' all fibre piezoelectric textiles for energy harvesting applications," *Energy Env. Sci.*, **7**(5), 1670–1679, 2014, doi: 10.1039/C3EE43987A.

45. Xiahou X., Wu S., Guo X., Li H., Chen C., Xu M., "Strategies for enhancing low-frequency performances of triboelectric, electrochemical, piezoelectric and dielectric elastomer energy harvesting: Recent progress and challenges," *Sci. Bull.*, S2095927323004085, 2023, doi: 10.1016/j.scib.2023.06.025.

46. Ferson N. D., Uhl A. M., Andrew J. S., "Piezoelectric and magnetoelectric scaffolds for tissue regeneration and biomedicine: A review," *IEEE Trans. Ultrason. Ferroelectr. Freq. Control*, **68**(2), 229–241, 2021, doi: 10.1109/TUFFC.2020.3020283.

47. Silva I., Guedes L. A., Portugal P., Vasques F., "Reliability and availability evaluation of wireless sensor networks for industrial applications," *Sensors*, **12**(1), 806–838, 2012, doi: 10.3390/s120100806.

48. Zhao Y., Liao Q., Zhang G., Zhang Z., Liang Q., Liao X., Zhang Y., "High output piezoelectric nanocomposite generators composed of oriented $BaTiO_3$ NPs@PVDF," *Nano Energy*, **11**, 719–727, 2015, doi: 10.1016/j.nanoen.2014.11.061.

49. Chen X., Shao J., An N., Li X., Tian H., Xu C., Ding Y., "Self-powered flexible pressure sensors with vertically well-aligned piezoelectric nanowire arrays for monitoring vital signs," *J. Mater. Chem. C*, **3**(45), 11806–11814, 2015, doi: 10.1039/C5TC02173A.

50. Sun Y., Du Z., "A flexible and highly sensitive pressure sensor based on AgNWs/NRLF for hand motion monitoring," *Nanomaterials*, **9**(7), 945, 2019, doi: 10.3390/nano9070945.

51. Pan X., Wang Z., Cao Z., Zhang S., He Y., Zhang Y., Chen K., Hu Y., Gu H., "A self-powered vibration sensor based on electrospun poly(vinylidene fluoride) nanofibres with enhanced piezoelectric response," *Smart Mater. Struct.*, **25**(10), 105010, 2016, doi: 10.1088/0964-1726/25/10/105010.

52. Bai S., Zhang L., Xu Q., Zheng Y., Qin Y., Wang Z. L., "Two dimensional woven nanogenerator," *Nano Energy*, **2**(5), 749–753, 2013, doi: 10.1016/j.nanoen.2013.01.001.

53. Sinha T. K., Ghosh S. K., Maiti R., Jana S., Adhikari B., Mandal D., Ray S. K., "Graphene-silver-induced self-polarized PVDF-based flexible plasmonic nanogenerator toward the realization for new class of self powered optical sensor," *ACS Appl. Mater. Interfaces*, **8**(24), 14986–14993, 2016, doi: 10.1021/acsami.6b01547.

54. Saravanakumar B., Soyoon S., Kim S.-J., "Self-powered pH sensor based on a flexible organic–inorganic hybrid composite nanogenerator," *ACS Appl. Mater. Interfaces*, **6**(16), 13716–13723, 2014, doi: 10.1021/am5031648.

55. Fujino, Y., Siringoringo D. M., Ikeda Y., Nagayama T., Mizutani T., "Research and implementations of structural monitoring for bridges and buildings in Japan," *Engineering*, 5(6), 1093–1119, 2019, doi: 10.1016/j.eng.2019.09.006.

56. Nechibvute A., Chawanda A., Luhanga P., "Piezoelectric energy harvesting devices: An alternative energy source for wireless sensors," *Smart Mater. Res.*, **2012**, 1–13, 2012, doi: 10.1155/2012/853481.

57. Murugu Nachippan N., Venkatesh A. P., Muniyappan M., "Modelling and analysis of piezoelectric microgripper for unmanned aerial vehicle," *Mater. Today Proc.*, **5**(9), 19456–19462, 2018, doi: 10.1016/j.matpr.2018.06.306.

58. Esmaeili A., Sousa J. M. M., "Flow-driven piezoelectric energy harvester on a full-span wing for micro-aerial-vehicle (MAV) application," *Arab. J. Sci. Eng.*, **45**(7), 5713–5728, 2020, doi: 10.1007/s13369-020-04526-2.

59. Guo S., Li D., Wu J., "Theoretical and experimental study of a piezoelectric flapping wing rotor for micro aerial vehicle," *Aerosp. Sci. Technol.*, **23**(1), 429–438, 2012, doi: 10.1016/j.ast.2011.10.002.

60. Fath A., Xia T., Li W., "Recent advances in the application of piezoelectric materials in microrobotic systems," *Micromachines*, **13**(9), 1422, 2022, doi: 10.3390/mi13091422.

61. Ortiz Villasuso H. F., Robador E. M., Pedre S., D'Ovidio C., "Modified Aerial Acoustic Piezoelectric Sensor for Underwater Robots in Closed Environments," Springer International Publishing, Cham, 285–292, 2021, doi: 10.1007/978-3-030-60372-4_32.

62. Pyun J. Y., Kim Y. H., Park K. K., "Design of piezoelectric acoustic transducers for underwater applications," *Sensors*, **23**(4), 1821, 2023, doi: 10.3390/s23041821.

63. Abdul B., Mastronardi V. M., Qualtieri A., Guido F., Algieri L., Rizzi F., De Vittorio M., "Design, fabrication and characterization of piezoelectric cantilever MEMS for underwater application," *Micro Nano Eng.*, **7**, 100050, 2020, doi: 10.1016/j.mne.2020.100050.

64. Zou H.-X., Li M., Zhao L.-C., Gao Q.-H., Wei K.-X., Zuo L., Qian F., Zhang W.-M., "A magnetically coupled bistable piezoelectric harvester for underwater energy harvesting," *Energy*, **217**, 119429, 2021, doi: 10.1016/j.energy.2020.119429.

65. Tiwari S., Gaur A., Kumar C., Maiti P., "Electrospun hybrid nanofibers of poly(vinylidene fluoride) and functionalized graphene oxide as a piezoelectric energy harvester," *Sustain. Energy Fuels*, **4**(5), 2469–2479, 2020, doi: 10.1039/D0SE00033G.

66. Sezer N., Koç M., "A comprehensive review on the state-of-the-art of piezoelectric energy harvesting," *Nano Energy*, **80**, 105567, 2021, doi: 10.1016/j.nanoen.2020.105567.

67. Kulkarni A. A., Gaikwad N. K., Salunkhe A. P., Dahotre R. M., Bhat T. S., Patil P. S., "An ensemble of progress and future status of piezo-supercapacitors," *J. Energy Storage*, **65**, 107362, 2023, doi: 10.1016/j.est.2023.107362.

68. Mishra S., Supraja P., Sankar P. R., Kumar R. R., Prakash K., Haranath D., "Controlled synthesis of luminescent ZnS nanosheets with high piezoelectric performance for designing mechanical energy harvesting

device," *Mater. Chem. Phys.*, **277**, 125264, 2022, doi: 10.1016/j.matchemphys.2021.125264.

69. Mondal S., Thakur S., Maiti S., Bhattacharjee S., Chattopadhyay K. K., "Self-charging piezo-supercapacitor: One-step mechanical energy conversion and storage," *ACS Appl. Mater. Interfaces*, **15**(6), 8446–8461, 2023, doi: 10.1021/acsami.2c17538.

70. Lv P., Qian J., Yang C., Liu T., Wang Y., Wang D., Huang S., Cheng X., Cheng Z., "Flexible all-inorganic Sm-doped PMN-PT film with ultrahigh piezoelectric coefficient for mechanical energy harvesting, motion sensing, and human-machine interaction," *Nano Energy* **97**, 107182, 2022, doi: 10.1016/j.nanoen.2022.107182.

71. Rufato F. C. T., De Sousa L. G., Scalize P. H., Gimenes R., Regalo I. H., Rosa A. L., Beloti M. M., De Oliveira F. S., Bombonato-Prado K. F., Regalo S. C. H., Siéssere S., "Texturized P(VDF-TrFE)/BT membrane enhances bone neoformation in calvaria defects regardless of the association with photobiomodulation therapy in ovariectomized rats," *Clin. Oral Investig.*, **26**(1), 1053–1065, 2022, doi: 10.1007/s00784-021-04089-y.

72. Xia W., Zhou J., Hu T., Ren P., Zhu G., Yin Y., Li J., Zhang Z., "Enhanced magnetoelectric coefficient and interfacial compatibility by constructing a three-phase CFO@BT@PDA/P(VDF-TrFE) core-shell nanocomposite," *Compos. Part Appl. Sci. Manuf.*, **131**, 105805, 2020, doi: 10.1016/j.compositesa.2020.105805.

73. Chen S., Zhu P., Mao L., Wu W., Lin H., Xu D., Lu X., Shi, J., "Piezocatalytic medicine: An emerging frontier using piezoelectric materials for biomedical applications," *Adv. Mater.*, **35**(25), 2208256, 2023, doi: 10.1002/adma.202208256.

74. Sood A., Desseigne M., Dev A., Maurizi L., Kumar A., Millot N., Han S. S., "A comprehensive review on barium titanate nanoparticles as a persuasive piezoelectric material for biomedical applications: Prospects and challenges," *Small*, **19**(12), 2206401, 2023, doi: 10.1002/smll.202206401.

75. Jarkov V., Allan S. J., Bowen C., Khanbareh H., "Piezoelectric materials and systems for tissue engineering and implantable energy harvesting devices for biomedical applications," *Int. Mater. Rev.*, **67**(7), 683–733, 2022, doi: 10.1080/09506608.2021.1988194.

76. Li J., Feng Y., Chen W., Zhang S., Ma J., Chen S., Liu S., Cao C., Zhang, Y., "Electroactive materials: Innovative antibacterial platforms for biomedical applications," *Prog. Mater. Sci.*, **132**, 101045, 2023, doi: 10.1016/j.pmatsci.2022.101045.

77. Najjari A., Mehdinavaz Aghdam R., Ebrahimi S. A. S., Suresh K S., Krishnan S., Shanthi C., Ramalingam M., "Smart piezoelectric biomaterials for tissue engineering and regenerative medicine: A review," *Biomed. Eng. Biomed. Tech.*, **67**(2), 71–88, 2022, doi: 10.1515/bmt-2021-0265.

78. Du Y., Du W., Lin D., Ai M., Li S., Zhang L., "Recent progress on hydrogel-based piezoelectric devices for biomedical applications," *Micromachines*, **14**(1), 167, 2023, doi: 10.3390/mi14010167.

Chapter 13

Applications of Smart Piezoelectric Materials for Medical Devices

Yashpal U. Rathod,[a] Shyam W. Dafare,[b] Sunil B. Zanje,[b] N. B. Singh,[c] Rashmi Dubey,[d] and Wasudeo B. Gurnule[d]

[a]*Department of Chemistry, C. J. Patel College, Tirora 441911, India*
[b]*Department of Chemistry, J. M. Patel Arts, Commerce and Science College, Bhandara 441904, India*
[c]*Department of Chemistry and Biochemistry, Sharda University, Greater Noida 441911, India*
[d]*Department of Chemistry, Kamala Nehru Mahavidyalaya, Nagpur 440024, India*
wbgurnule@gmail.com

Due to very distinctive characteristics, smart piezoelectric materials are of remarkable attention. Using piezoelectric materials, mechanical energy may be changed into electrical energy and vice versa. In the category of piezoelectric materials, there are mono and polycrystals, composites, and polymers. Due to its biocompatibility and biodegradability, piezoelectric materials have made progress in recent years, and their uses in biomedical devices, tissue engineering, actuators, sensors, and other medical devices

Piezoelectric Materials: Design and Applications
Edited by N. B. Singh and Dev Kumar Mahato

ISBN 978-981-5129-40-3 (Hardcover), 978-1-003-59897-8 (eBook)
www.jennystanford.com

are all being investigated. Piezoelectric sensors and actuators can transform flow rate, pressure, and other variables to produce or consume energy. This chapter explores the use of smart materials in the design of medical equipment and sheds light on the current use of the piezoelectric effect in the medical sector.

13.1 Introduction

It is essential to regularly monitor the physical and biological changes that occur in human organs and tissues, such as bone, skin, and ligaments, as a result of injuries or healing processes, in order to maintain the sound health of living things at clinical stages. Different recently developed smart materials, such as piezoelectric materials, have wonderful uses in observing the integrity of these tissues and dental implants. In this study, we focus on the most recent materials and technology for effectively harvesting biomechanical energy that might be utilized to power implanted medical devices.

Due to the features of piezoelectric materials as sensors and actuators as well as their capacity to produce and detect ultrasound, they are used in a wide range of medical applications. In hospitals, a lot of piezoelectric devices are frequently employed for both medical treatment and diagnosis. As piezoelectric materials have high potential therefore these gain interest for worldwide researchers. Some of the gadgets are still under research, while others are commercially available [1–3].

Electrical components for wearable health monitoring devices must be intelligent and flexible. Intelligent materials, sometimes referred to as smart or responsive materials, play a significant part in our lives because of their reaction to environmental stimuli including stress, light, temperature, moisture or pH, and electric or magnetic fields. They encompass a variety of sensing and actuation uses in the medical field. These materials can produce biomechanical energy by using the motions of humans, animals, and the environment. Due to their great sensitivity, higher resonance frequency, and higher steadiness, which are essential for electronic devices like sensors, actuators, ultrasonic motors, transducers, accelerators, filters, and resonators as well as micro-electromechanical systems

(MEMS), piezoelectric materials have attracted a lot of attention in comparison to other smart materials.

When subjected to mechanical stress, piezoelectric materials produce an electric current. We are aware that a variety of materials, including bone, proteins, crystals (like quartz), and ceramics (like lead, zirconate, and titanate), exhibit piezoelectric properties. Known as piezoelectric materials, a family of organic and inorganic compounds, predominantly polymers, these have the capacity to transform electrical energy into mechanical force and the opposite. In the crystal structures of dielectric materials, piezoelectricity coexists with ions [4]. The polarization of the materials, which alters linearly with applied force, is what creates the electrical field in the material. The piezoelectric effect in organic materials is primarily caused by the orientation and molecular structure of piezoelectric polymers [5].

Diverse mammalian tissues, hair, hooves, wool, and horns, which are made of keratin and aligned helical structures, exhibit piezoelectricity. Many components of the muscles and skeletal tissues are made of collagen. Collagen is distinguished by spiral and helical fibril structures. Every collagen fibril displays a lateral piezoresponse along the fibril axis. As a result, numerous tissues in nature, including bones, ligaments, cartilage, and tendons, are piezoelectric. An estimated 20 billion euros are spent annually on piezoelectric medical devices, a substantial portion of which are piezoelectric sensors and actuators. It is feasible to utilize the mechanical energy to support tiny-sized devices. Applications for piezoelectric technology, which incorporate biological structure interactions, constitute a groundbreaking rapid development [6, 7].

Organic materials are the best options for functional materials in medical applications even if they have lower piezoelectricity than inorganic materials. Various technologies, including micro- and nano-scaled medical devices, can use organic smart materials [8, 9].

This chapter contains in-depth details on numerous piezoelectric materials, including actuators, biosensors, tissue engineering, catheter applications, and development of piezoelectric materials for medical uses. This chapter discusses the background and method of developing piezoelectric materials for use in the medical field. It covers the principles of piezoelectricity as well as the characteristics

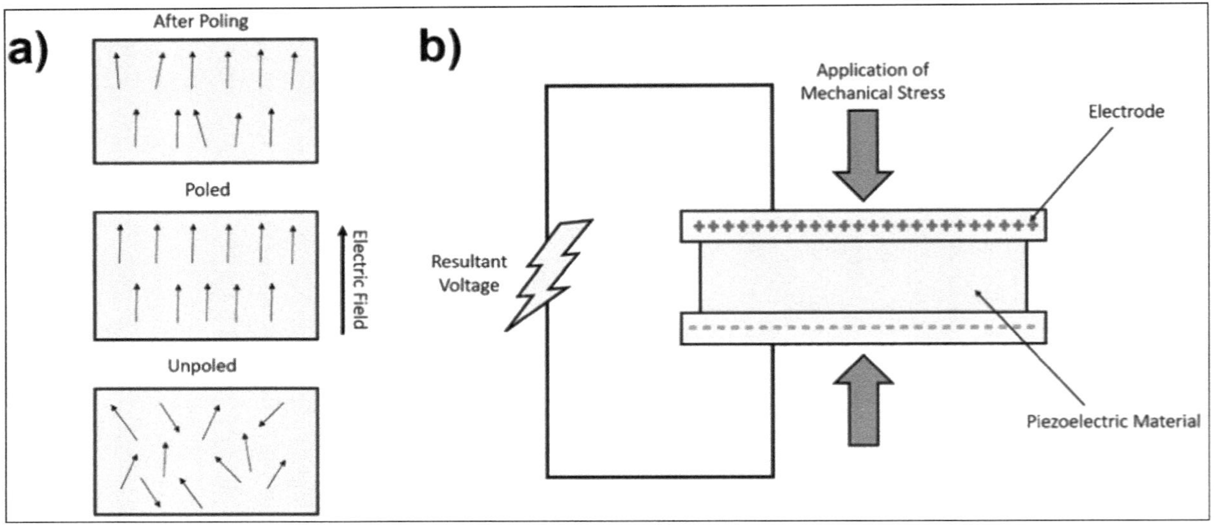

Figure 13.1 (a) After applying an electric field above its coercive field (Ec), piezoelectric dipoles in a ferroelectric material align, leaving behind a residue polarization (Pr). (b) Illustration of the direct piezoelectric effect, in which a charge is produced by a mechanical load that is applied. Courtesy: Jarkov et. al., *International Materials Reviews,* 2022, **67**(7), 683–733, https://doi.org/10.1080/09506608.2021.1988194.

of materials that may be found inside the human body, such as blood vessels, skin, and bones. These materials, which also include piezoelectric materials, naturally possess the capacity to cure illness discussing the uses of artificial and piezoelectric materials as sensors in relation to their use in medical implants.

13.2 Piezoelectricity

A family of materials known as piezoelectric materials undergo an electrical polarization while under stress; this phenomenon is known as the direct piezoelectric effect. The polarization direction of ferroelectric materials, a subclass of piezoelectric materials, may be changed by an externally applied electric field. A large electric field must be applied to support the dipoles in a ferroelectric-centered piezoelectric compound in order to trigger this response in scaffolds and ferroelectric-made biomedical equipment. Therefore, materials are exposed to such a poling procedure before application. In order to astound the abundance of arbitrarily orientated domains within the material, it is finished with the aid of exposing the materials to an electric powered subject above the coercive subject of the material, usually at accelerated temperatures, to facilitate dipole and domain alignment inside the electric powered subject course, as proven in Fig. 13.1a. When the fabric has reached room temperature, the poling field is withdrawn; however, the electric field is kept in location. This poling method produces net piezoelectric response and bulk polarization. As a result of the circuit's exchange in polarization after poling, strain application causes a shift in polarization that allows a charge and a brief current to flow in an external circuit.

13.3 Medical Applications of Piezoelectric Materials

13.3.1 Blood Pressure Monitor

During cardiac surgery, the pressure sensor may be utilized to continually monitor blood pressure in addition to measuring it. Thin films of polyvinylidene fluoride (PVDF) or lead zirconate titanate

(PZT) are employed as the piezoelectric material in the sensor. The sensor is positioned on the wrist or mid-arm to hit upon blood pressure for suitable backing and a strap with Velcro is furnished to safely preserve the sensor across the wrist or the arm. The sensor's components are shown in Fig. 13.2.

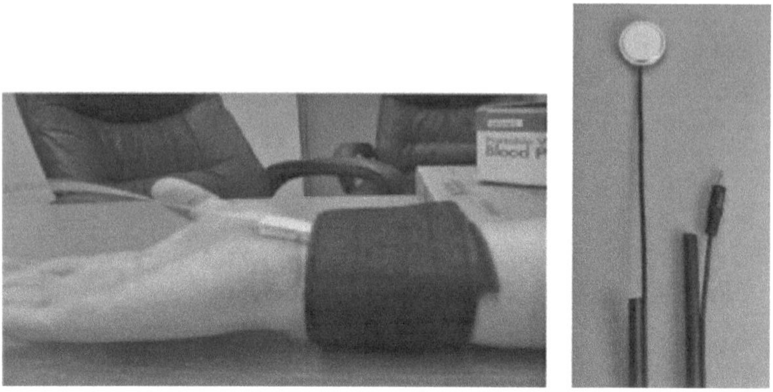

Figure 13.2 (a) Piezoelectric blood pressure monitor fixed on the wrist. Courtesy: Wade D. Peterson, David A. Skramsted, and Daniel E. Glumac. Reproduced from the website: www.phoenix.tc-ieee.org. (b) Pro-Wave's piezoelectric blood pressure sensor. Courtesy: Pro-Wave Electronic Corp. (website: http://www.prowave.com.tw/english/products/pp/bps.htm).

13.3.2 Piezoelectric Heartbeat Monitor

The medical industry uses the piezoelectric sound sensor, which is simply a piezoelectric microphone, as a heartbeat monitor. It is possible to use the piezoelectric microphone as a heartbeat monitor. Either an internal or external amplifier may be built into the gadget. A small disk-shaped sensor is firmly fastened to the patient's chest and the output can be seen on chart recorder or recorded on a strip chart recorder.

13.3.3 Biosensors

Using piezoelectric biosensors, different chemical, biological, or microbial structures can be found. Piezoelectric biosensors were prepared using constituents with extraordinary acoustic velocities.

Examples include an ethionamide sensor that is extremely selective and sensitive and is guided by molecular modeling [10], acoustic biosensors made of aluminum nitride (AlN) that offer present time reaction and computed data [11], and an AlN biosensor that is highly sensitive for detecting pesticide residues [12]. Low detection limits, a strong linear response, and exceptional repeatability are all features of piezoceramic sensors. A range of specific biological processes, such as genetic hybridization, may be observed using AlN piezoelectric biosensors, which may provide details on the kinetics of the process. AlN sensors can also be used to monitor protein-ligand interfaces or antigen-antibody binding.

13.3.4 Tissue Engineering

The literature thoroughly assesses and tests piezoelectric actuator-sensor systems for tissue engineering. Due to their stress, conventional tissue engineering actuators and sensors must be regularly removed from the frame. To avoid having to delay the instrument, it is possible to use short digital biological sensors with excellent mechanical characteristics [13]. Breast cancer cell survival may be impacted by piezoelectric nanoparticles exposed to ultrasonic frequencies [14].

13.3.5 Healthcare Monitoring

The physiological processes of human organisms can be dynamically measured in real time using piezoelectric sensors for medical monitoring. Non-invasive biosensors transmit data from interstitial fluids, perspiration, tears, or saliva for illness detection [15]. The development of electronic skin (or "e-skin") for use in medical monitoring systems is another development. E-skin has been utilized in numerous medical applications, including wearable sensors and illness diagnosis, because of its exceptional qualities [16]. Some devices' color can change thanks to pressure biosensors that mimic the chameleon's skin [17]. E-skin, for instance, may communicate with another pressure sensor [18]. A composite made of reduced graphene and PVDF was utilized to create piezoelectric artificial skin. High-conductivity PVDF/GO films were prepared and annealed to obtain a high proportion of phase.

13.3.6 Actuators in Tissue Engineering and Devices

A significant class of piezoelectric motion-controlling devices are medical actuators (Fig. 13.3). Advanced research is used to create the designs of piezoelectric actuators [19]. Despite the fact that piezoelectric forces are frequently quite weak, piezoelectric actuators have been modified for a variety of applications with minimal displacement. Flexible tissue actuators for piezoelectric devices can be utilized to encourage tissue regeneration and are also biocompatible [20]. According to certain reports, certain intelligent actuators may be able to restore an organ that has been lost in human organisms [21–24]. Shape-memory polymers (SMP) and shape-memory alloys (SMA) can be stimulated to activate them, while textile actuators can replicate the motion of the human body [25–30].

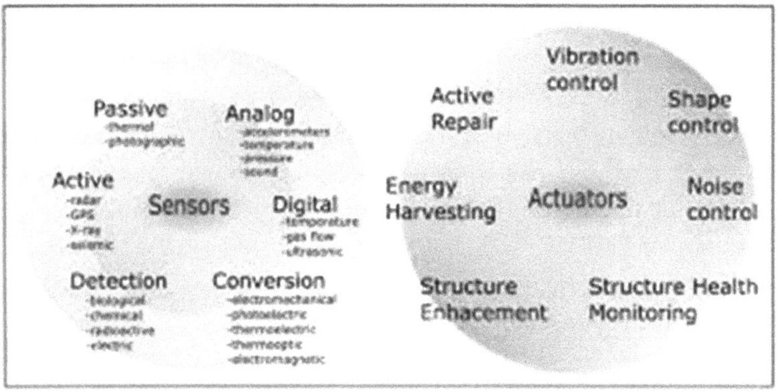

Figure 13.3 Applications of actuators and sensors. Courtesy: Ref. [29].

A number of articles have discussed using PLLA tweezers to treat thrombus. The tweezers show a great capacity to capture silica compounds when they are injected into a blood artery in the form of biodegradable fibers. By using jet spinning and alternating current (AC) voltage, fibers are created. These piezoelectric tweezers appear to be strong prospects for nanomedicine and tissue engineering applications because of the aforementioned characteristics and great sensitivity. Nanocomposites with oriented polymer fibers, which serve as the actuators and sensors in piezoelectric materials, can be created.

Silver nanoparticles can be added to small sensor–actuator systems to provide strain sensing. These devices can also be used to produce signals, ultrasonic energy, drilling equipment, and to convert vibration into electrical energy and electrical energy into mechanical energy. Numerous industrial products use piezoelectric actuators, including components for phones, sophisticated music systems, and musical instruments. Piezoelectric actuators are extensively researched in the medical field for use in endoscope lenses, tiny pumps, and other applications.

Additionally, three-dimensional (3D) fibrous piezoelectric scaffolds can support osteogenic differentiation and be employed with stimulation during in vitro experiments. Piezoelectric materials with low-voltage output, however, encourage chondrogenic differentiation. The study concludes by demonstrating improved chondrogenic differentiation by electromechanical stimulation. In comparison to mechanical actuation, electromechanical actuation is more effective.

According to reports, the electrospinning process's settings have a significant impact on the characteristics of nanofiber actuators. The method may be improved to produce intelligent PVDF bio-actuators for energy harvesting. Results revealed that fibroblasts grew exactly in the direction of the fibers. Due to their electric impact, such intelligent scaffolds can be used in neural tissue engineering for the regeneration of the nervous system.

Another PVDF-based nanocomposite system with barium titanate ($BaTiO_3$) nanoparticles added has been put to the test in an ultrasonic environment. The electrospinning procedure was used to create the material. Results indicated that stimulation improved cell viability. The most recent testing has revealed data on how using smart PVDF scaffolds might encourage cell communication. During stimulation, electro-active cardiomyocytes were put to the test. The differentiation and proliferation of in vitro cells were impacted by electrical stimulation.

13.3.7 Ultrasonic Imaging

The 1920s saw the introduction of piezoelectric ultrasonography, which was later applied to the assessment of mitral valve disease in the 1950s. Today, echocardiography is the gold standard for a quick assessment of the heart muscle and valves. Living beings have had endocardial piezoelectric crystals implanted along a number of the coronary heart's axis on the way to locate altered left ventricular-arterial coupling, which takes place before pump malfunction in early coronary heart failure. Now that surgeons are using transducer-tipped catheters for ultrasound biomicroscopy, and it is reasonable to assume that piezotechnology will ultimately lead to a complete ultrasound-based virtual heart display. Last but not least, paintings are being performed on a piezo-based sensor that may convert heartbeats into visuals.

13.3.8 Piezoelectric Cardiac Assistance Devices and Micropumps for Drug Delivery

Due to their high efficiency and ease of miniaturization, piezoelectric gadgets are best candidates for usage of implantable medicinal drug pumps. A peristaltic piezoelectric micropump device prototype was advanced and produced. This piezoelectric micropump has an adjustable stroke quantity and can minimize energy while maintaining a regular glide rate across extensive backpressure variety. PZT micropumps may be utilized to supply specific volumes of higher awareness medicines to the systemic flow due to the fact they are able to displace unique amounts of extraordinarily small volumes.

Piezoelectric materials had been employed to create VADs or cardiac assistance devices, due to their effectiveness and length. In the case of excessive coronary heart failure, those devices are utilized to pump blood in series or in parallel with the local heart. Piezo-VADs have now gone through in vivo and in vitro checking out.

13.3.9 Piezoelectric Pump-Drug Delivery

These micropumps are highly helpful in the promising field of drug delivery, where a lot of research is now being done. In the literature, numerous inventive piezoelectric micropump designs for drug delivery have been documented. Animals have undergone in vivo testing of biocompatible micropumps for the transfer of insulin that can be embedded in the body. The majority of piezoelectric micropumps that are commercially available are mostly utilized for biochemical investigation, including DNA and protein investigation, genomics, and environmental testing. "Lab-on chip" is a microfluidic chip that uses a micropump to pump fluids through a microchannel.

13.3.10 Bone Density Measurement Using Ultrasound

Age-related bone tissue density loss is a prevalent condition that affects many elderly persons. Osteoporosis is the name of the condition, which affects more women than males. Fracture risks are decreased by early disease detection and treatment. X-ray absorptiometry and quantitative ultrasound technique are two frequently utilized techniques for illness diagnosis. The latter method, which makes use of an ultrasonic transducer, is gaining popularity since it is less expensive, simple to use, and provides diagnostic values that are fairly trustworthy.

The ultrasonic approach relies on two principles: (1) measurement of ultrasonic wave velocity in the bone and (2) measurement of the intensity of the sound as it travels through the bone. In both procedures, an ultrasonic transmitter transmits ultrasonic pulses through the bone, and an ultrasonic receiver detects the transmitted pulse. The heel bone is typically used for measurement. A transportable ultrasonic device with a piezoelectric ultrasonic transmitter and receiver is utilized for both procedures. On either side of the patient's heel, the transmitter and receiver are permanently attached. The foot is submerged in water or a gel is placed between the transducer and the skin for optimal transmission at the border. The single-element transducer serves as a good example of the fundamental construction of the ultrasonic transducers that are used. Considering that focusing is not necessary, the piezoelectric element need not be concave. Inside a plastic box is affixed a piezoelectric disk with a backing and layer that match. Typically, frequencies between 200 and 1000 kHz are used. The piezoelectric disk's thickness is chosen to provide resonance with a broadband output throughout this frequency range.

Broadband ultrasonic attenuation (BUA) is the term for the parameter that is measured in the attenuation measuring method. With increasing frequency, the ultrasonic energy is attenuated more when it reaches the bone. The bone is penetrated by ultrasonic pulses that have a frequency in the 200–1000 kHz range. Through the detector, the attenuation as a function of frequency is monitored.

With frequency, the attenuation rises linearly. The term BUA refers to the slope of the linear attenuation against the frequency curve, which is used as a metric to assess the severity of osteoporosis. A healthy bone has a steeper slope than an osteoporotic bone. An indicator of osteoporosis is the ratio of the measured BUA to the BUA of healthy bone. The velocity of the sound wave is measured along with the amount of time that passed between the transmitted pulse and the pulse that was received at the receiver in the velocity technique. In contrast to osteoporotic bone, healthy bone has a greater velocity. The velocity difference is used to gauge the severity of osteoporosis.

13.3.11 Ablation of Tumor Cells Using High-Intensity Focused Ultrasound

Numerous cancer forms have been effectively treated with high-intensity focused ultrasound (HIFU). Due to absorption, when high-intensity ultrasound is concentrated at a location on human tissue, the tissue can quickly reach a high temperature. This results in tissue destruction or necrosis in the targeted location without damaging the tissues around it. HIFU can therefore be used to eliminate tumors.

Single-element PZT transducers are not ideal for producing high-intensity focused ultrasound since the PZT element has to be concave in form and rather big in size. Large PZT plates are carved into deep grooves to create array transducers, which are capable of producing high intensity and precise focusing.

In the frequency range of 200 kHz to 10 MHz and with an acoustic output of 10–30 W/cm^2, high-intensity piezoelectric 1-3 composite transducers have been created. The number of components present in the transducer varies from 64 to more than 200 components. The transducers are capable of producing high temperatures up to roughly 85°C with sharp focusing. The characteristics of HIFU transducers, including their frequency and the input power, focal length, and transducer head, define the volume of heat produced in that region. Standard HIFU transducers function in the 1–10 MHz range and have F-numbers between 1 and 1.5.

Higher frequency ultrasound raises the temperature at the focus point because the absorption of ultrasonic energy rises with frequency. However, high-frequency sound waves will not penetrate as deeply due to excessive absorption. When deep penetration is required, low-frequency transducers are utilized. High temperatures capable of ablating tumors can be attained, depending on the transducer's power and the length of exposure. The transducer needs to be continually moved in order to completely remove huge tumors. The movement of the transducer is observed using the live picture of the tumor.

13.3.12 Ultrasound for Drug Delivery

Following the surgical excision of the primary tumor, the remaining malignant cells in cancer patients must be treated with a systematic medication delivery regimen. Intravenous medication administration is the standard technique. This strategy has not proven particularly effective at getting enough medication to the right place in time to destroy all the sick cells. The following are a few causes of inefficient medication delivery:

- Tumors' vascular endothelia are not well-developed. Hence, it is possible that the medication cannot adequately reach the needed location.
- Drug penetration to the tumor area through diffusion is not effective because of the high interstitial pressure of tumors.
- The blood-brain barrier (BBB), which guards the brain from toxins and brain from germs and viruses in the blood.
- The medicine may not be given in sufficient quantities to destroy all the cancer cells since it cannot reach all the cells that can rebuild tumors.
- It is possible that the medicine does not reach the cancerous cells quickly enough to keep up with their pace of development.

The ability of anticancer medications to reach more tumors has been improved via the exploration of several methods. Some of the methods include implanting drug carriers, improving permeability and retention, and using nanoparticles. The approaches used, particularly in the treatment of brain tumors, include transient rupture of the BBB, interstitial drug administration through catheters, convection-enhanced drug delivery, and ultrasonic-enhanced drug delivery.

It is a relatively new technology that several researchers are looking into: using an ultrasonic field to improve medication delivery to the brain tumor. The usefulness of ultrasonic fields in enhancing medication delivery has been shown in a number of in vitro and a few in vivo investigations that have been described in the literature [19–22]. The non-inertial cavitation effect, which produces a high number of microbubbles that vibrate at the ultrasonic frequency, is the mechanism thought to be responsible for the enhancement. Without endangering the tissues, the oscillating bubbles put mechanical force on the walls of blood arteries.

13.3.13 Ultrasonic-Induced Transdermal Drug Delivery

Medication administration through the skin is done using a non-invasive method called transdermal medication delivery. Patches that are applied to the skin serve as the medication delivery method. The method offers benefits over injectable or oral drug administration since it allows the medication to enter the body without passing through the first-pass metabolic step and is non-invasive. Only a few medications with relatively tiny molecular sizes may be applied topically using patches. Due to the skin's poor permeability, the approach has limits even for these medications.

The use of ultrasound is one of the potential methods for improving medication delivery through the skin. The process, known as sonophoresis, is being employed extensively in hospitals. Ultrasound is produced by piezoelectric ultrasonic transducers. Microlysis, Sonoderm, and Sonoprep are a few of the sonophoresis devices that have been created for the commercial market. The ultrasound utilized for transdermal medication administration has a frequency in the 20–100 kHz range. Continuous ultrasound tends to heat up the region, and hence pulsed ultrasound is employed instead. Depending on the type of medicine and the desired dosage, the ultrasound's strength and pulse frequency must be appropriately chosen. There have been instances of using single or array transducers.

13.3.14 Therapeutic Ultrasound-Treatment of Injury, Muscular Pain, and Bone Fracture

For the treatment of wounds and bone repair, a piezoelectric transducer produces a low-intensity ultrasonic field. Researchers have made an effort to provide empirical support for ultrasound's capacity to mend broken bones [27–29]. But no solid study demonstrating the healing properties of ultrasonography has been published as of yet. In the majority of real-world instances, using ultrasound to repair fractures or ease muscle problems is done in conjunction with other traditional therapies.

The ultrasound employed in the procedures has a frequency of 1–3 MHz and a low intensity of less than 3 W/cm^2. Both pulsed and continuous ultrasound have been applied. The therapy lasts for a short while. Some of the hypotheses put up to support the therapeutic effects of ultrasound include the following:

- The moderate warmth caused by high-frequency vibrations traveling through the tissue reduces pain and speeds up the healing process. Ultrasound improves collagen's extensibility, which can aid in the repair of damaged tissues. Ultrasound also enhances metabolism and boosts blood flow.
- Sports-related injuries necessitate quick recovery. In these situations, ultrasonic therapy is used in the hopes of lowering swelling and imitating the healing process. Ultrasonic devices are used by physiotherapists to treat arthritic, joint, and muscle discomfort. It is thought that the ultrasound-induced vibrations provide a massaging and gentle heating impact.

13.3.15 Phacoemulsification-Cataract Surgery

The most famous method of cataract surgery is referred to as phacoemulsification, which involves shattering the damaged lens into portions with an ultrasonic subject. A tiny piezoelectric transportable ultrasonic transducer with a pointed titanium microtip, an irrigation-aspiration gadget with tiny channels, a field for fluids, and the vital electrical manipulating machine make up the instrument. Through a very little corneal incision, the handheld transducer is placed into the eye. Around 40 kHz is the operating frequency of the transducer. To create ultrasonic discipline bursts at the titanium tip, the ultrasonic area is pulsed with pulse frequencies between 4 and 400 Hz. Small portions of the cataract-affected lens are broken off by way of the vibrating microtip. A fluid is created to enter the attention concurrently.

13.3.16 Energy Harvesting

The creation of several implanted medical electronics (IMEs), which can enhance human life quality and perhaps even lengthen longevity, has been sparked by the significant improvements in microelectronics over the past few decades. As diagnostic and therapeutic tools, IMEs may now be implanted in a variety of human body areas. These implanted electronics might provide a variety of body component diagnostic tools and support therapy for such disorders [31–38].

Despite tremendous advancements in IME manufacture and use since Siemens' invention of the first pacemaker in 1958, there are still a number of difficulties that must be overcome.

To lessen their impact on human activities, it is strongly desired to create IMEs with reduced dimensions and weights. The most dependable power source for IMEs is widely acknowledged to be batteries. The highest amount of battery system miniaturization is challenging to attain, nevertheless, due to present technical restrictions. Another significant obstacle is the short battery life. For instance, the deep brain stimulators and cardiac pacemakers' electric pulse-generating batteries need to be changed or maintained after 3–5 years of operation.

Recent studies have shown that the physiological milieu and bodily activities may harvest energy in vivo and charge internal devices. Energy-autonomous implanted electronic devices are another way to miniaturize technology. These gadgets draw their power directly from the patient or from nearby artificial or natural sources. There have been reports of a number of methods for converting energy from thermal, mechanical, and chemical sources in living organisms, including electrical potentials from glucose oxidation, inner ear, muscular contraction, and organ vibration [39–42].

Energy harvesting is the process of obtaining electrical energy from several sources and storing it for use by tiny, wireless, autonomous devices. Multiple sources of energy, including thermal and mechanical ones, can be harnessed. Radio frequency and ambient light can also be gathered, both naturally and artificially. For usage in biomedical electronics, piezoelectric energy harvesting devices that are based on piezoelectricity appear to be the best options. Alongside, several types of piezo-energy harvesters were utilized to stock energy via body motions, heartbeats, and other sources. These devices are built of constituents like zinc oxide (ZnO), PZT, and barium titanate ($BaTiO_3$) imperfections lead magnesium niobate-lead titanate (PMN-PT).

For example, it is described as an energy harvester made of PZT that can store energy from the normal beating of the heart and other internal organs. Chromium (Cr) and gold (Au) layers on ribbons serve as electrodes in this particular device. Another illustration may be a medical device energy harvester that uses gold (Au) electrodes atop lead magnesium niobate-lead titanate (PMN-PT) films to safeguard medical equipment.

13.4 Conclusions

We have examined a number of uses for intelligent piezoelectric materials in electronics and medical equipment, including headphones, phones, and sophisticated electrical systems. It is obvious that piezoelectric materials are a class of contemporary materials. They can efficiently convert mechanical energy to electrical energy and vice versa, and after specific extra processing, they can become biocompatible, earning them the status of piezoelectric biomaterials with a promising future for use in numerous medical devices. Organic biomaterials have arisen as a relatively new class of contemporary materials exhibiting unique features, notably in sensing applications due to their flexibility, despite the fact that inorganic piezoelectric materials have been long and well researched. As a result, piezoelectric materials are extensively researched for use in actuators, sensors, and medical equipment. By providing instances of the difficulties, we intend to increase awareness of how crucial piezoelectricity is to the industry – mostly the medical sector – as well as of the possibilities for future advancement.

References

1. Nye JF. *Physical Properties of Crystals: Their Representation by Tensors and Matrices*. Oxford: Clarendon Press; 1985.
2. Heywang W, Lubitz K, Wersing W. *Piezoelectricity: Evolution and Future of a Technology*. New York, NY: Springer Science & Business Media; 2008.
3. Topolov VY, Bowen CR, Bisegna P. New aspect-ratio effect in three-component composites for piezoelectric sensor, hydrophone and energy-harvesting applications. *Sensors and Actuators A*, **229**, 94–103, 2015.
4. Ramadan KS, Sameoto D, Evoy S. A review of piezoelectric polymers as functional materials for electromechanical transducers. *Smart Materials and Structures*, **23**, 033001, 2014.
5. Mark HF, Kroschwitz JI. *Encyclopedia of Polymer Science and Engineering*. New York, NY: Wiley; 1985.
6. Liu Y, Huang Z, Gao Y. A three-dimensional and bi-objective topological optimization approach based on piezoelectric energy harvester. *Applied Science*, **10**, 6772, 2020.
7. Atif R, Khaliq J, Combrinck M, Hassanin A.H, Shehata N, Elnabawy E, Shyha I. Solution blow spinning of polyvinylidene fluoride based fibers for energy harvesting applications: A Review. *Polymers*, **12**, 1304, 2020.
8. Sciarra FM. A nonlocal model with strain-based damage. *International Journal of Solids Structure*, **46**, 4107–4122, 2009.
9. Barretta R, Fabbrocino F, Luciano R, de Sciarra FM. Closed-form solutions in stress-driven two-phase integral elasticity for bending of functionally graded nano-beams. *Physica E: Low Dimensional Systems and Nanostructures*, **97**, 13–30, 2018.
10. Kushwaha A, Gupta N, Srivastava J, Singh AK, Singh M. Development of highly sensitive and selective sensor for ethionamide guided by molecular modelling via electropolymerized molecularly imprinted films. *Microchemical Journal*, **152,** 104–355, 2020.
11. Perumal V, Hashim U. Advances in biosensors: Principle, architecture and applications. *Journal of Applied Biomedicine*, **12**, 1–15, 2014.
12. Liu S, Zheng Z, Li X. Advances in pesticide biosensors: Current status, challenges, and future perspectives. *Analytical and Bioanalytical Chemistry*, **405**, 63–90, 2013.

13. Lee G, Kang SK, Won SM, Gutruf P, Jeong YR, Koo J, Lee SS, Rogers JA, Ha JS. Fully biodegradable microsupercapacitor for power storage in transient electronics. *Advanced Energy Material*, **7**, 157, 2017.

14. Marino A, Battaglini M, De Pasquale D, Degl'Innocenti A., Ciofani G. Ultrasound-activated piezoelectric nanoparticles inhibit proliferation of breast cancer cells. *Scientific Reports*, **8**, 625, 2018.

15. Kim J, Campbell AS, de Ávila BEF, Wang J. Wearable biosensors for healthcare monitoring. *Natural Biotechnology*, **37**, 389–406, 2019.

16. Rodrigues D, Barbosa AI, Rebelo R, Kwon IK, Reis RL, Correlo VM. Skin-integrated wearable systems and implantable biosensors: A comprehensive review. Biosensors, **10,** 79, 2020.

17. Chou HH, Nguyen A, Chortos A, To JW, Lu C, Mei J, Kurosawa T, Bae WG, Tok JBH, Bao Z. A chameleon-inspired stretchable electronic skin with interactive colour changing controlled by tactile sensing. *Nature Communication*, **6**, 8011, 2015.

18. Zang Y, Zhang F, Di C, Zhu D. Advances of flexible pressure sensors toward artificial intelligence and health care applications. *Materials Horizons*, **2**, 140–156, 2015.

19. Liu YZ, Hao ZW, Yu JX, Zhou XR, Lee PS, Sun Y, Mu ZC, Zeng FL. A high-performance soft actuator based on a poly(vinylidene fluoride) piezoelectric bimorph. *Smart Materials and Structures*, **28**, 055011, 2019.

20. Mostafavi E, Medina-Cruz D, Kalantari K, Taymoori A, Soltantabar P, Webster TJ. Electroconductive nanobiomaterials for tissue engineering and regenerative medicine. *Bioelectricity*, **2**, 120–149, 2020.

21. Chen S, Pang Y, Yuan H, Tan X, Cao C. Smart soft actuators and grippers enabled by self-powered tribo-skins. *Advance Materials and Technology*, **5**, 1901075, 2020.

22. Cianchetti M, Laschi C, Menciassi A, Dario P. Biomedical applications of soft robotics. *Nature Review Materials*, **3**, 143–153, 2018.

23. Shih B, Shah D, Li J, Thuruthel TG, Park YL, Iida F, Bao Z, Kramer-Bottiglio R, Tolley MT. Electronic skins and machine learning for intelligent soft robots. Science Robotics, **5,** 9239, 2020.

24. McCracken JM, Donovan BR, White TJ. Materials as machines. *Advance Materials*, **32**, 1906564, 2020.

25. Persson NK, Martinez JG, Zhong Y, Maziz A, Jager EWH. Actuating textiles: Next generation of smart textiles. *Advance Materials Technologies*, **3**, 1700397, 2018.

26. Jager EWH, Martinez JG, Zhong Y, Persson NK. *Wearable Bioelectronics.* Elsevier, Amsterdam, The Netherlands: Chapter 7: Soft actuator materials for textile muscles and wearable bioelectronics. (201–218) 2020.

27. Maziz A, Concas A, Khaldi A, Stålhand J, Persson NK, Jager EWH. Knitting and weaving artificial muscles. *Science Advances*, **3**, e1600327, 2017.

28. Mokhtari F, Cheng Z, Raad R, Xi J, Foroughi J. Piezofibers to smart textiles: A review on recent advances and future outlook for wearable technology. *Journals of Materials Chemistry A.*, **8**, 9496–9522, 2020.

29. Zaszczyńska A, Gradys A, Sajkiewicz P. Progress in the applications of smart piezoelectric materials for medical devices. *Polymers (Basel)*, **12**(11), 2754, 2020.

30. Zaszczyńska A, Sajkiewicz P, Gradys A. Piezoelectric scaffolds as smart materials for neural tissue engineering. *Polymers*, **12**, 161, 2020.

31. Hwang GT, Im D., Lee SE, Lee J, Koo M, Park SY, Kim S, Yang K, Kim SJ, Lee K, Lee KJ. In vivo silicon-based flexible radio frequency integrated circuits monolithically encapsulated with biocompatible liquid crystal polymers. *ACS Nano*, **7**, 4545–4553, 2013.

32. Condon B, Hadley D. Cardiac pacing systems and implantable cardiac defibrillators (ICDs): A radiological perspective of equipment, anatomy and complications. *Clinical Radiology*, **59**, 1145, 2004.

33. Perlmutter JS, Mink JW. Deep brain stimulation. *Annual Review of Neuroscience*, 29, 229–257, 2006. https://doi.org/10.1146/annurev.neuro.29.051605.112824.

34. Cheng A, Tereshchenko LG. Evolutionary innovations in cardiac pacing. *Journal of Electrocardiology*, **44**, 611–615, 2011.

35. Bonawitz SC. Management of exposure of cardiac pacemaker systems. *Annals of Plastic Surgery*, **69**, 292–295, 2012.

36. Majerus SJA, Garverick SL, Suster MA, Fletter PC, Damaser MS. Wireless, ultra-low-power implantable sensor for chronic bladder pressure monitoring. *ACM Journal on Emerging Technologies in Computing Systems*, **8**, 1–13, 2012.

37. Mason RJ. Gastric electrical stimulation. *Archives of Surgery*, **140**, 841, 2005.

38. Horlbeck FW, Mellert F, Kreuz J, Nickenig G, Schwab JO. Real-world data on the lifespan of implantable cardioverter-defibrillators depending on manufacturers and the amount of ventricular pacing. *Journal of Cardiovascular Electrophysiology*, **23**, 1336–1342, 2012.

39. Mallela VS, Ilankumaran V, Rao NS. Trends in cardiac pacemaker batteries. *Indian Pacing Electrophysiology Journal*, **4**, 201–212, 2004.

40. Zurbuchen A, Pfenniger A, Stahel A, Stoeck CT, Vandenberghe S, Koch VM, Vogel R. Energy harvesting from the beating heart by a mass imbalance oscillation generator. *Annals of Biomedical Engineering*, **41**, 131–141, 2013.

41. Platt SR, Farritor S, Haider H. On low-frequency electric power generation with PZT ceramics. *IEEE/ASME Transactions on Mechatronics*, **10**, 240–252, 2005.

42. Wong LSY, Hossain S, Ta A, Edvinsson J, Rivas DH, Nääs H. A very low power CMOS mixed-signal IC for implantable pacemaker applications. *IEEE Journal of Solid-State Circuits*, **39**, 2446–2456, 2004.

Chapter 14

Piezoelectric Materials and Their Applications in Civil Engineering

R. R. Dubey,[a] W. B. Gurnule,[a] Y. U. Rathod,[b] and N. B. Singh[c]
[a]*Department of Chemistry, Kamala Nehru Mahavidyalaya, Nagpur 440024, India*
[b]*Department of Chemistry, C. J. Patel College, Tirora 441911, India*
[c]*Department of Chemistry and Biochemistry, Sharda University, Greater Noida 201310, India*
wbgurnule@gmail.com

Both natural and manufactured materials frequently exhibit the piezoelectric effect. Piezoelectric materials are desirable for applications including sensors, actuators, and energy harvesting devices because they can directly transduce electrical and mechanical energy. Through power harvesting and self-sustaining smart sensing in buildings, this property offers prospects for the implementation of renewable and sustainable energy. Piezoelectric materials are advantageous because they respond quickly and are highly precise. In MEMS and optoelectronic devices, they are

Piezoelectric Materials: Design and Applications
Edited by N. B. Singh and Dev Kumar Mahato

ISBN 978-981-5129-40-3 (Hardcover), 978-1-003-59897-8 (eBook)
www.jennystanford.com

often utilized. Despite being the most popular building material, piezoelectricity is not effective in capturing electrical energy from the background vibrations of a building system. Understanding how various building components' piezoelectric qualities work as well as the most recent innovations and building industry implementations are essential for the effective deployment of piezoelectric materials for sustainable building development. Piezoelectric materials have several uses in the field of monitoring the health of civil structures, including those involving concrete and structural engineering, bridge and pipeline engineering, earthquake and geological hazard prevention engineering, and so forth.

14.1 Introduction

The majority of researchers nowadays are concentrating on the use of smart materials in engineering constructions. These intelligent materials have some properties that may be advantageously changed in a controlled setting by external stimuli including temperature, stress, and an electric or magnetic field. Due to their peculiar nature and unique and intriguing features, piezoelectric materials are being explored more and more [1]. Buildings can employ piezoelectric materials as energy producers to provide operating energy for ventilation and water heating [2]. In addition to the energy issue, earthquakes and the vibrations caused by the energy released as a result of the movement of subterranean faults seriously harm the buildings. Some guidelines for earthquake-proofing buildings have also been accepted in writing form. These guidelines, which are based on a building's ability to survive without collapsing, have been included in today's standards. Some examples of environments where piezoelectric vibration energy harvesting can be used include civil infrastructure such as buildings and bridges [3, 4].

On the other hand, this chapter reports on the piezoelectric components employed as commonplace components in smart actuator systems, including small ultrasonic motors, precise positioners, and adaptive mechanical dampers. The major goals of piezoelectric and related ceramic actuators are to enhance the actuator's materials, form, design, structure, and applications. A new field of electrical and structural ceramics has emerged as a result of piezoelectric actuators. The use of piezoelectric actuators to reduce vibration in military vehicles and space constructions is another prospective application area. Shape memory alloy temperature-controlled solid-state displacement transducers have recently been investigated; however, piezoelectric ceramics perform better [5]. The use of the piezoelectric actuator goes beyond industrial machinery, it may be used for portable electronics like mobile phones and digital cameras [6].

Piezoelectric materials have gained significant attention in civil engineering owing to their distinctive capability to convert mechanical energy into electrical energy and vice versa. These materials exhibit the piezoelectric effect, where an electric charge accumulates in response to applied mechanical stress, and conversely, they deform when subjected to an electric field. This exceptional property has opened up new possibilities for various applications in the field of civil engineering, enabling the development of smart structures, monitoring systems, and energy harvesting technologies. Historically, the piezoelectric effect was first discovered by Pierre and Jacques Curie in 1880 when they observed the phenomenon in certain crystals such as quartz. Since then, the understanding and development of piezoelectric materials have evolved, leading to the discovery of various synthetic piezoelectric materials like piezoceramics, piezopolymers, and piezo composites.

Smart Materials and Adaptive Structures: Piezoelectric materials are considered "smart" materials because of their ability to sense and respond to external stimuli (Fig. 14.1). They can be incorporated into adaptive structures that can change their properties dynamically based on the environmental conditions and loads, optimizing their performance and resilience. Piezoelectric transducers are widely used in ultrasonic testing and non-destructive evaluation of civil structures. These transducers emit ultrasonic waves, which are used to inspect and assess the integrity of materials, detect hidden flaws, and evaluate the quality of welds and joints.

The rapid growth in industrialization and the increasing demand for sustainable energy sources have driven the exploration of innovative technologies for energy harvesting. Piezoelectric materials and energy harvesters have emerged as promising solutions for capturing mechanical energy and converting it into electrical power in various industrial applications.

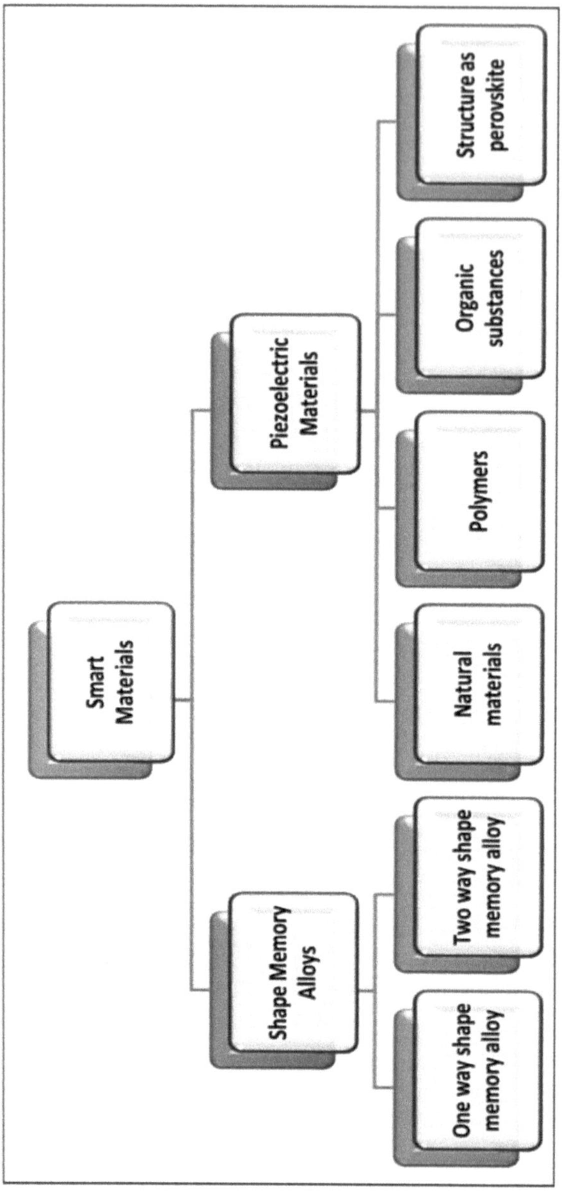

Figure 14.1 Most common smart materials [7].

In the current era of rapid technological advancements and growing concerns about sustainable energy sources, there is a burgeoning interest in the development and utilization of energy harvesting technologies. These technologies aim to capture and convert ambient energy from various sources into usable electrical power, thereby offering a promising pathway toward a more sustainable and self-sufficient future.

Among the various energy harvesting methods, piezoelectric energy harvesting has emerged as a particularly promising and versatile approach. Piezoelectric materials possess the unique property of generating an electrical charge in response to mechanical stress or vibration (Table 14.1). This characteristic makes them ideal candidates for extracting energy from the environment in the form of mechanical vibrations, oscillations, or strain. As a result, piezoelectric energy harvesters have found diverse applications in both consumer electronics and industrial sectors.

Table 14.1 Piezoelectric coefficients of piezoelectric materials [9]

Materials	Typical	d(pC/N)	g(Vm/N $\cdot 10^{-3}$)
Quartz	Crystal	2.3	50
Rochelle salt	Crystal	2300	Very low
Barium titanate	Ceramic	191	12.6
PZT	Ceramic	289	25.9
PVDF	Polymer	−33	−339

The potential for piezoelectric energy harvesters to power remote sensors, wireless devices, and low-power electronic systems in industrial settings is particularly attractive. Their ability to scavenge energy from everyday mechanical vibrations in machinery, rotating equipment, and structural elements has led to increased interest in exploring their practical applications within the industrial landscape. These applications not only offer opportunities for energy conservation but also enable the development of autonomous, self-powered systems, reducing maintenance costs and minimizing the reliance on conventional energy sources. Despite the growing interest and extensive research in the field of piezoelectric materials and energy harvesters, there is a need for a comprehensive and systematic review that can summarize and critically evaluate the progress made so far [7, 8].

This review objectives to afford a complete analysis of the state-of-the-art piezoelectric materials, energy harvesting methods, and their uses in industrial environments. By synthesizing existing literature, the review seeks to identify the strengths and limitations of various piezoelectric materials, their energy conversion efficiencies, and the challenges faced in practical implementations. Energy harvesting, also known as energy scavenging, refers to the process of capturing and converting ambient energy from the surrounding environment into usable electrical power. This concept has garnered significant attention due to its potential to provide a sustainable and eco-friendly alternative to traditional energy sources, reducing dependence on fossil fuels and contributing to a greener future. A family of intelligent materials known as piezoelectric materials has the singular capacity to produce an electrical charge in response to mechanical stress or deformation (Table 14.2). Since Pierre and Jacques Curie's first discovery of the piezoelectric effect in 1880, a great deal of study has been done on these materials for a variety of uses, including sensors, actuators, transducers, and energy harvesting devices (Fig. 14.2).

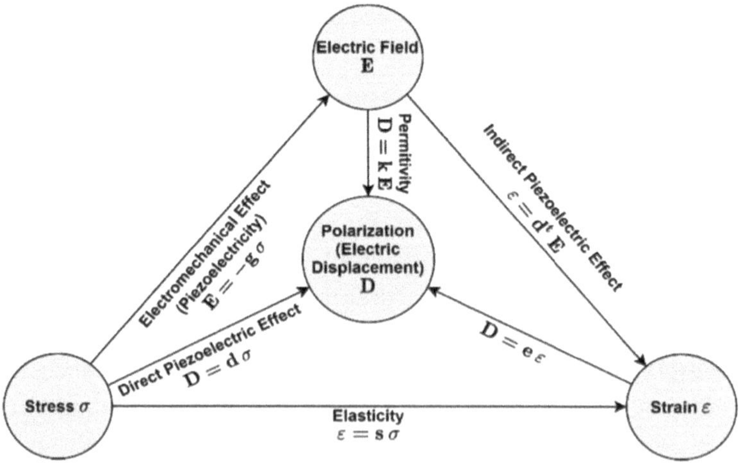

Figure 14.2 Effect of piezoelectric material [8].

Table 14.2 Advantages and disadvantages of piezoelectric materials relative to each other

Parameter	**Barium titanate [27, 28]**	**Lead zirconate titanate [29–32]**	**Quartz [33–35]**	**PVDF [36–38]**	**Rochelle salt [39]**
Range of frequency	0–70	0–100	Very low	0–10	Very low
Curing temperature	120	200	Very high	105	225
Environmental impact	Affected	Affected	Not affected	Affected	Not affected
Negative impact on the environment	No	Yes	No	No	No
Mechanical property	Rigid	Rigid	Rigid	Flexible	Rigid
Usage area	Sensor/actuator				
Energy production	High	Very high	Very low	Normal	Normal
Density (g/cm^3)	6.02	6.53	2.65	1.78	1.79

14.2 History

Piezoelectric materials play a crucial role in civil engineering due to their unique ability to convert mechanical stress or pressure into electrical energy and vice versa. This phenomenon, known as the piezoelectric effect, has been exploited in a wide range of civil engineering applications. Here's a detailed review and history of piezoelectric materials in civil engineering:

Piezoelectric Effect and Discovery: The piezoelectric effect was first observed in 1880 by Jacques and Pierre Curie, who discovered that certain crystals, such as quartz and tourmaline, generate electric charges when subjected to mechanical stress. They found that applying pressure to these crystals could polarize their molecular structure, leading to the separation of positive and negative charges, thus creating an electric potential.

Early Civil Engineering Applications: In the early 20th century, piezoelectric materials started finding applications in civil engineering, mainly in the form of sensors and actuators. These materials were used to measure stress and strain in structures like bridges, dams, and buildings. Additionally, they were utilized in the development of vibration-damping systems and ultrasonic devices for non-destructive testing of materials.

Rise of Smart Materials: The late 20th century witnessed significant advancements in piezoelectric materials and their integration into smart structures. Researchers began to explore the concept of piezoelectric actuators and sensors in active vibration control systems for civil engineering structures. These smart materials were capable of actively responding to dynamic loads and environmental changes, enhancing the safety and longevity of infrastructure.

Energy Harvesting Applications: As sustainable energy became an increasing concern, researchers realized the potential of piezoelectric materials for energy harvesting in civil engineering. The concept of using piezoelectric materials to convert ambient mechanical vibrations, such as traffic-induced vibrations or bridge movements, into electrical energy gained traction. This led to the development of piezoelectric energy harvesting systems in smart infrastructure.

Piezoelectric Transducers for Structural Health Monitoring (SHM): The 21st century brought further innovations in piezoelectric materials for SHM. Piezoelectric transducers were integrated into civil engineering structures to monitor their health continuously. These transducers can detect changes in structural properties, such as cracks, deformations, and material degradation, allowing for timely maintenance and preventing catastrophic failures [9].

The Curie brothers, Pierre and Jacques, discovered the piezoelectric effect in 1880. They used tourmaline, quartz, topaz, cane sugar, and Rochelle salt to show the effect by combining what they understood about pyroelectricity and crystal formations. They discovered that these crystals generated electricity in response to mechanical stress, and the voltage of these electrical charges was proportionate to the stress. Gabriel Lippmann later identified the converse effect by the mathematical application of the idea in 1881. The Greek word "piezein", which means to press or squeeze, was used to name these behaviors as the piezoelectric effect and the inverse piezoelectric effect, respectively. The first uses of piezoelectric ultrasonic transducers were created during World War I. Piezoelectricity is a modern technology utilized frequently. For instance, the airbag sensor in a car senses a change in the vehicle's acceleration by transmitting an electrical signal, which activates the airbag [10].

Piezoelectric materials are straightforward, inexpensive, lightweight, and controllable smart materials used in structural actuation. Piezoelectric materials are renowned for their versatility in a wide variety of applications in diverse structures; they can be easily shaped into numerous forms such as patches, thin films, cylinders, and fibers. Presently, lead zirconate titanate (PZT), lead titanate (LT), sodium potassium niobate (SPN), lead magnesium niobate (PMN), and lead metal niobate (LMN) are the piezoelectric materials that are most often utilized in automotive and aerospace engineering. Due to its wide range of uses as a pyroelectric material, PZT, one of the five types of piezoelectric materials described above, is one of the most often investigated ferroelectric materials [11, 12].

Piezoelectricity is a phenomenon in which certain materials generate an electric charge when subjected to mechanical stress or pressure. The effect was first discovered in 1880 by the French physicists Pierre and Jacques Curie, who found that applying mechanical pressure to certain crystals, such as quartz, produced an electric charge on their surfaces. In the decades that followed, researchers explored the potential applications of piezoelectric materials. During World War I and World War II, piezoelectric quartz crystals were used in early sonar devices, providing the foundation for modern underwater acoustic sensing technologies. The development of synthetic piezoelectric materials in the 1950s and 1960s further expanded the applications of piezoelectricity. These materials were more versatile and offered improved properties compared to natural crystals. With the advent of advanced materials science and manufacturing techniques, piezoelectric materials found applications in various fields, including sensors, actuators, and transducers. In recent years, there has been increasing interest in piezoelectric energy harvesting for industrial applications [13].

Piezoelectric energy harvesters can convert mechanical vibrations and movements present in the industrial environment into usable electrical energy. This concept is especially relevant as industries look for sustainable and efficient ways to power their devices and sensors. The systematic review mentioned would likely cover the various types of piezoelectric materials used for energy harvesting, such as PZT and other piezoceramics, as well as organic and flexible piezoelectric materials. The review will also explore the different configurations and designs of piezoelectric energy harvesters, their performance metrics, challenges, and potential solutions [14].

14.3 Piezoelectric–Mechanical Constitutive Equations

Piezoelectric constitution equations and Preisach-type models were used to define the hysteretic path-dependent strain-field relationship in piezoelectric actuator models. Nonlinear dielectric, elastic, and piezoelectric relationships were well established in piezoelectric ceramics. The constitutive equation for piezoelectric materials is often expressed as follows:

$$S = C^E \cdot T + [d]t \cdot E \tag{14.1}$$

$$D = d \cdot T + \varepsilon^T \cdot E \tag{14.2}$$

where d is the piezoelectric constant matrix, T is the constant stress, E is the constant field, C is the stiffness coefficient, D is the constant electrical displacement, S is the constant strain, and T is the dielectric permittivity. For the compliance matrix being examined, the superscripts E and T stand for a constant electric field and charting time, respectively. Usually, finite element (FE) analysis is modeled using these matrix relations. Only a few of the linkages are often helpful for theoretical approaches that further reduce the issue [15]. In order for the piezoelectric materials to show these properties, they must be polarized as shown in Fig. 14.3.

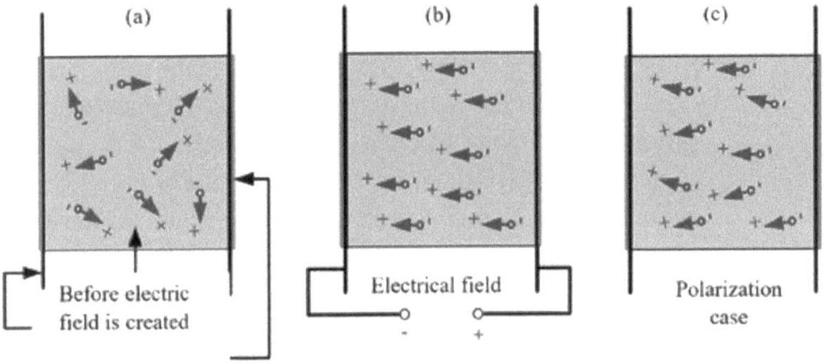

Figure 14.3 Polarization stages of piezoelectric material: (a) condition before the electric field is created, (b) effect of the electric field, (c) polarization condition [16].

14.4 Applications of Piezoelectric Materials

Piezoelectric materials are unique substances that exhibit the piezoelectric effect, which means they can generate an electric charge in response to mechanical stress or vice versa. This property has led to numerous applications of piezoelectric materials in various fields, including civil engineering. Here are some specific applications of piezoelectric materials in civil engineering

1. **Structural Health Monitoring (SHM)**: Piezoelectric sensors and actuators are used to monitor the health and integrity of civil infrastructure such as bridges, buildings, and dams. These sensors can detect changes in the structural behavior and vibrations, allowing early detection of damage or structural defects (Fig. 14.4).

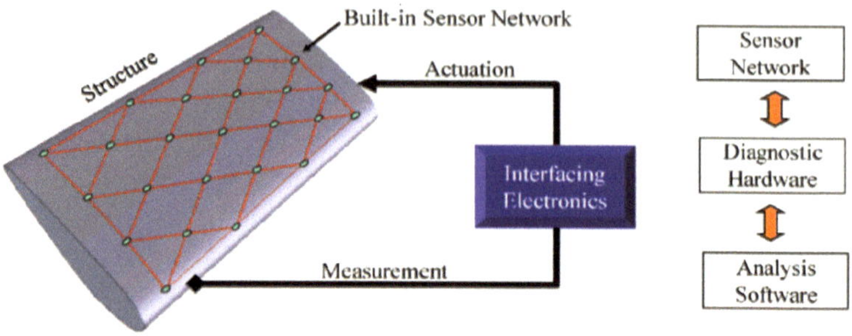

Figure 14.4 Composition of piezoelectric sensor network-based SHM system [16].

- Energy Harvesting: Piezoelectric materials can be used to convert ambient mechanical vibrations and movements from structures into electrical energy. This energy harvesting approach is particularly beneficial for powering low-power sensors or wireless sensor networks in remote locations.
- Aerospace Industry: In aerospace applications, piezoelectric sensors are used to monitor the health of aircraft structures, such as wings, fuselage, and tail components. By embedding or attaching piezoelectric sensors to critical areas, engineers can detect and monitor fatigue, cracks, and delamination, ensuring safe flight operations and reducing maintenance costs.
- Wind Turbine Monitoring: Piezoelectric sensors are employed to monitor the blades of wind turbines for any signs of damage or degradation. They can detect blade vibrations, leading edge erosion, and structural faults, enabling timely maintenance and preventing catastrophic failures.
- Automotive Industry: In automotive applications, piezoelectric sensors can be integrated into vehicle components, such as chassis, suspension systems, and engine parts. They help detect mechanical stress, wear and tear, and other structural issues, enabling predictive maintenance and improving vehicle safety.

- Civil Infrastructure: SHM using piezoelectric sensors is also applied to monitor the health of civil structures, including bridges, buildings, and dams. By placing sensors at critical points, engineers can monitor the structural integrity, identify potential issues like cracks or displacement, and take corrective actions to prevent accidents [17].

2. **Vibration Control and Damping**: Piezoelectric actuators can be integrated into civil engineering structures to actively control and reduce vibrations caused by external forces or environmental factors, such as wind and earthquakes.
    - Active Vibration Control: Piezoelectric actuators are integrated into structures or mechanical systems experiencing vibrations. These actuators can be bonded or attached to specific locations on the structure, such as beams, plates, or panels. By applying carefully controlled voltages to the piezoelectric actuators, they undergo controlled deformation, which introduces forces that counteract and suppress the undesired vibrations in real time.
    - Adaptive Vibration Control: Piezoelectric actuators are capable of adjusting their response dynamically based on the feedback from vibration sensors or accelerometers placed on the structure. This adaptive control system allows the actuators to continuously adapt their response to changing vibration conditions, making it particularly effective in dealing with varying vibrational loads or frequencies.
    - Vibration Damping: The inherent properties of piezoelectric materials, including their ability to convert electrical energy into mechanical motion and vice versa, allow them to act as damping elements. By applying voltage signals to the piezoelectric actuators that are out of phase with the vibration frequency, the actuators create out-of-phase motion that helps dampen the vibrations, dissipating the vibrational energy as heat [17].

3. **Smart Materials**: Piezoelectric materials are utilized as smart materials in adaptive civil engineering structures. These materials can change their shape or mechanical properties in response to electrical signals, enabling the construction of adaptive and morphing structures.
   - Impact and Load Sensing: Piezoelectric sensors can be embedded in pavements, roads, and bridge decks to monitor traffic loads, vehicle movements, and impact events. This information can be used for traffic management, load monitoring, and assessing the structural integrity of the infrastructure [18].
4. **Seismic Energy Dissipation**: Piezoelectric materials can be used in seismic energy dissipation devices to absorb and dissipate seismic energy during earthquakes, reducing structural damage and enhancing the seismic resilience of buildings and bridges.
   - Tunable Energy Dissipation: Piezoelectric materials offer tunability in their response by varying the applied voltage. This tunable characteristic enables engineers to adjust the seismic energy dissipation based on the intensity and frequency content of the seismic waves. By optimizing the response of the piezoelectric devices, the system can efficiently handle earthquakes of different magnitudes.
   - Retrofitting Existing Structures: Piezoelectric energy dissipation systems can also be used to retrofit existing structures that were not originally designed to withstand seismic events. Retrofitting involves adding piezoelectric dampers or actuators to vulnerable parts of the building, reinforcing its resilience against seismic forces without the need for extensive structural modifications.
5. **Vibration Reduction**: Piezoelectric materials can be incorporated into noise barriers and building facades to reduce noise pollution by converting sound energy into electrical energy or mechanical vibrations.
6. **Shape Control and Morphing Structures**: Piezoelectric materials can be applied to actively control the shape and deformation of civil engineering structures, allowing for shape morphing and adaptive functionalities.

7. **Bridge Weigh-in-Motion (WIM)**: Piezoelectric sensors can be used to estimate the weight and axle load of vehicles passing over a bridge (Fig. 14.5), providing valuable data for bridge load rating and maintenance.
   - Structural Health Monitoring: The data collected from the piezoelectric sensors can be used for SHM of the bridge. By continuously monitoring the traffic loads, engineers can assess the bridge's response to the dynamic loads over time, identify potential damage, and plan for maintenance and repair.
   - Traffic Management: WIM data provides valuable information for traffic management and planning. It helps transportation authorities analyze traffic patterns, identify peak periods, and make informed decisions about road maintenance and capacity improvements [19–25].

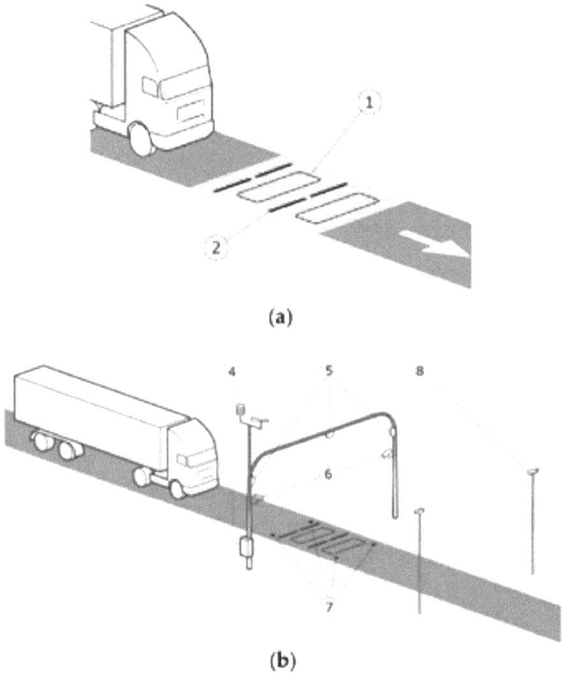

Figure 14.5 Diagram of sensor installation in WIM systems: (a) basic system configuration and (b) extensive system configuration.

8. **Self-Sensing Materials:** Some piezoelectric materials can also act as self-sensing elements, allowing them to monitor their own deformation and mechanical stress, which is valuable for structural integrity monitoring. Self-sensing materials, particularly those incorporating piezoelectric properties, are a class of smart materials that can sense and respond to external stimuli without the need for separate sensing devices. These materials have the unique ability to generate an electric signal in response to applied mechanical stress or deformation, which enables them to act as both sensors and actuators simultaneously. This self-sensing capability makes piezoelectric materials highly valuable in various engineering applications.
    - Self-Sensing Mechanism: In self-sensing materials, piezoelectric elements are integrated into the structure or component that requires sensing capabilities. When the material is subject to external forces, such as pressure, strain, or vibrations, it generates electrical signals proportional to the applied stress. These signals can be measured and analyzed to obtain information about the material's mechanical state and environmental conditions.

14.5 Conclusions

Some inferences have been drawn from the current investigations:

- Under the impact of mechanical stress, piezoelectric energy harvesters employ the direct piezoelectric effect to produce electrical energy. They have been included in certain contemporary self-powered gadget trends and methodologies.
- The design of the extra circuitry, the materials chosen, the type of arrangement, the mechanical limitations of the materials and structure, and many other variables affect how well the piezoelectric energy harvesters operate.
- It has been determined that piezoelectric energy harvesters are a recent trend, and the area of study, the methods employed, and their limits have been compiled.

- In order to prevent the impending collapse of structures based on the aerospace and civil sectors, such as offshore platforms, homes, bridges, and other structures, SHM is also proving to be extremely important.

References

1. Abdul A, Raheman Md A, Yasser EI, Asraar A, Meftah H, Bisma P, Nagma P, Jalal MZ. A systematic review of piezoelectric materials and energy harvesters for industrial applications. *Sensors*, 21, 4145, 2021.
2. Chen J, Qiu Q, Han Y, Lau D. Piezoelectric materials for sustainable building structures: Fundamentals and applications. *Renewable and Sustainable Energy Reviews*, 101, 14–25, 2019.
3. Ali SF, Friswell MI, Adhikari S. Analysis of energy harvesters for highway bridges. *Journal of International Materials System and Structures*, 22, 1929–1938, 2011.
4. Erturk A. Piezoelectric energy harvesting for civil infrastructure system applications: Moving loads and surface strain fluctuations. *Journal of International Materials System and Structures*, 22, 1959–1973, 2011.
5. Uchino, K. Introduction to Piezoelectric Actuators and Transducers; University Park: State College, PA, USA, 2003.
6. Zhu J, Gao C, He L. Piezoelectric-based crack detection techniques of concrete structures: Experimental study. *Journal of Wuhan University and Technology Materials*, 27(2), 346–352, 2012.
7. Yoichi M. Applications of piezoelectric actuator. *NEC Technology Journal*, 1, 82–86, 2006.
8. De J, Chen M, Geerlings W, Asta H, Persson MKA. A database to enable discovery and design of piezoelectric materials. *Science Data*, 2, 1–13, 2015.
9. Akgün M. Piezoelektriközellikli, polimer nano kompozitmalzemeg eliştirilmesivetitreşimsensörüolarakkullanilabilirliğininincelenme si. Masters Thesis; UludağÜniversitesi Fen BilimleriEnstitüsü, Nilüfer, Turkey. (151) 2018.
10. Panda P, Review: Environmental friendly lead-free piezoelectric materials. *Journal of Material Science*, 44, 5049–5062, 2009.
11. Park S, Ahmad S, Yun C-B, Roh Y. Multiple crack detection of concrete structures using impedance-based structural health monitoring techniques. *Experimental Mechanics*, 46, 609–618, 2016.

12. Zhu J, He L, Piezoelectric actuator/sensor wave propagation based nondestructive active monitoring method of concrete structures. *Journal of Wuhan University of Technology Materials*, 26(3), 541–547, 2011.
13. Anugonda P, Wiehn JS, Turner JA. Diffusion of ultrasound in concrete. *Ultrasonic Journal A*, 39(6), 429–435, 2021.
14. Ramegowda PC, Ishihara D, Takata R, Niho T, Horie T. Hierarchically decomposed finite element method for a triply coupled piezoelectric, structure, and fluid fields of a thin piezoelectric bimorph in fluid. *Computational Methods of Applied Mechanical Engineering*, 365, 113006, 2020.
15. Chen J, Qiu Q, Han Y, Lau D. Piezoelectric materials for sustainable building structures: Fundamentals and applications. *Renewable and Sustainable Energy Reviews*, 101, 14–25, 2019.
16. Hagood NW, von Flotow A. Damping of structural vibrations with piezoelectric materials and passive electrical networks. *Journal of Sound and Vibration*, 146(2), 243–268, 1991.
17. Qing X, Li W, Wang, Yishou SH. Piezoelectric transducer-based structural health monitoring for aircraft applications. *Sensors*, 19(3), 545, 2019.
18. Cai J, Shi L, Yuan S, Shao Z. High spatial resolution imaging for structural health monitoring based on virtual time reversal. *Smart Materials and Structures*, 20, 055018, 2011.
19. Xiong H, Zhang Y. Feasibility study for using piezoelectric-based weigh-in-motion (WIM) system on public roadway. *Applied Sciences*, 9(15), 3098, 2019.
20. Nia EM, Zawawi NAWA, Singh, BSM. A review of walking energy harvesting using piezoelectric materials. *IOP Conference Series: Materials Science and Engineering*, 012026, 2018.
21. Sodano HA, Inman DJ, Park G. A review of power harvesting from vibration using piezoelectric materials. *Shock and Vibration Digestion*, 36, 197, 2004.
22. Lee J, Lee M, Oh J, Kim K. Study on the energy conversion from the dynamic load of vehicles on the road using piezoelectric materials. *Materials Science Forum*, 658, 57, 2010. doi: 10.4028/www.scientific.net/MSF.658.57.
23. Hashemi VS, Haas CR, Haas C, Rothenburg L. Investigation of the effects of air temperature and speed on performance of piezoelectric weigh-in motion systems. *Canadian Journal of Civil Engineering*, 40, 935–944, 2013.

24. Burnos P, Gajda J, Sroka R, Wasilewska M, Dolega C. High accuracy weigh-in-motion systems for direct enforcement. *Sensors*, 21(23), 2021.

25. Liseli JB, Agnus J, Lutz P, Rakotondrabe M. An overview of piezoelectric self-sensing actuation for nanopositioning applications: electrical circuits, displacement and force estimation. *IEEE Transactions on Instrumentation and Measurement*, 1–14, 2019.

26. Wada S, Yako K, Kakemoto H, Tsurumi T, Kiguchi T. Enhanced piezoelectric properties of barium titanate single crystals with different engineered-domain sizes. *Journal of Applied Physics*, 98, 014109, 2005.

27. Lertcumfu N, Pengpat K, Eitssayeam S, Tunkasiri T, Rujijanagul G. Electrical properties of BZT/mullite ceramic composites. *Ceramics International*, 41, S447–S452, 2015.

28. Elvin NG, Lajnef N, Elvin AA. Feasibility of structural monitoring with vibration powered sensors. *Smart Materials and Structures*, 5(4), 977–986, 2006. doi: 10.1088/0964-1726/15/4/011.

29. Daniels A, Zhu M, Tiwari A. Evaluation of piezoelectric material properties for a higher power output from energy harvesters with insight into material selection using a coupled piezoelectric-circuit-finite element method. *IEEE Transactions on Ultrasonics, Ferroelectrics, and Frequency Control*, 60, 2626, 2013. doi: 10.1109/TUFFC.2013.2861.

30. Ottman GK, Hofmann HF, Bhatt AC, Lesieutre GA. Adaptive piezoelectric energy harvesting circuit for wireless remote power supply, *IEEE Transactions on Power Electronics*, 17, 669–676, 2002.

31. Shimamura K, Takeda H, Kohno T, Fukuda T. Growth and characterization of lanthanum gallium silicate $La_3Ga_5SiO_{14}$ single crystals for piezoelectric applications. *Journal of Crystal Growth*, 163(4), 388– 392, 1996.

32. Han JS. Investigation of PZT ceramics produced by powder processing. In: *Proceedings of 2014 Powder Metallurgy World Congress*, Compiled by Russell A, Chernenkoff W, Brian J. Orlando (FL, USA: Metal Powder Industries Federation) p. 04-28–34, 2014.

33. Bischur E, Schwesinger N. Energy harvesting from floor using organic piezoelectric modules. In: *Proceedings of the Power Engineering and Automation Conference*, PEAM2012; IEEE 2012.

34. Illgen U, Scheve J, Feldhaus R, Anders K. Influence of the silica support of iron-antimony oxides on the oxidation and ammoxidation of

propene. *Reaction Kinetics and Catalysis Letters*, 20(1–2), 203– 205, 1982.

35. Ramadan KS, Sameoto D, Evoy S. A review of piezoelectric polymers as functional materials for electromechanical transducers. *Smart Materials and Structures*, 23, 033001, 2014.
36. Jaitanong N, Yimnirun R, Zeng HR, Li GR, Yin QR, Chaipanich A. Piezoelectric properties of cement based/PVDF/PZT composites. *Materials Letters*, 130, 2014.
37. Ghosh SK, Sinha TK, Mahanty B, Mandal D. Self-poled efficient flexible "ferroelectretic" nanogenerator: a new class of piezoelectric energy harvester. *Energy Technology*, 3, 1190, 2015. doi: 10.1002/ente.201500167.
38. Chen C-D, Chiu Y-C, Huang Y-H, Wang P-H, Chien R-D. Assessments of structural health monitoring for fatigue cracks in metallic structures by using lamb waves driven by piezoelectric transducers. *Journal of Aerospace Engineering*, 34(1), 2021. doi: 10.1061/(ASCE)AS.1943-5525.0001212.
39. Tressler JF, Alkoy S, Newnham RE. Piezoelectric sensors and sensor materials. *Journal of Electroceramics*, 2(4), 257–272, 1998. doi: 10.1023/A:1009926623551.

Chapter 15

Lead-Free Piezoelectric Materials and Their Applications

Kartik Masson, Salma Shahni, and A. K. Shukla
Department of Physics, Amity Institute of Applied Sciences, Amity University, Noida 201313, India
akshukla@amity.edu

Recently developed some new lead-free piezoelectric materials have drawn extensive attention due to their environment-friendly nature and electromechanical response at par with the lead-based piezoelectrics. Lead zirconate titanate (PZT) and lead lanthanum zirconium titanate (PLZT) exhibit outstanding piezoelectric properties and are mostly used in actuators, sensors, and transducers. However, these materials utilize lead oxide, which is being restricted in many countries due to its hazardous nature. As such, it is important to develop lead-free piezoelectric materials which can supplement lead-based piezoelectric materials due to growing environmental concerns. This chapter reviews the progress in the development of some lead-free piezoelectric materials and discusses their structure-property correlation for device applications.

Piezoelectric Materials: Design and Applications
Edited by N. B. Singh and Dev Kumar Mahato

ISBN 978-981-5129-40-3 (Hardcover), 978-1-003-59897-8 (eBook)
www.jennystanford.com

15.1 Introduction

Piezoelectric materials are the class of materials with non-centrosymmetric crystal structures that have the capability of transforming mechanical energy to electrical energy and vice versa [1]. These abilities make them valuable in a wide range of applications like automotive fuel injections, motors, vibration control, ultrasound sensors, generators, civil systems (healthcare and network mobility), sensors, actuators, and piezoelectric energy harvesters [1–4]. Piezoelectric materials with perovskite structures show strong electromechanical response, are admired in academic circles and industry, and find numerous commercial applications. The chronological development of piezoelectric materials is depicted in Fig. 15.1.

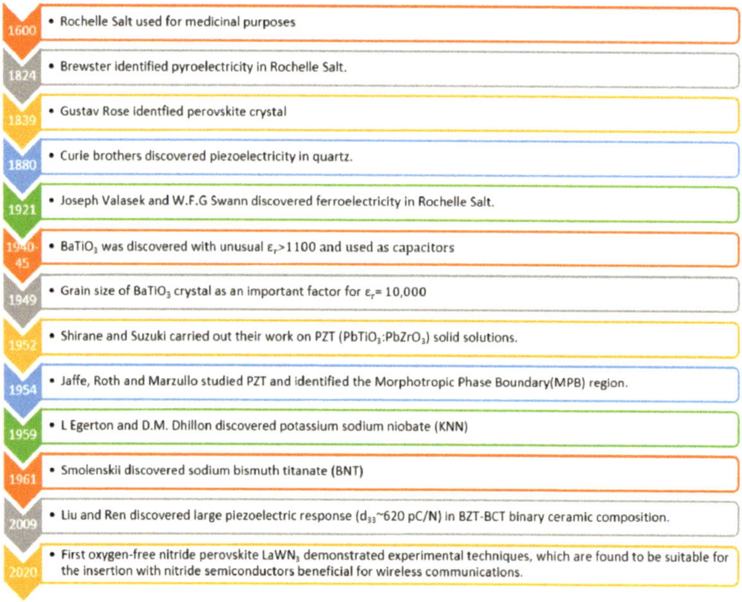

Figure 15.1 Gradual development of piezoelectric materials.

Perovskites, the blackish or grayish naturally occurring minerals, were found in metamorphic rocks by a Prussian mineralogist Gustav Rose in 1839 [5]. Calcium titanate ($CaTiO_3$) is the primary mineral on which perovskites are named which have general stoichiometry ABX_3 [6, 7], where A is a large cation, divalent and occupies corner positions, B is tetravalent which occupies the unit cell's body center position, and X is an anion which occupies the unit cell's face center positions. The ideal structure of perovskites is a cubic lattice, and the oxides prefer lower symmetry generally hexagonal and orthorhombic. Goldschmidt tolerance factor, $t = \frac{r_A + r_o}{\sqrt{2}(r_B + r_o)}$, determines the deviations from the ideal cubic structure [8], where r_A is the ionic radius of the A-site cation, r_B is the ionic radius of the B-site cation, and r_O is the ionic radius of oxygen (or halide) anion. Normally the ABO_3-type perovskite ferroelectrics show a tolerance factor of ~1 in the cubic phase and slightly >1 in the ferroelectric phase [8]. A breakthrough was the discovery of barium titanate ($BaTiO_3$) and PZT-based ceramics which showed a large piezoelectric response in comparison to the quartz crystal and became popular for technical and commercial applications [9]. Lead-based piezoelectric ceramics have been widely used for various applications due to their excellent piezoelectric coefficient and relatively high Curie temperature. PZT ceramics [10–13] are high-performance piezoelectric materials that are broadly used in sensors, actuators, and other electronic devices but contain more than 60 weight percent lead which is toxic to the environment and health. At the 9th International Meeting on Ferroelectricity (August 24–29, 1997, Seoul, Korea), Prof. Ding-Quan Xiao raised the issue of environmental concerns caused by lead-based materials [14]. In the last two decades, legislation governing the use of lead-containing compounds has been enforced in the EU and the State of California, including the draft Directives on Waste from Electrical and Electronic Equipment (WEEE), Restriction of Hazardous Substances (RoHS), and End-of-Life Vehicles (ELV) [15]. Therefore, it is desired to use lead-free materials for environmental protection. Lead-free piezoelectrics such as barium titanate, bismuth titanate, and potassium sodium niobate and their modified compositions have been widely explored to replace PZT. In this chapter, progress in the development of lead-free piezoelectric barium titanate, potassium sodium niobate, and bismuth sodium titanate systems is discussed. The effect of doping separately at A-site, B-site, and simultaneously both at A- and B-site in the above perovskites on the piezoelectric response has also been presented. Moreover, improvements in the electromechanical response due to the combination of different lead-free piezoelectric families in the form of a binary/ternary system are also included. This chapter also discusses the ongoing efforts for technical applications of these material systems.

15.2 History of Piezoelectricity

The phenomenon of piezoelectricity was discovered in the late 19th century which led to many innovations in industrial and commercial applications of piezoelectric transducers. From the Rochelle Salt period and demonstration of piezoelectricity in quartz by Curie Brothers in 1888 [9, 16] to the advancement in the perovskite family that includes nitrides perovskite $LaWN_3$ [17] which are used in telecommunication technology and play an important role in bringing 5G network, witnessed a tremendous change in the piezoelectric material fabrication methods, characterization techniques, and their innovative applications. Between 1600 to 1800s, the Rochelle Salt (sodium potassium tartrate tetrahydrate) was used mostly for medicinal purposes. Brewster [18] identified pyroelectricity (thermal-polar effect) in such crystals in 1824. The term *piezoelectricity* was first devised by Hankel in 1881 [19, 20]. A scientific definition of piezoelectricity was given by Cady as "electric polarization produced by mechanical strain in crystals belonging to certain classes, the polarization being proportional to the strain and changing sign with it" [21]. In 1921, Valasek and Swann [9, 22–24] discovered ferroelectricity in the Rochelle Salt. The identification of piezoelectricity in ceramics was introduced around the 1940s and the discovery of the effect in $BaTiO_3$ opened a new era for the use of piezoelectric ceramics in place of single crystals such as quartz and tourmaline due to their ease of fabrication and good piezoelectric response compared to the natural crystals. Prior to the discovery of the piezoelectric effect in barium titanate, it was used as a capacitor because of the high dielectric constant ($\varepsilon_r > 1100$) [25]. Gray in 1946 reported that the barium titanate ceramic can be made piezoelectric by poling in which the electrical dipoles of a multi-domain ferroelectric material are aligned in a preferred direction under application of a unipolar electric field [19].

Researchers have explored innovative applications of piezoelectric materials, such as piezoelectric fabrics for wearable energy harvesting and piezoelectric roadways to generate electricity from vehicular traffic. Today, piezoelectricity plays a crucial role in modern technology and continues to advance with ongoing research into new materials and applications.

15.3 Piezoelectric Realm

The characteristic feature of piezoelectric materials is the appearance of voltage across the material upon mechanical stress (generator effect) or conversely the material will get strained upon the action of an electric field (motor effect). Piezoelectric ceramics are used in various fields like actuating and sensing devices, ultrasound imaging, hydrophones, underwater devices, etc. [1, 3, 9, 19]. A lack of center of symmetry in a unit cell is a necessary condition to have piezoelectricity in the material.

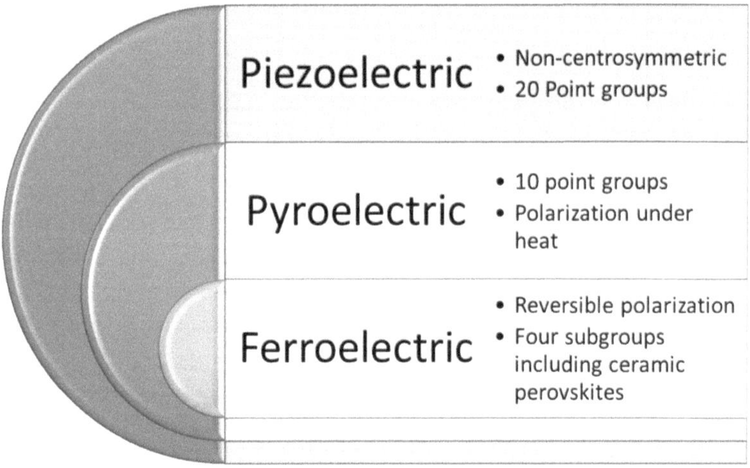

Figure 15.2 Realm of piezoelectricity.

The family of crystals can be divided into one of 32-point groups. These point groups are sub-classes of the seven-crystal systems, namely, cubic, rhombohedral, tetragonal, monoclinic, triclinic, hexagonal, and orthorhombic. Among 32-point groups, 21 exhibit a lack of center of symmetry in which 20 are piezoelectric (Fig. 15.2). In 20-point groups, 10 are pyroelectrics and divided into subgroups of ferroelectrics. Thus, all the ferroelectrics are piezoelectric but all piezoelectric are not ferroelectric. Ferroelectric ceramics are divided into four classes: (a) oxygen octahedral, i.e., ABO_3 perovskites, (b) tungsten bronze group, (c) pyrochlores, and (d) bismuth layer-structure group. PZT, lead titanate (PT), BT, lead lanthanum zirconate (PLZT), and potassium sodium niobate are the most common examples of ferroelectric ceramics that have been manufactured for commercial applications.

15.4 Piezoelectric Characteristics

The general parameters used to characterize a piezoelectric material are described below [1, 3, 7, 19, 26].

I. Dielectric constant (ε_r)

It signifies the ability of the dielectric material to hold the electric flux within it and is mathematically expressed as

$$\varepsilon_r = \frac{\varepsilon}{\varepsilon_0} \tag{15.1}$$

Here, ε_r is the dielectric constant of the material and ε and ε_0 are the permittivities of the material and the free space, respectively.

II. Curie temperature (T_c)

It is the temperature corresponding to the phase transformation of a material from a low temperature ferroelectric phase to a high temperature paraelectric phase by the method of heating. Above T_c, the material loses its ferroelectric behavior. That is why the material with higher T_c is preferred for device applications.

III. Remnant polarization (P_r)

The remnant polarization refers to the residual electric polarization in a ferroelectric material after the removal of the electric field (Fig. 15.3). This polarization can be switched between two states under the application of the alternating electric field. This switching property is useful for ferroelectric random-access memory (FeRAM) applications.

IV. Piezoelectric strain constant (d_{33})

The piezoelectric charge constant, also known as the strain constant, relates the strain produced in the material due to the influence of the electrical field at constant stress. The high value of charge constant is desirable for motional or vibrational devices such as sonars.

The unit of piezoelectric charge constant is pC/N or pm/V. It is an important figure of merit for actuator applications.

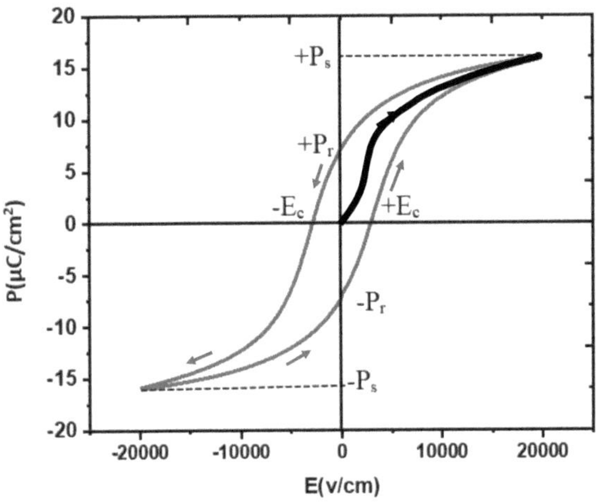

Figure 15.3 Typical P-E hysteresis loop.

V. Electromechanical coupling factor (k_p)

The electromechanical coupling factor provides a convenient way to describe the strength of the electromechanical effect. It is the energy conversion from mechanical energy to electrical energy and vice versa.

$$k^2 = \frac{\text{Output mechanical energy}}{\text{Input electrical energy}} \tag{15.2}$$

or

$$k^2 = \frac{\text{Output electrical energy}}{\text{Input mechanical energy}} \tag{15.3}$$

The term k_p is known as the planar mode piezoelectric coupling factor of the circular plate.

15.5 Equations Governing Piezoelectricity

The piezoelectric effect is of two types:

Direct effect where the polarization is directly proportional to the stress [1, 19].

$$\frac{\partial D}{\partial T} = d \tag{15.4}$$

Direct effect is responsible for the energy conversion from mechanical to electrical. The equation governing the direct effect is expressed as [1]

$$D = dE + \varepsilon^{T}E \tag{15.5}$$

This effect is utilized in piezoelectric generator applications such as pressure sensors, microphones, gas lighters, and ultrasonic detectors.

The indirect piezoelectric effect relates the strain produced in the material to the applied electric field and refers to the energy conversion from electrical to mechanical. The relation between the strain produced, the applied electric field, and the piezoelectric charge constant is given as

$$\frac{\partial S}{\partial E} = d \tag{15.6}$$

The equation governing the direct effect is expressed as [1]

$$S = dE + s^{\varepsilon}T \tag{15.7}$$

This effect is utilized in piezoelectric motor applications such as crystal oscillators, crystal speakers, actuators, ultrasonic generators, and record players. Here, D is the electric displacement, T is the stress applied, S is the strain, E is the electric field, ε is the dielectric constant, and d is the piezoelectric charge constant.

In superscript notations, the variable in the superscript is kept constant during the measurement of a particular parameter. Like ε^{T} denotes the value of the dielectric permittivity of the material under constant stress.

15.6 Phenomenon of Ferroelectricity

Ferroelectricity is an analogous term with ferromagnetism. Ferromagnetism was discovered many years before ferroelectricity. Around 1921, Joseph Valasek and his supervisor W.F.G. Swann discovered ferroelectricity in crystals, a phenomenon in which dielectric material got spontaneously polarized and they have the potential to reverse the polarization process by applying the electric field just before the electric breakdown of the material [9, 22–24,

27]. The three important characteristics of ferroelectric materials are [28]:

i. **Spontaneous polarization:** Ferroelectric materials show spontaneous polarization even after removing the external electric field. This polarization is caused by the displacement of electric charges within the crystal structure of the material.
ii. **Hysteresis loop:** Ferroelectric materials exhibit a hysteresis loop in their polarization versus electric field (P-E) curve, which is known as the fingerprint of ferroelectricity.
iii. **Paraelectric to ferroelectric phase transition**: Ferroelectricity is a closely related phenomenon with phase transitions, particularly at a critical temperature known as Curie temperature (T_c). Below T_c, the ferroelectric phase prevails and above T_c, the material possesses a paraelectric phase (Fig. 15.4).

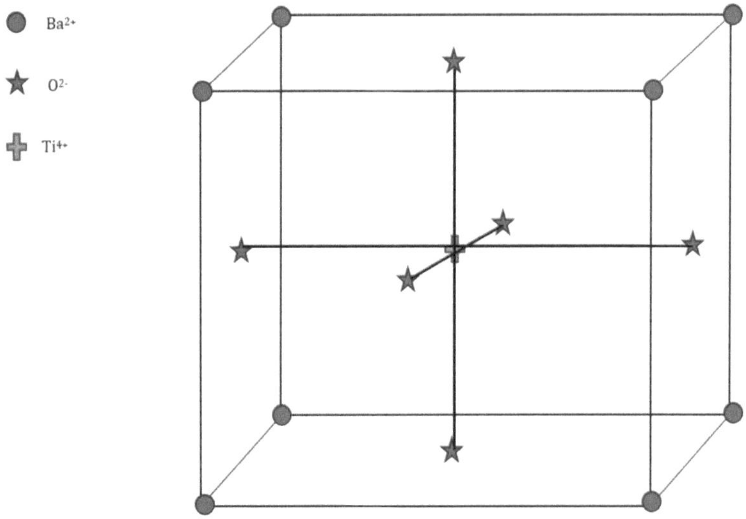

Figure 15.4 Cubic (paraelectric) phase of $BaTiO_3$ crystal.

15.7 Toxic Effects of Lead

Lead-based piezoelectric ceramics like lead titanate zirconate have dominated the world since its discovery. They surpassed barium titanate in terms of electrical and mechanical properties of materials like higher T_c values, high electromechanical coupling factor, large dielectric constant [1, 19], but contain lead around 60 wt% of the total material. Evaporation of lead during sintering and improper disposal of lead-based materials demonstrates stern health risks, from mild symptoms of lethargy and anorexia, intermediately causing headache and anemia, to severely affected persons such as cases of coma and convulsions and renal failure [29]. It can cause different aches in organs and abdominal discomfort also. Improper disposal of lead contaminates the five utmost important nature reserves, causing serious concerns and issues to the environment. Considering these severe issues, efforts are being made to produce lead-free piezoelectric materials that can replace lead-based material systems.

15.8 Synthesis of Lead-Free Piezoelectric Ceramics

Various methods have been devised to make piezoelectric materials like sol-gel, solid-state reaction method, and hydrothermal processing to obtain good quality piezoelectric ceramics with enhanced electromechanical parameters for practical applications. Some of these methods are discussed in the following sections.

15.8.1 Sol-Gel Method

The sol-gel method is used for the chemical stability of the solution for a variety of materials as they possess better stoichiometry control, atomic level mixing, and resulting in fine particle sizes. A tentative flow diagram for a typical sol-gel processing of piezoelectric ceramics is shown in Fig. 15.5. The domain orientation and configuration are usually done to get better piezoelectric properties, poling in ferroelectric materials [30, 31]. The advantage of the sol-gel method is to lower the processing temperature by fraternizing the ingredients at the atomic scales. Mixing reactants on a molecular level allows for better control of stoichiometry, and easy formation of the ultra-fine crystalline powders with controlled morphology and grain size in the final sintered ceramics [32, 33].

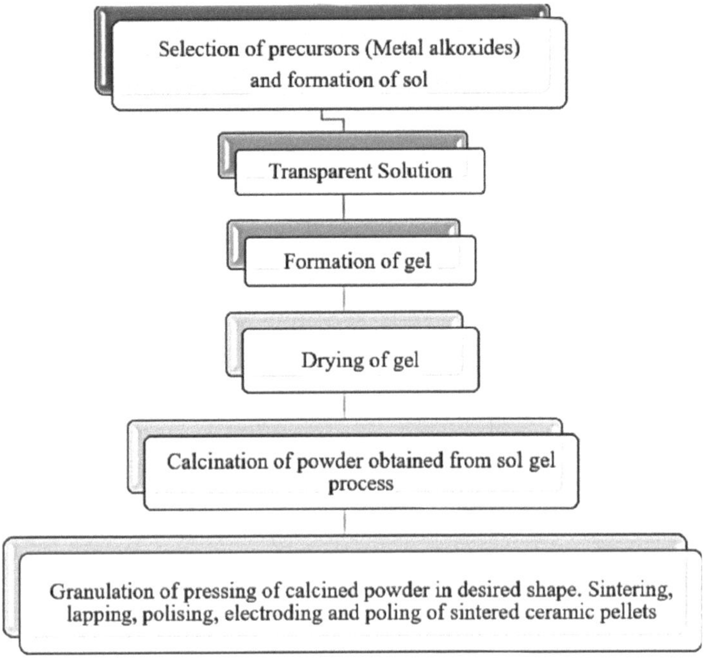

Figure 15.5 Piezoelectric material synthesis by sol-gel method.

In the sol-gel method, starting compounds (precursors) for the preparation of colloids are metal or metalloid surrounded by various legends. Metal alkoxides are the most common precursors in sol-gel processing. The metal alkoxides are a member of the family of metalorganic compounds that have an organic ligand to a metal or metalloid atom. Organometallics have direct metal-carbon bonds, whereas metal alkoxides have metal-oxygen-carbon linkage. There are three major steps in sol-gel processing: (i) formation of sol, (ii) formation of gel from sol, and (iii) drying and firing of gel. Initially, the dissolution of the desired compound in the solution occurs. The ligand phase of this solution is water, a short chain of alcohol, or an organic acid. Solutes are basically inorganic nitrates, inorganic-organic chlorides, or metal-organic compounds. Precursors should be soluble in the reaction media. Mostly all alkoxide and alkoxide salts have been used as precursors. An alkoxide typically has the formula unit $M(OR)_x$, where M represents a metallic ion, R is an alkyl group, and x denotes the valency state of the metal. The hydrolysis reaction happens as $M(OR)_x + xH_2O \rightarrow M(OH)_x + xROH$.

Conversion of a sol into the gel is observed with continuous stirring of sol at some elevated temperature. A gel represents a colloidal or polymeric solid containing a fluid component that has internal network structures such that both solid and liquid components are highly dispersed. The polymeric gel is formed through hydrolysis and condensation. Condensation reaction can occur through dehydration and dealcoholization. Gelation can be achieved through a cross-linking of M2Ox of the polymer specimen. Shaping of the viscous gel can be done in the desired form such as spheres, fibers, powder, and thin films. The final ceramic product can be obtained after drying and firing of gel. The major advantages of the sol-gel method are (i) easy purification of compounds, (ii) ultra homogenization due to atomic level mixing, (iii) densification to inorganic solids at comparatively low temperatures, and (iv) synthesis of active powders.

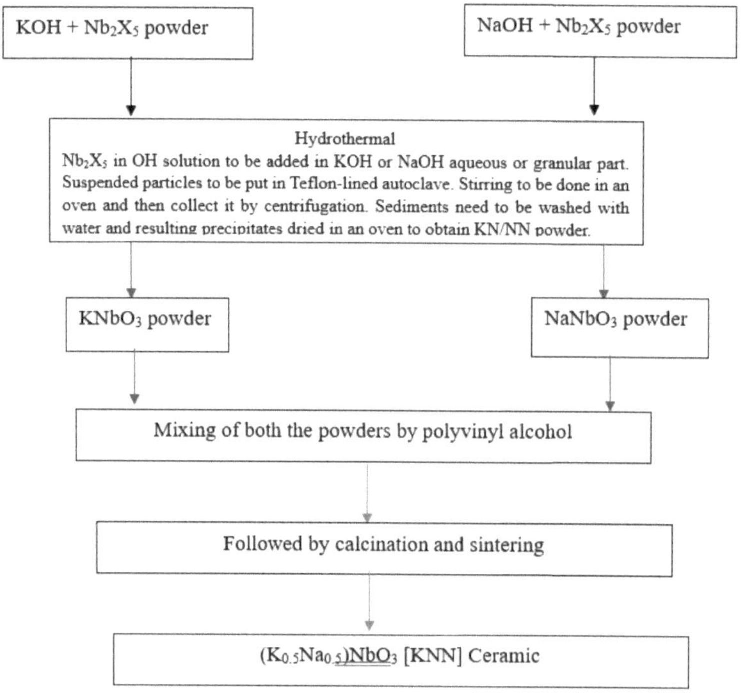

Figure 15.6 Typical KNN synthesis by hydrothermal method.

15.8.2 Hydrothermal Method

The hydrothermal method is a promising method that has been used for making nanomaterials. It uses the chemical reaction of solutions at temperatures less than 200°C [34]. Highly purified metal oxide powders can control the crystal morphological structure, in which the hydroxide ions play an important role. Water is used as the solvent in the hydrothermal process and is usually incorporated into the resultant titania product [35]. Hydrothermally derived ferroelectric particles are potentially useful because, when properly oriented and polarized, such particles can undergo almost complete polarization. High yielding quality of nanomaterials possessed by the hydrothermal method and inclusion of the hydrothermal method with a magnetic field led to the making of semicontinuous materials [36–38]. A typical hydrothermal processing method for a lead-free piezoelectric is depicted in Fig. 15.6.

15.8.3 Solid-State Reaction Method

One of the most popular methods for synthesizing piezoelectric materials is the solid-state reaction method. This method involves mixing the reactants in a solid state and heating them to high temperatures. The resulting product is then cooled and ground into a fine powder. It is a simple and cost-friendly method that can be used to synthesize a wide variety of piezoelectric materials. Piezoelectric materials synthesized using the solid-state reaction method possess high piezoelectric coefficients, good thermal stability, and high mechanical strength. Figure 15.7 shows the basic processing steps for synthesis of a typical piezoelectric ceramic using the solid-state reaction method.

15.9 Lead-Free Materials

There are various materials that have shown their capabilities that can be related to PZT materials and can replace lead-based materials. The following sections discuss the composition and properties of a few important classes of lead-free piezoelectric materials such as barium titanate, potassium sodium niobate, bismuth sodium titanate, and some of the combinations.

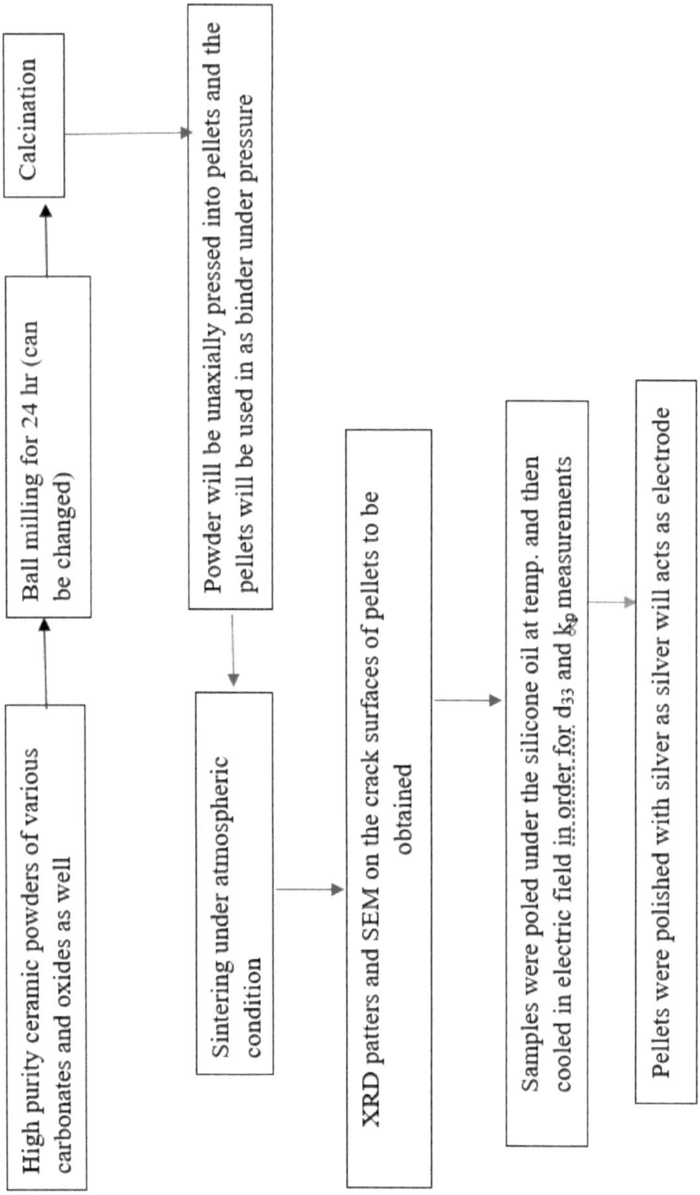

Figure 15.7 Piezoelectric ceramic synthesis by solid-state reaction method.

15.9.1 Barium Titanate

Barium titanate is one of the most promising lead-free ferroelectric materials which has been used for decades. Dominating lead-based piezoelectric materials like PZT and others are being used in every commercial and industrial application [1, 19], but their toxic and hazardous effects are alarming [29, 39]. The discovery of $BaTiO_3$ in the 1940s during the Second World War led to ceramic transducers and high dielectric constant capacitors [1, 9]. Barium titanium is the first polycrystalline material that exhibits ferroelectric and piezoelectric properties and belongs to the perovskite (ABO_3) family, which is characterized by a cubic unit cell at high temperatures. The coordination number of Ba^{+2} (barium) is 12 while the coordination number of Ti^{+4} (titanium) is 6. The perovskite structure of BT is one in which the atoms of Ba, Ti, and oxygen are arranged in a specific pattern. Ba atoms are at corner positions, Ti atoms are arranged at a central position, and oxygen atoms are at a face-centered position of the cubic unit cell. Barium titanate exhibits interesting phase transitions at different temperatures. Above the Curie temperature ($T_c \sim 120°C$), barium titanate is in a paraelectric phase and exists in a cubic perovskite phase with a centrosymmetric crystal structure and the material does not exhibit any spontaneous polarization in this phase [40]. As the temperature of barium titanate decreases below the Curie temperature (T_c), it undergoes a paraelectric to ferroelectric phase transition and has a tetragonal structure, leading to spontaneous electric polarization. Upon decreasing the temperature, the barium titanate exhibits a tetragonal ferroelectric phase to orthorhombic ferroelectric phase transition ($T \sim 5°C$). Further decreasing temperature around ($T \sim -90°C$), the material exhibits orthorhombic ferroelectric to rhombohedral ferroelectric phase transition and the polarization direction changes [41].

A tentative list of barium titanate-based compositions with their dielectric and electromechanical properties is presented in Table 15.1. Grain size plays an important role in determining the properties of barium titanate ceramics. In 1949, Walter et al. reported a dielectric constant of 10,000 at T_c [42]. It was a great achievement showing ε_r of BT could go up to 10,000. In 1958, Jaffe et al. showed ε_r = 1700 with d_{33} = 190 pC/N [43]. In 1971, Jaffe et al. observed ε_r = 3181 with k_p = 0.45 and d_{33} = 419 pC/N. High piezoelectric coefficients are excellent for use in device applications [44]. In 1984, Cross et al. examined the barium titanate at room temperature (RT) and got ε_r = 6000 [45]. Hennings et al. with grain size ~25 μm at ~130°C found ε_r ~ 9000 [46]. This showed the immense capabilities of BT to store electric charge. Advancements happened and a high piezoelectric coefficient was observed in 2008 by Shao et al., who found k_p = 0.461 with ε_r ~ 3162 and d_{33} = 405 pC/N by conventional sintering of fine grains of BT [47]. Lu et al. found that both A and B sites doped BT with La and Ce having ε_r > 9000 [48]. Praveen et al. synthesized the composition $50Ba(Zr_{0.2}Ti_{0.8})O_3$-$50(Ba_{0.7}Ca_{0.5})TiO_3$(50BZT-50BCT) by sol-gel method and reported the coupling factor k_p = 0.596, piezoelectric charge constant d_{33} = 637 pC/N and remnant polarization P_r = 12.2 μC/cm^2 [49]. Wang et al. reported a high piezoelectric constant d_{33} of $BaTi_{1-x}Sn_xO_3$ at x = 0.11 using the solid-state reaction method by practically applying the Landau theory [50–55] in which a high d_{33} can be obtained by two ways: coexistence of multiple phases (CMP) and local structural heterogeneity (LSH). By first principles calculations, the total energy of rhombohedral (R), tetragonal (T), orthogonal (O), and cubic (C) was measured. Then increasing the Sn content, the energy E was reduced to zero with x = 0.11, suggesting the CMP behavior, i.e., R-O-T-C from non-polar cubic to polar R, T, and O. Sn doping is responsible for high d_{33} = 1120 pC/N and high k_p = 0.55 [56]. Kaur et al. [57] attributed the increased grain growth and reduced porosity behind the observed high value of dielectric constant ~17000 around the transition temperature 125oC in $Ba_{0.85}Ca_{0.15}Zr_{0.05}Ti_{0.95}O_3$ composition prepared by hydrothermal method. Liu et al. [58] prepared the $Ba_{0.77}Ca_{0.23}TiO_3$ single crystal by Czochralski method and obtained the charge constant d_{33} = 130 pc/N and remnant polarization 17.93 microC/cm^2. The crystal growth was due to the calcium added to the BT ceramic. BZT-BCT-based compositions have excellent piezoelectric response but have low Curie point and coercive field, making it suitable for sensors and detectors for room temperature applications.

Table 15.1 Tentative list of barium titanate (BT) based piezoelectric compositions

Compositions	ε_r @ 1 kHz	T_c (°C)	d_{33} (pC/N)	k_p	Ref.
$BaTiO_3$	1700	115	190	0.37	[43]
$BaTiO_3$	3181		419	0.45	[44]
$BaTiO_3$	3140		315	0.37	[44]
$BaTiO_3$: a. Conventional sintering	3088		193	0.27	[59]
b. Spark plasma	4301		416	0.41	[59]
c. Conventional sintering (fine grain)	3162		405	0.46	[47]
d. Microwave sintering	1570		374		[60]
$BaTiO_3$	3181		419	0.45	[47]
$BaTiO_3$ (size: micro)		120	190		[61]
$BaTiO_3$ (size: nano)		120	240		[62]
$BaTiO_3$-4 mol% LiF (size: nano)		73	270	0.45	[63]
$BaTiO_3$-0.05 mol% Li_2O (size: nano)		120	200		[40]
$BaTiO_3$-3 at% Li_2CO_3 (size: nano)		130	260	0.43	[64]
$BaTiO_3$	10,000	120			[42]
$BaTiO_3$		125	300	0.4	[65]
$BaTiO_3$ (grain size ~ 1.5 μm)	1315	121			[66]
$BaTiO_3$ (grain size ~ 25 μm)	~9000				[46]
$BaTiO_3$	6000				[45]
$BaTiO_3$ (0.1%) filler hydroxyapatite/barium titanate grouped with polyetherimide (PEI) film	**(100 Hz)** ~4.35		~182		[67]
5 mol HA/95 mol BT nanocomposites ($Ca_{10}(PO_4)_6(OH)_2$ as HA)	~800				[68]

(*Continued*)

Table 15.1 (*Continued*)

Compositions	ε_r @ 1 kHz	T_c (°C)	d_{33} (pC/N)	k_p	Ref.
$Ba_{0.77}Ca_{0.23}TiO_3$	~1000	106	130		[58]
$Ba_{0.84}Ca_{0.16}TiO_3$ Temp ~ 140°C	**100 kHz** 5200				[69]
$BaTi_{0.9625}Zr_{0.0375}O_3$ + CuO	1546		300	0.49	[70]
$BaTi_{0.9625}Zr_{0.0375}O_3$	1322		260	0.42	[70]
$Ba_{0.85}Ca_{0.15}(Ti_{1-x}Zr_x)O_3$ BCZT	**100 kHz** 6927	87	510	0.56	[71]
$Ba(Ti_{0.96}Sn_{0.04})O_3$ [BTS]	3826		205	0.24	[72]
CuO-modified BTS	2421		390	0.49	[72]
CuO-modified $Ba(Ti_{0.96}Sn_xZr_{0.04-x})O_3$ [BTSZ] (x = 0.04)	3450	82.1	400	0.50	[73]
$BaTiO_3$	3587	125	300	0.42	[65]
$BaTiO_3$		130	140		[74]
$BaTi_{0.9625}Hf_{0.05}O_3$		109.2	305		[75]
$Ba(Zr_xTi_{1-x})O_3$					[76]
X = 0.00	>10,000	117			
X = 0.02	~10,000	111			
X = 0.05	~12,000	102			
X = 0.08	>12,000	90			
$BaTi_{1-x}Sn_xO_3$ (x = 0.11)	40,000	40	~1120	0.55	[56]
$50Ba(Zr_{0.2}Ti_{0.8})O_3$-$50(Ba_{0.7}Ca_{0.5})TiO_3$			637	0.59	[49]
$Ba(Ti_{0.8}Zr_{0.2})O_3$-$50(Ba_{0.7}Ca_{0.3})TiO_3$	4050	93	620	0.53	[77]
$(Ba_{0.92}Ca_{0.08})(Ti_{0.95}Zr_{0.05})O_3$		110	365	0.48	[78]
$(Ba_{0.84}Ca_{0.16})(Ti_{0.9}Zr_{0.1})O_3$	4800		328	0.37	[79]
$(Ba_{0.99}Ca_{0.01})(Ti_{0.98}Zr_{0.02})O_3$		115	375	0.44	[80]

Compositions	ε_r @ 1 kHz	T_c (°C)	d_{33} (pC/N)	k_p	Ref.
$(Ba_{0.96}Ca_{0.04})(Ti_{0.95}Zr_{0.05})O_3$	2070		338	0.36	[81]
$(Ba_{1-x}La_x)(Ti_{1-x/4-y}Ce_y)O_3$ X = 0.03 and y = 0.05	>10,000				[48]
$Ba_{0.85}Ca_{0.15}Zr_{0.05}Ti_{0.95}O_3$	~17,000	140			[57]

15.9.2 Potassium Sodium Niobate

Potassium sodium niobate (KNN) exhibits piezoelectric properties and has a relatively high Curie temperature than barium titanate based materials. It has gained significant attention for its potential applications like sensors, transducers, and actuators. It is discovered in 1959 by Egerton et al. [82], and found to be a promising lead-free ceramic which could replace lead-based materials. The stoichiometric composition of potassium sodium niobate is represented as $Na_xK_{1-x}NbO_3$, where potassium (K), sodium (Na), and niobium (Nb) ions form a complex lattice in which oxygen ions (O) occupy the spaces between these cations. Pure KNN has three-phase transition temperatures. At around 100°C, it shows the transition from ferroelectric rhombohedral to orthorhombic ($T_{R\text{-}O}$) phase [83].

When heated above the temperature ~ 200°C, it shows the transition from orthorhombic to tetragonal ($T_{O\text{-}T}$). In this phase, the spontaneous polarization also disappears, and the crystal structure becomes tetragonal. This transition is associated with a change in the symmetry of the crystal lattice. Upon further increasing temperature, typically above 440°C, KNN undergoes another phase transition from ferroelectric tetragonal to paraelectric cubic phase. In the cubic phase, the crystal symmetry changes again, and it becomes non-polar [84]. Advancements have been made in KNN ceramics over the years and materials with high piezoelectric responses have been developed. Table 15.2 contains a tentative list of KNN-based ceramics along with their dielectric, ferroelectric, and piezoelectric properties. In 2004, Saito et al. compared textured KNN $[(K_{0.44}Na_{0.52}Li_{0.04})(Nb_{0.84}Ta_{0.1}Sb_{0.06})O_3]$ with PZT and achieved results that could replace PZT (k_p = 0.61 and ε_r = 1570) in coming future [85]. Shrout et al. showed the KNN-LF4 having dielectric constant ε_r = 1560 with d_{33} = 271 pC/N [4]. In 2008, Lin et al. reported KNN with other compositions $K_{0.5}Na_{0.5}NbO_3$ + 0.75 mol% $K_{5.4}Cu_{1.3}Ta_{10}O_{29}$ and reported k_p = 0.41 with ε_r = 360 and T_c = 383°C [86]. KNN ceramic prepared by the hot-pressed (Jaeger–Egerton) method exhibited the value of kp = 0.45 and d33 = 160 pC/N [87]. Irregular grain growth and lowering of the transition temperature are seen in KNN ceramics when doped with antimony and tantalum [14]. Zheng et al. reported KNN with different compositions like $0.96(K_{0.48}Na_{0.52})(Nb_{0.95}Sb_{0.05})O_3$-$0.04(Bi_{0.5}(Ag_{0.5},(Na_{0.82}K_{0.18})_{0.5}))ZrO_3$ (KNNS-BANKZ) sintered at temperature 1065°C for 3 h in air and reported k_p ~ 0.46 with ε_r = 2750 and d_{33} = 450 pC/N [88]. In 2023, Pang et al. reported high electromechanical coupling factor k_p ~ 0.49 and d_{33} ~ 210 pC/N in $(K_{0.5}Na_{0.5})NbO_3$ – $LiSbO_{3-x}CaTiO_3$ (x = 0.01) composition [89]. They had attributed the high piezoelectric response to the evolution of nano-polar domains and enhanced phase boundary in this composition [89]. Xing et al. prepared KNN with $BiAlO_3$ and $BiNaZrO_3$ and reported piezoelectric coefficient d_{33} = 306 pC/N, k_p = 0.48, and T_c = 336°C suitable for high frequency ultrasound imaging transducer [90]. Maeda et al. [91] obtained $K_{0.48}Na_{0.52}NbO_3$ ceramic material by sintering the KN and NN at 1100°C for 2 h. The control of stoichiometric ratio between potassium and niobium in KNN ceramics has been emphasized for reducing the conductivity and obtaining the maximum piezoelectric response in the material [92, 93]. For high power transducers applications, large piezoelectric response with low mechanical loss are required in KNN ceramics. KNN-based single crystals show strong anisotropy and high electromechanical coupling factors which are advantageous for improving the piezoelectric device performance [94]. Li et al. synthesized $(0.96(K_{0.5}Na_{0.5})(Nb_{0.965}Sb_{0.035})$-$0.01CaZrO_3$-$0.03(Bi_{0.5}K_{0.5})HfO_3$ + x mol% $NaNbO_3$ (x = 3)). They further reported that the low temperature rhombohedral to orthorhombic transition was favorable for achieving high piezoelectric response and resulted a large value of charge constant d_{33}~700 pC/N and k_p = 0.76 making it suitable for use in sensors or actuators. High piezoelectric properties in KNN based compositions are achieved due to formation of nanodomains and occurrence of intermediate monoclinic phase between rhombohedral and orthorhombic structures [95].

Table 15.2 List of potassium sodium niobate (KNN) based compositions

Compositions	ε_r @ 1 kHz	T_c (°C)	d_{33} (pC/N)	k_p	Ref.
$K_{0.48}Na_{0.52}NbO_3$	~600	~410	160		[96]
$K_{0.48}Na_{0.52}NbO_3$	**(100 Hz)** ~2000	420	114	0.34	[91]
$K_{0.5}Na_{0.5}NbO_3$ (SPS)	606	395	148	0.39	[97]
$K_{0.5}Na_{0.5}NbO_3$ + 1 mol% CuO	~6000	421	82	0.39	[98]
$K_{0.5}Na_{0.5}NbO_3$+0.75mol% $K_{5.4}Cu_{1.3}Ta_{10}O_{29}$	360	383	90	0.41	[99]
$K_{0.5}Na_{0.5}NbO_3$ + 1 wt% $K_4CuNb_8O_{23}$			100	0.40	[100]
$K_{0.5}Na_{0.5}NbO_3$	240	393	160		[101]
0.5 % MnO_2 doped $K_{0.5}Na_{0.5}NbO_3$	730	416	270		[102]
$(K_{0.5}Na_{0.5})NbO_3$	540		46 pm/V		[103]
$(K_{0.5}Na_{0.5})NbO_3$–$LiSbO_{3-x}$ $CaTiO_3$ (x = 0.01)	1170	330	210	0.49	[89]
$(K_{0.5}Na_{0.5})NbO_3$	420		160	0.45	[87]
$(K_{0.5}Na_{0.5})NbO_3$	**10 kHz**- 522		132	0.38	[104]
$(K_{0.5}Na_{0.5})_{1-x}Li_xNbO_{3-x}$ $BaZrO_3$ (x = 0.08)	2600		317	0.43	[89]
$K_{0.5}Na_{0.5}(Nb_{0.7}Ta_{0.3})O_3$			61		[105]
$K_{0.44}Na_{0.52}Li_{0.04}Nb_{0.84}$ $Ta_{0.1}Sb_{0.06}O_3$			90		[106]
$K_{0.5}Na_{0.5}(Mn_{0.005}Nb_{0.995})O_3$	330		53		[107]

(*Continued*)

Table 15.2 (*Continued*)

Compositions	ε_r @ 1 kHz	T_c (°C)	d_{33} (pC/N)	k_p	Ref.
$(K_{0.44}Na_{0.52}Li_{0.04}Nb_{0.86}Ta_{0.1}Sb_{0.04})O_3$			21		[108]
$K_{0.4425}Na_{0.52}Li_{0.0375}Nb_{0.8825}Ta_{0.0375}Sb_{0.08}O_3$	341				[109]
$Li_{0.04}(K_{0.5}Na_{0.5})_{0.96}NbO_3$			48.1		[110]
$0.95K_{0.5}Na_{0.5}NbO_3-0.05LiNbO_3$	259		83.1		[111]
$Li_{0.06}(K_{0.5}Na_{0.5})_{0.94}NbO_3$	360		105		[112]
$0.96(K_{0.48}Na_{0.52})(Nb_{0.95}Sb_{0.05})O_3$-$0.04(Bi_{0.5}(Ag_{0.5},(Na_{0.82}K_{0.18})_{0.5}))ZrO_3$ (KNNS-BANKZ)	2750	~215	450	~0.46	[88]
$0.96(K_{0.48}Na_{0.52})(Nb_{0.95}Sb_{0.05})O_3$-$0.04Bi_{0.5}Ag_{0.5}ZrO_3$ (KNNS-BAZ)	~2600	~220	475	~0.46	[88]
$[(K_{0.44}Na_{0.52}Li_{0.04})(Nb_{0.84}Ta_{0.1}Sb_{0.06})O_3]$ KNN-LF4 textured	1570	253	416	0.61	[85]
$[(K_{0.44}Na_{0.52}Li_{0.04})(Nb_{0.84}Ta_{0.1}Sb_{0.06})O_3]$ KNS-2	1560	262	270		[4, 113]
$0.96(K_{0.5}Na_{0.5})(Nb_{0.965}Sb_{0.035})$-$0.01CaZrO_3$-$0.03(Bi_{0.5}K_{0.5})HfO_3$ + x mol% $NaNbO_3$ (x = 3) [KNN-xT] {Textured}		242	~700/ ~980 pm/V	0.76	[95]
$(K_{0.4}Na_{0.52})(Nb_{0.84}Sb_{0.08})O_{3-(0.08-x)}LiTaO_3$-$xBaZrO_3$ (x = 0.025)	~3200	175	~325	0.45	[114]

15.9.3 Bismuth Sodium Titanate

The bismuth sodium titanate (BNT) material was discovered by Smolenskii in 1961 [115]. Bismuth-based materials are among the most promising lead-free piezoelectric ceramics that have a perovskite crystal structure with a rhombohedral crystallographic symmetry at room temperature. The sodium (Na)/bismuth (Bi) ions are arranged in the eight corners in equal proportions, oxygen (O) ions occupy face-centered positions, and titanium (Ti) ions are in the center of the unit cell. Bismuth sodium titanate $(Bi_{0.5}Na_{0.5})TiO_3$ is one of the main compositions for replacing the lead-based materials [116]. BNT has a cubic perovskite structure at high temperatures (>320°C). As the temperature decreases, it undergoes a phase transition to a ferroelectric tetragonal structure. However, at a specific temperature range (200 to 300°C), it can undergo a transition to an antiferroelectric state. Further decreasing the temperature below 200°C, BNT undergoes another phase transition from tetragonal to orthorhombic. This phase transition further modifies the crystal structure and affects the material's properties. Table 15.3 shows some of the BNT-based ceramics that are showing capabilities to replace lead-based materials. Li et al. used the combination of BNT with sodium niobate (NN) and got the piezoelectric coefficient d_{33} = 88 pC/N [117]. Herabut et al. reported a maximum of the transition temperature T_c~370°C and d_{33}~80 pC/N for La doped BNT ceramic composition $(Bi_{0.5}Na_{0.5})_{(1-1.5x)}La_xTiO_3$ for x = 1 at% [118]. Bismuth sodium titanate has been found to be useful for piezoelectric device applications in the rhombohedral phase [14]. Xiao et al. [119, 120] reported that BNT exhibits d_{33} = 231 pC/N and k_p = 0.41 maintaining high depolarization temperature to produce lead-free middle frequency filter with performance comparable to PZT-based filter [121]. Yin et al. reported that doping of Li and Sr at B-site in the BNT ceramic composition $([Bi_{0.5}(Na_{0.82-x}K_{0.18}Li_x)_{0.5}]_{1-y}Sr_y)TiO_3$ with x = 0.10 and y = 0.05 in ε_r = 1480, P_r ~ 36 μC/cm^2, d_{33} = 271 pC/N, and planar coupling factor k_p = 0.38 [122]. No doubt BNT system is showing its amazing dielectric and piezoelectric properties but its low depolarization temperature minimizes its applications [94]. High T_c materials are promising for high temperature sensing applications. However, BNT based materials are promising for high power applications because of the large coercive field which can withstand with high voltage under hard drive conditions. Sidak et al. annealed the BNT ceramics 400 to 500oC in vacuum which significantly enhanced the maximum dielectric constant ~50000 at a measurement frequency of 0.5 kHz [123]. Chaou et al. showed that the doping of Sn in BNT $(Bi_{0.5}Na_{0.5})(Ti_{1-x}Sn_x)O_3$ increases the randomness in the B-site of BNT and reported the conductivity increases ~61 times than that of the pure BNT which will be useful for electrochemical oxide fuel cell application [124]. Mesrar et al. sintered BNT hydrothermally at 1000°C for 4 h and reported ε_r = 2250 [125]. Very recently in February 2023, Lai et al synthesized the textured Nb doped BNT ceramics using reactive template grain growth (RTGG) method in which $Bi_4Ti_3O_{12}$ particles were used as templates [126]. They observed that the dnor doping reduces the concentration of oxygen vacancies and defects which significantly enhances the electrostrain ~0.65% and field induced strain constant d33*~1083 pm/V at room temperature. Such a large values of electro-strain are promising for development of environmental friendly actuators [126].

Table 15.3 List of bismuth sodium titanate (BNT) based compositions

Compositions	**ε_r @ 1 kHz**	**T_c (°C)**	**d_{33} (pC/N)**	**k_p**	**Ref.**
Mn:BNT			120		[127]
0.7BNT-0.3BKT		365	160		[128]
BNT-BT	1230		283		[129]
$(Bi_{0.5}Na_{0.5})TiO_3$-$Ba(Zr,Ti)O_3$	8814	244	147		[130, 131]
Mn:BNT-BT			486		[132]
$(Bi_{0.5}Na_{0.5})TiO_3$	302.6	310	64		[130]
$(1-x)(Bi_{1/2}Na_{1/2})TiO_{3-x}$ $Bi(Mg_{2/3}Nb_{1/3})O_3$ (x=0-1.5 mol%)	469 **(x=0.5%)**		94 **(x=0.7%)**		[133]
$Na_{0.5}Bi_{0.5}TiO_3$	**0.5 kHz** 50,000				[123]
$(Bi_{0.5}Na_{0.5})TiO_3$	2250				[125]
$(Bi_{0.5}Na_{0.5})TiO_3$	~130				[134]

Compositions	ε_r @ 1 kHz	T_c (°C)	d_{33} (pC/N)	k_p	Ref.
$(Bi_{0.5}Na_{0.5})_{(1-1.5x)}La_xTiO_3$ (x = 1.72)	550		92		[118]
$0.08(Bi_{0.5}Na_{0.5})TiO_3$-$0.02NaNbO_3$ [BNT-NN]	624 (10 kHz)		88	0.179	[117]
$(Bi_{0.51}Na_{0.49})Sc_{0.02}Ti_{0.98}O_3$	431	358	74.7		[135]
$\{[Bi_{0.5}(Na_{0.82-x}K_{0.18}Li_x)_{0.5}]_{1-y}Sr_y\}TiO_3$ x = 0.10 and y = 0.05	~1480		271	0.38	[122]
$(Bi_{0.5}Na_{0.5})(Ti_{1-x}Sn_x)O_3$ (x = 1)	>2750 (1 MHz)				[124]
$[Bi_{0.5}(Na_{0.8}K_{0.2})_{0.5}](Ti_{1-x}Nb_x)O_3$ (x = 0.015)	4834		1083 pm/V		[126]

15.9.4 Combinations of Different Ferroelectric Families

Better piezoelectric properties can be achieved by combining different compositions like barium titanate with bismuth ferrate ($BaTiO_3$-$BiFeO_3$), bismuth sodium titanate with bismuth potassium titanate (BNT-BKT), and many more. Table 15.4 shows tentative list of several different combinations that showed better properties rather than their pure or doping forms.

Zang et al. reported appreciable peizoelectric and ferroelectric properties like dielectric constant ~1370, transition temperature T_c ~385°C, and high coupling factor k_p= 0.51 of KNN ceramic with lithium antimony [KNN-$LiSbO_3$]. [136]. The electromechanical properties of 0.9BNT-0.05BKT-0.05BT composition are good enough to replace PZT in certain applications such as ultrasonic transducers, wire-bonding, and accelerometers [137, 138]. There have been combinations related to BNT-BKT or BNT-BT on which ultrasonic transducers were developed. Wu et al. [124] used the KNN material doped with lithium, tantalum, and antimony $K_{0.38}Na_{0.58}Li_{0.04}Nb_{0.91}Ta_{0.05}Sb_{0.04}O_3$ and obtained ε_r = 1327, T_c = 337°C, d_{33} = 306 pC/N, and high coupling factor k_p = 0.48 [139]. Lin et al. synthesized the composition KNN with barium zirconate BZT, $0.94K_{0.5}Na_{0.5}NbO_3$-

$0.06Ba(Zr_{0.05}Ti_{0.95})O_3$, obtaining ε_r = 1191, T_c = 318°C, d_{33} = 234 p/N, and high coupling factor k_p = 0.49 [140]. Zuo et al. used the BNT-BT composition with tantalum oxide 0.995(0.94BNT-0.06BT) -$0.005Ta_2O_5$ and observed dielectric constant ε_r = 1861, T_c = 273°C, d_{33} = 171 pC/N, and k_p = 0.33 [141]. Du et al. fabricated KNN ceramic with $BiScO_3$ $0.9825Na_{0.5}K_{0.5}NbO_3$-$0.0175BiScO_3$, achieving ε_r = 1000, T_c = 351°C, d_{33} = 253 pC/N, and coupling factor k_p = 0.48 [142]. Zhang et al. synthesized KNN material doped with lithium and tantalum $(K_{0.55}Na_{0.45})_{0.965}Li_{0.035}Nb_{0.8}Ta_{0.2}O_3$ and achieved ε_r = 1290, T_c = 325°C, d_{33} = 262 pC/N, and coupling factor k_p = 0.53 [143]. Wang et al. reported calcium niobium titanate with bismuth doping $CaBi_2NbTiO_9$ and reported ε_r = 97 with Curie temperature T_c = 936°C which can be used as fire sensors [144]. Sun et al. fabricated BNT with BKT with 96/4 mole ratio NKBT96/4 [$0.96(Na_{0.5}Bi_{0.5})TiO_3$-$0.4(K_{0.5}Bi_{0.5})TiO_3$] and reported dielectric constant at room temperature and the maximum constant value $\varepsilon_{RT}/\varepsilon_{max}$ =766/3656, respectively, with P_r = 10.46 μC/cm^2 and d_{33} = 120 pC/N [145]. Yin et al. in 2017 reported BNT-BLT with $SrTiO_3$ (ST) achieving ε_r ~ 1480 with P_r ~ 36 μC/cm^2 and d_{33} ~ 271 pC/N [122]. Xue et al. reported d_{33} value of ~600 pC/N by using the BNT-BT single crystal system with Fe doping. Doping of Fe results in making nano domains crystal systems leading to higher domain density and getting high piezoelectric response (d_{33}) with high remnant polarization P_r ~ 40 μC/cm^2. The nanodomains were measured using a piezoelectric force microscope (PFM) and performed in AC resonance tracking PFM modes which reduced morphological deviation while looking out for electromechanical properties [146]. Wang et al. reported the BF-BT composition with BKT+Mn $0.725BF$-$0.25BT$-$0.025Bi_{0.5}K_{0.5}TiO_3$+Mn and obtained ε_r = 135 and high Curie temperature T_c = 708°C [147]. Kunnath et al. synthesized boron sodium gadolinium niobate (BNGN) at morphotropic phase boundary (MPB) $(1\text{-}x)B_{0.5}Na_{0.5}GdO_3$ – $xB_0.5Na_{0.5}NbO_3$ and obtained ε_r = 1733 ± 8 and d_{33} = 556 ± 7 pC/N [148]. Chen et al. fabricated the BNT ceramic composition with $SrLaTiO_3$ $(1–x)(0.9Bi_{0.5}Na_{0.5}TiO_3$-

$0.1Na_{0.8}Sr_{0.1}NbO_3$)-x($Sr_{0.85}La_{0.1}TiO_3$) [BNT-NSN-xSLT] by a viscous polymer process and reported dielectric constant ε_r = 2600 and remnant polarization $P_r \sim 45\ \mu C/cm^2$ [149].

Table 15.4 List of several combinations-based compositions

Compositions	ε_r @ 1 kHz	T_c (°C)	d_{33} (pC/N)	k_p	Ref.
$0.66BiFeO_3$-$0.3BaTiO_3$-$0.04Bi(Mg_{0.5}Ti_{0.5})O_3$		482	184	0.28	[150]
0.85BNT-$0.15(Bi_{0.5}Li_{0.5})TiO_3$			110	0.18	[151]
$0.995(Bi_{0.5}Na_{0.5})TiO_3$ [BNT] -$0.005Ba(Cu_{0.5}W_{0.5})O_3$	328	308	80	0.18	[152]
0.98BNT-$0.02NaNO_3$	624		88	0.18	[117]
0.82BNT-0.18 $(Bi_{0.5}K_{0.5})TiO_3$ [BKT]	893		144	0.29	[153]
0.77BNT-0.23BKT+1 mol% Bi_2O_3	1261	330	207	0.29	[154]
0.88BNT-0.12BKT+0.2 wt% CeO_2			132	0.27	[155]
0.82BNT-0.16BKT-0.02KN	1437		215	0.35	[156]
0.805BNT-0.18BKT-$0.015BiCrO_3$			168		[157]
0.7BNT-0.2BKT-$0.1(Bi_{0.5}Li_{0.5})TiO_3$	119		231	0.36	[120]
0.7BNT-0.2BKT-$0.1(Bi_{0.5}Li_{0.5})TiO_3$+$Na^+$		334	171	0.30	[158]
0.9(0.79BNT-0.21BKT)-$0.1(Bi_{0.5}Ag_{0.5})TiO_3$	1110		160	0.30	[159]
0.94BNT-0.06BT	826	288	155	0.36	[160]
0.94BNT-0.06BT+0.4 wt% CeO_2	652		152	0.34	[161]

(*Continued*)

Table 15.4 (*Continued*)

Compositions	ε_r @ 1 kHz	T_c (°C)	d_{33} (pC/N)	k_p	Ref.
0.92BNT-0.08BT+0.3 wt% MnO	1596		153	0.36	[162]
0.93BNT-0.07BT+0.8 wt% CoO			137	0.23	[163]
0.94BNT-0.06BT+0.5 wt% Co_2O_2	1200		139	0.27	[164]
0.985(0.94BNT-0.06BT)-0.0075La_2O_5	1520	270	133	0.29	[165]
0.93BNT-0.07BT+0.16 wt% In_2O_3	1046		205		[166]
0.92BNT-0.08BT+Nb_2O_5	1230	250	149		[135]
0.995(0.94BNT-0.06BT) -0.005Ta_2O_5	1861	273	171	0.33	[141]
0.865BNT-0.06BT-0.075$(Bi_{0.5}Li_{0.5})TiO_3$		260	208	0.36	[86]
0.94[0.94BNT-0.06$(Bi_{0.5}Ag_{0.5})$ TiO_3]-0.06BT		265	168	0.31	[167]
0.92BNT-0.06BT-0.02$(Y_{0.5}Na_{0.5})TiO_3$		267	159	0.28	[168]
0.91BNT-0.09$Ba(Ti_{0.942}Zr_{0.058})O_3$	881	244	147		[169]
0.92BNT-0.08$Ba(Ti_{0.95}Hf_{0.05})O_3$		259	135		[170]
0.95BNT-0.02BT-0.03$NaNO_3$			110		[171]
0.9BNT-0.05BKT-0.05BT	700		148	0.34	[172]
0.895BNT-0.07BKT-0.035BT			150	0.29	[173]
0.85BNT-0.12BKT-0.03BT			149	0.28	[174]
0.85BNT-0.12BKT-0.03BT+Co_2O_3			~143	0.25	[175]
$(Bi_{0.98}Na_{0.755}K_{0.15}Li_{0.075})_{0.5}Ba_{0.02}TiO_3$	1040		205	0.29	[176]
$(Bi_{0.95}Na_{0.75}K_{0.15}Li_{0.05})_{0.5}Ba_{0.02}TiO_3$	1510		194	0.17	[177]
0.92$Na_{0.535}K_{0.48}NbO_3$-0.08$LiNbO_3$ (domain engineering)			324		[178]

Compositions	ε_r @ 1 kHz	T_c (°C)	d_{33} (pC/N)	k_p	Ref.
$(K_{0.55}Na_{0.45})_{0.965}Li_{0.035}Nb_{0.8}Ta_{0.2}O_3$	1290	325	262	0.53	[143]
$0.948K_{0.5}Na_{0.5}NbO_3$-$0.052LiSbO_3$	1372	385	286	0.51	[136]
$K_{0.38}Na_{0.58}Li_{0.04}Nb_{0.91}Ta_{0.05}Sb_{0.04}O_3$	1327	337	306	0.48	[139]
$0.91K_{0.5}Na_{0.5}NbO_3$-$0.09AgSbO_3$+0.75mol% MnO_2	1400	215	216	0.44	[179]
$K_{0.42}Na_{0.52}Li_{0.04}Ag_{0.02}Nb_{0.91}Ta_{0.05}Sb_{0.04}O_3$	1478	353	263	0.45	[180]
$(K_{0.47}Na_{0.47}Li_{0.06})_{0.985}Bi_{0.005}NbO_3$	1020	455	185	0.43	[181]
$0.98Na_{0.475}K_{0.475}Li_{0.05}NbO_3$-$0.02Bi_{0.48}Na_{0.48}Ba_{0.04}TiO_3$	702	415	328	0.48	[182]
$0.97Na_{0.5}K_{0.5}NbO_3$-$0.03Bi_{0.5}Na_{0.5}TiO_3$.		375	195	0.43	[183]
$0.9825Na_{0.5}K_{0.5}NbO_3$-$0.0175BiScO_3$.	~1000	351	253	0.48	[142]
$0.94K_{0.5}Na_{0.5}NbO_3$-$0.06Ba(Zr_{0.05}Ti_{0.95})O_3$	1191	318	234	0.49	[140]
$0.94K_{0.5}Na_{0.5}NbO_3$-$0.06BaTiO_3$+1mol% CuO	230	314	193	0.43	[184]
$0.95(K_{0.95}Na_{0.5})$ NbO_3-$0.05LiNbO_3$		426	405		[185]
$Ca_{1-x}(KCe)_{x/2}Bi_2Nb_2O_9$ (x = 0.1)	~250	868	16	0.086	[144]
$Ca_{0.9}(K,Ce)_{0.05}Bi_2NbTiO_9$	118	868	16		[144]
$Bi_4Ti_{2.98}V_{0.02}O_{12.01}$	170	674	20		[186]
$Bi_4Ti_{2.975}W_{0.025}O_{12.025}$+0.2wt% Cr_2O_3	178	658	22		[187]
$Na_{0.5}Bi_{4.5}Ti_4O_{15}$	153	668	18		[188]
$Na_{0.5}Bi_{4.5}Ti_4O_{15}$+0.5wt% CeO_2	146	655	28		[188]
$Na_{0.5}Bi_{4.5}Ti_4O_{15}$+0.3wt% CoO	152	663	30		[189]
$CaBi_4Ti_4O_{15}$	143	789	7		[190]
$CaBi_4Ti_{3.975}W_{0.025}O_{15}$	158	789	10		[190]

(Continued)

Table 15.4 (*Continued*)

Compositions	ε_r @ 1 kHz	T_c (°C)	d_{33} (pC/N)	k_p	Ref.
$Ca_{1-x}Sr_xBi_4Ti_4O_{15}$+1.5 mol% MnO_2	122	677	21		[191]
$CaBi_4Ti_{3.95}V_{0.05}O_{15}$		791	14		[192]
$CaBi_{3.75}Nd_{0.25}Ti_4O_{15}$	159	756	12		[193]
$Bi_7Ti_4NbO_{21}$	150	841	13		[194]
$Bi_{6.25}La_{0.75}Ti_4NbO_{21}$	180	700	17		[194]
$(Bi_{0.5}Na_{0.5})_{0.94}Ba_{0.06}TiO_3$+ x%$Ce_2O_3$+ y%$La_2O_3$ + z%Y_2O_3 [BNT-BT-x/y/z]	At T_m/ RT	Tm			[195]
0.5/0.25/0.25	2874/ 1050	316	95		
(1-x) $B_{0.5}Na_{0.5}GdO_3$ – $xB_{0.}5Na_{0.5}NbO_3$	1733 ± 8		556 ± 7		[196]
$BiFeO_3$-x$BaTiO_3$, Mn doped					[147]
X = 0.31	704	598	331		
$BiFeO_3$-x$BaTiO_3$, undoped					
X = 0.33	795	605	151		
[$(Na_{0.5}Bi_{0.5})TiO_3$-$(K_{0.5}Bi_{0.5})TiO_3$-$BaTiO_3$]	730	290	173		[197]
[0.96$(Na_{0.5}Bi_{0.5})TiO_3$-0.4$(K_{0.5}Bi_{0.5})TiO_3$]	766_{RT}/ 3656_m		120		[145]
0.96$(K_{0.5}Na_{0.5})$0.95$Li_{0.05}$ $Nb_{1-x}Sb_xO_3$-0.04$BaZrO_3$		197	425	0.50	[198]
0.5 wt% Mn-$(K_{0.5}Na_{0.5})NbO_3$		416	350		
(Na,K)(Nb,Ta)O_3		291	200		
0.95$(Na_{1/2}Bi_{1/2})TiO_3$-0.05$BaTiO_3$			360	0.72	
Fe-doped 0.95$(Na_{1/2}Bi_{1/2})TiO_3$-0.05$BaTiO_3$			600		[146]
0.83$Na_{0.5}Bi_{0.5}TiO_3$-0.17$Bi_{0.5}K_{0.5}TiO_3$			161		
0.94$(Bi_{0.5}Na_{0.5})TiO_3$-0.06$BaTiO_3$-x$Bi_4Ti_3O_{12}$(BNT-BT-xBiT)		675	152		

Compositions	ε_r @ 1 kHz	T_c (°C)	d_{33} (pC/N)	k_p	Ref.
$BaBi_4Ti_4O_{15}$-xCeO_2		407	24.2		
$CaBi_4Ti_{3.95}Nb_{0.05}O_{15}$+ 0.2 wt% MnO_2		790	23	0.047	
$BaBi_4Ti_4O_{15}$		395	24	0.036	
$Ca_{1-x}(Li,Ho)_{x/2}Bi_4Ti_4O_{15}$		814	10.2		
$(Na_{0.5}Bi_{0.5})_{0.92}Ba_{0.08}TiO_3$		280	125		[199]
$(Na_{0.5}Bi_{0.5})_{0.92}Ba_{0.08}TiO_3$+ x$Nb_2O_5$		250	149		[200]
$(Na_{0.5}Bi_{0.5})$0.94-6BaTiO_3+0.5 mol% CeO_2+0.5 mol% La_2O_3			162		[199]
$(Na_{0.5}Bi_{0.5})_{0.92}$-$Ba_{0.08}TiO_3$+x mol% $MnCO_3$		243	160		
$(K_{0.5}Bi_{0.5})_{0.92}$-$Ba_{0.08}TiO_3$+x mol% $MnCO_3$		174	117		[201]
$(Bi_{0.5}Na_{0.5})TiO_3$-$(Bi_{0.5}K_{0.5})TiO_3$-$BaTiO_3$ (1-3x)BNT-2xBKT-BT		125	150		[202]
$(Bi_{0.5}Na_{0.5})TiO_3$-Ba(Ti,Zr)O_3		244	147		[130]
$Bi_{0.5}(Na_{0.84}K_{0.16})_{0.5}TiO_3$+xSr TiO_3		292	185		[203]
$Bi_{1-z}(Na_{1-x-y-z}K_xLi_y)]_{0.5}Ba_zTiO_3$ ($_{BN-x/y/z}$) BN-0.15/0.075/0.02		210	205		[204]
$(K_{0.5}Na_{0.5})_{1-2y}AE_yNbO_3$ [AE=Mg,Ca,Sr, Ba]		400	95		[205]
(1-x)$BaTiO_3$-x$(Bi_{0.5}K_{0.5})$ TiO_3+$MnCO_3$ [BTBK-20+Mn]		223	58.9		[201]
(1-x)$BaTiO_3$-x$(Bi_{0.5}K_{0.5})$ TiO_3+$MnCO_3$ [BTBK-5+Mn]		174	117		[201]
0.75BiFeO_3-0.25BaTiO_3		574	47		[147]
0.67$B_{1.05}F_{0.97}Ga_{0.03}$-0.33T		454	402		
0.95(0.72BNT0.18BKT-0.10BLT)-0.05ST	~1480		~271	~0.38	[122]
(1–x)(0.9$Bi_{0.5}Na_{0.5}TiO_3$-0.1$(Na_{0.8}Sr_{0.1}NbO_3)$-x$(Sr_{0.85}La_{0.1}TiO_3)$ [BNT-NSN-xSLT] (x = 0.15)	~2600				[149]

15.10 Applications

We have seen in Sections 15.9.1–15.9.4 that the lead-free piezoelectric materials are being improvised by doping of different elements/compositions and extensive efforts have been put to replace lead-based piezoelectrics. Applications of lead-free piezoelectric materials are extensive and pervasive, dominating from industrial and commercial applications to the growing need of the digital world. They are proving their abilities of low cost, high reliability, better response to electrical and mechanical effects, and miniaturized size as well. Inorganic oxide ceramics are the promising ones due to their cationic nature and oxygen bonding effects. The first perovskite piezoelectric ceramic, $BaTiO_3$, showed unusually high ε_r and has been used in multi-layer ceramic capacitors (MLCCs). BNT-based ceramics showed a large value of the coercive field and high T_c which are suitable for high power applications such as actuators and transducers. KNN-based ceramics showed high k_p and strong anisotropy lead advantages for acoustic transducer applications and are being used in the healthcare sector as well [94]. These materials can be further improved by adding dopants and achieving the coexistence of multiphase (CMP). In the race for development of lead-free materials, KNN and BNT are more focused. Piezoelectric polymers like polyvinylidene fluoride (PVDF) nanocomposite and poly(lactic acid) (PLA) are being used as foil sensors and transducers, respectively [206]. As our global physical world is transforming into a digital era, so the need for extensive storage of data, sensor applications, and cumulative energy harvesting procedures are also in increasing demand. The potential of lead-free piezoelectric ceramics, including both inorganic and organic types, is also evolving and being tested [207]. In addition to the huge demand for piezoelectric materials in sonars, medical ultrasound, and nano-positioners, they are also being explored for futuristic applications such as mini robotic parts to miniaturize the size of the devices as well as in fulfilling the requirement of micro-electro-mechanical systems (MEMS) [107, 112, 208, 209]. The future of digitization is to produce low-cost materials with excellent properties and small sizes. The piezoelectric industries are trying to miniaturize the electrical components to its extent and leading the world into a nanotechnological era, thus producing nano-generators, nano-transducers, and even IC chips (integrated circuits) in every electrical and/or mechanical device application. A few applications of lead-free piezoelectric/ferroelectric ceramics are discussed in the following sections.

15.10.1 Capacitor Applications

A capacitor is the fundamental component in an electrical circuit and is characterized by its charge (energy) storing capacity. As digitization needs are growing day by day, data is increasing every second, which requires storing that data in any digital format whether cloud drive or hard drive. Ceramics based on barium titanate dominate the capacitor industries. Barium titanate is known for its high ε_r (ε_r > 1100) since its discovery in the 1940s and its low dielectric loss characteristics. BT-based multi-layer ceramic capacitors (MLCCs) are mostly fabricated by tape casting procedures which result in miniaturized size capacitors. One of the prominent applications of ferroelectric ceramics is the ferroelectric random access memory (FeRAM) or the dynamic random-access memory (DRAM). A "Y5V" capacitor made of BT doped with La and Ce exhibited ε_r >10,000, which is accompanied and considered to be used for capacitors and DRAM-like applications [48]. It has already been reported that the BT-based ceramics show the matching properties with the commercial PZT [85]. Li et al. fabricated the BNT-0.45SBT multi-layer ceramic capacitors with the help of the tape casting method. Sintering at 1100–1200°C by the conventional solid-state method, they reported the high energy density acquiring capacitor ~9.5 J/cm^3 with an efficiency over 95% and the dielectric permittivity of 2000. The tape casting method is found to be beneficial in producing ceramic tapes with lower porosity and enhanced grain size and leads to high dielectric breakdown strength, which is needed in MLCC. Owing to its high energy storing density, the BNT-0.45SBT can be used for high power electronic devices and high energy storage applications [210]. Li et al. fabricated the BNT ceramic doped with SZSHTN-xNN (1−x) $(0.75Na_{0.5}Bi_{0.5}TiO_3)$-$0.25Sr(Zr_{0.2}Sn_{0.2}Hf_{0.2}Ti_{0.2}Nb_{0.2})O_3$-$xNaNbO_3$

exhibited good piezoelectric properties in pseudo cubic phase with low loss factor tanδ < 0.01 in the temperature range of 48.2–303.4°C under the influence of 293 kV/m field. Results suggested that increasing of NN doping also increases the ε_r value, possessing high energy density ~2.4 J/cm^3 with the efficiency of 71%. It was found that the operating temperature range –35.4–224.3°C of (1–x) $(0.75Na_{0.5}Bi_{0.5}TiO_3)$-$0.25Sr(Zr_{0.2}Sn_{0.2}Hf_{0.2}Ti_{0.2}Nb_{0.2})O_3$-$xNaNbO_3$ ceramics satisfied the "Y9P" standard capacitors and can be used in high temperature capacitors. Moreover, its discharging process makes it useful for application in pulse powered devices [211].

15.10.2 Piezoelectric Generators

Generators work on the direct piezoelectric effect, which can transform mechanical energy to electrical energy (Fig. 15.8). Gas ignitors, accelerometers, microphones, biosensors, ultrasound imagers, knock sensors, and acoustic emission sensors are some of the prominent applications of piezoelectric generators [19, 212]. Piezoelectric materials with high d_{33} and with low dielectric loss factor are considered useful for generator applications.

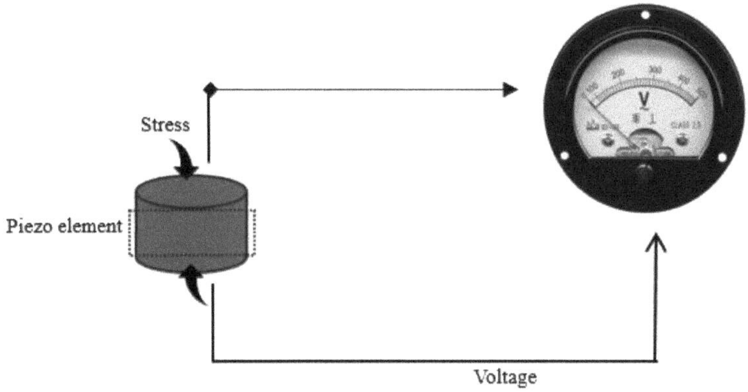

Figure 15.8 Direct piezoelectric effect.

In the knock sensors, piezoelectric material is placed between the vibrating structure and a seismic mass introducing the vibrational forces into the piezo element. The piezo element converts the vibrations into an electric charge proportional to the applied force.

These sensors need to have high temperature absorption/control power to work efficiently. Organic materials like PVDF are proving their abilities to be included in the nanogenerator applications leading into the IoT sector [213]. Won et al. demonstrated a new way of high voltage and high power generation with flexible devices by using BNKT nanoparticle filler polymethyl siloxane (BNKT-PDMS) composite on cellulose paper [214]. The BNKT-PDMS showed $d_{33} = 15$ pC/N, which is lower than the pure BNKT ceramics, and a maximum strain value of 0.17% obtained at electric field 60 kV/cm. Comparing with other ceramics compositions, BNKT-PDMS demonstrated large output voltage of 100 V, with an average current of ~20 μA through tapping stimulation. These findings suggested that paper-based BNKT-PDMS composites can be used for high power generators with improved power output and efficiency. The BNKT-PDMS can also be used for developing self-powered wearable devices as it possesses a large value of piezoelectric voltage constant $g_{33} \sim 80$ $mVmN^{-1}$, which is four times higher than pure BNKT ceramics [214]. Kim et al. prepared a lead-free fiber generator by melt spinning process of BCZT-PVDF composite fiber [215]. Piezoelectric constant $d_{33} \sim 1200$ pC/N with $k_p = 0.28$ were obtained for BCZT-PVDF fiber generator which gives output voltage 1.9 V, which was 1.6 times higher than that of pure PVDF generator. The results show that the technology of flexible fiber generators using lead-free ceramics as additives has great potential [215]. The discovery of nanogenerators in 2006 paved the path to harvest energy and use in the other forms [216]. Triboelectric generators (TENG) are a new alternative of energy harvesting owing to their capabilities of converting mechanical energy into electrical power [217]. Today, the world is surrounded by smart wireless devices that require a sustainable and ecological source of energy. Piezoelectric generators are emerging as highly efficient, cost-effective, and easy-to-produce power sources.

15.10.3 Piezoelectric Motors

It is one of the piezoelectric applications which work on the indirect (or converse) effect (Section 15.5). There are many applications in which it is used like miniature ultrasonic motors, actuating purposes, positioners, and inkjet printers, etc. (Fig. 15.9). The parameters in

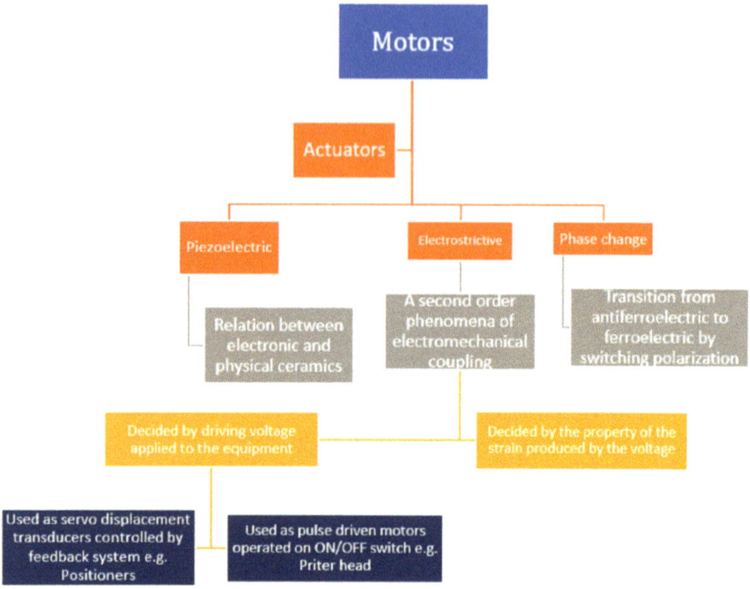

Figure 15.9 Piezoelectric motors paradigm.

converse effect that have been checked for measuring capability of the material to do desired work are k_p and d_{33} (in pm/V). BNT-BKT-BT systems show large d_{33}, k_p, and coercive field and can be used in high power actuator applications. Piezoelectric devices have become crucial for smart actuating devices. Piezoelectric actuators are reliable to be used in tracking magnetic heads, micro-positioners for robotic technology, different machines, commercial tools, and adaptive optic systems [218]. Piezoelectric actuators are unique in their own sense as they can produce displacements up to several hundreds of microns with high accuracy (±0.01 mm), large generative force (400 kg/cm2) with short response time (~10 ms), and utilizes much less power than the electromagnetic motors. Hence the piezoelectric materials with large actuating strain are found to be useful in vibration suppressors, motors, and positioner devices. The barium stannate titanate system (Ba(Sn,Ti) O_3) showed linear strain behavior at higher electric field and the reorientation of domains is observed only at low electric fields [219]. Such unusual strain response of barium stannate titanate has led to the development of a new family of actuators. Micromotors based on multilayer piezoelectric actuators are more reliable as they have shorter stabilization time. This technique is utilized in diesel injection valve systems.

Zheng et al. fabricated the KNN material with other dopants for large d_{33} and high strain with wide temperature range for commercial and industrial applications. They have reported high d_{33}* of 599–1553 pm/V (depending upon different dopants) with high strain of 0.18–0.46% accompanied with phase boundary of rhombohedral to tetragonal (R-T), making it a suitable material for electromechanical actuator applications. High value of d_{33} was obtained at the R-T phase boundary and in the absence of Li^+. High piezoelectric strain of 0.46% with an electric field of ~3 kV/mm and high T_c (~250°C) are appreciable factors to be considered for practical applications [88].

15.10.4 Energy Harvesting

The introduction of MEMS and improvisation in semiconductor industry led to miniaturized components for portable device like smartphones, printers, laptops, sensors, and transducers. Nowadays, every electronic device is required to be interconnected through a network. This network system needs to be synced with real time information through internet. Such kind of ecosystem is known as Internet of Things (IoT) [220–222]. A power source is necessary to run the components of such ecosystem round the clock. Batteries are one of the options to provide constant power supply. However, it was found to be unfeasible due to high cost and availability in difficult terrains. Batteries are also non-replenishable; hence, we need to harvest the energy from any external source. Extracting and storing the energies from the solar, thermal, vibrational, or radio frequency (RF) sources can provide good sustainability of such devices (Fig. 15.10). One of the primary objectives of energy harvesting is to make the self-powered wireless sensor networks (WSN). Considering that the energy conversion efficiency of pure polymer materials is quite low, which prevents their use as mechanical-to-electrical transducers, its insertion as a piezoelectric active secondary phase to create organic/inorganic composites is a promising strategy for the design of new piezoelectric energy harvesters [206, 223–225]. The piezoelectric potential generated inside a piezoelectric material under external stress will generate a transient charge flow from the external load. Efficiency and power density of a piezoelectric vibrational energy harvesting device are strongly frequency dependent because piezoelectric generates maximum power at its resonance frequency. The fundamental resonance frequency determines the size of the piezoelectric element of a piezoelectric energy harvesting unit [226].

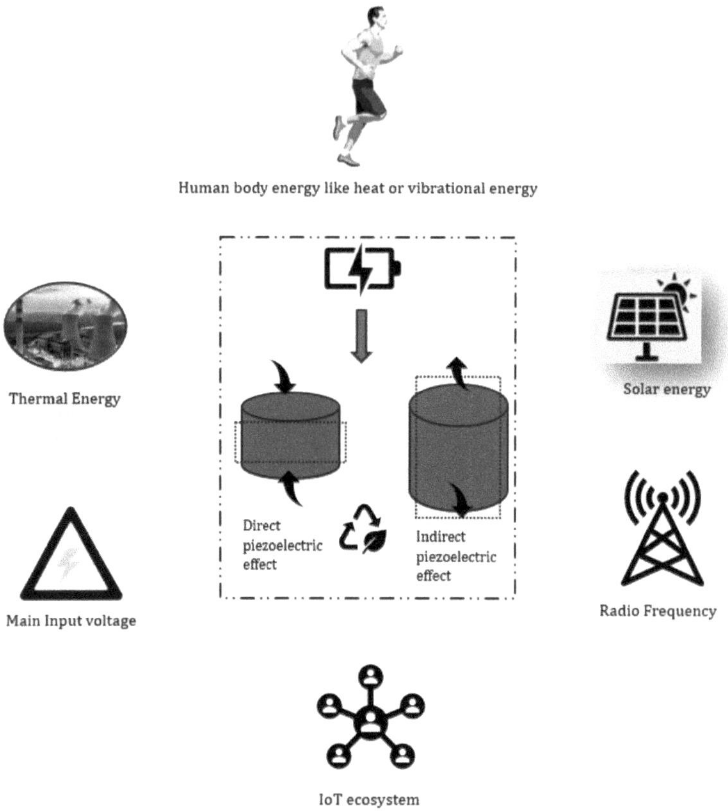

Figure 15.10 Factors replacing batteries in IoT devices.

Wang et al. developed a nanocomposite of KNN-epoxy resin with carbon-fiber-reinforced-polymer (CFRP) electrodes for piezoelectric generator. The CFRP composites have high electrical conductivity and mechanical stability and hence can be used in place of Au electrodes. The KNN-resin nanocomposite-based generator with CFRP electrodes showed a power generation of ~0.8 μW/cm^3 at 60 Hz with a d_{33} of 5.8 pC/N. The KNN piezoelectric/ CFRP nanogenerators are found to be promising for energy harvesting, and particularly for structural health monitoring of aircraft and spacecraft [227].

15.11 Conclusions

Recent advances in the development and characterization of BT, KNN, BNT based perovskite materials with excellent dielectric and piezoelectric properties have been discussed. There is still a long way ahead for lead-free piezoelectric materials to overcome the lead-based piezoelectric materials regarding their electromechanical response and temperature range of application. The barium titanate-based compositions exhibited good piezoelectric properties but were constrained by a low Curie temperature. Sodium potassium niobate (KNN) and bismuth sodium titanate (BNT) that come under the perovskite type ceramics seem to be suitable for actuator and high-power applications because of their high Curie temperature; however, it requires improvement in piezoelectric properties. Lead-free piezoelectric materials are believed to take place of lead-based piezoelectric materials in the future by improving the piezoelectric performance and enhancing the ferroelectric to paraelectric phase transition temperature of lead-free piezoelectric materials through compositional engineering and processing techniques.

References

1. Haertling G H 1999 Ferroelectric ceramics: History and technology *J. Am. Ceram. Soc.* **82** 797–818.
2. Brennecka G, Sherbondy R, Schwartz R and Ihlefeld J 2020 Ferroelectricity – A revolutionary century of discovery *Am. Ceram. Soc. Bull.* **99** 24–30.
3. Uchino K 2009 *Ferroelectric Devices* (CRC Press).
4. Shrout T R and Zhang S J 2007 Lead-free piezoelectric ceramics: Alternatives for PZT? *J. Electroceramics* **19** 111–24.

5. Rose G 1839 Beschreibung einiger neuen Mineralien des Urals *Ann. der Phys. und Chemie* **124** 551–73.
6. Tilley R J D 2016 *Perovskites Structure–Property Relationships* (Wiley).
7. Zheng T, Wu J, Xiao D and Zhu J 2018 Recent development in lead-free perovskite piezoelectric bulk materials *Prog. Mater. Sci.* **98** 552–624.
8. Goldschmidt V M, Barth T, Lunde G and Zachariasen W 1926 Geochemische verteilungsgesetze der elemente VII. *Skrifter der Norske Videnskaps Akademi Klasse. Matematisk Naurvidenskaplig Klasse.*
9. Cross L E and Newnham R E 1987 History of ferroelectrics *Ceram. Civiliz.* **III** 289–305.
10. Bajpai K K, Sreenivas K, Gupta A K and Shukla A K 2019 Cr-doped lead lanthanum zirconate titanate (PLZT) ceramics for pyroelectric and energy harvesting device applications *Ceram. Int.* **45**(11) 14111–20.
11. Shukla A K, Agrawal V K, Das I M L, Singh J, Singh D P and Sood K N 2006 Structural, dielectric and electromechanical properties of Cr-doped PLZT close to the morphotropic phase boundary region *Phase Transitions* **79** 875–87.
12. Shukla A K, Agrawal V K, Das I M L, Singh J and Srivastava S L 2011 Dielectric response of PLZT ceramics x/57/43 across ferroelectric-paraelectric phase transition *Bull. Mater. Sci.* **34** 133–42.
13. Shukla A K, Agrawal V K, Das I M L, Singh J and Srivastava S L 2010 Temperature dependence of electromechanical properties of PLZT x/57/43 ceramics *Bull. Mater. Sci.* **33** 383–90.
14. Lu Y-Q and Li Y-X 2011 A review on lead-free piezoelectric ceramics studies in China *J. Adv. Dielectr.* **01** 269–88.
15. Konoval G J 2006 Electronic waste control legislation: Observations on a new dimension in state environmental regulation *58 A. F. L. Rev.* **147-174**.
16. Curie J and Curie P 1880 Development by pressure of polar electricity in hemihedral crystals with inclined faces *Bull. soc. min. de France* **3** 90.
17. Talley K R, Perkins C L, Diercks D R, Brennecka G L and Zakutayev A 2020 Synthesis of ferroelectric $LaWN_3$ – the first nitride perovskite arxiv (Cornell University) 1–26, https://arxiv.org/pdf/2001.00633.
18. Brewster 1824 Observation of pyroelectricity of minerals, *Edinburgh J. Sci.* **1** 208–14.
19. Shukla A K and Srivastava S L 2012 Smart Materials: Piezoelectrics, in *Synthesis, Characterization and Application of Smart Materials*, ed R Rai, (Nova Science Publishers Inc.) pp 155–200.

20. Hankel W G 1881 *Über die aktino- und piezoelektrischen Eigenschaften des Bergkrystalles und ihre Beziehung zu den Thermoelektrischen* (S. Hirzel).

21. Cady W G 1946 *Piezoelectricity: An Introduction to the Theory and Applications of Electromechancial Phenomena in Crystals* (New York and London: McGraw-Hill).

22. Valasek J 1921 Piezo-electric and allied phenomena in Rochelle salt *Phys. Rev.* **17** 475–81.

23. Valasek J 1922 Piezo-electric activity of Rochelle salt under various conditions *Phys. Rev.* **19** 478–91.

24. Valasek J 1924 Dielectric anomalies in Rochelle salt crystals *Phys. Rev.* **24** 560–8.

25. Thurnauer H 1977 Reflections *Am. Ceram. Soc. Bull.* **56** 861–6.

26. Rödel J, Jo W, Seifert K T P, Anton E M, Granzow T and Damjanovic D 2009 Perspective on the development of lead-free piezoceramics *J. Am. Ceram. Soc.* **92** 1153–77.

27. Valasek J 1922 Properties of Rochelle salt related to the piezo-electric effect *Phys. Rev.* **20** 639–64.

28. Rangarajan G and Vijaya M S 2014 *Materials Science* (McGraw Hill Education).

29. Gordon J N, Taylor A and Bennett P N 2002 Lead poisoning: Case studies *Br. J. Clin. Pharmacol.* **53** 451–8.

30. Chauhan A K S, Gupta V and Sreenivas K 2006 Dielectric and piezoelectric properties of sol-gel derived Ca doped $PbTiO_3$ *Mater. Sci. Eng. B* **130** 81–8.

31. Chilibon I and Marat-Mendes J N 2012 Ferroelectric ceramics by sol-gel methods and applications: A review *J. Sol-Gel Sci. Technol.* **64** 571–611.

32. Praveen J P, Karthik T, James A R, Chandrakala E, Asthana S and Das D 2015 Effect of poling process on piezoelectric properties of sol-gel derived BZT-BCT ceramics *J. Eur. Ceram. Soc.* **35** 1785–98.

33. Kwok K W, Tsang R C W, Chan H L W and Choy C L 2004 Effects of niobium doping on the piezoelectric properties of sol-gel-derived lead-zirconate-titanate films *J. Appl. Phys.* **95** 1372–6.

34. Morita T 2010 Piezoelectric materials synthesized by the hydrothermal method and their applications *Materials (Basel)* **3** 5236–45.

35. Moon J, Li T, Randall C A and Adair J H 1997 Low temperature synthesis of lead titanate by a hydrothermal method *J. Mater. Res.* **12** 189–97.

36. Kanie K, Numamoto Y, Tsukamoto S, Sasaki T, Nakaya M, Tani J, Takahashi H and Muramatsu A 2011 Size-controlled hydrothermal synthesis of bismuth sodium and bismuth potassium titanates fine particles and application to lead-free piezoelectric ceramics *Mater. Trans.* **52** 1396–401.

37. Mimura K, Liu Z, Itasaka H and Kato K 2020 Hydrothermal synthesis of A-site substituted $BaTiO_3$ nanocubes *J. Ceram. Soc. Jpn.* **128**(8) 475–480.

38. Tan I C K 2013 *Synthesis, Structure and Properties of Lead-Free Piezoelectric Ceramics,* Doctoral Thesis, Nanyang Technological University, Singapore. https://doi.org/10.32657/10356/54814.

39. Iqra L, Rashid M S, Ali Q, Latif I and Malik A 2020 Evaluation for Na^+/K^+ ratio under salt stress condition in wheat *Life Sci J.* **17**(7) 43–7.

40. Vijatovic M M, Bobic J D and Stojanovic B D 2008 History and challenges of barium titanate: Part II *Sci. Sinter.* **40** 235–44.

41. Tewatia K, Sharma A, Sharma M and Kumar A 2021 Factors affecting morphological and electrical properties of barium titanate: A brief review *Mater. Today Proc.* **44** 4548–56.

42. Merz W J 1949 The electric and optical behavior of $BaTiO_3$ single-domain crystals *Phys. Rev.* **76** 1221–5.

43. Jaffe H 1958 Piezoelectric ceramics *J. Am. Ceram. Soc.* **41** 494–8.

44. Jaffe B, William R, Cook J and Hans J 1971 *Piezoelectric Ceramics* (Elsevier).

45. Cross L E and Moulson A J 1984 The effect of grain size on the permittivity of $BaTiO_3$ *Ferroelectrics* **54** 147–50.

46. Hennings D 1987 Barium titanate based ceramic materials for dielectric use *Int. J. High Technol. Ceram.* **3** 91–111.

47. Shao S, Zhang J, Zhang Z, Zheng P, Zhao M, Li J and Wang C 2008 High piezoelectric properties and domain configuration in $BaTiO_3$ ceramics obtained through the solid-state reaction route *J. Phys. D. Appl. Phys.* **41** 125408.

48. Lu D-Y, Sun X-Y and Toda M 2007 A novel high-k 'Y5V' barium titanate ceramics co-doped with lanthanum and cerium *J. Phys. Chem. Solids* **68** 650–64.

49. Praveen J P, Karthik T, James A R, Chandrakala E, Asthana S and Das D 2015 Effect of poling process on piezoelectric properties of sol–gel derived BZT–BCT ceramics *J. Eur. Ceram. Soc.* **35** 1785–98.

50. Li F, Cabral M J, Xu B, Cheng Z, Dickey E C, LeBeau J M, Wang J, Luo J, Taylor S, Hackenberger W, Bellaiche L, Xu Z, Chen L-Q, Shrout T R and Zhang S 2019 Giant piezoelectricity of Sm-doped $Pb(Mg_{1/3}Nb_{2/3})O_3$-$PbTiO_3$ single crystals *Science (80-.)* **364** 264–8.

51. Li F, Lin D, Chen Z, Cheng Z, Wang J, Li C, Xu Z, Huang Q, Liao X, Chen L-Q, Shrout T R and Zhang S 2018 Ultrahigh piezoelectricity in ferroelectric ceramics by design *Nat. Mater.* **17** 349–54.

52. Haun M J, Furman E, Jang S J and Cross L E 1989 Thermodynamic theory of the lead zirconate-titanate solid solution system, part I: Phenomenology *Ferroelectrics* **99** 13–25.

53. Fu H and Cohen R E 2000 Polarization rotation mechanism for ultrahigh electromechanical response in single-crystal piezoelectrics *Nature* **403** 281–3.

54. Damjanovic D 2009 Comments on origins of enhanced piezoelectric properties in ferroelectrics *IEEE Trans. Ultrason. Ferroelectr. Freq. Control* **56** 1574–85.

55. Landau L 1937 On the theory of phase transitions *Zh. Eksp. Teor. Fiz* **7** 19–32.

56. Wang D, Fan Z, Rao G, Wang G, Liu Y, Yuan C, Ma T, Li D, Tan X, Lu Z, Feteira A, Liu S, Zhou C and Zhang S 2020 Ultrahigh piezoelectricity in lead-free piezoceramics by synergistic design *Nano Energy* **76** 0–17.

57. Anit Kaur G, Kumar S, Sharma V, Kainthla I, Thakur S, Thakur S, Rai R and Shandilya M 2023 Enhancement in the dielectric and ferroelectric behaviour by interface between the electrode and grain bulk boundaries of Ca, Zr-doped barium titanate *Inorg. Chem. Commun.* **151** 110644.

58. Liu L, Yin F, Zhao G, Zheng L, Fu X, Jia Z and Tao X 2023 High quality crystal growth and characterization of ferroelectric $Ba_{0.77}Ca_{0.23}TiO_3$ single crystal *J. Adv. Dielectr.* **13** 1–6.

59. Shen Z-Y and Li J-F 2010 Enhancement of piezoelectric constant d_{33} in $BaTiO_3$ ceramics due to nano-domain structure *J. Ceram. Soc. Japan* **118** 940–3.

60. Zhu K, Ji Hand Chen Y 2010 Microwave sintering of lead-free barium titanate piezoelectric ceramics from barium titanate powders *JCPR* **11** 293–296..

61. Ma N, Zhang B-P and Yang W-G 2012 Low-temperature sintering of Li_2O-doped $BaTiO_3$ lead-free piezoelectric ceramics *J. Electroceramics* **28** 275–80.

62. Yang W-G, Zhang B-P, Ma N and Zhao L 2012 High piezoelectric properties of $BaTiO_3$–xLiF ceramics sintered at low temperatures *J. Eur. Ceram. Soc.* **32** 899–904.
63. Hu S, Luo C, Li P, Hu J, Li G, Jiang H and Zhang W 2017 Effect of sintered temperature on structural and piezoelectric properties of barium titanate ceramic prepared by nano-scale precursors *J. Mater. Sci. Mater. Electron.* **28** 9322–7.
64. Kimura T, Dong Q, Yin S, Hashimoto T, Sasaki A, Aisawa S and Sato T 2013 Synthesis and piezoelectric properties of Li, Ca and Mn-codoped $BaTiO_3$ by a solvothermal approach *IOP Conf. Ser. Mater. Sci. Eng.* **47** 012018.
65. On D V, Vuong L D, Chuong T Van, Quang D A, Tuyen H Van and Tung V T 2021 Influence of sintering behavior on the microstructure and electrical properties of $BaTiO_3$ lead-free ceramics from hydrothermal synthesized precursor nanoparticles *J. Adv. Dielectr.* **11** 2150014.
66. Kinoshita K and Yamaji A 1976 Grain-size effects on dielectric properties in barium titanate ceramics *J. Appl. Phys.* **47** 371–3.
67. Wu X and Tan D Q 2023 Enhanced energy density of polyetherimide using low content barium titanate nanofillers *J. Phys. Conf. Ser.* **2500** 012008.
68. Swain S, Bhaskar R, Narayanan K B, Gupta M K, Sharma S, Dasgupta S, Han S S and Kumar P 2023 Physicochemical, mechanical, dielectric, and biological properties of sintered hydroxyapatite/barium titanate nanocomposites for bone regeneration *Biomed. Mater.* **18**(2) 025016.
69. Krishna P S R, Pandey D, Tiwari V S, Chakravarthy R and Dasannacharya B A 1993 Effect of powder synthesis procedure on calcium site occupancies in barium calcium titanate: A Rietveld analysis *Appl. Phys. Lett.* **62** 231–3.
70. Zheng P, Zhang J L, Shao S F, Tan Y Q and Wang C L 2009 Piezoelectric properties and stabilities of CuO-modified $Ba(Ti,Zr)O_3$ ceramics *Appl. Phys. Lett.* **94**(3) 1–4.
71. Panda P K, Sahoo B, Sureshkumar V and Politova E D 2021 Effect of Zr^{4+} on piezoelectric, dielectric and ferroelectric properties of barium calcium titanate lead-free ceramics *J. Adv. Dielectr.* **11** 1–9.
72. Zhang J, Zhang Z, Shao S, Zheng P and Wang C 2011 High piezoelectric performance and relevant physical mechanism of CuO-modified $Ba(Ti_{0.96}Sn_{0.04})O_3$ ceramics *J. Adv. Dielectr.* **01** 79–84.
73. Tan Y Q, Zhang J L and Wang C L 2013 High piezoelectric properties and good temperature stabilities of CuO -modified $Ba(Ti_{0.96}Sn_xZr_{0.04-x})O_3$ ceramics *J. Adv. Dielectr.* **03** 1350014.

74. Matsuo K, Xie R-J, Akimune Y and Sugiyama T 2002 Special issue ceramics integration. Preparation of lead-free $Sr_{2-x}Ca_xNaNb_5O_{15}$ (x=0.1)-based piezoceramics with tungsten bronze structure. *J. Ceram. Soc. Japan* **110** 491–4.

75. Tian H Y, Wang Y, Miao J, Chan H L W and Choy C L 2007 Preparation and characterization of hafnium doped barium titanate ceramics *J. Alloys Compd.* **431** 197–202.

76. Nanakorn N, Jalupoom P, Vaneesorn N and Thanaboonsombut A 2008 Dielectric and ferroelectric properties of $Ba(Zr_xTi_{1-x})O_3$ ceramics *Ceram. Int.* **34** 779–82.

77. Liu W and Ren X 2009 Large piezoelectric effect in Pb-free ceramics *Phys. Rev. Lett.* **103** 257602.

78. Li W, Xu Z, Chu R, Fu P and Zang G 2010 Piezoelectric and dielectric properties of $(Ba_{1-x}Ca_x)(Ti_{0.95}Zr_{0.05})O_3$ lead-free ceramics *J. Am. Ceram. Soc.* **93** 2942–4.

79. Li W, Xu Z, Chu R, Fu P and Zang G 2010 Polymorphic phase transition and piezoelectric properties of $(Ba_{1-x}Ca_x)(Ti_{0.9}Zr_{0.1})O_3$ lead-free ceramics *Phys. B Condens. Matter* **405** 4513–6.

80. Li W, Xu Z, Chu R, Fu P and Zang G 2010 High piezoelectric d_{33} coefficient in $(Ba_{1-x}Ca_x)(Ti_{0.98}Zr_{0.02})O_3$ lead-free ceramics with relative high Curie temperature *Mater. Lett.* **64** 2325–7.

81. Zhang S-W, Zhang H, Zhang B-P and Yang S 2010 Phase-transition behavior and piezoelectric properties of lead-free $(Ba_{0.95}Ca_{0.05})(Ti_{1-x}Zr_x)O_3$ ceramics *J. Alloys Compd.* **506** 131–5.

82. Egerton L and Dillon D M 1959 Piezoelectric and dielectric properties of ceramics in the system potassium—sodium niobate *J. Am. Ceram. Soc.* **42** 438–42.

83. Tellier J, Malic B, Dkhil B, Jenko D, Cilensek J and Kosec M 2009 Crystal structure and phase transitions of sodium potassium niobate perovskites *Solid State Sci.* **11** 320–4.

84. Sharma J P, Kumar D and Sharma A K 2021 Structural and dielectric properties of pure potassium sodium niobate (KNN) lead free ceramics *Solid State Commun.* **334–335** 114345.

85. Saito Y, Takao H, Tani T, Nonoyama T, Takatori K, Homma T, Nagaya T and Nakamura M 2004 Lead-free piezoceramics *Nature* **432** 84–7.

86. Lin D, Kwok K and Chan H 2008 Structure and electrical properties of $Bi_{0.5}Na_{0.5}TiO_3$–$BaTiO_3$–$Bi_{0.5}Li_{0.5}TiO_3$ lead-free piezoelectric ceramics *Solid State Ionics* **178** 1930–1937.

87. Panda P K 2009 Review: Environmental friendly lead-free piezoelectric materials *J. Mater. Sci.* **44** 5049–62.

88. Zheng T, Wu J, Xiao D, Zhu J, Wang X and Lou X 2015 Potassium-sodium niobate lead-free ceramics: Modified strain as well as piezoelectricity *J. Mater. Chem. A* **3** 1868–74.

89. Pang Q, Han L and Yu X 2023 Doping modification in lead-free piezoelectric ceramics *Highlights in Science, Engineering and Technology* **55** 166–75.

90. Xing J, Jiang L, Zhao C, Tan Z, Xu Q, Wu J, Chen Q, Xiao D and Zhu J 2020 Potassium sodium niobate based lead-free ceramic for high-frequency ultrasound transducer applications *J. Mater.* **6** 513–22.

91. Maeda T, Takiguchi N, Morita T, Ishikawa M and Hemsel T 2010 Hydrothermal $(K_{1-x}Na_x)NbO_3$ lead-free piezoelectric ceramics *J. Korean Phys. Soc.* **57** 924–8.

92. Ishikawa M, Takiguchi N, Hosaka H and Morita T 2008 Nondoped potassium niobate ceramics synthesized by hydrothermal method with optimum temperature condition *Jpn. J. Appl. Phys.* **47** 3824–8.

93. Ishikawa M, Yazawa K, Fujisawa T, Yasui S, Yamada T, Hasegawa T, Morita T, Kurosawa M and Funakubo H 2009 Growth of epitaxial $KNbO_3$ thick films by hydrothermal method and their characterization *Jpn. J. Appl. Phys.* **48** 09KA14.

94. Zhang S, Malič B, Li J F and Rödel J 2021 Lead-free ferroelectric materials: Prospective applications *J. Mater. Res.* **36** 985–95.

95. Li P, Zhai J, Shen B, Zhang S, Li X, Zhu F and Zhang X 2018 Ultrahigh piezoelectric properties in textured $(K,Na)NbO_3$-based lead-free ceramics *Adv. Mater.* **30** 1–32.

96. Dai Y-J, Zhang X-W and Chen K-P 2009 Morphotropic phase boundary and electrical properties of $K_{1-x}Na_xNbO_3$ lead-free ceramics *Appl. Phys. Lett.* **94** 042905.

97. Li J-F, Wang K, Zhang B-P and Zhang L-M 2006 Ferroelectric and piezoelectric properties of fine-grained $Na_{0.5}K_{0.5}NbO_3$ lead-free piezoelectric ceramics prepared by spark plasma sintering *J. Am. Ceram. Soc.* **89** 706–9.

98. Lin D, Kwok K W and Chan H L W 2007 Double hysteresis loop in Cu-doped $K_{0.5}Na_{0.5}NbO_3$ lead-free piezoelectric ceramics *Appl. Phys. Lett.* **90** 232903.

99. Lin D, Kwok K W and Chan H L W 2008 Piezoelectric properties and hardening behavior of $K_{5.4}Cu_{1.3}Ta_{10}O_{29}$-doped $K_{0.5}Na_{0.5}NbO_3$ ceramics *J. Appl. Phys.* **103**(6) 064105.

100. Chen Q, Chen L, Li Q, Yue X, Xiao D, Zhu J, Shi X and Liu Z 2007 Piezoelectric properties of $K_4CuNb_8O_{23}$ modified $(Na_{0.5}K_{0.5})NbO_3$ lead-free piezoceramics *J. Appl. Phys.* **102**(10) 104109.

101. Lin D, Li Z, Zhang S, Xu Z and Yao X 2009 Dielectric/piezoelectric properties and temperature dependence of domain structure evolution in lead free single crystal *Solid State Commun.* **149** 1646–9.

102. Lin D, Li Z, Zhang S, Xu Z and Yao X 2010 Influence of MnO_2 doping on the dielectric and piezoelectric properties and the domain structure in $(K_{0.5}Na_{0.5})NbO_3$ single crystals *J. Am. Ceram. Soc.* **93** 941–4.

103. Nakashima Y, Sakamoto W, Maiwa H, Shimura T and Yogo T 2007 Lead-free piezoelectric $(K,Na)NbO_3$ thin films derived from metal alkoxide precursors *Japanese Journal of Applied Physics* **46**(14) L311–3.

104. Li H 2023 Enhanced piezoelectric properties for potassium sodium niobate lead-free piezoelectric ceramics prepared by microwave technology *Mater. Lab* **2** 1–6.

105. Lee S Y, Ahn C W, Kim J S, Ullah A, Lee H J, Hwang H-I, Choi J S, Park B H and Kim I W 2011 Enhanced piezoelectric properties of Ta substituted-$(K_{0.5}Na_{0.5})NbO_3$ films: A candidate for lead-free piezoelectric thin films *J. Alloys Compd.* **509** L194–8.

106. Abazari M, Choi T, Cheong S-W and Safari A 2010 Nanoscale characterization and local piezoelectric properties of lead-free KNN-LT-LS thin films *J. Phys. D. Appl. Phys.* **43** 025405.

107. Won S S, Lee J, Venugopal V, Kim D-J, Lee J, Kim I W, Kingon A I and Kim S-H 2016 Lead-free Mn-doped $(K_{0.5},Na_{0.5})NbO_3$ piezoelectric thin films for MEMS-based vibrational energy harvester applications *Appl. Phys. Lett.* **108**(23) 232908.

108. Kim J S, Ahn C W, Ullah A, Chae S A and Kim I W 2016 Ferroelectric and electrical properties of lead-free $(K_{0.44}Na_{0.52}Li_{0.04})(Nb_{0.86}Ta_{0.10}Sb_{0.04})O_3$ thin films *J. Korean Phys. Soc.* **68** 1461–6.

109. Wang L, Zuo R, Liu L, Su H, Shi M, Chu X, Wang X and Li L 2011 Preparation and characterization of sol–gel derived (Li,Ta,Sb) modified $(K,Na)NbO_3$ lead-free ferroelectric thin films *Materials Chemistry and Physics* **130** 165–9.

110. Lin C-C, Chen C-C, Weng C-M, Chu S-Y, Hong C-S and Tsai C-C 2015 Effects of lithium doping on microstructure, electrical properties, and chemical bonds of sol-gel derived NKN thin films *J. Appl. Phys.* **117** 085307.

111. Chin Goh P, Yao K and Chen Z 2011 Lithium diffusion in $(Li,K,Na)NbO_3$ piezoeletric thin films and the resulting approach for enhanced performance properties *Appl. Phys. Lett.* **99**(9) 092902.

112. Bu X, Yang C, Fan M, Wang W, Lin X and Huang S 2023 Ferroelectric and piezoelectric properties of lead-free $Li_{0.06}(K_{0.5}Na_{0.5})_{0.94}$ NbO_3 thin films *J. Adv. Dielectr.* **13** 1–8.

113. Lang S B, Zhu W and Cross L E 2006 Piezoelectric and pyroelectric properties of $(K_{0.5}Na_{0.5})_{1-x}$ $(Nb_{1-y}Ta_y)O_3$ ceramics *Ferroelectrics* **336** 15–21.

114. Zuo R and Fu J 2011 Rhombohedral-tetragonal phase coexistence and piezoelectric properties of $(NaK)(NbSb)O_3$-$LiTaO_3$-$BaZrO_3$ lead-free ceramics *J. Am. Ceram. Soc.* **94** 1467–70.

115. Smolenskii G 1961 New ferroelectrics of complex composition. IV *Sov. Physics-Solid State* **2** 2651–4.

116. Quan N D, Huu Bac L, Thiet D Van, Hung V N and Dung D D 2014 Current development in lead-free $Bi_{0.5}(Na,K)_{0.5}TiO_3$-based piezoelectric materials *Adv. Mater. Sci. Eng.* **2014** 1–14.

117. Li Y, Chen W, Zhou J, Xu Q, Sun H and Xu R 2004 Dielectric and piezoelecrtic properties of lead-free $(Na_{0.5}Bi_{0.5})TiO_3$–$NaNbO_3$ ceramics *Mater. Sci. Eng. B* **112** 5–9.

118. Herabut A and Safari A. 1997 Processing and electromechanical properties of $(Bi_{0.5}Na_{0.5})_{(1-1.5x)}LaxTiO_3$ ceramics *Journal of the American Ceramic Society* **80**(11) 2954–58.

119. Xiao D Q, Lin D M, Zhu J G and Yu P 2008 Studies on new systems of BNT-based lead-free piezoelectric ceramics *J. Electroceramics* **21** 34–8.

120. Lin D, Xiao D, Zhu J and Yu P 2006 Piezoelectric and ferroelectric properties of $[Bi0.5(Na_{1-x-y}K_xLi_y)_{0.5}]TiO_3$ lead-free piezoelectric ceramics *Appl. Phys. Lett.* **88**(6) 062901.

121. Coondoo I, Panwar N and Kholkin A 2013 Lead-free piezoelectrics: Current status and perspectives *J. Adv. Dielectr.* **03** 1330002.

122. Yin J, Zhao C, Zhang Y and Wu J 2017 Composition-induced phase transitions and enhanced electrical properties in bismuth sodium titanate ceramics *J. Am. Ceram. Soc.* **100** 5601–9.

123. Sidak V M, Trubitsyn M P and Panchenko T V 2022 Dielectric relaxation induced by oxygen vacancies in $Na_{0.5}Bi_{0.5}TiO_3$ ceramics *Condens. Matter Phys.* **25** 1–10.

124. Chaou F, Bendahhou A, Chourti K, Jalafi I, Yahakoub E H, Barkany S El and Abou-salama M 2023 Optimisation of electrical conductivity and dielectric properties of Sn^{4+}-doped $(Na_{0.5}Bi_{0.5})TiO_3$ perovskite *Ceramics International* **49**(11) 17940–52.

125. Mesrar M, Elbasset A, Mrabet I El, Zaitan H, Abdi F, Echatoui N and Lamcharfi T 2023 Hydrothermal synthesis and characterization of sodium bismuth titanate for photocatalytic applications **24** 185–97.

126. Lai L, Zhao Z, Tian S, Ou B, Liang G, Li B and Dai Y 2023 Ultrahigh electrostrain with excellent fatigue resistance in textured Nb^{5+}-doped $(Bi_{0.5}Na_{0.5})TiO_3$-based piezoceramics *J. Adv. Ceram.* **12** 487–97.

127. Ge W, Li J, Viehland D and Luo H 2010 Influence of Mn doping on the structure and properties of $Na_{0.5}Bi_{0.5}TiO_3$ single crystals *Journal of the American Ceramic Society* **93**(5) 1372–77.

128. Yi X, Chen H, Cao W, Zhao M, Yang D, Ma G, Yang C and Han J 2005 Flux growth and characterization of lead-free piezoelectric single crystal $[Bi_{0.5}(Na_{1-x}K_x)_{0.5}]TiO_3$ *J. Cryst. Growth* **281** 364–9.

129. Zhang Q, Zhang Y, Wang F, Lin D, Li X, Zhao X and Luo H 2010 Growth and electric properties of $0.96Na_{0.5}Bi_{0.5}TiO_3$–$0.04BaTiO_3$ single crystal *J. Cryst. Growth* **312** 457–60.

130. Yoo J, Oh D, Jeong Y, Hong J and Jung M 2004 Dielectric and piezoelectric characteristics of lead-free $Bi_{0.5}(Na_{0.84}K_{0.16})_{0.5}TiO_3$ ceramics substituted with Sr *Mater. Lett.* **58** 3831–5.

131. Panda P K 2009 Review: environmental friendly lead-free piezoelectric materials *J. Mater. Sci.* **44** 5049–62.

132. Zhang Q, Zhang Y, Wang F, Wang Y, Lin D, Zhao X, Luo H, Ge W and Viehland D 2009 Enhanced piezoelectric and ferroelectric properties in Mn-doped $Na_{0.5}Bi_{0.5}TiO_3$–$BaTiO_3$ single crystals *Appl. Phys. Lett.* **95**(10) 102904.

133. Zhou C and Liu X 2008 Dielectric and piezoelectric properties of bismuth-containing complex perovskite solid solution of $Bi_{1/2}Na_{1/2}TiO_3$–$Bi(Mg_{2/3}Nb_{1/3})O_3$ *J. Mater. Sci.* **43** 1016–9.

134. Alaoui I H, Marrec F Le, Cantaluppi A, Favry D, Lahmar A and Scholar G 2023 Effect of oxygen annealing atmosphere on structural, electrical and energy storage properties of $Bi_{0.5}Na_{0.5}TiO_3$ polycrystalline thin film *Quantum Beam Science* **7**(3) 29.

135. Chu B-J, Chen D-R, Li G-R and Yin Q-R 2002 Electrical properties of $Na_{1/2}Bi_{1/2}TiO_3$–$BaTiO_3$ ceramics *J. Eur. Ceram. Soc.* **22** 2115–21.

136. Zang G-Z, Wang J-F, Chen H-C, Su W-B, Wang C-M, Qi P, Ming B-Q, Du J, Zheng L-M, Zhang S and Shrout T R 2006 Perovskite $(Na_{0.5}K_{0.5})_{1-x}(LiSb)_x Nb_{1-x}O_3$ lead-free piezoceramics *Appl. Phys. Lett.* **88**(21) 212908.

137. Choy S H, Wang X X, Chong C P, Chan H L W, Liu P C K and Choy C L 2006 $0.90(Bi_{1/2}Na_{1/2})TiO_3$–$0.05(Bi_{1/2}K_{1/2})TiO_3$– $0.05BaTiO_3$ transducer for ultrasonic wirebonding applications *Appl. Phys. A* **84** 313–6.

138. Choy S H, Wang X X, Chan H L W and Choy C L 2006 Study of compressive type accelerometer based on lead-free BNKBT piezoceramics *Appl. Phys. A* **82** 715–8.

139. Wu J, Xiao D, Wang Y, Zhu J, Wu L and Jiang Y 2007 Effects of K/Na ratio on the phase structure and electrical properties of $(K_xNa_{0.96-x}Li_{0.04})(Nb_{0.91}Ta_{0.05}Sb_{0.04})O_3$ lead-free ceramics *Appl. Phys. Lett.* **91**(25) 252907.

140. Lin D, Kwok K W and Chan H W L 2007 Dielectric and piezoelectric properties of $(K_{0.5}Na_{0.5})NbO_3$–$Ba(Zr_{0.05}Ti_{0.95})O_3$ lead-free ceramics *Appl. Phys. Lett.* **91** 143513.

141. Zuo R, Ye C, Fang X and Li J 2008 Tantalum doped $0.94Bi_{0.5}Na_{0.5}TiO_3$–$0.06BaTiO_3$ piezoelectric ceramics *J. Eur. Ceram. Soc.* **28** 871–7.

142. Du H, Zhou W, Luo F, Zhu D, Qu S, Li Y and Pei Z 2008 Design and electrical properties' investigation of $(K_{0.5}Na_{0.5})NbO_3$–$BiMeO_3$ lead-free piezoelectric ceramics *J. Appl. Phys.* **104**(3) 034104.

143. Zhang J L, Zong X J, Wu L, Gao Y, Zheng P and Shao S F 2009 Polymorphic phase transition and excellent piezoelectric performance of $(K_{0.55}Na_{0.45})_{0.965}Li_{0.035}Nb_{0.80}Ta_{0.20}O_3$ lead-free ceramics *Appl. Phys. Lett.* **95**(2) 022909.

144. Wang C, Wang J, Zhang S and Shrout T R 2009 Piezoelectric and electromechanical properties of ultrahigh temperature $CaBi_2Nb_2O_9$ ceramics *Phys. status solidi – Rapid Res. Lett.* **3** 49–51.

145. Sun R, Lin D, Zhang Q, Fang B, Zhang H, Li X, Wang S, Wang D, Zhao X and Luo H 2011 Growth and characterization of $Na_{0.5}Bi_{0.5}TiO_3$–$K_{0.5}Bi_{0.5}TiO_3$ lead-free piezoelectric crystal by the TSSG method *J. Adv. Dielectr.* **01** 237–42.

146. Xue S, Ma J, Zhao X, Wang F, Sun D, Wang T, Shi W, Fu Z, Zhou H and Luo H 2017 Nanoscale insight into the giant piezoelectric response in lead-free Fe-doped $0.95(Na_{1/2}Bi_{1/2})TiO_3$-$0.05BaTiO_3$ single crystal *Appl. Phys. Lett.* **111**(16) 162902.

147. Wang D, Wang G, Murakami S, Fan Z, Feteira A, Zhou D, Sun S, Zhao Q and Reaney I M 2018 $BiFeO_3$-$BaTiO_3$: A new generation of lead-free electroceramics *J. Adv. Dielectr.* **08** 1830004.

148. Kunnath N and Philip J 2020 Dielectric and piezoelectric perfomance of lead-free ceramics of boron sodium gadolinate niobate at morphotropic phase boundary *J. Adv. Dielectr.* **10** 1–9.

149. Chen F, Yang L, Feng H, Li Q, Zeng X, Yu K, Song C, Yan Y, Jin L and Zhang D 2023 An ultrahigh energy storage efficiency and recoverable density in $Bi_{0.5}Na_{0.5}TiO_3$ with the modification of $Sr_{0.85}La_{0.1}TiO_3$ via viscous polymer process *Journal of Materiomics* 1–42.

150. Zhu L F, Zhang B P, Li S, Zhao L, Wang N and Shi X C 2016 Enhanced piezoelectric properties of $Bi(Mg_{1/2}Ti_{1/2})O_3$ modified $BiFeO_3$-$BaTiO_3$

ceramics near the morphotropic phase boundary *J. Alloys Compd.* **664** 602–8.

151. Lu W, Fan G, Wang X and Liang F 2006 Dielectric and piezoelectric properties of $[Bi_{0.5}(Na_{1-x}Li_x)_{0.5}]TiO_3$ lead-free ceramics *Jpn. J. Appl. Phys.* **45** 8763.

152. Wang X, Chan H L-W and Choy C-L 2003 $(Bi_{1/2}Na_{1/2})TiO_3$-$Ba(Cu_{1/2}W_{1/2})O_3$ Lead-Free Piezoelectric Ceramics *J. Am. Ceram. Soc.* **86** 1809–11.

153. Yang Z, Liu B, Wei L and Hou Y 2008 Structure and electrical properties of $(1-x)Bi_{0.5}Na_{0.5}TiO_3$–$xBi_{0.5}K_{0.5}TiO_3$ ceramics near morphotropic phase boundary *Mater. Res. Bull.* **43** 81–9.

154. Zhang Y-R, Li J-F, Zhang B-P and Peng C-E 2008 Piezoelectric and ferroelectric properties of Bi-compensated $(Bi_{1/2}Na_{1/2})TiO_3$–$(Bi_{1/2}K_{1/2})TiO_3$ lead-free piezoelectric ceramics *J. Appl. Phys.* **103**(7) 074109.

155. Li Y, Chen W, Xu Q, Zhou J, Wang Y and Sun H 2007 Piezoelectric and dielectric properties of CeO_2-doped $Bi_{0.5}Na_{0.44}K_{0.06}TiO_3$ lead-free ceramics *Ceram. Int.* **33** 95–9.

156. Fan G, Lu W, Wang X and Liang F 2007 Morphotropic phase boundary and piezoelectric properties of $(Bi_{1/2}Na_{1/2})TiO_3$–$(Bi_{1/2}K_{1/2})TiO_3$–$KNbO_3$ lead-free piezoelectric ceramics *Appl. Phys. Lett.* **91**(20) 202908.

157. Zhou C, Liu X, Li W and Yuan C 2009 Dielectric and piezoelectric properties of $Bi_{0.5}Na_{0.5}TiO_3$–$Bi_{0.5}K_{0.5}TiO_3$–$BiCrO_3$ lead-free piezoelectric ceramics *J. Alloys Compd.* **478** 381–5.

158. Zhang Y, Chu R, Xu Z, Hao J, Li G and Yin Q 2010 Dielectric and piezoelectric properties of low temperature sintering lead free $(Bi_{0.5}Na_{0.7+x}K_{0.2}Li_{0.1})0.5TiO_3$ piezoelectric ceramics *Phys. B Condens. Matter* **405** 1228–32.

159. Lu Y, Li Y, Wang D and Yin Q 2007 Morphotropic phase boundary and electrical properties of $(Bi_{1/2}Na_{1/2})TiO_3$-$(Bi_{1/2}K_{1/2})TiO_3$-$(Bi_{1/2}Ag_{1/2})TiO_3$ ceramics *Ferroelectrics* **358** 109–16.

160. Xu C, Lin D and Kwok K W 2008 Structure, electrical properties and depolarization temperature of $(Bi_{0.5}Na_{0.5})TiO_3$–$BaTiO_3$ lead-free piezoelectric ceramics *Solid State Sci.* **10** 934–40.

161. Wang X, Chan H L-W and Choy C 2003 Piezoelectric and dielectric properties of CeO_2-added $(Bi_{0.5}Na_{0.5})_{0.94}Ba_{0.06}TiO_3$ lead-free ceramics *Solid State Commun.* **125** 395–9.

162. Zhu M, Liu L, Hou Y, Wang H and Yan H 2007 Microstructure and electrical properties of MnO-doped $(Na_{0.5}Bi_{0.5})_{0.92}Ba_{0.08}TiO_3$ lead-free piezoceramics *J. Am. Ceram. Soc.* **90** 120–4.

163. Xu Q, Chen M, Chen W, Liu H-X, Kim B-H and Ahn B-K 2008 Effect of CoO additive on structure and electrical properties of $(Na_{0.5}Bi_{0.5})_{0.93}Ba_{0.07}TiO_3$ ceramics prepared by the citrate method *Acta Mater.* **56** 642–50.

164. Li H, Feng C and Yao W 2004 Some effects of different additives on dielectric and piezoelectric properties of $(Bi_{1/2}Na_{1/2})TiO_3$–$BaTiO_3$ morphotropic-phase-boundary composition *Mater. Lett.* **58** 1194–8.

165. Li H-D, Feng C-D and Xiang P-H 2003 Electrical Properties of La^{3+}-Doped $(Na_{0.5}Bi_{0.5})_{0.94}Ba_{0.06}TiO_3$ Ceramics *Jpn. J. Appl. Phys.* **42** 7387–91.

166. Zhou T-S, Huang R-X, Shang X-Z, Peng F, Guo J-Y, Chai L-Y, Gu H-S and He Y-B 2007 Lead-free In_2O_3-doped $(Bi_{0.5}Na_{0.5})_{0.93}Ba_{0.07}TiO_3$ ceramics synthesized by direct reaction sintering *Appl. Phys. Lett.* **90**(18) 182903.

167. Wu L, Xiao D-Q, Lin D-M, Zhu J-G and Yu P 2005 Synthesis and Properties of $[Bi_{0.5}(Na_{1-x}Ag_x)_{0.5}]_{1-y}Ba_yTiO_3$ Piezoelectric Ceramics *Jpn. J. Appl. Phys.* **44** 8515.

168. Lin D and Kwok K W 2010 Structure and electrical properties of $Bi_{0.5}Na_{0.5}TiO_3$-$Y_{0.5}Na_{0.5}TiO_3$-$BaTiO_3$ lead-free piezoelectric ceramics *J. Mater. Sci. Mater. Electron.* **21** 291–7.

169. Peng C, Li J-F and Gong W 2005 Preparation and properties of $(Bi_{1/2}Na_{1/2})TiO_3$–$Ba(Ti,Zr)O_3$ lead-free piezoelectric ceramics *Mater. Lett.* **59** 1576–80.

170. Tian H Y, Wang D Y, Lin D M, Zeng J T, Kwok K W and Chan H L W 2007 Diffusion phase transition and dielectric characteristics of $Bi_{0.5}Na_{0.5}TiO_3$–$Ba(Hf,Ti)O_3$ lead-free ceramics *Solid State Commun.* **142** 10–4.

171. Anon Wu, Y., Zhang, H., Zhang, Y., Ma, J., and Xie, D. (2003). Lead-free piezoelectric ceramics with composition of $(0.97\text{-}x)Na_{1/2}Bi_{1/2}TiO_3$-$0.03NaNbO_3$-$xBaTiO_3$. *J. Mater. Sci.*, 38(5), 987–994. https://doi.org/10.1023/A:1022333427521.

172. Wang X X, Tang X G and Chan H L W 2004 Electromechanical and ferroelectric properties of $(Bi_{1/2}Na_{1/2})TiO_3$–$(Bi_{1/2}K_{1/2})TiO_3$–$BaTiO_3$ lead-free piezoelectric ceramics *Appl. Phys. Lett.* **85** 91–3.

173. Li Y, Chen W, Xu Q, Zhou J, Gu X and Fang S 2005 Electromechanical and dielectric properties of $Na_{0.5}Bi_{0.5}TiO_3$–$K_{0.5}Bi_{0.5}TiO_3$–$BaTiO_3$ lead-free ceramics *Mater. Chem. Phys.* **94** 328–32.

174. Li Y, Chen W, Xu Q, Zhou J and Gu X 2005 Piezoelectric and ferroelectric properties of $Na_{0.5}Bi_{0.5}TiO_3$–$K_{0.5}Bi_{0.5}TiO_3$–$BaTiO_3$ piezoelectric ceramics *Mater. Lett.* **59** 1361–4.

175. Hu H, Zhu M, Xie F, Lei N, Chen J, Hou Y and Yan H 2009 Effect of Co_2O_3 additive on structure and electrical properties of $85(Bi_{1/2}Na_{1/2})TiO_3$-$12(Bi_{1/2}K_{1/2})TiO_3$-$3BaTiO_3$ lead-free piezoceramics *J. Am. Ceram. Soc.* **92** 2039–45.

176. Lin D, Xiao D, Zhu J and Yu P 2005 Electrical properties of $[Bi_{1-z}(Na_{1-x-y-z}K_xLi_y)]_{0.5}Ba_zTiO_3$ multi-component lead-free piezoelectric ceramics *Phys. status solidi* **202** R89–91.

177. Wang X Y, Wang C L, Zhao M L, Wang J F, Yang K and Li J C 2007 Ferroelectric properties of lithia-doped $(Bi_{0.95}Na_{0.75}K_{0.20})_{0.5}Ba_{0.05}TiO_3$ ceramics *Mater. Lett.* **61** 3847–50.

178. Wang K and Li J-F 2010 Domain engineering of lead-free Li-modified $(K,Na)NbO_3$ polycrystals with highly enhanced piezoelectricity *Adv. Funct. Mater.* **20** 1924–9.

179. Lin D, Kwok K W and Chan H L W 2009 Dielectric and piezoelectric properties of $K_{0.5}Na_{0.5}NbO_3$–$AgSbO_3$ lead-free ceramics *J. Appl. Phys.* **106**(3) 034102.

180. Wu J, Wang Y, Xiao D, Zhu J and Pu Z 2007 Effects of Ag content on the phase structure and piezoelectric properties of $(K_{0.44-x}Na_{0.52}Li_{0.04}Ag_x)(Nb_{0.91}Ta_{0.05}Sb_{0.04})O_3$ lead-free ceramics *Appl. Phys. Lett.* **91**(13) 132914.

181. Du H, Zhou W, Luo F, Zhu D, Qu S and Pei Z 2007 Perovskite lithium and bismuth modified potassium-sodium niobium lead-free ceramics for high temperature applications *Appl. Phys. Lett.* **91**(18) 182909.

182. Zuo R and Ye C 2007 Structures and piezoelectric properties of $(NaKLi)_{1-x}(BiNaBa)_xNb_{1-x}Ti_xO_3$ lead-free ceramics *Appl. Phys. Lett.* **91**(6) 062916.

183. Zuo R, Fang X and Ye C 2007 Phase structures and electrical properties of new lead-free $(Na_{0.5}K_{0.5})NbO_3$–$(Bi_{0.5}Na_{0.5})TiO_3$ ceramics *Appl. Phys. Lett.* **90**(9) 092904.

184. Lin D, Kwok K W and Chan H L W 2007 Structure, dielectric, and piezoelectric properties of CuO-doped $K_{0.5}Na_{0.5}NbO_3$–$BaTiO_3$ lead-free ceramics *J. Appl. Phys.* **102**(7) 074113.

185. Chen K, Xu G, Yang D, Wang X and Li J 2007 Dielectric and piezoelectric properties of lead-free $0.95(K_{0.5}Na_{0.5})NbO_3$–$0.05LiNbO_3$ crystals grown by the Bridgman method *J. Appl. Phys.* **101**(4) 044103.

186. Tang Q-Y, Kan Y-M, Li Y-G, Zhang G-J and Wang P-L 2006 Effect of vanadium doping on fabrication and property of $Bi_4Ti_3O_{12}$ ceramics *Scr. Mater.* **54** 2075–80.

187. Hou J, Qu Y, Vaish R, Varma K B R, Krsmanovic D and Kumar R V 2010 Crystallographic evolution, dielectric, and piezoelectric properties

of $Bi_4Ti_3O_{12}$:W/Cr ceramics *Journal of the American Ceramic Society* **93**(5) 1414–21.

188. Zhao L, Xu J-X, Yin N, Wang H-C, Zhang C-J and Wang J-F 2008 Microstructure, dielectric, and piezoelectric properties of Ce-modified $Na_{0.5}Bi_{4.5}Ti_4O_{15}$ high temperature piezoceramics *Phys. status solidi – Rapid Res. Lett.* **2** 111–3.
189. Wang C-M, Zhao L, Wang J-F, Zhang S and Shrout T R 2009 Enhanced piezoelectric properties of sodium bismuth titanate ($Na_{0.5}Bi_{4.5}Ti_4O_{15}$) ceramics with B-site cobalt modification *Phys. status solidi - Rapid Res. Lett.* **3** 7–9.
190. Zeng J, Wang Y, Li Y, Yang Q and Yin Q 2008 Ferroelectric and piezoelectric properties of tungsten doped $CaBi_4Ti_4O_{15}$ ceramics *J. Electroceramics* **21** 305–8.
191. Li G, Zheng L, Yin Q, Jiang B and Cao W 2005 Microstructure and ferroelectric properties of MnO_2-doped bismuth-layer (Ca,Sr)$Bi_4Ti_4O_{15}$ ceramics *J. Appl. Phys.* **98**(6) 064108.
192. Zeng J, Li Y, Yang Q and Yin Q 2005 Ferroelectric and piezoelectric properties of vanadium-doped $CaBi_4Ti_4O_{15}$ ceramics *Mater. Sci. Eng. B* **117** 241–5.
193. Zeng J, Li Y, Wang D and Yin Q 2005 Electrical properties of neodymium doped $CaBi_4Ti_4O_{15}$ ceramics *Solid State Commun.* **133** 553–7.
194. Yi Z G, Li Y X, Wang Y and Yin Q R 2006 Dielectric and ferroelectric properties of intergrowth Bi_{7-x}LaxTi$_4$NbO$_{21}$ ceramics *Appl. Phys. Lett.* **88**(15) 152909.
195. Chaouchi A, Saidi M, D'astorg S, Rguiti M and Courtois C 2012 Dielectric and piezoelectric properties of lead-free $(Bi_{0.5}Na_{0.5})_{0.94}Ba_{0.06}TiO_3$ + x%Ce_2O_3 + y%La_2O_3 + z%Y_2O_3 compounds synthesized by sol–gel technique *J. Adv. Dielectr.* **02** 1250022.
196. Kunnath N and Philip J 2020 Dielectric and piezoelectric perfomance of lead-free ceramics of boron sodium gadolinate niobate at morphotropic phase boundary *J. Adv. Dielectr.* **10** 2050014.
197. Zuo R, Fu J, Lu S and Xu Z 2011 Normal to relaxor ferroelectric transition and domain morphology evolution in (K,Na)(Nb,Sb)O_3-$LiTaO_3$-$BaZrO_3$ lead-free ceramics, ed D W Johnson *J. Am. Ceram. Soc.* **94** 4352–7.
198. Wei H, Wang H, Xia Y, Cui D, Shi Y, Dong M, Liu C, Ding T, Zhang J, Ma Y, Wang N, Wang Z, Sun Y, Wei R and Guo Z 2018 An overview of lead-free piezoelectric materials and devices *J. Mater. Chem. C* **6** 12446–67.
199. Wang X X, Chan H L W and Choy C L 2005 $(Bi_{0.5}Na_{0.5})_{0.94}Ba_{0.06}TiO_3$ lead-free ceramics with simultaneous addition of CeO_2 and La_2O_3 *Appl. Phys. A* **80** 333–6.

200. Takenaka T, Sakata K O and Toda K O 1990 Piezoelectric properties of $(Bi_{1/2}Na_{1/2})TiO_3$-based ceramics *Ferroelectrics* **106** 375–80.

201. Takenaka T and Nagata H 2005 Current status and prospects of lead-free piezoelectric ceramics *J. Eur. Ceram. Soc.* **25** 2693–700.

202. Lin D M, Xiao D Q, Zhu J G, Yu P, Yan H J, Li L Z and Zhang W 2004 The relations of sintering conditions and microstructures of $[Bi_{0.5}(Na_{1-x-y}K_xLi_y)_{0.5}]TiO_3$ piezoelectric ceramics *Cryst. Res. Technol.* **39** 30–3.

203. Yoo J, Oh D, Jeong Y, Hong J, and Jung M 2004 Dielectric and piezoelectric characteristics of lead-free $Bi_{0.5}(Na_{0.84}K_{0.16})_{0.5}TiO_3$ ceramics substituted with Sr *Materials Letters* **58**(29) 3831–35.

204. Sasaki A, Chiba T, Mamiya Y and Otsuki E 1999 Dielectric and piezoelectric properties of $(Bi_{0.5}Na_{0.5})TiO_3$–$(Bi_{0.5}K_{0.5})TiO_3$ systems *Jpn. J. Appl. Phys.* **38** 5564.

205. Wang Y, Wu J, Xiao D, Wu W, Zhang B, Wu L and Zhu J 2008 Microstructure and electrical properties of $[(K_{0.50}Na_{0.50})_{0.95-x}Li_{0.05}Ag_x](Nb_{0.95}Ta_{0.05})O_3$ lead-free ceramics *J. Am. Ceram. Soc.* **91** 2772–5.

206. Ali M, Bathaei M J, Istif E, Karimi S N H and Beker L 2023 Biodegradable piezoelectric polymers: Recent advancements in materials and applications *Adv. Healthc. Mater.* **12**(23) 2300318 (1-32).

207. Zaszczyńska A, Gradys A and Sajkiewicz P 2020 Progress in the applications of smart piezoelectric materials for medical devices *Polymers (Basel)* **12** 1–19.

208. Samoei V K and Jayatissa A H 2021 Zinc oxide-based piezoelectric pressure sensor **1** 1–15. https://arxiv.org/pdf/2110.13643.

209. Lu F, Lee H P and Lim S P 2004 Modeling and analysis of micro piezoelectric power generators for micro-electromechanical-systems applications *Smart Mater. Struct.* **13** 57–63.

210. Li J, Li F, Xu Z and Zhang S 2018 Multilayer lead-free ceramic capacitors with ultrahigh energy density and efficiency *Adv. Mater.* **30**(32) 1802155 (7 pages).

211. Li H, Zhou S, Zhao J, Yan T, Du Y, Zhou H, Pu Y and Wang D 2023 Dielectric temperature stability and energy storage performance of NBT-based lead-free ceramics for Y9P capacitors *J. Adv. Dielectr.* **13**(01) 2242007 (7 pages).

212. Jürgen Nuffer T B and Fraunhofer 2006 Application of piezoelectric materials in transportation industry *Design* **4** 12.

213. Jin C C, Liu D M and Zhang L X 2023 An emerging family of piezocatalysts: 2D piezoelectric materials *Small* **19**(44) 2303586.

214. Won S S, Kawahara M, Ahn C W, Lee J, Lee J, Jeong C K, Kingon A I and Kim S H 2020 Lead-free $Bi_{0.5}(Na_{0.78}K_{0.22})TiO_3$ nanoparticle filler–

elastomeric composite films for paper-based flexible power generators *Adv. Electron. Mater.* **6** 1–10.

215. Kim K-B and Lee J 2022 A lead-free piezoelectric fiber generator with a high energy conversion constant material *Energies* **15** 6787.
216. Wang Z L and Song J 2006 Piezoelectric nanogenerators based on zinc oxide nanowire arrays *Science* **312** 242–6.
217. Kim D W, Lee J H, Kim J K and Jeong U 2020 Material aspects of triboelectric energy generation and sensors *NPG Asia Mater.* **12**(6) 1–17.
218. Damjanovic D and Newnham R E 1992 Electrostrictive and piezoelectric materials for actuator applications *J. Intell. Mater. Syst. Struct.* **3** 190–208.
219. Cieminski J von and Beige H 1991 High-signal electrostriction in ferroelectric materials *J. Phys. D. Appl. Phys.* **24** 1182–6.
220. Stuber V L, Deutz D B, Bennett J, Cannel D, de Leeuw D M, van der Zwaag S and Groen P 2019 Flexible lead-free piezoelectric composite materials for energy harvesting applications *Energy Technol.* **7** 177–85.
221. Vázquez-Rodríguez M, Jiménez F J, Pardo L, Ochoa P, González A M and de Frutos J 2019 A new prospect in road traffic energy harvesting using lead-free piezoceramics *Materials (Basel)* **12** 3725.
222. Shirvanimoghaddam M, Shirvanimoghaddam K, Abolhasani M M, Farhangi M, Zahiri Barsari V, Liu H, Dohler M and Naebe M 2019 Towards a green and self-powered internet of things using piezoelectric energy harvesting *IEEE Access* **7** 94533–56.
223. Chen S, Zhu P, Mao L, Wu W, Lin H, Xu D, Lu X and Shi J 2023 Piezocatalytic medicine: An emerging frontier using piezoelectric materials for biomedical applications *Adv. Mater.* **35**.
224. Pohanka M 2017 The piezoelectric biosensors: Principles and applications, a review *Int. J. Electrochem. Sci.* **12** 496–506.
225. Mahdi S M and Habeeb M A 2023 Low-cost piezoelectric sensors and gamma ray attenuation fabricated from novel polymeric nanocomposites *AIMS Mater. Sci.* **10** 288–300.
226. Li H, Tian C and Deng Z D 2014 Energy harvesting from low frequency applications using piezoelectric materials *Appl. Phys. Rev.* **1**(4) 041301.
227. Wang Z, Kurita H, Nagaoka H and Narita F 2020 Potassium sodium niobate lead-free piezoelectric nanocomposite generators based on carbon-fiber-reinforced polymer electrodes for energy-harvesting structures *Compos. Sci. Technol.* **199** 108331.

Chapter 16

Lithium Doped Lead-Free Piezoelectric Materials

Preeti Gupta
Department of Chemistry, DDU Gorakhpur University, Gorakhpur, India
preeti17_nov@yahoo.com

In the past few decades, tremendous work has been done in the area of lead-free piezoelectric materials to replace traditional lead-based piezoelectric materials for environmental protection. A number of piezoelectric materials such as $BaTiO_3$, $Bi_{0.5}Na_{0.5}TiO_3$, $(K,Na)NbO_3$, bismuth-layered structured ferroelectric materials, and alkali-based perovskites have been developed. Such materials have high electrical, dielectric, and piezoelectric properties along with durability because of the formation of morphotropic phase boundary (MPB). These properties make the material suitable for medical and communication instruments such as ultrasonic cleaners, buzzers, sonars, motors, transducers, microphones, structure-health monitoring devices, and energy harvesting devices. Lithium-ion dopants effectively lower the sintering temperature of piezoelectric

Piezoelectric Materials: Design and Applications
Edited by N. B. Singh and Dev Kumar Mahato

ISBN 978-981-5129-40-3 (Hardcover), 978-1-003-59897-8 (eBook)
www.jennystanford.com

materials. Doping of Li ion enhances the piezoelectric properties, which are important in industrial applications. This chapter covers the important features of lead-free piezoelectric materials with Li dopant. It includes the preparation of samples, properties, and application of materials in recent years.

16.1 Introduction

For the advancement of science and technology, smart materials with multifunctional properties, viz., pyroelectric, dielectric, and piezoelectric, are very crucial. Among various types of smart materials, piezoelectric is a significant class that is applicable to fabricating microphones and buzzers [1], transducers [2–7], actuators [2, 3], energy harvesting devices [1, 7], health monitoring devices [1], as well as in communication system and medical instruments such as echo graphic pictures, new imaging strategies, portable amplifiers, and miniature sensors [8].

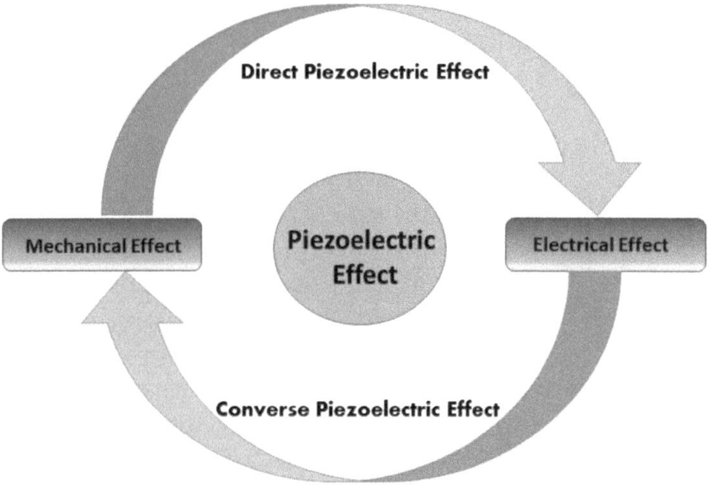

Figure 16.1 Direct and converse piezoelectric effect.

Piezoelectric materials can produce an electric field on applying stress/pressure. This is called the direct piezoelectric effect. However, in the converse piezoelectric effect, an electric field is applied to the material and it produces mechanical strain (Fig. 16.1).

In the unstressed condition, the electric dipoles of piezoelectric materials cancel each other, and thus no electricity is obtained. On applying the stress, the relative position of the electrical charges gets changed with the alteration in dipole moment. It creates a potential difference and the material undergoes deformation and builds the potential difference. The result shows the flow of electricity. Thus, one can state that applied stress is the ratio of force and cross-sectional area of the material [8].

These piezoelectric materials are categorized into various classes, i.e., piezoelectric ceramics, semiconductors, composites, glass ceramics, and single crystalline materials [9, 10]. On the basis of crystal system, piezoelectric materials are of four types: tungsten bronze, pyrochlore, bismuth layer, and perovskite structure.

There are several naturally occurring piezoelectric materials that belong to the hexagonal crystal system, viz., enamel, silk, wood dentin, and tendon. In 1880, Curie brothers had discovered the piezoelectricity in tourmaline, quartz, Rochelle salt, etc. In the year 1940, remarkable attention was drawn by the discovery of $BaTiO_3$ as an artificial piezoelectric material. Other potential piezoelectric materials are zinc oxide, lead titanate, potassium niobate, sodium niobate, and lead zirconate titanate (PZT). Among all the mentioned materials, lead zirconate titanate, a perovskite, has excellent piezoelectric, dielectric, ferroelectric, and electrochemical properties. It is the most widely used material in different sectors. However, lead is an acute toxic element and may cause damage to various organs of the human body. It has also a bad impact on the environment and climate change [11–13].

Taking account of environmental awareness, extensive researches have been done to replace lead zirconate titanate by lead-free piezoelectric materials, i.e., ZnO, barium titanate (BT), Rochelle salt, lithium niobate, and potassium niobate-sodium niobate (KNN) [14]. Other lead-free piezoelectric materials have been synthesized by using copolymer, biopolymer, and polyvinylidene difluoride (PVDF) and its composites [15]. These materials are tremendous piezoelectric and ferroelectric producers and form a morphotropic phase boundary (MPB). Doping of various metal ions such as lithium, antimony, and tantalum crucially affects the polymorphic phase boundary and thus enhances the piezoelectric property. Guo et al. [16] have extensively discussed the effect of the doping of multi-element in (K,Na)NbO_3 ceramics and studied their piezoelectric property. Literature has uncovered the studies of Li^+ doped piezoelectric material with improved properties [17–44].

Looking at the significance of lead-free piezoelectric material with the additional doping of metal ions, this chapter is focused on the preparation of various materials, their properties, and their applications in diverse fields.

16.2 Li^+ Ion Doped Lead-Free Piezoelectric Materials

The most promising lead-free piezoelectric materials are barium titanate (BT), sodium bismuth titanate (NBT), and potassium niobate-sodium niobate (KNN). They all have perovskite structures. Because of its good piezoelectric properties, BT is also one of the promising materials. However, it shows a low Curie temperature (~120°C) [45] and has a high storage capacity application along with a good dielectric coefficient of ~7000 [46].

Niobium bismuth titanate is an important piezoelectric material with a Curie temperature of ~320°C and remnant polarization is 38 $\mu C/cm^2$ [47–50]. This material is a promising candidate for electromechanical equipment works at high temperature, viz., solid oxide fuel cells, oxygen sensors, oxygen pumps, chemical sensors, and electrolyzers [51, 52]. It has major applications in energy storage devices such as electromechanical and electrostatic capacitors, Li^+/Na^+ ion batteries, and fuel cells [53]. Major drawbacks of this material are large dielectric losses, leakage of current, and high conductivity. Moreover, moderate piezoelectric coefficient (d_{33} = 57–73 pC/N) and large coercive field (Ec = 73 kV/cm) affect the property and applications.

KNN is a perovskite formed by two alkalines, potassium niobate ($KNbO_3$) and sodium niobate ($NaNbO_3$), which are ferroelectric and antiferroelectric, respectively. It shows good piezoelectric properties, high Curie temperature, and good dielectric properties [54–57]. BCZT is calcium and zirconium-based barium titanate. It exhibits significant piezoelectric properties because of the morphotropic phase boundary. Ji et al. [58] have reported the electrical properties of BCZT ceramics synthesized by microwave-assisted sol-gel hydrothermal method. The details about the Curie temperature, polarization coefficient, and other factors are provided in Table 16.1.

Table 16.1 Various electrical properties of lead-free piezoelectric materials [9, 58]

Ferroelectric Material	T_c	d_{33} (pC/N)	Dielectric	P_r ($\mu C/cm^2$)	E_c (kV/cm)
BT	~120	700	7000	-	-
NBT	320	-	-	38	73
KNN	~420	8–160	-	-	-
BCZT	83–100	500–600	-	25.22	7.52

In recent years, remarkable researches have been done in the field of lead-free piezoelectric materials with the addition of dopant ions. It has been found in the literature that lithium (Li^+) ion is used in various materials as a dopant because it is cost-effective and easily available. There are certain other significant purposes for the addition of Li^+ ion in lead-free piezoelectric materials. It has been reported that addition of Li_2CO_3 and LiF decreases the temperature involved in sintering and thus shows great enhancement in the electrical property of $(Na_{0.52}K_{0.435}Li_{0.045})Nb_{0.87}Sb_{0.08}Ta_{0.05}O_3$ lead-free piezoelectric [59]. This dopant ion also improves the density of the materials by the additional formation of liquid phase, which increases the sinterability, reduces sintering temperature, and densifies the material. Incremental doping of lithium ion results in a significant loss in dielectric property, a decrease in tan δ, and an increase in electrical field-induced strain [60, 61]. The addition of lithium increases the internal bias field value and improves the mechanical quality [62, 63]. Chen et al. [62] have also reported that there is coexistence of orthorhombic-tetragonal phase structures in the KNN-based material at room temperature. This improves the electrical and piezoelectrical properties. In $(K_{0.5}Na_{0.5})NbO_3$-$BiAlO_3$ lead-free piezoelectric ceramics, when lithium ion is added in different amounts, it forms orthorhombic-tetragonal phase boundary with the largest value of electrical properties. Van et al. [22] have reported that doping of lithium ion in $0.97Bi_{0.5}Na_{0.4}$-$xLi_xK_{0.1}TiO_3$-

$0.03CaZrO_3$ (BNKTCZ-xLi) decreases the band gap. When the Li ion is doped in piezoelectrical materials, it distorts the regular structure of the materials and thus reduces the band gap. This may shift the polymorphic phase transition to room temperature and the piezoelectric properties get enhanced [64, 65].

16.3 Method of Preparation of Lead-Free Piezoelectric Ceramic Materials

There are several methods for preparation of piezoelectric materials which are broadly classified into two categories: (i) physical methods and (ii) chemical methods. Physical methods involve two main methods: mixed oxide method and mechanochemical synthesis. In chemical methods, there are three major methods: coprecipitation, sol-gel technique, and hydrothermal synthesis. Other methods are spark plasma sintering, combustion, etc.

The most convenient and relatively easy method for preparation is mixed oxide technique. In this method, first the mixing of oxides in an appropriate amount is performed in liquid medium (distilled water/organic solvents). Then, calcination is done at a required temperature in order to complete the solid-state reaction and the formation of the desired product. The last step is milling of the calcined powder. This brings homogeneity in the material and reduces the particle size. Table 16.2 represents the list of materials prepared by the conventional mixed oxide method. Most of the lithium doped lead-free piezoelectric materials have been synthesized by solid-state method. The precursor materials were taken in stoichiometry ratio and dried in the oven at 120^{0}C for at least 10–12 h. Once the material gets dried, it is properly ground in mortar and again calcined at the said temperature (700–900°C), and the sample is ball-milled with gradual addition of ethanol. Thus, materials were ground in mortar and dried for 12–24 h at 80–100°C and again calcined at 850–900°C for 2–4 h. The ball-milling process was done for 24 h. Pellets of materials were formed by the hydraulic press and sintering was performed at 1160–1180^{0}C for 1–2 h. The solid-state reaction method needs high temperature that densifies the materials and increases the grain size, and hence the piezoelectric property improves.

Table 16.2 Various lead-free piezoelectric materials and their method of synthesis [9]

Method of synthesis	Li doped lead-free piezoelectric material	Composition (x) and experimental temperature
Solid state	$Na_{0.484}Bi_{0.456}Ba_{0.06}Ti_{0.97}Nb_{0.03}O_3$-xLi	x = 0, 0.005, 0.010, and 0.015
	$K_{0.48}Na_{0.48}Li_{0.04}(Nb_{0.8}Ta_{0.2})O_3$ (KNN-LT)	1178°C (sintering)
	$(0.7BiFeO_3\text{-}0.3BaTiO_3) + 0.025Bi\ (Zn_{0.5}Ti_{0.5})O_3 + xLi_2CO_3 + 0.0035MnO_2$ (BFBT-BZT-xLC)	0%, 0.1%, 0.2%, 0.3%, 0.4% mol 970°C (sintering)
	$Ba_{0.85}Ca_{0.15}(Zr_{0.10}Ti_{0.90})O_3 + xLi_2TaO_3$ (BCZT-xLT)	
	$(1\text{–}x)Na_{0.47}Bi_{0.47}Ba_{0.06}TiO_3$ – xLN (NBT–BT–xLN)	0, 0.02, 0.04, 0.05, 0.06, and 0.08
	$Ba_{0.95}Ca_{0.05}(Ti_{0.93}Sn_{0.07})O_3$ – yLiF (BCTS-yLiF)	y = 0, 0.01, 0.04, 0.07, and 0.1 wt%
	$Ba_{0.85}Ca_{0.15}(Ti_{0.9}Zr_{0.1})O_3$ + x wt% Li_2CO_3	x = 0, 0.1, 0.3, 0.5, 0.7, and 1.0
	$0.96(K_{0.48}Na_{0.52-x}Li_x)NbO_3\ \text{–}0.04Bi_{0.5}Na_{0.5}ZrO_3$ (KNLxN-BNZ)	x = 0, 0.01, 0.02, 0.03, and 0.05
	$0.96(Na_{0.4}K_{0.1}Bi_{0.5}TiO_3)$ – 0.04Na1 – xLi_xNbO_3	x = 0, 0.25, 0.5, 0.75, and 0.10 [76] (x = 0.75)
	$Bi_{0.5}(Na_{0.4}\text{-}xK_{0.1}Li_x))0.96Sr_{0.04}Ti_{0.975}Ta_{0.025}O_3$	x = 0, 0.025, 0.050, 0.075, and 0.100
	$BaTiO_3 - xLi_2O$	x = 0. 0.3, 0.5, 0.7, 1.0, 2.0, 3.0, and 4.0 mol% [70] 1050°C (sintering)
	$Ba_{0.99}Ca_{0.01}Zr_{0.02}Ti_{0.98}O_3$ – xLiF	x = 0, 0.4, 0.8, 1.2, 2.0, and 4.0 mol%
	$Ba_{0.85}Ca_{0.15}Ti_{0.9}Zr_{0.1}O_3$ + 0.3 wt% Li_2CO_3	x = 0.8 (BCTZL)
	$Ba_{0.85}Ca_{0.15}Zr_{0.1}Ti_{0.9}O_3$ – xLiF	x = 0.00, 0.02, 0.04, 0.06, 0.10, and 0.15 mol (sintered at 1350°C)

(Continued)

Table 16.2 (*Continued*)

Method of synthesis	Li doped lead-free piezoelectric material	Composition (x) and experimental temperature
	$(K_{0.48}Na_{0.52})_{1-x}Li_xNbO_3$-$0.015CuO$ (KNN-Li-CuO)	x = 0.00, 0.01, 0.02, 0.03, 0.04, 0.05, 0.06 0.07, and 0.08
	$0.97Bi_{0.5}Na_{0.4}$-$xLi_xK_{0.1}TiO_3$-$0.03CaZrO_3$ (BNKTCZ-xLi)	x = 0.00, 0.01, 0.02, 0.03, 0.04, and 0.05
	$Li_{0.058}(Na_{0.535}K_{0.480})$ $0.966(Nb_{0.9}Ta_{0.1})O_3$	x wt% LiF and y wt% HfO_2 [43] (x = 0.2, y = 0.1)
Conventional mixed oxide method	$0.94(K_{0.45}Na_{0.55})_{1-x}Li_x$ $(Nb_{0.85}Ta_{0.15})O_3$-$0.94AgNbO_3)$	x = 0, 0.02, 0.04, 0.06, 0.08, and 0.10
	$K_{0.5}Na_{0.5})_{1-x}LixNbO_3$	x = 0, 2, 4, and 6 mol%
	$(Na_{0.52}K_{0.435}Li_{0.045})$ $Nb_{0.87}Sb_{0.08}Ta_{0.05}O_3$ + x wt% LiF	x = 0, 2, 4, and 6 930°C (sintering)
	$(1\text{-}x)K_{0.5}Na_{0.5}NbO_{3\text{-}x}LiBiO_3$	x = 0%, 0.05%, 0.1%, 0.2%, 0.4%, 0.6%, 0.8%, and 1.0%
	$(Na_{0.5}K_{0.5})_{(1-x)}Li_xNbO_3$	x = 6.6 and 8 mol% x = 6 mol%
	$(1\text{-}x)(Na_{0.5}K_{0.5})NbO_{3\text{-}x}$ $LiSbO_3$	x = 0.01, 0.02, 0.04, and 0.06 1100°C (sintered)
	$(1\text{-}x)K_{0.48}Na_{0.52}NbO_3$ -$xLi_{0.8}Ni_{0.2}Nb_{0.96}O_3$ (KNN-LNNx)	x = 0, 0.005, 0.010, 0.015, 0.020, 0.025, 0.030, 0.035, and 0.040
Solid sintering	$0.85Bi_{0.5}Na_{0.5}$ $(1-x)Li_{0.5x}TiO_3$ -$0.11Bi_{0.5}K_{0.5}TiO_3$-$0.04BaTiO_3$	x = 0.00, 0.05, 0.10, and 0.15
Sol-gel	$0.852[Bi_{0.5}(Na_{(1\text{-}x)}Li_x)_{0.5}]$ TiO_3-0.110 $(Bi_{0.5}K_{0.5})TiO_3$-$0.038BaTiO_3$	x = 0, 0.02, 0.05, 0.10, and 0.15
	$0.75BaTiO_3$-$0.15CaTiO_3$-$0.1BaZrO_3$-1% Li (BCZT-Li)	700°C (calcined) 1500°C (sintered)

The sol-gel process involves the formation of a homogeneous solution of precursor in a suitable solvent. This is followed by the hydrolysis process to form a gel. This gel is calcined to yield a nanopowder of that specific material followed by sintering at a somewhat higher temperature than the calcination temperature. The chemical reactions are influenced by several parameters such as the amount of solvent, water, and precursor and temperature. The benefit of the sol-gel process is that it yields small particle sizes, and the stoichiometry of the material can be controlled as compared to the solid-state reaction [66, 67].

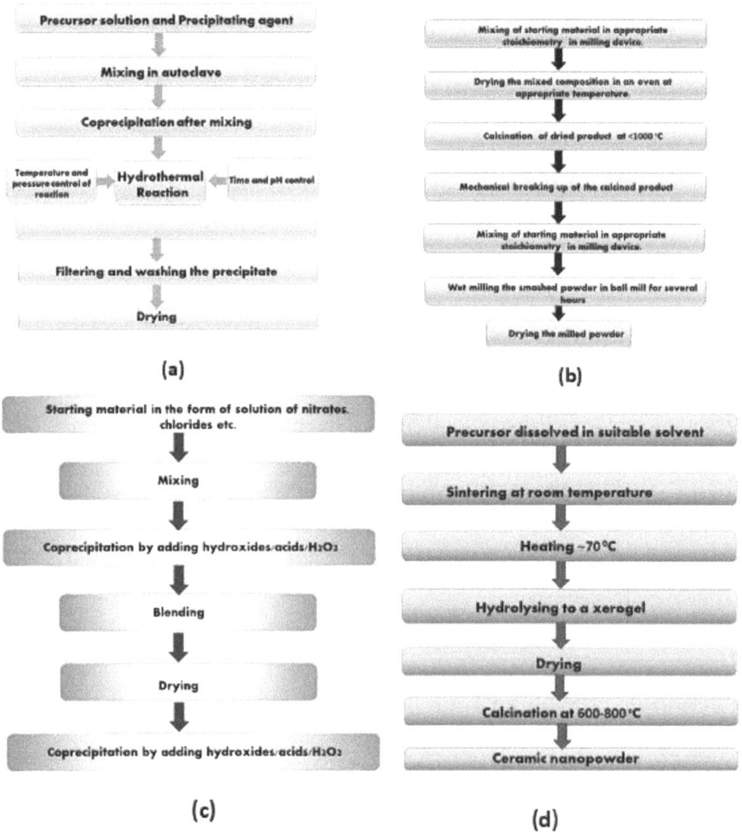

Figure 16.2 Steps involved in the synthesis of lead-free piezoelectric material: (a) hydrothermal method; (b) mixed oxide technology; (c) coprecipitation method; and (d) sol-gel method.

Hydrothermal processing includes mixing the solutions of precursors with some precipitating agent in an autoclave at a reasonable temperature and pressure up to several hours. For the hydrothermal process, different ratios of sodium hydroxide/potassium hydroxide have been used in precursors. Hydrothermal treatment for 20 h at 200°C was done before overnight drying at 80°C [68]. Figure 16.2 represents the flow chart of the process of synthesis of various piezoelectric materials and Table 16.2 summarizes the materials synthesized by different methods.

16.4 Properties

16.4.1 Structure and Microstructure

The structure of piezoelectric material has been analyzed by X-ray diffraction and it has been found to belong to perovskite structure (ABO_3). Some important structural changes have been noticed on the basis of the study of peak position, intensity, and bandwidth of the XRD pattern. It is reported in the literature that on the addition of Li^+ to lead-free piezoelectric ceramics, the material becomes more crystalline. The crystallinity of the structure is also achieved by applying suitable calcination and sintering temperature. On the incremental addition of Li_2CO_3 in KNN material, the lattice parameter decreases. Another feature of this material is the coexistence of orthogonal and tetragonal phases. Another conclusion is also drawn on the basis of the study of doping of $LiSbO_3$ that decreasing and increasing of lattice parameter are related to Li^+ size that enters A-site or B-site [9].

The microstructure of lead-free piezoelectric material has been studied by SEM, FESEM, and TEM. Studies have revealed that on increasing the doping of Li^+ in the piezoelectric material, there is an increase in sintering temperature. Other studies report that there is a decrease in grain size with addition of a dopant ion. It inhibits the growth of the grain. An increase in sintering temperature offers additional energy for the growth. Hence, it is concluded that optimal use of Li as a dopant produces grains of high-density material and the electrical properties will enhance.

16.4.2 Piezoelectric Property

Piezoelectric property can be assessed by piezoelectric constant represented by d_{33}, electromechanical coupling coefficient, k, and mechanical quality factor (Qm). The piezoelectric constant is the polarization produced in the direction of the z-axis on the application of mechanical stress in the same direction. The electromechanical coupling coefficient (k) is the longitudinal vibration of piezoelectric and is recorded by the resonance-antiresonance method employing an impedance analyzer [27, 69, 70]. Another parameter, Qm, shows the ratio of bandwidth of the resonator and its center frequency. It is a dimensionless factor [71]. It has been reported that the piezoelectric property of material is enhanced by the addition of Li^+. When Li_2CO_3 (0.3% mol) doped $BiFeO_3$-$BaTiO_3$-$Bi(Zn_{0.5}Ti_{0.5})O_3$, d_{33} is 184 pC/N at 970^{0}C (sintering temperature) [35]. The piezoelectric constant of Li^+ doped $(K_{0.48}Na_{0.48}Li_{0.04})(Nb_{0.8}Ta_{0.2})O_3$ is 165 pC/N when the material is sintered at 1178°C.

On the addition of Li_2TaO_3 in $Ba_{0.85}Ca_{0.15}(Zr_{0.10}Ti_{0.90})O_3$ material, the piezoelectric constant is reported the highest value; i.e., BNKLBT0.05 ceramics is sintered at 1110°C and shows piezoelectric constant ~184 pC/N, kp~0.30 [9].

The internal bias field (E_i) affects the mechanical quality factor. With an increase in the concentration of Li doping, the value of E_i increases. The formation of defect dipoles are good in numbers. It directly enhances the pinning effect to the domain wall, thus the Q_m improvises [72, 22]. The mechanical quality factor of KNN-Li-CuO has been studied in different composition range (x_{Li}) and it has been found that when $x < 0.04$, there is an increase in the Q_m and when x_{Li} is greater than 0.04, there is a decrease in Q_m because the pinning effect of domain wall gets weakened. It has also been reported that there is formation of secondary phase in which the production of oxygen vacancy can be enhanced. This oxygen vacancy inhibits the movement of ferroelectric domain wall and thus improves the Q_m [39]. The ferroelectric lattice instability reduces when the lithium doping enters the B-site. The result is a decrease in the planar coupling coefficient, k_p [73].

Table 16.3 Dielectric, piezoelectric, and ferroelectric properties of Li^{+} doped lead-free piezoelectric material [9]

Li doped lead-free piezoelectric material	Dielectric property		Piezoelectric property				Ferroelectric property	
	ε_r	tan δ	d_{33} (pC/N)	k_p	k_t	Q_m	P_r ($\mu C/cm^2$)	E_c (kV/cm)
$Na_{0.484}Bi_{0.456}Ba_{0.06}Ti_{0.97}Nb_{0.03}O_3$-xLi	–	–	–	–	–	–	–	–
$(K_{0.48}Na_{0.48}Li_{0.04})(Nb_{0.8}Ta_{0.2})O_3$ (KNN-LT)	–	–	165	0.363	0.438	–	17.651–13.917	14.514–12.5
$(0.7BiFeO_3$-$0.3BaTi_03)$ + 0.025Bi $(Zn_{0.5}Ti_{0.5})O_3$ + xLi_2CO_3 + $0.0035MnO_2$ (BFBT-BZT-xLC)	–	–	184	0.357	–	35.687	21.6	–
$Ba_{0.85}Ca_{0.15}(Zr_{0.10}Ti_{0.90})O_3$ + xLi_2TaO_3 (BCZT-xLT)	–	–	433	44.6	–	–	9.16	0.32 kV/mm
(1–x) $Na_{0.47}Bi_{0.47}Ba_{0.06}TiO_3$ –xLN (NBT–BT–xLN)	–	–	–	–	–	–	–	–
$Ba_{0.95}Ca_{0.05})(Ti_{0.93}Sn_{0.07})O_3$ – yLiF (BCTS-yLiF)	–	–	413	0.48	–	–	–	–
$Ba_{0.85}Ca_{0.15})(Ti_{0.9}Zr_{0.1})O_3$ + x wt% Li_2CO_3	–	–	530	0.43	–	206	Decrease	Decrease
$0.96(K_{0.48}Na_{0.52}- xLi_x)NbO_3$ $-0.04Bi_{0.5}Na_{0.5}ZrO_3$ (KNLxN-BNZ)	–	–	252 (x = 0.03)	–	–	–	–	–

Li doped lead-free piezoelectric material	Dielectric property		Piezoelectric property				Ferroelectric property	
	ε_r	tan δ	d_{33} (pC/N)	k_p	k_t	Q_m	P_r (μC/cm^2)	E_c (kV/cm)
$0.96(Na_{0.4}K_{0.1}Bi_{0.5}TiO_3)$ –$0.04Na_{1-x}Li_xNbO_3$	–	–	–	–	–	–	Decrease	Decrease
$Bi_{0.5}(Na_{0.4}$-$xK_{0.1}Li_x)$ $0.96Sr_{0.04}Ti_{0.975}Ta_{0.025}O_3$	–	–	–	–	–	–	–	–
$BaTiO_3$-xLi_2O	1977	Lowest	200	–	–	–	2.0	8.0
$Ba_{0.99}Ca_{0.01}Zr_{0.02}Ti_{0.98}O_3$ – xLiF			340	0.44	–	–	9.8	220
$Ba_{0.85}Ca_{0.15}Ti_{0.9}Zr_{0.1}O_3$ + 0.3 wt% Li_2CO_3	2250	1.35	512	0.49	–	190	10	1.8
$Ba_{0.85}Ca_{0.15}Zr_{0.1}Ti_{0.9}O_3$ – xLiF	4394	1.2	380	39.5	–	–	–	–
$K_{0.48}Na_{0.52})_{1-x}Li_xNbO_3$ – 0.015CuO (KNN-Li-CuO)	–	–	>110	–	–	350	16	–
$0.97Bi_{0.5}Na_{0.4}$ – $xLi_xK_{0.1}TiO_3$-$0.03CaZrO_3$ (BNKTCZ-xLi)	–	–	–	–	–	–	–	0.34
$Li_{0.058}$ $(Na_{0.535}K_{0.480})$ 0.966 $(Nb_{0.9}Ta_{0.1})$ O_3	2701	0.013	347	0.48	–	–	–	–
$(Na_{0.52}K_{0.435}Li_{0.045})$ $Nb_{0.87}Sb_{0.08}Ta_{0.05}O_3$ + x wt% LiF	–	–	D31-41	0.27	0.40	–	–	–

(*Continued*)

Table 16.3 (*Continued*)

Li doped lead-free piezoelectric material	Dielectric property		Piezoelectric property				Ferroelectric property	
	ε_r	tan δ	d_{33} (pC/N)	k_p	k_t	Q_m	P_r (μC/cm^2)	E_c (kV/cm)
(1-x)$K_{0.5}Na_{0.5}NbO_3$-xLiBiO_3	871.8	Decrease with increasing x	115	0.27		93.1	–	–
(1-x)($Na_{0.5}K_{0.5}$)NbO_3-xLiSbO_3	Highest	–	143	57.8	–	–	–	–
(1-x)$K_{0.48}Na_{0.52}NbO$–	–	–	110	27.15	27.75	195	23.92	20.76
$BaTiO_3$–xLiF	–	–	270	–	–	–	9.5	3.89
0.85$Bi_{0.5}Na_{0.5}$(1-x)$Li_{0.5x}TiO_3$ – 0.11$Bi_{0.5}K_{0.5}TiO_3$-0.04$BaTiO_3$	–	–	200	–	–	33	–	–
0.852[$Bi_{0.5}(Na_{(1-x)}Li_x)_{0.5}$]$TiO_3$-0.110 ($Bi_{0.5}K_{0.5}$)$TiO_3-0.038BaTiO_3$	1239	0.033	184	0.30	–	–	–	–

16.4.3 Dielectric Property

Dielectric property is concerned with Curie temperature (T_c), dielectric constant (ε_r), and dielectric loss (tan δ). It has been studied by Zhang et al. [74] that on the addition of Li_2CO_3 to the lead-free (K,Na)NbO_3 ceramic material, the T_c is found to be 450°C. There is a positive correlation between T_c and concentration of Li^+. It has been reviewed by many researchers that as the Li^+ doping enhances, there is an increase in T_c and the phase transition from orthorhombic to tetragonal decreases [31]. When Li_2O is doped in (1-x) (99.6 $K_{0.5}Na_{0.5}$ NbO_3 – 0.4 $MnBiO_3$), the material shows highest dielectric constant, i.e., 420. In pure KNN ceramic, the reported T_c is 420°C, but in f (1-x) ($Na_{0.535}K_{0.480}$)$NbO_{3-x}LiNbO_3$ the T_c is 450°C [75]. Table 16.3 summarizes various electrical properties of lithium-ion doped lead-free piezoelectrical material.

16.4.4 Ferroelectric Property

Various ferroelectric properties, viz., values of remnant polarization, P_r, coercive field, E_c, and saturated polarization, P_s, have been shown by the P-E hysteresis loop. In order to have good ferroelectric properties, materials should have a higher value of P_r and small E_c to maintain their orientation. Better ferroelectric properties can be achieved by lithium doping. All the electrical properties of Li^+ doped lead-free piezoelectric material are listed in Table 16.3.

16.5 Applications

There are various applications of piezoelectric materials such as transducers, actuators, nanogenerators, and bioimaging devices. Barium titanate (BT) KNN, ($Bi_{0.5}Na_{0.5}$)TiO_3-$BaTiO_3$, $LiNbO_3$, and polymer-based piezoelectric materials have potential applications in ultrasound imaging. Lead zirconate titanate (PZT)/polyvinylidene difluoride (PVDF), poly(methylmethacrylate) (PMMA), polar poly(γ-benzyl α,l-glutamate) (PBLG), etc. [12] play an important part in the fabrication of piezoelectric biosensors to detect the target molecules. KNN-based ceramics possess outstanding piezoelectric constant and high T_c. It can replace Pb-based piezoelectrics. With the enhanced piezoelectric properties of KNN-based materials, they are widely used in the fabrication of high-sensitivity ultrasound transducers for biomedical applications. Hydrothermal process is employed for the syntehsis of such materials. This synthesis process enhances the properties of ZnO-based PVDF material which has application in the field of wearable nanogenerator [76]. Because of its high power density, BT is also used as nanogenerator [77]. Lead-free piezoelectric materials are also used in transducers and they require large bandwidth, high frequency, and good electrochemical coupling coefficient. High Curie temperature, large electric field-induced strain, and blocking force are needed for the application as an actuator. Sahibata et al. [78] have reported that $0.96(K_{0.5}Na_{0.5})NbO_3 - 0.04CaZrO_3 + 0.03ZrO_2 + 0.05MnCO_3$ is used to develop multilayer (12 piezoceramics layers) actuators, which have maximum strain of order 360 pm/V at 2 kV/mm. Knocking sensor has also been developed by using KNN and $KTiNbO_3$ (KNN-KTN).

16.6 Conclusions

Research of piezoelectric material is a diverse field and huge work has been reported in this area. This chapter tried to summarize an idea about lead-free piezoelectric material and their applications in various fields. In the past few years, extensive research has been done in lithium doped lead-free piezoelectric material. The chapter provides knowledge about the synthesis methods for lead-free piezoelectric material. It also covers various important properties, viz., piezoelectric, dielectric, and ferroelectric properties, of Li^+ doped ceramics and also discusses the major effects of their doping. Lithium doped lead-free piezoelectric materials have wide application areas such as medical and health applications, communication, and technology. Considerable work has to be done in this field focusing on improved performance in electrical and piezoelectrical devices. Lithium will be a good and cost-effective alternative dopant for such lead-free piezoelectric ceramics.

References

1. Jiang M, Zhang J, Rao G, Li D, Randall CA, Li T, Peng B, Li L, Gu Z, Liu X, Huang H, Ultrahigh piezoelectric coefficient of a lead-free

$K_{0.5}Na_{0.5}NbO_3$-based single crystal fabricated by a simple seed-free, *Solid-State Growth Meth.* **7**, 1445–14854, 2019.

2. Kushvaha DK, Rout SK, Tiwari B, Structural, piezoelectric and high-density energy storage properties of lead-free BNKT-BCZT solid solution, *J. Alloy. Compd.* **782**, 270–276, 2019.
3. Zhao R, Li Y, Zheng Z, Kang W, Phase structure regulation and enhanced piezoelectric properties of Li-doped KNN-based ceramics, *Mater. Chem. Phys.* **245**, 122806-122810, 2020.
4. Shi C, Ma J, Wu J, Wang X, Miao F, Huang Y, Chen K, Wu W, Wu B, Coexistence of excellent piezoelectric performance and high Curie temperature in KNN-based lead-free piezoelectric ceramics, *J. Alloy. Compd.* **846**, 156245–156253, 2020.
5. Xue H, Zheng T, Wu J, Understanding the nature of temperature stability in potassium sodium niobate-based ceramics from structure evolution under external field, *ACS Appl. Mater. Interfaces* **12**, 32925–32934, 2020.
6. Tan L, Sun Q, Wang Y, Outstanding piezoelectric properties of Al-substituted potassium-sodium niobate-based lead-free piezoceramics. *J. Alloys Compd.*, **836**, 155419(1-7), 2020.
7. Yan X, Zheng M, Zhu M, Hou Y, Enhanced electrical resistivity and mechanical properties in BCTZ-based composite ceramic, *J. Adv. Dielect.* **9**, 1950036–1950042, 2019.
8. Mayeen A, Kalarikkal N, Development of ceramic-controlled piezoelectric devices for biomedical applications. In: *Fundamental Biomaterials: Ceramics*. Elsevier Ltd. 2018.
9. Fadhlina H, Atiqah A, Zainuddin Z, A review on lithium doped lead-free piezoelectric materials, *Mater. Today Commun.* **33**, 104835–104851, 2022.
10. Wang H, Jasim A, Piezoelectric energy harvesting from pavement, In *Eco-Efficient Pavement Construction Material*. Elsevier Ltd. 2020.
11. Benyoussef M, Zannen M, Belhadi J, Monoun B, Dellis JL, Marssi ME, Dielectric, ferroelectric, and energy storage properties in dysprosium doped sodium bismuth titanate ceramics, *Ceram. Int.* **44**(16), 19451–19460, 2018.
12. Li S, Liu J, Guo T, Dong W, Bi K, Luo Y, Piezoelectricity and flexoelectricity of sodium bismuth titanate-based ceramics, *Ceram. Int.* **46**, 2049–2054, 2019.
13. He X, Han F, Liu M, Yuan Z, Jiang X, Hu C, Ren S, Lei X, Liu L, High temperature dielectric and relaxation behavior of tantalum-doped sodium bismuth titanate-barium titanate ceramics, *J. Elect. Mater.* **49**, 6643–6655, 2020.

14. Nan B, Olhero S, Pinho R, Vilarinho PM, Button TW, Direct ink writing macroporous lead-free piezoelectric $Ba_{0.85}Ca_{0.15}Zr_{0.1}Ti_{0.9}O_3$, *J. Am. Ceram. Soc.* **102**, 3191–3203, 2018.
15. Wei H, Wang H, Xia Y., Cui D, Shi Y, Dong M, Liu C, Ding T, Zhang J, Ma Y, Wang N, Wang Z, Sun Y, Wei R, Guo Z, An overview of lead-free piezoelectric materials and devices, *J. Mater. Chem.* C **6**, 12446–12467, 2018.
16. Gao X, Cheng Z, Chen Z, Liu Y, Meng X, Zhang X, Wang J, Guo Q, Li B, Sun H, Gu Q, Hao H, Shen Q, Wu J, Zhou X L, Ringer SP, Liu H, Zhang L, Chen W, Li F, Zhang X, The mechanism for the enhanced piezoelectricity in multi-elements doped $(K,Na)NbO_3$ ceramics, *Nat. Commun.* **12**, 881(1-9), 2021.
17. Gio PD, Phong ND, Effects of LiF on the structure and electrical properties of $(Na_{0.52}K_{0.435}Li_{0.045})Nb_{0.87}Sb_{0.08}Ta_{0.05}O_3$ lead-free piezoelectric ceramics sintered at low temperatures, *J. Mater. Sci. Chem. Eng.* **3**, 13–20, 2015.
18. Ma N, Zhang BP, Yang WG, Low-temperature sintering of Li_2O-doped $BaTiO_3$ lead-free piezoelectric ceramics, *J. Electroceram.* **28**, 275–280, 2012.
19. Chen C, Huang Y, Tan Y, Sheng Y, Structure and electric properties of Li-modified $(K_{0.48}Na_{0.52})NbO_3$-0.015CuO lead-free piezoceramics, *J. Alloys Compd.* **663**, 46–51, 2016.
20. Zhang RZ, Guo WP, Zhao GL, Improved Curie temperature of Li-doped lead-free $(K,Na)NbO_3$ ceramics analysis, *Adv. Mater. Res.* **1061**, 91, 2014.
21. Chen Y, Xue D, Chen Z, Jiang X, Gou J, Liu G, Liu X, Lithium-modified $(K_{0.5}Na_{0.5})NbO_3$–$BiAlO_3$ lead-free piezoelectric ceramics with high Curie temperature, *Ceram. Int.* **43**, 634–640, 2016.
22. Quyet NV, Bac LH, Dung DD, Effect of Li_2CO_3 addition on the structural, optical, ferroelectric, and electric-field-induced strain of lead-free BNKT-based ceramics, *J. Phys. Chem. Solids.* **85**, 148–184, 2015.
23. Tan CKI, Yao K, Ma J, Effects of LiF on the structure and properties of $Ba_{0.85}Ca_{0.15}Zr_{0.1}Ti_{0.9}O_3$ lead-free piezoelectric ceramics, *Inter. J. Appl. Ceram. Tech.* **10**, 701–706, 2013.
24. Li Q, Ning L, Wang C, Fan H, Gaint electric field-induced strain and ferroelectric behavior of $Bi_{0.5}Na_{0.4}K_{0.1}TiO_3$-$Na_{1-x}Li_xNbO_3$ lead-free ceramics, *Mater. Sci. Eng. B* **263**, 114819, 2021.
25. Wongsaenmai S, Ananta S, Yimnirun R, Effect of Li addition on phase formation behavior and electrical properties of $(K_{0.5}Na_{0.5})NbO_3$ lead free ceramics, *Ceram. Int.* **38**(1), 147–152, 2012.

26. Yan B, Fan H, Wang C, Zhang M, Yadav AK, Zheng X, Wang H, Du Z, Giant electro-strain and enhanced energy storage performance of $(Y_{0.5}Ta_{0.5})^{4+}$ co-doped $0.94(Bi_{0.5}Na_{0.5})TiO_3$-$0.06BaTiO_3$ lead-free ceramics, *Ceram. Int.* **46**(1), 281–288, 2020.

27. Du H, Huang Y, Tang H, Feng W, Qin H, Lu X, Structure and electrical properties of $[(K_{0.49}Na_{0.51})_{1-x}Li_x](Nb_{0.90}Ta_{0.04}Sb_{0.06})O_3$ lead-free piezoceramics, *Ceram. Int.* **39**, 5689–5694, 2013.

28. Liu X, Zhu M, Chen Z, Fang B, Ding J, Zhao X, Xu H, Luo H, Structure and electrical properties of Li-doped $BaTiO_3$-$CaTiO_3$-$BaZrO_3$ lead-free ceramics prepared by citrate method, *J. Alloys Compd.* **613**, 219–225, 2014.

29. Zhang Y, Sun HJ, Chen W, Li-modified $Ba_{0.99}Ca_{0.01}Zr_{0.02}Ti_{0.98}O_3$ lead-free ceramics with highly improved piezoelectricity, *J. Alloys Compd.* **694**, 745–751, 2017.

30. Feng W, Huang Y, Yang C, Lu J, Huang Y, Effects of the addition of $Li_{0.8}Ni_{0.2}Nb_{0.96}O_3$ on the structure and electrical properties of $K_{0.48}Na_{0.52}NbO_3$ lead-free piezoceramics, *Mater. Sci. Eng. B.* **189**, 7–12, 2014.

31. Li M, Chan NY, Wang D, Improved thermal stability of the piezoelectric properties of (Li, Ag)-co-modified (K, Na) NbO_3-based ceramics prepared by spark plasma sintering, *J. Am. Ceram. Soc.* **100**, 2984–2990, 2017.

32. Zhang C, Wu F, Wang J, Lin W, He X, Effect of sintering temperature on the electric properties of KNLNT ceramics, *Ferroelectrics* **562**, 1–9, 2020.

33. Kim J, Ji JH, Shin DJ, Koh JH, Improved Li and Sb doped lead-free (Na,K) NbO_3 piezoelectric ceramics for energy harvesting applications, *Ceram. Int.* **44**, 22219–22224, 2018.

34. Gao S, Yao Z, Ning L, Dong G, Fan H, Li Q, Enhanced bipolar strain response in lithium/niobium co-doped sodium–barium bismuth titanate lead-free ceramics, *Adv. Eng. Mater.* **19**, 1700125, 2017.

35. Sun Y, Yang H, Guan S, Cao Y, Jiang M, Liu X, Strong piezoelectricity of Li_2CO_3-doped $BiFeO_3$–$BaTiO_3$–$Bi(Zn_{0.5}Ti_{0.5})O_3$ lead-free piezoelectric ceramics with high Curie temperature and high temperature stability, *J. Alloys Compd.* **819**, 153058, 2019.

36. Yang W, Zhang B, Ma N, Zhao L, High piezoelectric properties of $BaTiO_3$–xLiF ceramics sintered at low temperatures, *J. Eur. Ceram. Soc.* **32**, 899–904, 2012.

37. Cheng C, Chen K, Lee D, Jong F, Chen M, Chang J, Dielectric, piezoelectric, and vibration properties of the LiF-doped $(Ba_{0.95}Ca_{0.05})(Ti_{0.93}Sn_{0.07})O_3$ lead-free. *Piezoceram. Sheets.***11**, 182 (1-9), 2018.

38. Li D, Jiang M, Han S, Jin Q, Xu Y, Yao X, Zhang K, Li L, Miao L, Zhou C, Rao G, Influence of trace lithium addition on the structure and properties of $K_{0.5}Na_{0.5}NbO_3$-based single crystals, *J. Mater. Sci. Mater. Electr.* **31**, 4857–4866, 2020.

39. Chen X, Ruan X, Zhao K, He X, Zeng J, Li Y, Zheng L, Park CH, Li G, Low sintering temperature and high piezoelectric properties of Li-doped (Ba,Ca)(Ti,Zr) O_3 lead-free ceramics, *J. Alloys Compd.* **632**, 103–109, 2015.

40. Feng Q, Huang K, Luo N, Yuan C, Zhou C, Wei Y, Li Q, Fujita T, Chen G, Formation mechanism, dielectric properties, and energy-storage density in $LiNbO_3$-doped $Na_{0.4}7Bi_{0.47}Ba_{0.06}TiO_3$ ceramics, *J. Mater. Sci. Mater. Electr.* **31**, 13368–13375, 2020.

41. Jiang MH, Deng MJ, Yang ZP, Fu ZX, Structure and piezoelectric properties of $(1\text{-}x)K_{0.5}Na_{0.5}NbO_3\text{-}x LiBiO_3$ lead-free piezoelectric ceramics, *Trans. Nonferrous Met. Soc.* **22**, s133–s137, 2012.

42. Yin DS, Zhao Z, Dai YJ, Electrical properties and relaxor phase evolution of Li-modified BNT-BKT-BT lead-free ceramics, *J. Am. Ceram. Soc.* **99**, 2354–2360, 2016.

43. Pei CH, Effect of LiF & HfO_2 on the LKNNT lead-free piezoelectric ceramics, *Int. Conf. Appl. Syst. Innov.* 653–656, 2017.

44. Sun T, Dai Y, Wang H, Sol-gel derived Li_2O doped BNT-BKT-BT ceramics: Microstructure and piezoelectric properties, *Key Eng. Mater.* **515**, 1355–1358, 2012.

45. Jamil NHBJ, Jumali MHH, Zainuddin ZB, Izzuddin IB, Pojprapai S, Determination of optimum parameter for electrical poling treatment on BaTiO3 piezo-ceramic using home-built system, *Sains Malays.* **48**(2), 425–433, 2019. DOI: 10.17576/jsm-2019-4802-21.

46. Adewale AA, Chik A, Mohd Zaki R, Electronic properties of calcium and zirconium co-doped $BaTiO_3$, *Mater. Sci. Forum* **1010**, 308–313, 2020.

47. Kornphom C, Laowanidwatana A, Bongkarn T, Influence of sintering temperature on properties of BNKLLT–6wt% BCTZ binary lead-free piezoelectric ceramic prepared through the solid-state combustion technique, *Phase Trans.* **90**, 317–324, 2017.

48. Tamrakar P, Bajpai PK, Structural and dielectric behaviour of doped bismuth sodium titanate: Lead free piezoelectric materials. *AIP Conf. Proc.* **1837**, 040062, 2017.

49. Ruth DEJ, Sundarakannan LVB, Structure–property relation to enhance the piezoelectric free piezoelectric ceramics, *J. Mater. Sci. Mater. Electr.* **29**, 5433–5438, 2018.

50. Verma A, Yadav AK, Khatun N, Kumar S, Jangir R, Srihari V, Reddy VR, Liu SW, Biring S, Sen S, Structural, dielectric and ferroelectric studies of thermally stable and efficient energy storage ceramic material: $(Na_{0.5-x}K_xBi_{0.5-x}La_x)TiO_3$, *Ceram. Int.* **1**, 2018, doi:10.1016/j.ceramint.2018.07.312.

51. Singh P, Pandey R, Singh P, Tailoring the electrical and structural properties of sodium bismuth titanate with sintering temperature, *Mater. Today Proc.* 44, 166, 2020.

52. Bhattacharyya R, Omar S, Electrical conductivity study of B-site Ga doped nonstoichiometric sodium bismuth titanate ceramics, *J. Alloys Compd.* **746**, 54–61, 2018.

53. Benyoussef M, Zannen M, Belhadi J, Monoun B, Dellis JL, Marssi ME, Dielectric, ferroelectric, and energy storage properties in dysprosium doped sodium bismuth titanate ceramics, *Ceram. Int.* **44**, 19451–19460, 2018.

54. Saito Y, Takao H, High performance lead-free piezoelectric ceramics in the $(K,Na)NbO_3$-$LiTaO_3$ solid solution system, *Ferroelectrics* **338**, 17–32, 2006.

55. Mahdi R., Al-bahnam NJ, Ajeel MA, Al-keisy A, Ali T, Majid WHA, High-performance $(K,Na)NbO_3$-based binary lead-free piezoelectric ceramics modified with acceptor metal oxide, *Ceram. Int.* **46**, 21762–21770, 2020.

56. Zhai Y, Du J, Chen C, Li W, Hao J, The photoluminescence and piezoelectric properties of Eu_2O_3 doped KNN-based ceramics, *J. Alloys Compd.* **829**, 154518, 2020.

57. Kornphom C, Udeye T, Thongbai P, Bongkarn T, Phase structures, PPT region and electrical properties of new lead-free KNLNTS-BCTZ ceramics fabricated via the solid-state combustion technique, *Ceram. Int.* **43**, s182–s192, 2017.

58. Xiang J, Chuanbin W, Harumoto T, Zhang S, Tu R, Shen Q, Shi J, Structure and electrical properties of BCZT ceramics derived from microwave-assisted sol–gel-hydrothermal synthesized powders, *Sci. Rep.* **10**, 20352, 2020.

59. Gio PD, Phong ND, Effects of LiF on the structure and electrical properties of $(Na_{0.52}K_{0.435}Li_{0.045})Nb_{0.87}Sb_{0.08}Ta_{0.05}O_3$ lead-free piezoelectric ceramics sintered at low temperatures, *J. Mater. Sci. Chem. Eng.* **3**, 13–20, 2015.

60. Li Q, Ning L, Wang C, Fan H, Gaint electric field-induced strain and ferroelectric behavior of $Bi_{0.5}Na_{0.4}K_{0.1}TiO_3$-$Na_{1-x}Li_xNbO_3$ lead-free ceramics, *Mater. Sci. Eng. B: Solid State Mater. Adv. Tech.* **263**, 114819, 2021.
61. Wongsaenmai S, Ananta S, Yimnirun R, Effect of Li addition on phase formation behavior and electrical properties of $(K_{0.5}Na_{0.5})NbO_3$ lead free ceramics, *Ceram. Int.* **38**, 147–152, 2012.
62. C. Chen, Y. Huang, Y. Tan, Y. Sheng, Structure and electric properties of Li modified $(K_{0.48}Na_{0.52})NbO_3$-0.015CuO lead-free piezoceramics, *J. Alloys Compd.* **663**, 46–51, 2016.
63. Wang C, Li Q, Zhang W, Yan B, Yadav AK, Peng H, Fan H, $[Bi_{0.5}(Na_{0.4-x}Li_xK_{0.1})]_{0.96}Sr_{0.04}Ti_{0.975}Ta_{0.025}O_3$ lead-free RELAXOR ceramics with the enhanced recoverable energy density, *Ceram. Int.* **46**, 715–721, 2020.
64. Chen Y, Xue D, Chen Z, Jiang X, Gou J, Liu G, Liu X, Lithium-modified $(K_{0.5}Na_{0.5})NbO_3$–$BiAlO_3$ lead-free piezoelectric ceramics with high Curie temperature, *Ceram. Int.* **43**, 634–640, 2016.
65. Tan CKI, Yao K, Ma J, Effects of LiF on the structure and properties of $Ba_{0.85}Ca_{0.15}Zr_{0.1}Ti_{0.9}O_3$ lead-free piezoelectric ceramics, *Int. J. Appl. Ceram. Tech.* **10**, 701–706, 2013.
66. Kumar S, Shandilya M, Thakur S, Thakur N, Kaur GA, Effect of sol–gel synthesis method on the structural, electrical, *J. Sol Gel Sci. Tech.* 2019, doi:10.1007/s10971-019-05077-1.
67. Chilibon I, Marat-Mendes JN, Ferroelectric ceramics by sol-gel methods and applications: A review, *J. Sol Gel Sci. Tech.* **64**, 571–611, 2012.
68. Ghasemian MB, Rawal A, Liu Y, Wang D, Approaching piezoelectric response of Pb-piezoelectrics in hydrothermally synthesized $Bi_{0.5}(Na_{1-x}K_x)_{0.5}TiO_3$ nanotube, *ACS Appl. Mater. Interfaces* **10**(24), 20816–20825, 2018.
69. Zhang Q, Shi S, Chen W, An electromechanical coupling model of a longitudinal vibration type piezoelectric ultrasonic transducer, *Ceram. Int.* **41**, S638–S644, 2015.
70. Zu H, Wu H, Wang QM, High-temperature piezoelectric crystals for acoustic wave sensor applications, *IEEE Trans. Ultrason. Ferroelectr. Freq. Control.* **63**, 486–505, 2016. DOI: 10.1109/TUFFC.2016.2527599.
71. Suna T, Daib YJ, Wang HQ, Sol-gel derived Li_2O doped BNT-BKT-BT ceramics: Microstructure and piezoelectric properties, *Key Eng. Mater.*, **512–515**, 1355–1358, 2012.
72. Chen C, Huang Y, Tan Y, Sheng Y, Structure and electric properties of Li modified $(K_{0.48}Na_{0.52})NbO_3$-0.015CuO lead-free piezoceramics, *J. Alloys Compd.* **663**, 46–51, 2016.

73. Chen X, Zeng J, Kim D, Zheng L, Lou Q, Hong Park C., G. Li, Understanding of role of Li in high-performance Pb-free Li-doped $(Ba_{0.85}Ca_{0.15})(Ti_{0.9}Zr_{0.1})O_3$ piezoceramics from theory and experiments, *Mater. Chem. Phys.* **231**, 173–180, 2019.

74. Zhang RZ, Guo WP, Zhao GL, Improved Curie temperature of Li-doped lead-free $(K,Na)NbO_3$ ceramics analysis, *Adv. Mater. Res.* **1061**, 91–95, 2014.

75. Pramanik S, Murphy BP, Azuan N, Osman A, Developments of immobilized surface modified piezoelectric crystal biosensors for advanced applications, *Int. J. Electrochem. Sci.* **8**, 8863–8892, 2013.

76. Kim M, Wu YS, Kan EC, Fan J, Breathable and flexible piezoelectric ZnO@PVDF fibrous nanogenerator for wearable applications, *Polymers* **10**, 745, 2018.

77. Wei H, Wang H, Xia Y, Cui D, Shi Y, Dong M, Liu C, Ding T, Zhang J, Ma Y, Wang N, Wang Z, Sun Y, Wei R, Guo Z, An overview of lead-free piezoelectric materials and devices, *J. Mater. Chem. C.* **6**, 12446–12467, 2018.

78. Shibata K, Wang R, Tou T, Koruza J, Applications of lead-free piezoelectric materials, *MRS Bull.* **43**, 612–616, 2018.

Chapter 17

Virus-Based Piezoelectric Energy Generation

Reena Saxena[a] and Bhaskar Pratap Chaudhary[b]
[a]*School of Applied Sciences, Suresh Gyan Vihar University, Jaipur, India*
[b]*Department of Applied Sciences, Chandigarh Engineering College, Jhanjeri, India*
saxena8284@rediffmail.com

Piezoelectric materials are well known for their ability to generate electrical charge after applied mechanical stress. Many researchers have developed various types of piezoelectric materials by using ceramics and natural biomaterials. Due to toxic nature, poor biocompatibility, and brittle characteristics, ceramic-based piezoelectric materials have limitations in wearable and biomedical applications. To overcome these limitations, recently researchers have been focused on synthesizing virus-based piezoelectric energy generators. On the basis of well-defined shape, biological and physical properties, as well as ability to show functional peptides using genetic and chemical modifications, M13 phage has been considered as most suitable virus for the synthesis of biomaterial-

Piezoelectric Materials: Design and Applications
Edited by N. B. Singh and Dev Kumar Mahato

ISBN 978-981-5129-40-3 (Hardcover), 978-1-003-59897-8 (eBook)
www.jennystanford.com

based piezoelectric energy generators. These M13 phages have unique structural properties that make them suitable to become piezoelectric materials. With the help of genetic engineering, these M13 phages can be easily modified at molecular structure and surfaces as per requirement. Genetically modified bacterial viruses have pulled attraction of many researchers to develop biopiezoelectric devices. Research has shown that due to liquid crystalline property, these phages are having capability of self-assembly into phage films. This assembly will result in simple fabrication of piezoelectric generators that will be capable to turn on a liquid crystal display (LCD). This chapter has been focused on the properties of viruses for the production of piezoelectric materials used for energy generation.

17.1 Introduction

Energy is the basic requirement to perform daily activities of human life. All industrial sectors also depend on energy sources. Increasing demand of energy consumption globally has become major threat for environment and has created impact on global warming [1]. Due to sustainable feature and eco-friendly nature, renewable energy sources have become popular worldwide. These sources are solar energy, bioenergy, kinetic energy, and sources based on traditional fossil fuels. Nowadays, batteries are widely used to store large amount of energy for further use but use of batteries as storage has some drawback such as short durability, large size, replacement issue, and side effect on environment [2]. To overcome all these issues, researchers have focused on utilizing energy sources like vibration, heat, light, and sound and converting them into low-power electrical energy to develop self-powered electric devices [3]. At present time, scientists are more interested to develop such flexible wearable electronic devices that can obtain energy from human body activity [4]. Keeping these properties in mind, piezoelectric materials have been synthesized by researchers to show electric responses by applying mechanical stress. Conventional piezoelectric materials have acquired great significance due to their unique properties such as biosafety, biodegradability, and biocompatibility. Hence, piezoelectric materials can be considered with the property of environmental friendliness [5–8]. Furthermore, organic-based piezoelectric biomaterials have weak piezoelectric property as compared to inorganic-based materials, but they have high biocompatibility and eco-friendly nature over conventional inorganic materials [9]. In 1880, Jacques and Pierre Curie discovered piezoelectric property in quartz material. After that, many other types of materials such as ceramics and naturally found biomaterials have been obtained with piezoelectric property. In case of ceramic materials, lithium niobate ($LiNbO_3$), lithium tantalite ($LiTaO_3$), barium titanate ($BaTiO_3$), lead zirconate titanate (PZT), etc. [10–13] have been utilized to develop piezoelectric devices. At present time, lead zirconate titanate has gained attraction of many researchers due to its practical applications. Additional to ceramics, natural biomaterials such as hair, bone, tendon, and wood have been considered as the major source of piezoelectric effect [14–19]. Biomolecules such as amino acids, peptides, protein, polysaccharides, cellulose, animal-derived polymer, and many more have been utilized for piezoelectric studies [32]. Proteins are large molecules made up of amino acids; an antibacterial enzyme lysozyme, found in egg white of birds, mammalian tears, saliva, and milk, has weak piezoelectric effect [33]. After measuring by piezometer, both tetragonal and monoclinic crystal structures of lysozyme were found with weak piezoelectric effect with an average piezoelectric coefficient of 3.16 pm V^{-1} in case of tetragonal and 0.94 pm V^{-1} for monoclinic one [34]. Amino acids are important organic compounds that are found in genetic code of human body. There are 20 kinds of amino acids, each with unique characteristics. The study of piezoelectric effect of 20 amino acids was performed by Lemanov et al. [35–37]. After the study, over 16 amino acids and their compounds were found with piezoelectric effect at room temperature. According to the study, L-methionine showed moderate piezoelectricity at $T < 210$ K while γ-glycine and DL-alanine were observed to have the best property of piezoelectricity, even better than quartz crystal. This study attracted focus of researchers on piezoelectric explorations of amino acids [36]. Other biomolecules such as polysaccharides, which are generally found in animals, plants, and microorganisms, exhibit piezoelectric effect due to their fibrous structure. Additional to that, some natural polysaccharides such as cellulose found in wood show piezoelectric effect [32]. Similar to cellulose, chitin is also considered

Table 17.1 Measurement of piezoelectricity in different biomolecules

Type of biomolecule			Source of biomolecule	Piezoelectric coefficient	Reference
Amino acids	Glycine	α-glycine	Food and animal protein	No net polarization	[32]
		β-glycine		d33: 10.4 pm V^{-1} d16: 195 pm V^{-1}	[32]
		γ-glycine		d11, d22, d33: 1.7, −1.1, and 9.93 pm V^{-1}, respectively	[32]
	DL-alanine			d3: 10.34 pm V^{-1}.	[43, 44]
	L-methionine			Moderate piezoelectricity	[43]
Diphenylalanine (FF) peptide nanostructure			Animal protein and plants	d33: 9.9–17.9 pm V^{-1}	[45, 46]
Lysozyme	Tetragonal crystal structure		Egg whites of birds, mammalian tears, saliva, and milk	3.16 pm V^{-1}	[47, 48]
	Monoclinic crystal structure			0.94 pm V^{-1}	
Polysaccharides	Cellulose		Natural wood	4.3–36.4 pm V^{-1}.	[32]
	Chitin		Exoskeletons and cell walls	~4 pm V^{-1}	[49]
Animal-derived polymer	Silk		Insect	~1 pC N^{-1}	[50, 51]
	Collagen		High-protein food	d33 of ~2.64 pC N^{-1}	

as an important biomolecule as its production is 100 billion tons/year and is used in the exoskeletons of crustaceans, insects, and cell wall of fungi [38]. Due to similarity in structure with cellulose, chitin also exhibits piezoelectricity. One important biomolecule is animal-derived polymer that exists in semi-crystalline and crystalline structure and is well known for piezoelectric effect. Collagen provides structure to our bodies, protecting and supporting the softer tissues and connecting them with the skeleton, considered as a good source of piezoelectricity [39–42]. Table 17.1 gives the value of piezoelectricity found in different biomolecules.

Although all these materials have good piezoelectric effect, they have some limitations when applied for practical applications. To overcome these limitations, a variety of self-assembled biomaterials, polysaccharides, and viruses have been considered as good replacement [20–24]. Along with the development of piezoelectric devices, these biomaterials can also be utilized for developing sensors, energy harvesting, energy storage, and many more [25–27]. Additional to all these types of materials, virus-based piezoelectric materials for the applications of energy harvesting devices have rarely been reported in the literature. This chapter has focused on various properties of viruses that make them superior materials for the applications of piezoelectric energy harvesting. This chapter has also explained the synthesis process of virus-based piezoelectric materials for the applications of energy harvesting devices.

17.2 Viruses: Building Blocks for Piezoelectric Energy Devices

Viruses are well-known biological materials due to their unique structures, physical properties, high reactivity, and a wide range of structural diversity.

On the basis of engineering applications, viruses have the following unique features [28]:

(i) They have higher similarity with collagen (M13 phages).
(ii) Mass productivity after bacterial infection.
(iii) They can be easily modified with the help of genetic engineering.

(iv) They can easily self-assemble to make highly ordered structure.

(v) Viruses have higher degree of piezoelectric effect.

Recently, researchers have used the virus called M13 bacteriophage for developing piezoelectric energy devices [29]. This virus has capability to replicate itself to produce millions of copies within few hours, so there is huge amount available for producing piezoelectric materials. Phages can be easily engineered genetically, so naturally they modify themselves into ordered shape. Viral population do not grow through cell division, because they are acellular. When infected, the host cell is forced to rapidly produce thousands of copies of the original viruses. M13 phage has a long rod-shaped structure with 880 nm length and 6.6 nm diameter (Fig. 17.1). Around 2700 copies of major coat proteins and five copies of minor coat proteins are present on either end of virus. Due to lack of inversion symmetry in M13 protein coat structure, it is capable to demonstrate intrinsic piezoelectric properties [29]. When an electric field is applied to the film of M13 viruses, it creates the protein into helical shape and this helical structure makes the virus twist and turn in response due to which piezoelectric effect is produced. Piezoelectric strength of viruses can be increased by genetically modifying it with addition of four negatively charged amino acid residues to one end of helical protein of virus. Addition of this molecule will enhance the charge difference between positive and negative ends of protein and will create voltage of virus. Bacteriophages, a type of viruses, exist in different shapes. Genome of viruses can be DNA or RNA, either single or double stranded (Fig. 17.1). Bacteriophages were discovered prior to archaea and some of the bacteriophages, mostly with tailed shape, can be found similar to both bacteria and archaea. Archaeal viruses are larger than bacterial viruses [52]. Most of the bacteriophages are found with covering of protein sheet but without lipid external envelop. Phages are found with different shapes like filamentous or icosahedral. Icosahedral phages are found in different forms and length along with tails and tail fibers. Previously, many phage structures have been identified with the help of crystallography and electronic spectroscopy technique [53–55] while in recent years cryo-electron microscopy has also become a key tool in structural study of viruses [56–58].

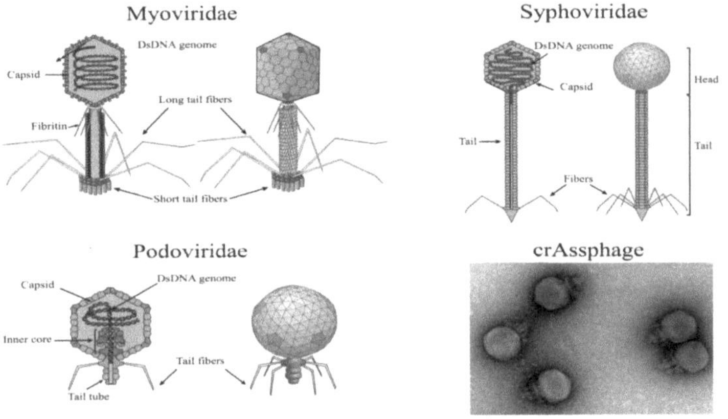

(a) Structure of ds DNA phages

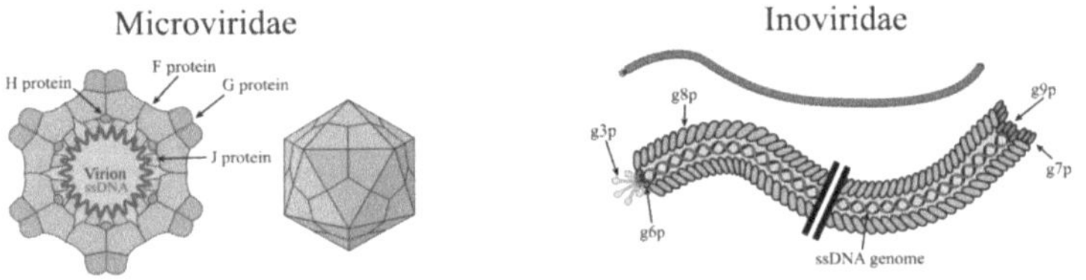

(b) Structure of ss DNA phages

Figure 17.1 Different shapes of bacteriophages.

17.3 Structural Characteristics of M13 Bacteriophage to Support Piezoelectricity

M13 bacteriophage is having a filamentous structure that is made up of single-stranded deoxyribonucleic acid (ssDNA) and capsid protein. Around 2700 copies of major coat proteins with the helical structure pVIII encapsulate ssDNA of M13 phage and approximately 5–7 copies of coat protein under minor category (pVI, pVII, pIX, and pIII) are found on the end of it. The length and diameter ratio of M13 phage is about 880 nm:6.6 nm, respectively [30]. The structure of M13 phage is found very similar to human collagen (Fig. 17.2), and hence M13 phage is capable to mimic nature based on collagen [31]. Group of M13 lysogenic infections are caused by phage viruses, which insert their genome into the bacterial genome. After inserting its genome, the coat protein of minor category pIII of M13 phage gets attached to the receptor present on host bacterial surface to spread infection. After insertion into bacterial genome, multiple copies of M13 phage are produced under metabolic reactions.

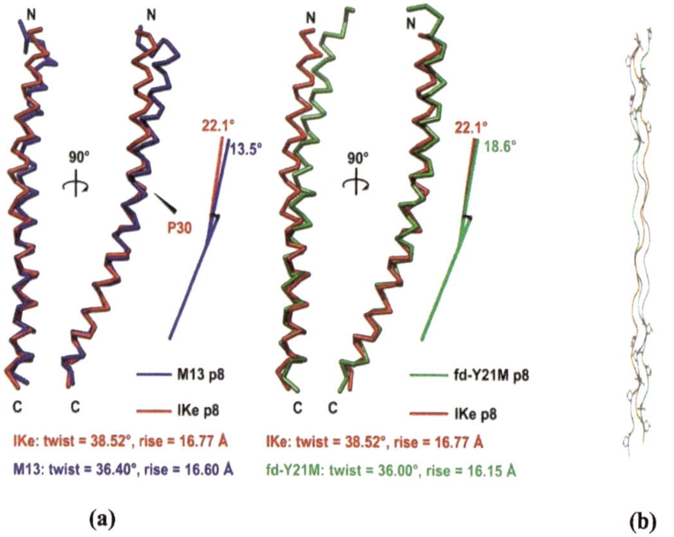

Figure 17.2 (a) Structure of M13 bacteriophage; (b) structure of human collagen.

Ju-Hyuck Lee et al. (2019) have genetically modified M13 phage to enhance piezoelectricity. In their study, M13 phages have been used to produce a variety of nanostructures for uses including the production of tissue, energy storage, and biosensors. Additionally, due to its twofold symmetry and the fivefold rotating arrangement of its coat proteins, the M13 phage itself exhibits piezoelectricity [60]. Radial and axial directional confirmation provides different piezoelectric response. Due to variation in these directional parameters, phages show different responses such as if angle of pVIII protein is tilted by 20° angle, then the expected piezoelectricity of M13 phages in axial direction is three times stronger as compared to radial direction [61–63]. The observed value of polarization of the radial and axial components of a single pVIII is expected to be 0.34 P_0 and 0.94 P_0, respectively. To improve the piezoelectric energy generation from phage-based devices, vertically aligned phage nanostructures with polarization control must be built [59]. In order to manage its geometry and mechanical properties, researchers genetically modified the M13 phage to express a number of functional peptides. The following steps have been approached for this purpose:

(i) To make phage binding to the substrate, more polarized hexa-histidine (6H) has been added to the minor coat protein (pIII) of the M13 phage. With a modified nickel-nitrilotriacetic acid (Ni-NTA) substrate, 6H can attach to it firmly [65].

(ii) To enhance piezoelectric effect, three glutamates (3E) have been fused at the N-terminal of pVIII major coat protein for the enhancement of dipole strength of the protein.

(iii) Tyrosine (Y) residue has been expressed at the N-terminal of the phage in order to provide Y-Y cross-linkage between adjacent phages, which would improve the mechanical stability of the vertically aligned phage nanostructure [64].

Naturally, M13 phages are found with such types of properties that support piezoelectric effect in phage. M13 phages are made up of single-stranded DNA and capsid proteins, which have dipole moments because they contain positively charged, neutrally charged, and negatively charged regions. The body coat proteins of phages, which are also known as 2700pVIII primary coat proteins, are organized in right-handed helical configurations on the ssDNA with a 200-tilt angle in tandem with the phage's long axis. In this form of arrangement, all major coat proteins assemble to form a pentagonal shape. This pentagonal shape indicates that M13 phages have no inversion center, twofold screw symmetry, and fivefold rational symmetry [66, 67]. On the basis of all these arrangements in M13 phages, it has been concluded that M13 phages can have piezoelectric properties. By applying phage tension along the phage-long axis, electrical polarization and net dipole moments are produced in the direction of the applied stress, establishing the piezoelectric effect in M13. In the second method, electrical polarization and net dipole moments are produced in two different directions if tension is applied along the phages' bodies. These distinctive qualities of M13 phages allow for the development of many types of piezoelectric energy harvesting systems [28].

17.4 Observation of Piezoelectric Electric Effect in M13 Phages

Based on conformational arrangement, it has been observed that M13 phages exist with strong piezoelectric properties because of their permanent axial polarization induced by net dipole moment in the pVIII properties [66, 67]. Piezoresponse force microscopy (PFM) was used by B.Y. Lee et al. in 2012 to examine the piezoelectric characteristics of M13 phages [68]. In order to conduct this experiment, a phage monolayer sample was created by quickly and vertically removing an octadecanethiol (ODT)/cysteine pattern substrate from the suspension of phage. After this arrangement, piezoelectric properties of phages can be observed by PFM. For the characterization of structural-based piezoelectric properties at the single-phage level, they prepared a self-assembled phage monolayer film of 1 mm wide lines of gold-based substrate. On the basis of PFM characterization, it was observed that the piezoelectric phage responded to an applied electric field in both axial and lateral directions. For the PFM in lateral direction, the films showed bright characteristics and texture with dark fibril. Opposite to that when lateral PFM was performed in perpendicular direction of long axes by rerating sample at 90°, the texture contrast of fibril disappeared. The observed contrast was probably due to net dipole moment of the phage along with its axial direction. Due to this net dipole moment, cantilever undergoes torsion in opposite directions. In the condition of tilt angle at 20°, alpha helical coat protein and helical screw advancement of dipoles create a radial component with orientation independence. According to these results, shear piezoelectricity can be observed. Along with that, compressive piezoelectricity can be observed due to axial response [68].

Piezoelectric effect of any material can be measured by measuring piezoelectric coefficient. Piezoelectric coefficient represents correlation between piezoelectric effect and the direction of applied external stimuli. Different types of piezoelectric coefficients can be utilized to represent the piezoelectric responses of choice; these parameters such as piezoelectric charge coefficients (d constant), the piezoelectric voltage coefficients (g), and the piezoelectric coupling factors (k). Additionally, the permittivity coefficient (ε) is the dielectric displacement measured in farad m^{-1} [69].

Identification of directions in piezoelectric materials is an important parameter to calculate piezoelectric response. For this purpose, three-axis coordinate system has been used by researchers to identify the directions of piezoelectric response. These three directions, X, Y, and Z, in the classic orthogonal coordinate system for normal piezoelectricity are defined as the directions 1, 2, and 3 while their shear directions are 4, 5, and 6 in case of shear piezoelectricity, respectively. On the basis of these observations, two subunits *i* and *j* are used to describe directions of piezoelectric coefficients. In these subunits, *i* subunit is used to describe direction of electrical field or applied voltage while *j* shows the direction of applied mechanical stress. For example, *d*33, *d*31, and *d*34, where "3" direction represents poling direction. From all these examples, it can be seen that electric field can be generated in the "3" direction by mechanically deforming piezoelectric materials along the "3" (*d*33) or "1" direction (*d*31) or applying shear stress along the "4" direction (*d*34) [69].

17.5 Genetically Modified M13 Bacteriophage to Enhance Piezoelectricity

As per previous study, M13 phage can be genetically modified with the help of recombinant DNA engineering method [70]. For the modification of M13 phage, 2, 3, and 4 glutamate (2E, 3E, and 4E peptides) were expressed at the N-terminus of the M13 major coat protein pVIII, with the replacement of 2-4 Ala-Glu-Gly-Asp-Asp with Ala-(Insert)-Asp). For enhancing the piezoelectric energy generation from the phage-based devices, it is hard to construct vertically aligned phase nanostructures with the control of polarization effect. Ju-Hyuck Lee et al. (2019) solved this limitation by genetically modifying M13 phage for the expression of multifunctional peptides for controlling their geometrical, mechanical, and electrical functions. In the first step of modification, they polarized the phage binding on substrate; for that purpose, they modified M13 phage with hexa-histidine (6H) at the N-terminal of the minor coat protein (pIII). 6H is very much specific to bind with nickel-nitrilotriacetic acid (Ni-NTA). In the second step, piezoelectrical performance was enhanced by fusing 3 glutamates (3E) at the N-terminal of pVIII major coat protein to enhance the protein's dipole strength. After second step, at last mechanical stability of the vertically aligned phage-based nanostructure was enhanced by using tyrosine (Y) residue at the N-terminal to form Y-Y cross-linkage. In the process of vertical alignment of the phage nanostructure, a micropatterned polydimethylsiloxane (PDMS) mold was utilized as a template. For further process, phage solution should be deposited on Ni-NTA-coated substrate. When phage solution gets evaporated, it produces a capillary force on upward direction. Due to this induced force, phages self-assemble into monolayer of vertically aligned bundles. Mechanical properties of genetically modified phage can be achieved for enhancing properties of nanostructures. Obtained vertically aligned modified nanostructure phage was tested for polarization effect and dipole orientation of phage using second harmonic generation (SHG). In the analysis result it was found that modified phage nanostructures exhibited strong signals in SHG, which directly indicates strong unidirectional polarization effect. While just opposite to that engineered phage with control samples and vertically aligned phage can show weak SHG signals. For the vertical alignment of phages, researchers have used polyethylene naphthalate (PEN) substrate using 15 mg mL^{-1} of phage solution and PDMS micropattern, which have been coated using NTA/Au with a width size of around 5 μm. Once fabrication procedure was over, a protecting layer of micropatterned polydimethylsiloxane (PDMS) was deposited on the vertically aligned phage. After polymerization, another layer of Cr/Au-coated PEN was deposited on PDMS to work as topmost electrode [71]. By following all these procedures, a piezoelectric device can be fabricated. M13 phages also have benefit of surface modification with the help of genetic engineering. For surface modification, it also follows recombinant DNA technology.

17.6 Developmental Process of Virus-Based Piezoelectric Energy Harvesting Devices

Piezoelectricity in materials was discovered by Pierre Curie and Jacques Curie in 1880 during the quartz, tourmaline, and Rochelle salt study [72]. The piezoelectric effects in materials have been divided into two categories, which are direct effect and inverse effect. In the first category effect that is direct, after polarization of material, a voltage is generated under the influence of compressive and tensile stress while in the case of inverse effect a mechanical displacement gets induced in the material after applying electric potential [73]. There are two kinds of operating modes: one is mode 33 in which stress is applied parallel to the polarization and the other is mode 31 in which applied stress is opposite to the polarization effect [74, 75]. The reason of piezoelectric effect generation is that some materials have effective distribution of ions in their crystalline structures; due to that, steady-state equilibrium originates between the positive and negative electric charge in the material and effect remains neutral. When compressive stress is applied to any square-shaped material, it creates equivalent center of charge at the same point, so observed polarization is considered with no change. In the case of energy harvesting devices, observed piezoelectric effect is due to direct effect while generation of electrical charge is due to applied force. For the utilization of energy harvesting at initial stage, the dipoles that are present in piezoelectric materials align randomly between two electrodes just because poling is absent [76]. When electrical field is applied to the energy harvester, the alignment of dipoles changes to the same direction as applied field. But when external force is zero, device takes possession of electrical signals due to presence of equilibrium state. In another case with the influence of compression force in the vertical direction, the material gets polarized just because of compressive strain and this applied compressive strain generates potential of piezoelectric effect between electrodes. After releasing of this applied force, there has been observed a modest tensile force due to which reverse piezoelectric effect generates into energy harvester. After discovery of piezoelectricity, different types of materials have been developed ranging from ceramics to biomaterials. Certain ceramic materials such as lithium niobate ($LiNbO_3$), lithium tantalite ($LiTaO_3$), barium titanate ($BaTiO_3$), and lead zirconate titanate (PZT) have been focused as superior materials with piezoelectric effect [78–80]. Among all these superior materials, PZT has been considered as the most common material due to its practical application and higher range of piezoelectricity. Although piezoelectric ceramic materials have very good piezoelectric effect, still they are poor in biocompatibility and brittleness, which limits their applications as a piezoelectric energy harvesting devices in the field of biomedical [77]. The limitation of ceramic materials leads to new approach of biomaterials as an alternative source to produce energy harvesting devices with piezoelectric effect. The development of M13 phage into piezoelectric energy harvesting devices was started in 2012 by Prof. Lee at UC Berkeley after discovery of its excellent piezoelectric effect. A self-assembled multilayered film using M13 phage was developed on Au-coated flexible base by researchers. At the time of evaporation period, self-assembly of M13 phage was utilized to form long-range ordered smectic-phase liquid crystalline film with the help of its chirality and monodispersive phase. After achievement of this Au-coated flexible substrate on the film and embedding this produced device in between two 2.5 mm thick polydimethylsiloxane (PDMS), they were able to synthesize phage-based energy generator. To modulate the generated power of this fabricated M13 phage device, it was required to modify the phage with the help of genetic engineering. After being genetically engineered, M13 phage-based device was recorded to turn on a liquid crystal display (LCD) [81]. Researchers have modified the power capacity of phage-based energy generating system by modification in the morphology of film. The observed outcomes are superior to the previous one after the installation of piezoelectric energy producers, which are based on hierarchically ordered film.

17.7 Conclusion

Viruses have been considered as promising materials to synthesize piezoelectric harvesting devices. The reason behind their popularity is their unique characteristics. Although various types of ceramics and biomaterial-based piezoelectric materials have been synthesized, they all have some drawback such as limitation for mass production, lower piezoelectric effect, and properties for modification. All these drawbacks have been overcome after development of M13 bacteriophages. Researchers find these M13 phase-based piezoelectric materials advantageous due to their exceptional attributes, which include their ability to be produced in large quantities, their surface modification capabilities, their ease of genetic engineering for improved qualities, and many more. Morphology and properties of M13 bacteriophages can be simply altered with the help of genetic engineering, controlling dipole alignment, etc. This chapter has focused on the working principle of the different types of biomaterials-based piezoelectric energy devices as well as M13 phage-based energy harvesting devices. Additional to that, a basic comparison between various piezoelectric biomaterials such as organic and inorganic substances has also been made with M13 phage-based piezoelectric devices. Methods for enhancing energy conversion efficiency of M13 phage-based piezoelectric material have been explored. The piezoelectric biomaterials are continuously attracting researchers' focus due to their great importance for future self-powered implantable medical devices as well as wearable electronics.

References

1. J. Krane, Climate change and fossil fuel: An examination of risks for the energy industry and producer states. *MRS Energy Sustain.* **4**, 2, 2017.
2. B.H. Robinson, E-waste: An assessment of global production and environmental impacts. *Sci. Total Environ.* **408**, 183–191, 2009.
3. Z.L. Wang, J. Chen, L. Lin, Progress in triboelectric nanogenerators as a new energy technology and self-powered sensors. *Energy Environ. Sci.* **8**, 2250–2282, 2015.
4. L. Dong, C. Xu, Y. Li, Z.-H. Huang, F. Kang, Q.-H. Yang, X. Zhao, Flexible electrodes and supercapacitors for wearable energy storage: A review by category. *J. Mater. Chem. A* **4**, 4659–4685, 2016.
5. M.J. Tan, C. Owh, P.L. Chee, A.K.K. Kyaw, D. Kai, X.J. Loh, Biodegradable electronics: Cornerstone for sustainable electronics and transient applications. *J. Mater. Chem. C* **4**, 5531–5558, 2016.
6. V.R. Feig, H. Tran, Z. Bao, Biodegradable polymeric materials in degradable electronic devices. *ACS Cent. Sci.* **4**, 337–348, 2018.
7. L.S. Nair, C.T. Laurencin, Biodegradable polymers as biomaterials. *Prog. Polym. Sci.* **32**, 762–798, 2007.
8. J.C. Middleton, A.J. Tipton, Synthetic biodegradable polymers as orthopedic devices. *Biomaterials* **21**, 2335–2346. 2000.
9. D.-M. Shin, S.W. Hong, Y.-H. Hwang, Recent advances in organic piezoelectric biomaterials for energy and biomedical applications. *Nanomaterials*, **10**, 123, 2020.

10. Y.P. Guo, K. Kakimoto, H. Ohsato, $(Na_{0.5}K_{0.5})NbO_3$-$LiTaO_3$ lead-free piezoelectric ceramics. *Mater. Lett.* **59**, 241, 2005.

11. R.T. Smith, F.S. Welsh, Temperature dependence of elastic, piezoelectric, and dielectric constants of lithium tantalate and lithium niobate. *J. Appl. Phys.* **42**, 2219, 1971.

12. N. Setter, D. Damjanovic, L. Eng, G. Fox, S. Gevorgian, S. Hong, A. Kingon, H. Kohlstedt, N.Y. Park, G.B. Stephenson, et al., Ferroelectric thin films: Review of materials, properties, and applications. *J. Appl. Phys.* **100**, 051606, 2006.

13. G.H. Haertling, Ferroelectric ceramics: History and technology. *J. Am. Ceram. Soc.* **82**, 797, 1999.

14. E. Fukada, Piezoelectric properties of biological polymers. *Q. Rev. Biophys.* **16**, 59, 1983.

15. M. Minary-Jolandan, M.F. Yu, Nanoscale characterization of isolated individual type I collagen fibrils: Polarization and piezoelectricity. *Nanotechnology* **20**, 085706, 2009.

16. V. Vivekananthan, N.R. Alluri, Y. Purusothaman, A. Chandrasekhar; S. Selvarajan, S.J. Kim, Biocompatible collagen nanofibrils: An approach for sustainable energy harvesting and battery-free humidity sensor applications. *ACS Appl. Mater. Interfaces* **10**, 18650, 2018.

17. W.S. Williams, L. Breger, Analysis of stress distribution and piezoelectric response in cantilever bending of bone and tendon. *Ann. N. Y. Acad. Sci.* **238**, 121, 1974.

18. A.R. Liboff, M.H. Shamos, Piezoelectric effect in dentin. *J. Dent. Res.* **50**, 516, 1971.

19. E. Fukada, Piezoelectricity of natural biomaterials. *Ferroelectrics* **60**, 285, 1984.

20. S. Maiti, S. Kumar Karan, J. Lee, A. Kumar Mishra, B. Bhusan Khatua, J. Kon Kim, Bio-waste onion skin as an innovative nature-driven piezoelectric material with high energy conversion efficiency. *Nano Energy* **42**, 282, 2017.

21. S.K. Ghosh, D. Mandal, High-performance bio-piezoelectric nanogenerator made with fish scale. *Appl. Phys. Lett.* **109**, 103–701.

22. S.K. Karan, S. Maiti, O. Kwon, S. Paria, A. Maitra, S.K. Si, Y. Kim, J.K. Kim, B.B. Khatua, Nature driven spider silk as high energy conversion efficient bio-piezoelectric nanogenerator. *Nano Energy* **49**, 655, 2018.

23. K. Heo, H.-E. Jin, H. Kim, J.H. Lee, E. Wang, S.-W. Lee, Transient self-templating assembly of M13 bacteriophage for enhanced biopiezoelectric devices. *Nano Energy* **56**, 716, 2019.

24. J.H. Lee, K. Heo, K. Schulz-Schonhagen, J.H. Lee, M.S. Desai, H.E. Jin, S.W. Lee, Diphenylalanine peptide nanotube energy harvesters. *ACS Nano* **12**, 8138, 2018.

25. M.T. Chorsi, E.J. Curry, H.T. Chorsi, R. Das, J. Baroody, P.K. Purohit, H. Ilies, T.D. Nguyen, Piezoelectric biomaterials for sensors and actuators. *Adv. Mater.* **31**, 1802084, 2019.

26. H. Yuan, T. Lei, Y. Qin, J.-H. He, R. Yang, Design and application of piezoelectric biomaterials. *J. Phys. D* **52**, 194002, 2019.

27. A. Ben Amar, A.B. Kouki, H. Cao, Power approaches for implantable medical devices. *Sensors*, **15**, 28889, 2015.

28. P. Woo, K.W. Kim, Y. Hong, H.J. Yoon, Y. Lee, D. Gwak, K. Heo, Recent developments and prospects of M13-bacteriophage based piezoelectric energy harvesting devices. *Nanomaterials* **10**, 93, 2020.

29. B.Y. Lee, J. Zhang, C. Zueger, W.J. Chung, S.Y. Yoo, E. Wang, J. Meyer, R. Ramesh, S.W. Lee, Virus-based piezoelectric energy generation, *Nat. Nanotechnol.* 2012, doi: 10.1038/NNANO.2012.69

30. W.J. Chung, J.W. Oh, K. Kwak, B.Y. Lee, J. Meyer, E. Wang, A. Hexemer, S.W. Lee, Biomimetic self-templating supramolecular structures. *Nature*, **478**, 364, 2011.

31. H.E. Jin, J. Jang, J. Chung, H.J. Lee, E. Wang, S.W. Lee, W.J. Chung, Biomimetic self-templated hierarchical structures of collagen-like peptide amphiphiles. *Nano Lett.* **15**, 7138, 2015.

32. J. Li, Y. Long, F. Yang, X. Wang, Degradable piezoelectric biomaterials for wearable and implantable bioelectronics. *Current Opinion in Solid State and Materials Science* **24**(1), 100806, 2020. https://doi.org/10.1016/J.COSSMS.2020.100806.

33. A. Stapleton, M.R. Noor, E.U. Haq, C. Silien, T. Soulimane, S.A. Tofail, Pyroelectricity in globular protein lysozyme films. *J. Appl. Phys.* **123**, 124701, 2018.

34. A. Stapleton, M. Noor, J. Sweeney, V. Casey, A. Kholkin, C. Silien, A. Gandhi, T. Soulimane, S. Tofail, The direct piezoelectric effect in the globular protein lysozyme. *Appl. Phys. Lett.* **111**, 142902, 2017.

35. V. Lemanov, Piezoelectric and pyroelectric properties of protein amino acids as basic materials of soft state physics. *Ferroelectrics* **238**, 211–218, 2000.

36. V. Lemanov, S. Popov, Phonon echo in L-alanine. *Phys. Solid State* **40**, 1921–1922, 1998.

37. V. Lemanov, S. Popov, Unusual electromechanical effects in glycine. *Phys. Solid State* **40**, 991–994, 1998.

38. A. Jardine, S. Sayed, Challenges in the valorisation of chitinous biomass within the biorefinery concept. *Curr. Opin. Green Sustain. Chem.* **2**, 34–39, 2016.
39. S.B. Lang, Pyroelectric effect in bone and tendon. *Nature* **212**, 704, 1966.
40. A.A. Marino, R.O. Becker, S.C. Soderholm, Origin of the piezoelectric effect in bone. *Calcif. Tissue Res.* **8**, 177–180, 1971.
41. V. Volkov, A.V. Ferreira, A. Cavaco-Paulo, On the routines of wild-type silk fibroin processing toward silk-inspired materials: A review. *Macromol. Mater. Eng.* **300**, 1199–1216, 2015.
42. C. Vepari, D.L. Kaplan, Silk as a biomaterial. *Prog. Polym. Sci.* **32**, 991–1007, 2007.
43. V. Lemanov, S. Popov, G. Pankova, Piezoelectricity in protein amino acids. *Phys. Solid State* **53**, 1191–1193, 2011.
44. D. Vasilescu, R. Cornillon, G. Mallet, Piezoelectric resonances in amino-acids. *Nature* **225**, 635, 1970.
45. V. Nguyen, K. Jenkins, R. Yang, Epitaxial growth of vertically aligned piezoelectric diphenylalanine peptide microrods with uniform polarization. *Nano Energy* **17**, 323–329, 2015.
46. K. Jenkins, S. Kelly, V. Nguyen, Y. Wu, R. Yang, Piezoelectric diphenylalanine peptide for greatly improved flexible nanogenerators. *Nano Energy* **51**, 317–323, 2018.
47. A. Stapleton, M. Noor, J. Sweeney, V. Casey, A. Kholkin, C. Silien, A. Gandhi, T. Soulimane, S. Tofail, The direct piezoelectric effect in the globular protein lysozyme. *Appl. Phys. Lett.* **111**, 142902, 2017.
48. A. Stapleton, M.R. Noor, E.U. Haq, C. Silien, T. Soulimane, S.A. Tofail, Pyroelectricity in globular protein lysozyme films. *J. Appl. Phys.* **123** 124701, 2018.
49. K. Kim, M. Ha, B. Choi, S.H. Joo, H.S. Kang, J.H. Park, B. Gu, C. Park, C. Park, J. Kim, Biodegradable, electro-active chitin nanofiber films for flexible piezoelectric transducers. *Nano Energy* **48**, 275–283, 2018.
50. T. Yucel, P. Cebe, D.L. Kaplan, Structural origins of silk piezoelectricity, *Adv. Funct. Mater.* **21**, 779–785, 2011.
51. Z. Zhou, D. Qian, M. Minary-Jolandan, Molecular mechanism of polarization and piezoelectric effect in super-twisted collagen. *ACS Biomater. Sci. Eng.* **2**, 929–936, 2016.
52. A. Goldbourt, Structural characterization of bacteriophage viruses by NMR. *Prog. Nucl. Magn. Reson. Spectrosc.* **114–115**, 192–210, 2019. https://doi.org/10.1016/J.PNMRS.2019.06.004.

53. D.A. Marvin, M.F. Symmons, S.K. Straus, Structure and assembly of filamentous bacteriophage. *Prog. Biophys. Mol. Biol.* **114**, 80–122, 2014.

54. R. Olsthoorn, J. van Duin, Bacteriophages with ssRNA, in: eLS, (Ed.). 2011. https://doi.org/10.1002/9780470015902.a0000778.pub3.

55. A. King, M. Adams, E. Carstens, E. Lefkowitz, Virus Taxonomy: Ninth Report of the International Committee on Taxonomy of Viruses, 2012.

56. M.G. Rossmann, Structure of viruses: A short history. *Q. Rev. Biophys.* **46**, 133–180, 2013.

57. K.R. Richert-Pöggeler, K. Franzke, K. Hipp, R.G. Kleespies, Electron microscopy methods for virus diagnosis and high resolution analysis of viruses. *Front. Microbiol.* **9**, 3255, 2019.

58. J. Xu, N. Dayan, A. Goldbourt, Y. Xiang, Cryo-electron microscopy structure of the filamentous bacteriophage IKe. *Proc. Natl. Acad. Sci.* **116**, 5493–5498, 2019.

59. J.H. Lee, J.H. Lee, J. Xiao, M.S. Desai, X. Zhang, S.W. Lee, Vertical self-assembly of polarized phage nanostructure for energy harvesting. *Nano Letters* **19**(4), 2019. https://doi.org/10.1021/acs.nanolett.9b00569.

60. B.Y. Lee, J.X. Zhang, C. Zueger, W.J. Chung, S.Y. Yoo, E. Wang, J. Meyer, R. Ramesh, S.W. Lee, Virus-based piezoelectric energy generation. *Nat. Nanotechnol.* **7**, 351–356, 2012.

61. M. Minary-Jolandan, M.F. Yu, Nanoscale characterization of isolated individual type I collagen fibrils: Polarization and piezoelectricity. *Nanotechnology* **20**, 085706, 2009.

62. D.M. Shin, H.J. Han, W.G. Kim, E. Kim, C. Kim, S.W. Hong, H.K. Kim, J.W. Oh, Y.H. Hwang, Bioinspired piezoelectric nanogenerators based on vertically aligned phage nanopillars. *Energy Environ. Sci.* **8**, 3198–3203, 2015.

63. S. Vasilev, P. Zelenovskiy, D. Vasileva, A. Nuraeva, V.Y. Shur, A.L. Kholkin, Piezoelectric properties of diphenylalanine microtubes prepared from the solution. *J. Phys. Chem. Solids* **93**, 68–72, 2016.

64. L. Sando, M. Kim, M.L. Colgrave, J.A.M. Ramshaw, J.A. Werkmeister, C.M. Elvin, Photochemical crosslinking of soluble wool keratins produces a mechanically stable biomaterial that supports cell adhesion and proliferation. *J. Biomed. Mater. Res. A*, **95a**, 901–911, 2010.

65. D.F. Kelly, D. Dukovski, T. Walz, Monolayer purification: A rapid method for isolating protein complexes for single-particle electron microscopy. *Proc. Natl. Acad. Sci. U.S.A.* **105**, 4703–4708, 2008.

66. A. Goldbourt, Structural characterization of bacteriophage viruses by NMR. *Prog. Nucl. Magn. Reson. Spectrosc.*, **114–115**, 192, 2019.

67. J. Xu, N. Dayan, A. Goldbourt, Y. Xiang, Cryo-electron microscopy structure of the filamentous bacteriophage IKe. *Proc. Natl. Acad. Sci. U.S.A.* **116**(12), 2019.

68. B.Y. Lee, J. Zhang, C. Zueger, W.J. Chung, S.Y. Yoo, E. Wang, J. Meyer, R. Ramesh, S.W. Lee, Virus-based piezoelectric energy generation. *Nat. Nanotechnol.* **7**, 351, 2012.

69. W. Yuan, H. Min, V. Jeffrey, L. Ting, D. Matthew, Expedient secondary functions of flexible piezoelectrics for biomedical energy harvesting. *Bioact. Mater.* **22**, 291–311, 2023.

70. W.-J. Chung, J.-W Oh, K. Kwak, B.Y. Lee, J. Meyer, E. Wang, A. Hexemer, S.-W. Lee, Biomimetic self-templating supramolecular structures. *Nature* **478**, 364–368, 2011. https://doi.org/10.1038/nature10513.

71. J.H. Lee, J.H., Lee, J. Xiao, M.S. Desai, X. Zhang, S.W. Lee, Vertical self-assembly of polarized phage nanostructure for energy harvesting. *Nano Letters*, **19**(4), 2019. https://doi.org/10.1021/acs.nanolett.9b00569.

72. H.S. Klickstein, P. Curie, An appreciation of his scientific achievements. *Journal of Chemical Education* **24**(6), 1947. https://doi.org/10.1021/ed024p278.

73. S.D. Mahapatra, P.C. Mohapatra, A.I. Aria,G. Christie, Y.K. Mishra, S. Hofmann, V.K. Thakur, Piezoelectric materials for energy harvesting and sensing applications: Roadmap for future smart materials, in: *Advanced Science* (Vol. 8, Issue 17) 2021. https://doi.org/10.1002/advs.202100864.

74. C. Jean-Mistral, S. Basrour, J.J. Chaillout, Comparison of electroactive polymers for energy scavenging applications. *Smart Materials and Structures*, **19**(8), 2010. https://doi.org/10.1088/0964-1726/19/8/085012.

75. T. Furukawa, Piezoelectricity and pyroelectricity in polymers. *IEEE Transactions on Electrical Insulation*, **24**(3), 1989. https://doi.org/10.1109/14.30878.

76. X. Ren, H. Fan, Y. Zhao, Z. Liu, Flexible lead-free $BiFeO_3$/PDMS-based nanogenerator as piezoelectric energy harvester. *ACS Applied Materials and Interfaces*, **8**(39), 2016. https://doi.org/10.1021/acsami.6b04497.

77. Z.L. Wang, Towards self-powered nanosystems: From nanogenerator nanopiezotronics. *Adv. Funct. Mater.* **18**, 3553, 2008.

78. Y. Guo, K.I. Kakimoto, H. Ohsato, $(Na_{0.5}K_{0.5})NbO_3$–$LiTaO_3$ lead-free piezoelectric ceramics. *Materials Letters*, **59**(2–3), 241–244, 2005. https://doi.org/10.1016/J.MATLET.2004.07.057.

79. R.T. Smith, F.S. Welsh, Temperature dependence of elastic, piezoelectric, and dielectric constants of lithium tantalate and lithium niobate. *J. Appl. Phys.* **42**, 2219, 1971.

80. N. Setter, D. Damjanovic, L. Eng, G. Fox, S. Gevorgian, S. Hong, A. Kingon, H. Kohlstedt, N.Y. Park, G.B. Stephenson, et al., Ferroelectric thin films: Review of materials, properties, and applications. *J. Appl. Phys.* **100**, 051606, 2006.

81. B.Y. Lee, J. Zhang, C. Zueger, W.J. Chung, S.Y. Yoo, E. Wang, J. Meyer, R. Ramesh, S.W. Lee, Virus-based piezoelectric energy generation. *Nat. Nanotechnol.* **7**, 351, 2012.

Chapter 18

Development in Lead-Free Perovskite Piezoelectric Bulk Materials

Asima Monjur,[a] **Md. Hasibul Hasan,**[a]
Tasnim Mahzabin Tanha,[a] **and Md. Abu Bin Hasan Susan**[a,b]
[a]*Department of Chemistry, University of Dhaka, Dhaka 1000, Bangladesh*
[b]*Dhaka University Nanotechnology Center, University of Dhaka, Dhaka 1000, Bangladesh*
susan@du.ac.bd

Piezoelectric materials possess unique ability to interconvert electrical and mechanical energy, enabling their application in a variety of important areas like harvesting energy, high-power applications, sensing and actuation, and more. In this regard, lead titanate and other lead-based perovskites have been commercialized due to their compelling dielectric and ferroelectric properties. However, growing environmental concerns regarding lead exposure during the synthesis and integration of lead-based materials in high-performance applications led to the enforcement of restrictive laws and regulations. A major part of the ongoing research in this field is more or less influenced by the desire to develop lead-free substitutes

Piezoelectric Materials: Design and Applications
Edited by N. B. Singh and Dev Kumar Mahato

ISBN 978-981-5129-40-3 (Hardcover), 978-1-003-59897-8 (eBook)
www.jennystanford.com

that have either comparable or, in some cases, better piezoelectric properties. As a potential substitute, perovskite ceramics have gained considerable interest because of their characteristic feature to respond to applied electrical fields as well as for the ease of tuning properties simply by making any compositional change. This chapter delves into the latest developments in lead-free piezoelectric perovskites (LFPP), providing a thorough overview of the crystallographic structures, polarization mechanisms, and material properties of $BaTiO_3$, $BiFeO_3$, $Bi_{0.5}Na_{0.5}TiO_3$, and (K, Na) NbO_3, which all contribute to their piezoelectric behavior. Finally, current challenges and prospects associated with LFPPs based on their performance have been discussed.

18.1 Introduction

18.1.1 Piezoceramics

Piezoelectric effect, a groundbreaking discovery in the early 19th century, opened up numerous possibilities for scientists and engineers, allowing them to develop new technologies and solutions that harness the unique properties of piezoelectric materials (PEMs). Direct and reverse piezoelectric effects are well-known phenomena where materials are capable of generating electricity when subjected to mechanical stress and vice versa [1]. The macroscopic electromechanical responses of piezoelectric material evolve from either reversible lattice expansion or contraction or irreversible domain wall motion [2]. Piezoelectric ceramics represent the progress and technological prowess of the exploitation of electromechanical effects for high-precision actuation, fuel injection, medical ultrasonic imaging, and green electric power generation [3].

Lead zirconate titanate (PZT), potassium sodium niobate (KNN), barium titanate (BT), sodium bismuth titanate (BNT), and bismuth ferrite (BFO) are some well-known piezoelectric ceramic materials belonging to the family of perovskite [4]. It is worth mentioning that ferroelectric (FE) perovskite ceramics are almost exclusively used in polycrystalline form due to the ease of preparation and less expense compared to their counterpart single crystals. Among ceramic compositions, the most important were solid solutions of

$PbZrO_3$, $PbTiO_3$, and PZT. Although PZT materials have a long and productive history, the current research focus is on piezoelectricity from lead-free materials rather than only optimizing PZT for better performances [5]. There has been a surge of interest in exploring the promising capabilities of piezoelectric perovskites in high-performance applications. Specific emphasis is placed on inducing a FE phase transition through modification of composition, which has the potential to yield significant advancements in this field. Despite the recognition of the exceptional piezoelectric properties of lead-based ferroelectrics, research on the development of lead-free materials capable of generating equivalent levels of piezoelectricity is still in the rudimentary stage. The phenomena of critical behavior induced by electric field and miniaturized isotropic domains induced by polarization have been elucidated. But there remains a pressing need for feasible method of synthesis of materials of this kind [6]. The use of piezoelectric ceramics devoid of lead has been increasingly popular owing to the growing demand for eco-friendly materials and the increased concerns about the toxicity of lead. These ceramics have garnered significant market momentum and are a viable solution to address the aforementioned apprehensions [7].

18.1.2 Perovskite Bulk Materials and Piezoelectric Perovskites

The general formula for the perovskites is ABO_3, where "A" and "B" are cations with differences in their ionic radii and with valency of +1, +2, or +3 and +3 to +6, respectively. The perovskite structures mainly consist of oxygen octahedrons, BO_6, with B cations positioned at interstices whereas A cations are packed in the dodecahedral holes. Ions of A site are often alkaline earth or rare-earth elements whereas ions of B site belong to transition metal elements in the *d*-block [3]. Distortion takes place in the ideal perovskite structure most prominently at low temperatures about the sublattice formed by A cation resulting in polar FE properties. The symmetrical distribution of constitutional atoms in perovskites is highly pronounced, which promotes easy deformation of the unit cell, resulting in monoclinic (M), orthorhombic (O), rhombohedral (R), and tetragonal (T) FE phases (Fig. 18.1).

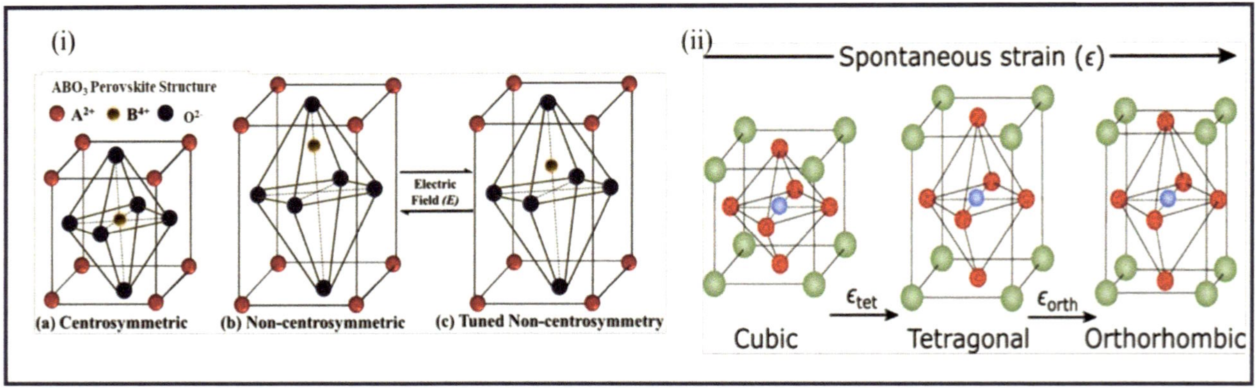

Figure 18.1 (i) Perovskite structure (a) centrosymmetric, (b) non-centrosymmetric arrangements, and (c) external stimulus-induced tuned non-centrosymmetric [8]. (ii) Phase transition among different perovskite structures [9]. Here, ϵ_{orth} and ϵ_{tet}, respectively, represent degenerate O and T symmetry-adapted strains emerging when the phase transitions occur.

Among them, non-centrosymmetric perovskite structured crystals lack a point of inversion which is the requirement to exhibit piezoelectric behavior. The versatility of perovskite structure stems from its ability to undergo diverse unit cell distortions. Several phases are produced by distortion of the structure (Table 18.1) [10].

Table 18.1 Phases produced by unit cell distortion of perovskite structure

Distorted dimension	Direction of distortion	Generated structure
One	[100]	T
Two	[110]	O
Three	[111]	R
Three	Arbitrary ([hko], [hkl])	M, Tr

M = monoclinic; O = orthorhombic; R = rhombohedral; T = tetragonal; Tr = triclinic

The varying degrees of complexity during phase transitions of perovskite solid solutions do not hinder them from exhibiting interesting dielectric, optical, and piezoelectric properties. Certain perovskite compositions experience a series of phase transitions into FE phases, whereas others display a relaxor phase preceding the FE phase [5].

18.1.3 Classification of Piezoelectric Perovskite

18.1.3.1 Lead-based piezoelectric perovskite

One of the earliest commercially employed PEMs is $PbTiO_3$ crystallized in the tetragonal structure at room temperature (RT). High temperatures tend to produce a significant amount of Pb vacancies in the single crystals from which the piezoelectric and FE behaviors are believed to be generated. The reduced piezoelectricity and dielectric behavior correspond to the low resistivity, a mechanical failure, and the substantial strain that occurs during the transition from the cubic to the tetragonal (C-T) phase. The subsequent developments in the modification of $PbTiO_3$ are targeted by introducing additives, mainly rare earths and alkaline elements, to achieve electrical resistivity and reduce spontaneous strain [11].

The advancement in piezoelectric research has been widely recognized as a momentous historical feat. PZT is incredibly versatile

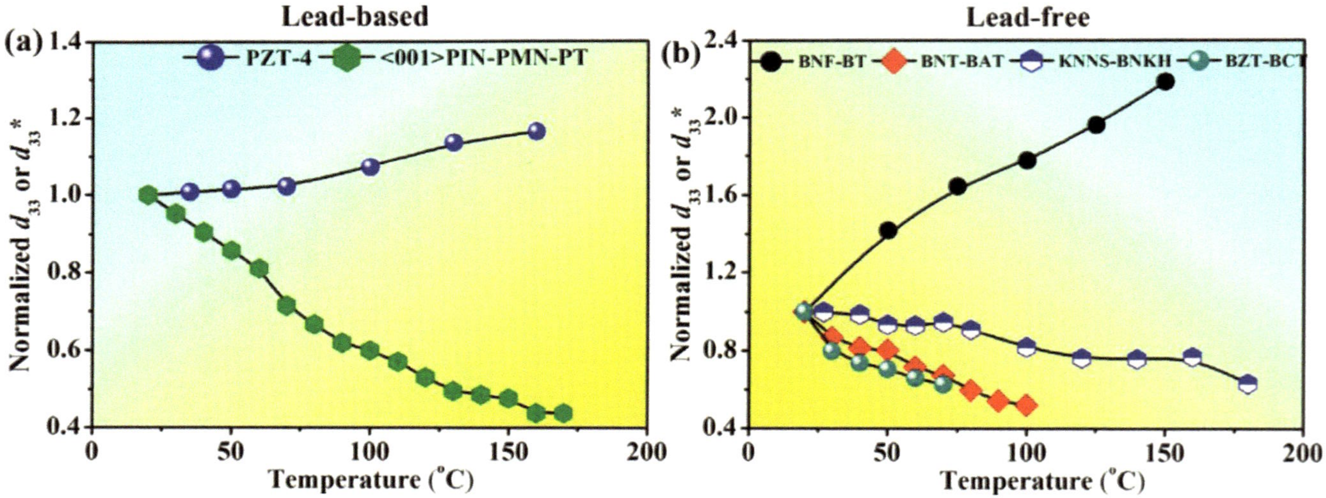

Figure 18.2 (a) Variation of d_{33} or d_{33}* for Ba and La modified commercial PZT (PZT-4) ceramic and ⟨011⟩ indium niobate and magnesium niobate modified lead titanate (PIN-PMN-PT) single crystals with temperature and (b) stability of d_{33} or d_{33}* with the variation of temperature in LFPP (neodymium modified BFO-BT (BNF-BT), sodium modified BFO-aluminum tantalum co-doped barium oxide (BNT-BAT), antimony doped KNN-sodium potassium co-doped bismuth Hoffmate (KNNS-BKNH), zirconium modified BT-calcium modified BT (BZT-BCT)) [15]. Here, d_{33} is the piezoelectric charge constant which represents polarization along the z-axis upon applying unit stress along the same direction, and d_{33}* is the inverse piezoelectric charge constant.

due to its ease of compositional engineering, enabling it to be used in a range of applications. The composition-induced phase transitions are thus crucial to piezoelectric behavior [3]. For the tetragonal to rhombohedral (T-R) phase transition of PZT materials, especially with a Zr/Ti ratio of 52:48, a high amount of piezoelectricity is produced. The high values of the piezoelectric coefficient and electromechanical coupling factor (k_{33}) indicate excellent electrical properties derived from the composition at or near the morphotropic phase boundary (MPB). It can have a wide variety of attainable properties due to the simplicity of forming solid solution compositions and sintering [12]. However, the phase transition at its critical temperature, T_c, of 350°C makes the operating temperature sufficiently low restricting its application in high-power transducer applications [13]. At a particular molar ratio (i.e., x = 0.47–0.50), the two phases can coexist [11]. The large d_{33} of relaxor-PT single crystals are considered to have originated from engineered domain configurations (Fig. 18.2) [13].

PZT contains toxic lead and the toxicity issue is further enhanced due to the enhanced volatility, particularly during the calcination or sintering process. Thus, LFPPs have been commercialized and their development has been promoted by targeting problems like low durability and d_{33} [14].

18.1.3.2 Lead-free piezoelectric perovskite

The environmental concerns about the lead-based perovskites act as a triggering force to look for lead-free substitutes with better piezoelectric performance and greater thermal stability but with PZT-like phase boundaries [16]. Phase boundary defines the junction of two different crystals with distinct compositions and lattice structures [17].

Polycrystalline ceramics are not that sufficient to produce remarkable piezoelectricity as they lack dipole orientation. A significant breakthrough in this aspect was achieved by constructing phase boundaries (Fig. 18.3) like polymorphic phase boundaries (PPBs) and MPBs usually progressed by composition designing strategy [18–20]. The boundary conditions of FE domains in piezoelectric oxides are determined by crystallographic phase and are often complex at phase boundaries. The concept of MPB was first introduced from the observation of a piezoelectric constant of PZT ceramics of a composition characterized by R (*R3m*) and T (*P4mm*) phase regions [19]. Dipole polarization or its projection along the electric field direction marks the polarity of the piezoceramic. Owing to randomly distributed grains, the generated piezoelectricity of the polycrystalline ceramics is insufficient. At the phase coexistence region, dipoles are more likely to take up a wide range of orientations upon application of an electric field owing to the low-energy barrier. Furthermore, the mobile domain walls make the material more susceptible to any change in the electric field during phase coexistence. Thus, building phase coexistence has become one of the efficient ways to improve piezoelectric response. In the early 20th century, the phenomenological theory of Landau explained this behavior based on the trends in Gibbs free energy changing, with electric field, composition, temperature, and other external parameters. Near MPB, flattening in the energy profile of Gibbs free energy facilitates the piezoelectricity by enhancing the polarization rotation.

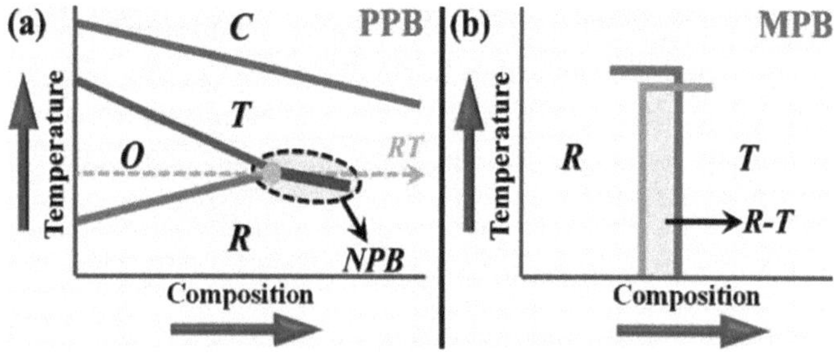

Figure 18.3 Traits in PPB and MPB. (a) New phase boundary construction in a phase diagram and (b) a typical MPB diagram [20].

The MPB in perovskites is constructed by bringing out necessary changes in composition to adjust the T_c down toward RT such as isovalent or aliovalent dopants, sintering aids, and ABO_3-type compounds [21]. Doping with electron donor elements is found to induce superior electromechanical characteristics along with a low T_c, coercive field E_c, remnant polarization P_r. Acceptor doping, on the other hand, introduces the converse effects than the former. The compositionally driven phase behavior is of significant practical interest as it imparts notable stability over a wide pressure, time, and temperature range [22].

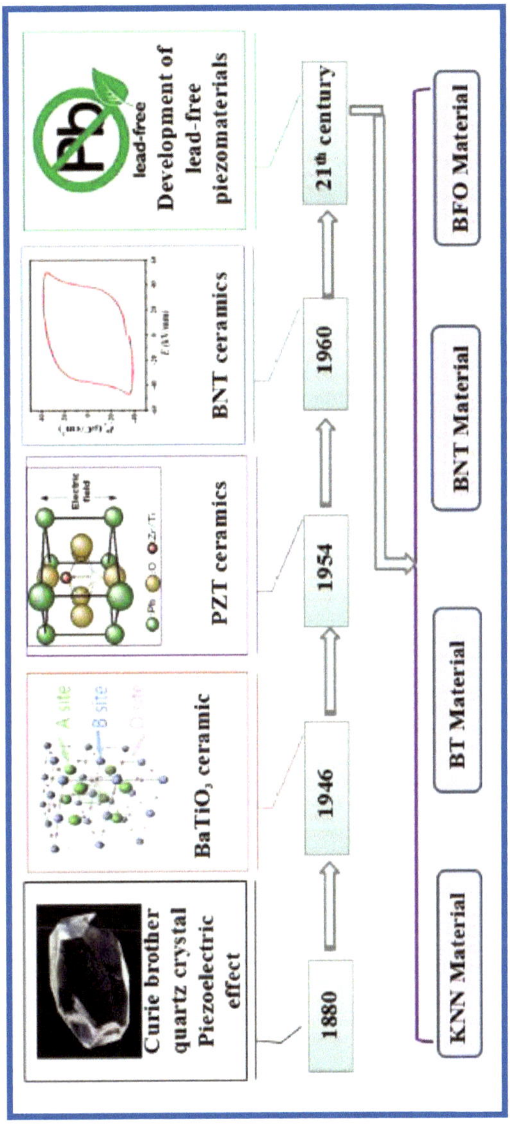

Figure 18.4 Development and advances in LFPPs [24].

PPBs involve a structural phase transition that is temperature driven. The polarization rotation at PPB is favored by the flattened energy profile, almost resembling MPB [23]. Additionally, any extrinsic contribution, e.g., domain wall motion, is considered upon approaching PPBs. Hence, the fundamental approach for enhancement of the piezoelectric properties is the construction of a PPB that ensures energetically favorable phase coexistence. Temperature sensitivity is, however, a major drawback of PEMs containing PPB, which hinders their integration in high-power applications. Figure 18.4 shows a schematic representation of development in LFPPs [24].

The following materials have been considered as potential substitutes for PZT:BT and its derivatives, KNN and their derivatives, BNT and its derivatives, and BFO or $BiFeO_3$. As shown in Fig. 18.5, these ceramics possess relatively larger d_{33}, and their properties suit high-power applications and actuators well [25, 26]. Table 18.2 summarizes piezoelectric and FE properties of various LFPPs at different compositions [27–64].

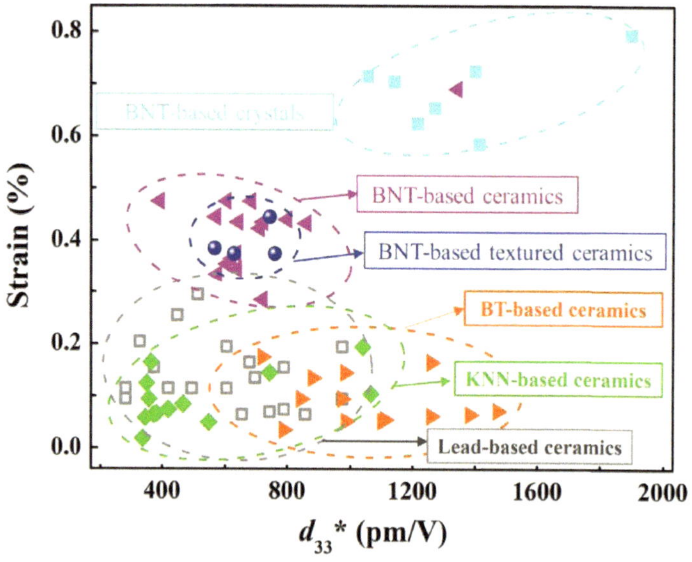

Figure 18.5 Comparison of variation in strain with d_{33}^* for BNT, KNN, PZT, and BT piezoceramics [25].

Table 18.2 Piezoelectric and ferroelectric properties of various LFPPs at different compositions

Material system	Composition	Phase structure	T_c (°C)	P_r (μC/cm^2	d_{33} (pC/N)	E_C (kV/cm)	Ref.
BT	$(Ba_{0.85}Ca_{0.15})(Ti_{0.9}Zr_{0.1})O_3$ + 0.06 wt% Pr_2O_3	T	80		460		[27]
	$(Ba_{0.99}Ca_{0.01})(Ti_{0.98}Zr_{0.02})O_3$ + 0.08 wt% Ga_2O_3	O–T	120		440		[28]
	$(Ba_{0.95}Ca_{0.05})(Ti_{0.90}Sn_{0.10})O_3$ + 3% Li_2CO_3	R–T	50		485		[29]
	$(Ba_{0.95}Ca_{0.05})(Ti_{0.9}Sn_{0.1})O_3$ – 2% CuO	R–O–T	50		683		[30]
	$Ba_{0.98}Ca_{0.02}Zr_{0.02}Ti0_{.98}O_3$ + 0.3 mol% MnO_2	O–T	120		308		[31]
	$Ba_{0.85}Ca_{0.15}Ti_{0.9}Zr_{0.1}O_3$ + 0.5 wt% Li_2CO_3	R–O–T	70		493		[32]
	$(Ba_{0.98}Ca_{0.02})(Ti_{0.94}Sn_{0.06})O_3$ + 0.1 wt% ZnO	T	89		428		[33
	$(Ba_{0.85}Ca_{0.15})(Ti_{0.9}Zr_{0.1})O_3$ + 0.04 wt% CeO_2	R–T	90		600		[34]
	$(Ba_{0.95}Ca_{0.05})(Ti_{0.90}Sn_{0.10})O_3$ + 4% Li_2O	R–T	50		578		[35]
	$(Ba_{0.98}Ca_{0.02})(Ti_{0.94}Sn_{0.06})O_3$ + 0.1 wt% HfO_2	O	80		430		[36]
	$(Ba_{0.85}Ca_{0.15})(Ti_{0.9}Zr_{0.1})O_3$ + 0.8 wt% $Ba(Cu_{0.5}W_{0.5})O_3$	R–T	78		541		[37]
	$(Ba_{0.7}Ca_{0.3})TiO_3$-$Ba(Zr_{0.2}Ti_{0.8})O_3$ + 0.04 wt% CuO	R–T	95		510		[38]

(*Continued*)

Table 18.2 (*Continued*)

Material system	Composition	Phase structure	T_c (°C)	P_r (μC/cm^2	d_{33} (pC/N)	E_C (kV/cm)	Ref.
KNN	(K,Na,Li)(Nb,Ta)O_3:Mn		235	3.45	630		[39]
	$(K_{0.45}Na_{0.55})_{0.96}Li_{0.04}NbO_3$		432	24.1	689	13.9	[38]
	$K_{0.5}Na_{0.5}NbO_3$	O			160		[40]
	(K, Na)(Nb,Ta)O_3		300		200		[41]
	$K_{0.47}Na_{0.53}NbO_3$	O		27	104		[42]
	(K,Na,Li)(Nb,Ta)O_3		276		255		[43]
	$Na_{0.5}K_{0.5}Nb_{0.995}Mn_{0.005}O_3$	O		106			[44]
	$K_{0.622}Na_{0.378}NbO_3$	O		18	148		[45]
	(K,Na,Li)(Nb,Ta)O_3		230	2.67	354		[46]
	$K_{0.8}Na_{0.2}NbO_3$	O		7.2	145		[47]
	$(K_{0.5}Na_{0.5})_{0.98}Li_{0.02}NbO_3$		430	21.4	280		[48]
	$K_{0.252}Na_{0.75}NbO_3$	O		12	161		[9]
	$K_{0.5}Na_{0.5}NbO_3$ + 0.5% MnO_2	O			270		[50]

Material system	Composition	Phase structure	T_c (°C)	P_r (μC/cm^2	d_{33} (pC/N)	E_C (kV/cm)	Ref.
BNT	$0.944Na_{0.5}Bi_{0.5}TiO_3$–$0.056BaTiO_3$	P_c–T		46.4	2500	28	[51]
	$0.93Na_{0.5}Bi_{0.5}TiO_3$–$0.07BaTiO_3$	R–T		41	450	22	[52]
	$0.9(Bi,Na)TiO_3$–$0.1BaTiO_3$ + MnO_2	T		45.3	483	2.91	[53]
	$0.95Na_{0.5}Bi_{0.5}TiO_3$–$0.05BaTiO_3$ + Fe	R–T		53.2	590	3.5	[54]
	$0.94Na_{0.5}Bi_{0.5}TiO_3$–$0.06BaTiO_3$	R–T		16.02	360	24.89	[55]
	$Bi_{0.5}Na_{0.5}TiO_3$	R		44			[56]
BFO	$Bi_{0.88}Sm_{0.12}FeO_3$	R		40–50	–	130	[57]
	$Bi_{1-x}La_xFeO_3$ (x = 0–0.20)	R–T–P_c		>9.8	>25		[58]
	$Bi_{0.9}Ho_{0.1}FeO_3$	R		13		63	[59]
	$Bi_{0.85}La_{0.15}FeO_3$	P_c		12.08	27.7	60	[60]
	$Bi_{0.875}Sm_{0.125}FeO_3$	T		15.09	29	90	[61]
	$Bi_{0.95}Dy_{0.05}FeO_3$	R		31		120	[62]
	$Bi_{0.875}Sm_{0.125}FeO_3$	R		~40	45	~150	[63]
	$Bi_{1-x}Dy_xFeO_3$ (x ≤ 0.08)	R		>18	>35	110	[64]

R-O-T = rhombohedral orthorhombic tetragonal phase structure; R-T-Pc = rhombohedral tetragonal primitive cubic phase structure; T_c: Curie temperature

18.2 BT Ceramics

The first ever polycrystalline perovskite structured ceramic exhibiting both ferroelectricity and piezoelectricity is BT, best known for its low dielectric loss (tanδ), high maximum polarization (P_m), high P_r, and high dielectric constant (ϵ_r). In-depth research is currently focused on its dielectric properties rather than its piezoelectric properties. Despite their good performance, the main concern is their low T_c and coercive field responsible for less stable polarization and challenges with poling treatments [65–67].

The BT crystal changes to a FE tetragonal (*4mm*) phase from a paraelectric cubic (*m3m*) phase below the Curie temperature. It then changes to a FE O (*mm2*) phase with further cooling below 0°C, and at 90°C, it changes to FE rhombohedral (*3m*) one. The material is found to show mainly three phase transitions: orthorhombic to rhombohedral ($T_{O\text{-}R} \sim -90°C$), tetragonal to orthorhombic ($T_{T\text{-}O} \sim 0°C$), and cubic to tetragonal ($T_{C\text{-}T} \sim 120°C$). At low T_c, this shows a normal FE state with tetragonal symmetry (*P4mm*).

Studies on $Ba(Zr,Ti)O_3$ concentrate on bringing the O and R phases into the RT range since the former two phases are below RT and therefore difficult to work with. The main approach to solve this problem has been ion substitution.

The polymorphic nature of BT-based ceramics restrains it from having an ideal MPB along the axis of temperature. Despite the fact that the improved piezoelectric response is typically achieved with a significant reduction in T_c, a major performance barrier presently exists as a consequence of the high phase coexistence [68]. BT-based ceramics with R-T and R-O-T phase boundaries have superior piezoelectricity to those with O-T boundaries as demonstrated in BZT-xBCT where the R–T starts a tricritical point, C–T–R [69]. The phase diagram for BZT-xBCT is shown in Fig. 18.6(i). The underlying cause is the expedited polarization rotation and the loss of polarization anisotropy from decreased domain wall energy and elastic softening between T and R states. However, there is an intermediate O phase in R–T which is believed to be coupled with a weak R-O and an O-T phase transition. According to Cai et al., the piezoelectric properties of calcium and zirconium modified barium titanate (BCZT) ceramics are enhanced by the interaction of grain size and phase boundaries

[70]. The FE phase is considered to be unstable at RT and the temperature corresponding to phase transition in BT relies on grain size.

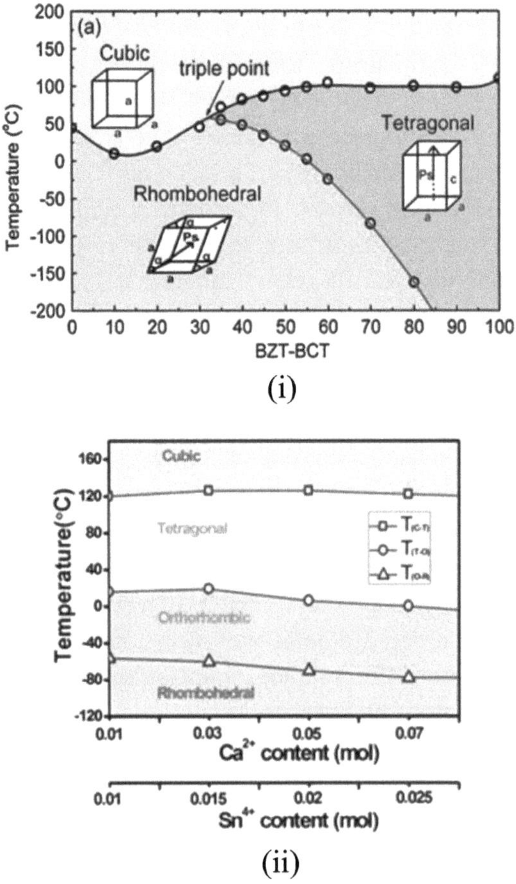

Figure 18.6 (i) Pseudobinary phase diagram of a BZT-BCT ferroelectric system [66] and (ii) phase diagram of $Ba_{1-x}Ca_xTi_{1-y}Sn_yO_3$ ceramics with the variation of Ca^{2+} and Sn^{4+} [67].

18.2.1 Chemical Modification: Doping with Different Materials

Ba and/or Ti ions at the A and B sites, respectively, can be suitably substituted to change the phase coexistence of the properties of

BT-based ceramics. Apart from radius difference, dopants can be those having low or high oxidation numbers from the lattice site. In addition to promoting densification, doping causes changes in cell parameters. The correct choice of dopant allows the ceramic to be soft to very hard. For example, the use of Nb in the doping of Co-doped BT compensates the charge defects created by Co, thus allowing the "hard" character. There have been several previous reports of BT piezoelectric ceramics that have been doped with different elements, including Ca, Sn, Zr, Co, and Cu. Dopants cause the shifting or disappearance of one or more specific phase transitions in addition to lattice distortion. Thermal energy also influences the occupancy of the site. At a given temperature, higher oxygen partial pressure causes a higher hole concentration and a lower electron concentration affects the occupancy of the site. Consequently, doping with two or more elements may establish phase boundaries at RT for BT ceramics, resulting in improved strain and piezoelectric features through changes in their $T_{R\text{-}O}$ and $T_{O\text{-}T}$.

Acceptor doping (Mn, Fe, Co, Ni, and Cu ions) at the Ti site is found to stabilize hexagonal structures kinetically rather than thermodynamically at RT. The oxygen vacancies in the structure must undergo diffusion to domain walls through octahedral sites in order to produce orientated dipoles with acceptor sites. By restraining domain walls, the defect dipoles diminish the extrinsic contribution of ferroelectricity [72]. The sample ages due to the induced dipolar effects at very low concentrations of the acceptor, but the tetragonal structure endures. Conversely, Ca isovalent doping in the perovskite lattice A site changes the increment of piezoelectric properties of the material to a noticeable extent. Low concentrations of dopants can cause the T-O phase transition to occur at low temperatures, despite having a small influence on T_c. The effect of doping with Ca^{2+} and Sn^{4+} simultaneously on the T_c is shown in Fig. 18.6(ii).

Ca^{2+}, Sr^{2+}, La^{3+}, and Zr^{4+}/Sn^{4+} are commonly used as substituents at the A and B sites of ABO_3 structures to improve piezoelectric, dielectric, and FE properties. Dopants of the same valence state are not expected to produce the same effect; e.g., doping with Ca is frequently used to lower the $T_{R\text{-}O}$ and/or $T_{O\text{-}T}$ values whereas the addition of Zr significantly increases both $T_{R\text{-}O}$ and $T_{O\text{-}T}$ values with greater chemical stability of the system as shown in Fig. 18.7a,d. When substitution is performed with the Zr ions, the cell parameter ratio begins to decrease, which causes the polarization to become incredibly erratic and the overall domain structure to eventually be disrupted. The result is reflected in a shift of T_c toward lower temperature; the same effect has been observed when doping with Hf (Fig. 18.7b) [73]. Aside from this, it was seen that substituting Zr^{4+} or Sn^{4+} for Ti^{4+} lowers the T_c, which makes their use in real-world applications impossible. Since the ionic radius of Li is small, it frequently settles at interstitial sites of BCZT ceramics. With the perovskite structure being loosened, compared to the undoped BCZT ceramics, easier rotation of FE domains becomes possible, improving the piezoelectric properties. According to a neutron scattering measurement by Jeong et al., Ca substitution on Ba sites increases the octahedral tilting of ZrO_6, which results in an irregular lattice distortion. The easy polarization rotation is articulated by the random local strain caused by either the ion size disparity or the octahedral tilting between TiO_6 and ZrO_6 [74].

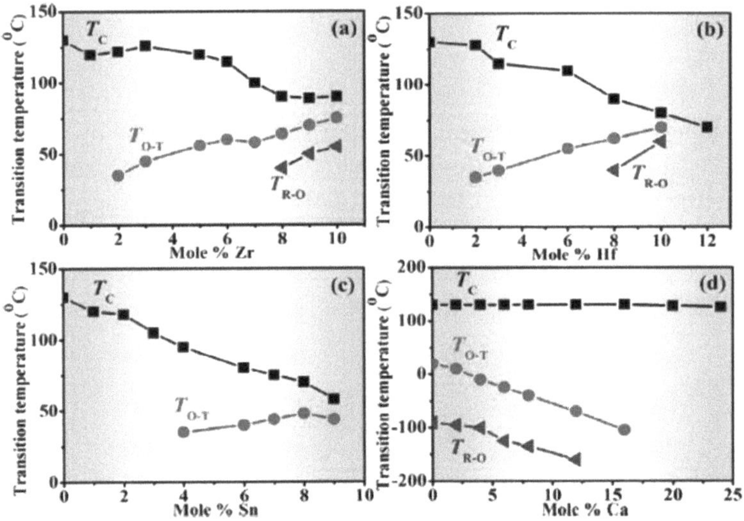

Figure 18.7 Representation of phase diagram for (a) $BaTi_{1-x}Zr_xO_3$, (b) $BaTi_{1-x}Hf_xO_3$, (c) $BaTi_{1-x}Sn_xO_3$, and (d) $Ca_xBa_{1-x}TiO_3$ piezoelectric ceramics [71].

The doping of BT ceramics with cations, especially of elements that have intermediate ionic radii of Ba^{2+} and Ti^{4+}/Zr^{4+} ions, emerges as a good substitute for A or B site. Dielectric properties are thus improved along with the stabilization of the perovskite structure and reduction of tanδ. The microstructure and dielectric characteristics of BT are changed by doping with lanthanum. The formation of oxygen vacancies brings about a decrease in the T_c [75]. In the same way, the choice of these trivalent lanthanides (Ln^{3+} = Sm^{3+} and La^{3+}) is justified by their particular electronic configuration, which can lead to their behavior being so specific as to be justified by their use. As cationic vacancy is required for the neutrality of charge in the perovskite structure, the dielectric permittivity value of BZT compounds doped with lanthanides Ln^{3+} = La^{3+} and Sm^{3+} is higher than that of non-doped BZT compounds. This has a detrimental impact on the T_c. In the Er^{3+} doped smart FE ceramics with Yb^{3+} doping, properties deteriorate with increasing Yb^{3+} doping.

The piezoelectricity in modified BT ceramics with MPB composition can be comprehensive while discussing it from the perspective of symmetry-adapted Landau polynomials. The energy profile will be flattened as a result of the grain correlation, causing the path of polarization rotation to diverge from its equilibrium. This will increase ε, which will ultimately increase the k_{33}. Textured polycrystalline ceramics can achieve a high k_{33} value in comparison to single crystals since the transition of (001) of T to (111) R state takes place with ease as polarization anisotropy disappears (Fig. 18.8). Ren et al. observed that a quasi-quadruple point exists in the $BaTiO_{3-x}BaSnO_3$ ceramics and considered it to be accountable for the gigantic strain response [76]. It should be obvious that for both polarization rotation and extension, the four phases coexisting together lowers the energy barrier.

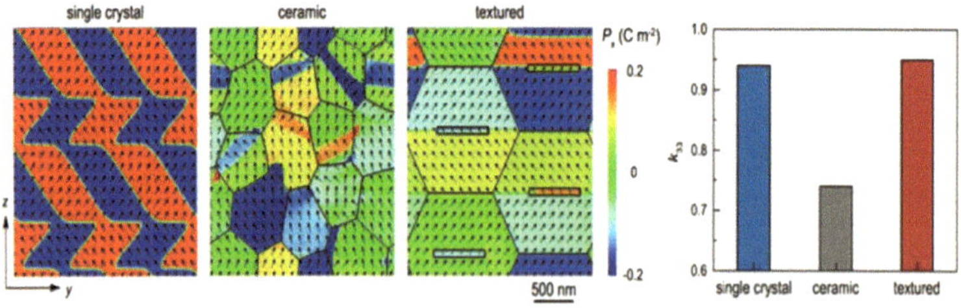

Figure 18.8 Structures of domain, and k_{33} factor of random ceramic, single crystal, and textured ceramic after poling in [001] direction electrically [77].

The $T_{O\text{-}T}$ and the $T_{R\text{-}O}$ rise as substitution of Sn^{4+} and Zr^{4+} takes place at the B site of $BaTiO_3$ (shown in Fig. 18.7a,c) while the T_c shows an abrupt decrease [78]. Insertion of donor or acceptor in the B site leads to "soft" or "hard" piezoelectric ceramics, respectively.

18.2.2 Microstructure Engineering

18.2.2.1 Controlling grain size

The improvement of ferro/piezoelectric properties depends, to a large extent, on grain and domain size. The grain size effect correlates electrical properties with grain size [79]. The dependence of the ratio of maximum strain (S_m) to maximum stress (E_m), S_m/E_m, d_{33}, k_p on grain size for BCZT is shown in Fig. 18.9b. Materials with grain size above a critical value are supposed to exhibit enhanced properties; ~10 μm is proved to be the critical size in BZT-50BCT. The effects that reduce domain reorientation or produce constrained motion of domain walls are mainly responsible for insufficient electrical properties. Alternative strategies might consider the grain size effect for tuning the electrostrictive strain in BT-based ceramics. Sintering temperature (T_s) and time influence the bulk density and grain boundary, while calcination temperature is widely used to regulate the grain size. The value of d_{33} between 350 and 500 pC/N and piezoelectric charge constant along the perpendicular direction of applied stress (d_{31}) to −185 pC/N at RT were found to be related to improved grain size. Below a threshold value of the particle diameter, even the FE phase becomes unstable at RT [80]. Recently evolved microstructural engineering techniques work on targeting both optimizations of grain size and their orientation and domain engineering to have control of the micro- to nanoscale structure of PEMs [81].

When mixed with other oxides, dopants can form solid solutions at low temperatures like that of liquid phase sintering, a technique that is used to promote densification in microstructure and reduce the temperature required for sintering. However, when the dopant concentration in the sample exceeds the solubility limit, the oxide additives start to inhibit grain growth by being segregated at grain boundaries like that of substitution of Pr^{4+}/Pr^{3+} at the A site of BZT-50 BCT ceramics. Exceeding the P_r content by more than about 0.06% tends to reduce both ε and grain size [82]. On the other hand, the microstructure of Li and La oxide-doped BCZT ceramics showed homogeneity, where nm- and μm-sized grains were obtained by doping the system with Li_2CO_3 and La_2O_3, respectively. The μm-sized grains deteriorated the ferro/piezoelectric properties of the BCZT ceramics [83].

Even the effect of strain induced by the electric field is also vivid. A significant difference in achieving large piezoelectric activity is observed in optimally prepared piezoceramics compared to microstructurally engineered piezoceramics. Lowering the domain and grain sizes is effective in the sense that it results in fewer recoverable non-180° domains because smaller grains have less restriction along their grain boundaries. The various sintering techniques include normal sintering, two-step sintering, and spark plasma sintering, which have produced grains with diameters between 0.4 and 32.2 μm. Apart from the coexistence of two or more FE phases around RT, doping imparts control over the growth of grains, which contributes to the piezoelectricity of a neoteric lead-free system [84].

18.2.2.2 Poling

The anisotropic properties of polycrystalline materials produce a direction-dependent electric field response. Crystallites in these materials are characterized by microscopic regions having dipoles oriented in a particular direction known as a domain. As a whole, the overall dipole of these materials gets canceled to maintain depolarization of the material. Both the domain walls and the domain sizes contribute to the enhanced piezoelectricity. The direction of polarization of the domain during phase transition is dictated by the symmetry of the crystal, while domain patterns rely on the symmetry among the phases. For instance, lamellar and wedge-shaped domains are commonly observed in T and R phases, respectively. Poling refers to the process of application of an electric field to switch domains and align dipoles toward the direction of the field. Poling by an external electric field is supposed to produce P_r in a particular direction. During the polarization, deformation occurs depending on the poling direction. Several factors, such as dimensions of the materials, doping concentration, temperature, field strength, time, porosity, and percentage of impurities, influence the alignment of the dipoles during poling. In addition to increasing piezoelectricity in the nonpolar direction and inhibiting domain wall motion, a modification of the engineered domain can also alter the macroscopic symmetry of the crystal [85]. The value of the d_{33} coefficient and electromechanical coupling factor (k_p) shows large dissimilarity for BZT-BCT and calcium doped BT-bismuth scandium oxide (BCT–xBS) system even if the high electric poling condition is applied.

18.2.2.3 Processing condition

Sintering is an essential step that involves heating a material at a high temperature below its melting point to fuse its particles. Electromechanical properties of polycrystalline ceramics can be influenced by sintering processes, such as sampling, grain size, porosity, density, volatilization of elements, and segregation of composition (Fig. 18.9) [86]. The properties of polycrystalline ceramics such as grain growth, reduction in energy loss, and improvement in mechanical quality factor (Q_m) are also affected significantly by the sintering process. Therefore, optimization of the T_s and time is crucial to gain improved piezoelectric properties. The low relative density is often accompanied by abnormal growth of grains which occurs at either very low or high T_s.

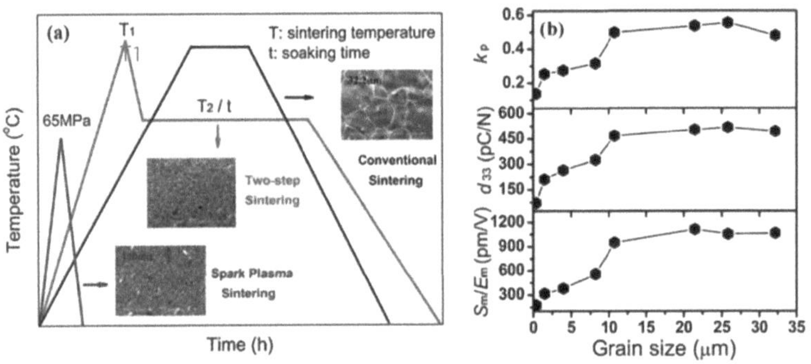

Figure 18.9 (a) Temperature programmed spark plasma, two-step, and conventional sintering (SEM images of BCZT ceramics sintered at different conditions). (b) The plot of S_m/E_m, d_{33}, k_p as a function of grain size for BCZT ceramics [86].

BT-based ceramics employed in high-performance applications could be synthesized using high T_s values (≥1450°C). In addition, successful attempts are taken to avoid fluctuation of composition such as lowering T_s or using aids in liquid phase sintering [71].

When sintering is being carried out, Ti starts to accumulate at the grain boundaries, simultaneously forming several impurities such as $BaTi_3O_7$ and $BaTi_2O_5$. The impurity phases, while converted into $BaTiO_3$ in subsequent cooling, generated numerous oxygen and cation vacancies. Rather than accumulating inside the grains, oxygen vacancies show the tendency to accumulate at grain boundaries. The growth inhibition contributes in part to the decent thermal stability of oxygen vacancies. Once Ce and Sb were used to dope ceramics, vacancies generated inside the grains facilitated the diffusion of ions. The restricted oxygen vacancy promotes Ba^{2+} at grain boundaries to consequently prevent the generation of the impurity phase. Nevertheless, excessive Ce doping would show the opposite effect [87]. However, there are shortcomings in using sintering aids; their effect on T_c is not always praiseworthy. For example, the reduction of T_c of BCZT ceramics is linear in increasing the content of $Ba(Cu_{0.5}W_{0.5})$. However, the reverse is observed in the CuO-modified BCTS ceramics with d_{33} = 683 pC/N [88]. The significant difference in crystallite and particle size at elevated temperatures is an indication of the C-T phase transition [89].

18.2.2.4 Texture engineering

Crystallographic texturing is traditionally used for the improvement of the performance of polycrystalline materials, especially when there is a limitation on the usage of single crystals. Single crystals lack mechanical integrity and are often expensive to produce in a wide range of sizes, geometry, and composition. Polycrystalline ceramics can simply overcome these limitations. Textured polycrystalline ceramics possess anisotropy, in addition to mechanical reliability and compositional versatility [90]. Textured ceramics with specific grain orientations are capable of generating anisotropy and possessing high dielectric and piezoelectric properties when template seed crystals are introduced. The difference in grain orientation in single crystal, random or nontextured ceramics, and textured ceramic and their effect on k_{33} are shown in Fig. 18.8.

BT-based perovskite materials, especially textured BZT-50BCT ceramics with different degrees of texture, have received increased attention in the last few years. As opposed to any other preferred method, it has been observed that control of orientation degree up to 90% can result in an approximately twofold improvement in the piezoelectric property. The perfect alignment of the polar vectors increases the poling efficiency and, in turn, has a drastic effect on the piezoelectricity generated by polycrystalline and single-crystal PEMs.

Two familiar texturing techniques are templated grain growth (TGG) and reactive TGG (RTGG), which involve applying external shear force, followed by heat treatment, to align the templates of anisotropic shape and crystallographic characteristics in a matrix [91]. The resultant textured ceramics are produced as a result of the nucleation and development of the required crystals on an aligned template [92]. Crystallographic texturing at $[001]_c$ of BZT-50BCT ceramics was found to be successful, as is evident from the increased value of piezoelectric coefficients. Ye et al. achieved this from a 5 mol% template with a texture fraction of about 82% and grain size of 18 μm. The RTGG method using plate-like BT templates made it possible to produce textured BCZT ceramics possessing a high degree of orientation. RTGG is used for texturing piezoceramics, especially with a perovskite structure exceeding the limitations of appropriate templates [93]. The firstly designed [94] texturing of piezoceramics by RTGG involves embedding the templates in equiaxed matrix powders. After heat treatment in this process, an in situ reaction will take place to form the expected texture.

The main challenge remaining is the degree of texturing. The textured ceramics rarely show a polarization rotation effect since after poling a "1T" domain configuration is formed. Template particles must possess an identical crystal structure to the matrix particles and for this specific purpose, needle-like, TiO_2 plate-like BT and plate-like $CaTiO_3$ (CT) are commonly considered. Although prepared BZT-5BCT was highly textured (98%), the d_{33} value reported was 455 pC/N only [74].

The development of BT-based ceramics also involves a new strategy considering domain characteristics, phase structure selection, and crystallographic texturing. Following this strategy, $(Ba_{0.94}Ca_{0.06})(Ti_{0.95}Zr_{0.05})O_3$ ceramics with $[001]_c$ grain orientation and texture degrees as high as ~98.6% could be successfully synthesized with an ultrahigh of d_{33}~755 pC/N. The piezoelectricity here inherits from reversible grain orientation-induced piezoelectric anisotropy [95]. Another method, called the vat photopolymerization (VPP), has recently been employed to develop BT-based ceramics that are highly textured and at a time 3D-structural [91].

18.3 KNN Ceramics

KNN is obtained by mixing $KNbO_3$ and $NaNbO_3$, which are, respectively, FE and antiferroelectric in nature. The KNN-based system has been one of the front-runners of the LFPP, widely regarded for its high d_{33}, high T_c (420°C), and low anisotropy of k_{33} [23]. The comparable piezoelectric properties of KNN with PZT make it the most promising environment-friendly substitute for PZT.

The common phase transitions of this system correspond to R-O, O-T, and R-T. They possess maximum piezoelectric activity [96] at (i) R-T and O-T boundaries because of low anisotropy and (ii) PPB composition with miniaturized nanodomain structures. A variety of experimental and theoretical methods are used to study the relationship between piezoelectric activity and phase boundaries. The shifting of $T_{R\text{-}O}$ or $T_{O\text{-}T}$ to RT is advantageous for formation of R-O or O-T phase boundaries using appropriate additives (Fig. 18.10).

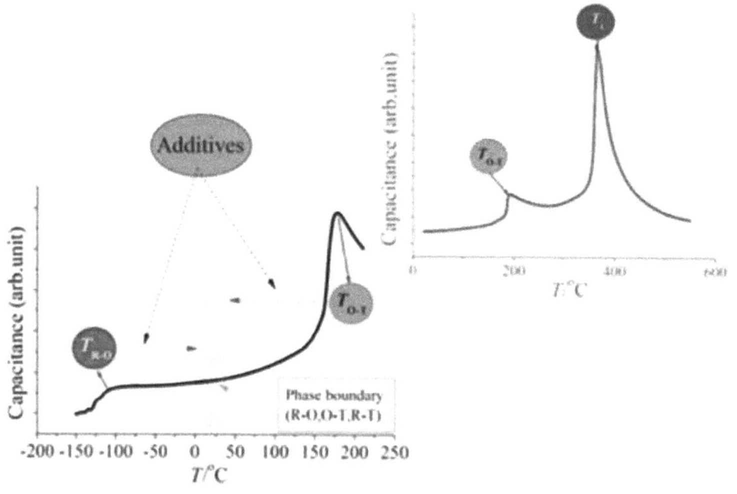

Figure 18.10 Phase boundaries of $K_{0.48}Na_{0.52}Nb_{0.98}Sb_{0.02}O_3$ piezoceramics [97].

The construction of new phase boundaries (R-T or R-O-T) is considered to overcome the problems of low piezoelectricity through the coexistence of nanoscale domains. The d_{33} value has been much higher even than that of KNN-based nontextured ceramics [98].

Any composition at the phase boundary that satisfies the requirements for simple polarization rotation among various symmetries can also be able to improve piezo/dielectric properties, regardless of the presence or absence of MPB or PPB. Increased domain wall density due to the new phase boundary facilitates domain switching, thereby promoting the piezoelectric effect. So far, the attempts to escalate the piezoelectric performance of these perovskites involving either O-T or R-O phase boundaries seem to produce insufficient piezoelectricity compared to PZT [97].

During sintering, high volatility of metal oxides hinders the ideal densification of KNN-based ceramics. As a result, poor density along with low piezoelectricity arises together, contributing to degradation of performance. The difficulties encountered in the synthesis of well-sintered KNN ceramics make it quite challenging to substitute PZT completely. The general approaches adopted by the scientists to resolve these issues are: firstly, lowering T_c to ambient or sub-ambient temperature and, secondly, producing highly textured ceramic samples. The key strategies to enhance piezoelectric properties of such materials include the development of phase boundaries, grain texturing, and domain engineering.

18.3.1 Chemical Modification: Doping with Different Materials

The densification and electric properties are, in general, accomplished by constructing phase boundaries. Perovskite materials differ in electrical properties for different phase transitions with changes in composition and temperature. Typically, monovalent Ag^+ and Li^+ can be substituted for the (K, Na) site, and pentavalent Ta^{5+} and Sb^{5+} cations often replace the Nb^{5+} site because of the nearly close values of valency and ionic radii. ABO_3-type compounds are also used. The oxides of elements like Zn, Cu, Ni, Co, Ce, Mn, B, La, V, Pr, Yb, Nd, and Dy are claimed to be efficient in tailoring the defect types. The incorporation of one or more elements to KNN affects phase boundaries with simultaneous shifts in $T_{R\text{-}O}$, $T_{O\text{-}T}$, and T_c, as it was for Sb illustrated in Fig. 18.11(i). The increased domain wall density achieved by this strategy makes domain switching easier under an external electric field [7].

The dopant type and amount can be adjusted to shift transitions of the R-T and R-O phases to RT. Phase transitions could be brought about by the distortion of A cations influenced by the size of B cations, which was experimentally supported by Ahart et al. [100]. Strategies to create O-T phase boundaries include ion substitutions (e.g., Li^+, Ta^{5+}, Sb^{5+}), adjustments in composition ratios, and innovative preparation techniques. The T_c of O-T can be shifted around RT while doping it with Li and Ta [101]. Li^+ and other small cations are specifically chosen for resubstitution if the target is to both increase T_c and decrease the $T_{O\text{-}T}$ [102].

Several equipollent metals are used as replacements for both A- and B-type cations in KNN. Li^+ is most commonly used in place of the A site and Sb^{5+} or Ta^{5+} is most commonly used in place of the B site. These additives are designed to affect the $T_{R\text{-}O}$, $T_{O\text{-}T}$, and T_c in a way that the formation of phase boundaries takes place easily [103]. In recent research, the R-T boundary has been induced in KNN materials through the addition of certain additives, such as Sb^{5+}, Ta^{5+}, and $Bi_{0.5}$(Na, K, Li)ZrO_3. With these additives, KNN ceramics with high d_{33} values (400–490 pC/N) and moderate values of T_c (217–304°C) have been achieved. Chemically modifying phase boundaries offers tremendous potential for optimizing piezoelectric properties in perovskite materials, as depicted in Fig. 18.11(ii). To achieve desired performance outcomes, it is crucial to engineer the right type of phase boundary [97].

Zr is a common candidate used for increasing the density of KNN. The addition of ZrO_2 in small amounts to KNN ceramics during air sintering inhibits grain growth [104]. The microstructure is refined by several conditions: first, ZrO_2 particles in the grain matrix junctions reduce grain boundary mobility, and second, enrichment of Zr at the grain boundary region introduces defects in the structure that restrict grain boundary mobility in the sintering process [104].

The addition of Cu to KNN can produce characteristic hard behaviors [105]. Hard ions like Cu^{2+} create oxygen vacancies and pin the domain walls. This leads to piezoelectric "hardening" effects [97]. In the case of Mn^{2+} doped KNN, it is the elimination of the free carrier holes that induces enhanced FE properties. According to Yao et al., Ti diffusion highly affects the structure, dielectric properties, and composition of doped KNN due to the evolution of holes and oxygen vacancies [106].

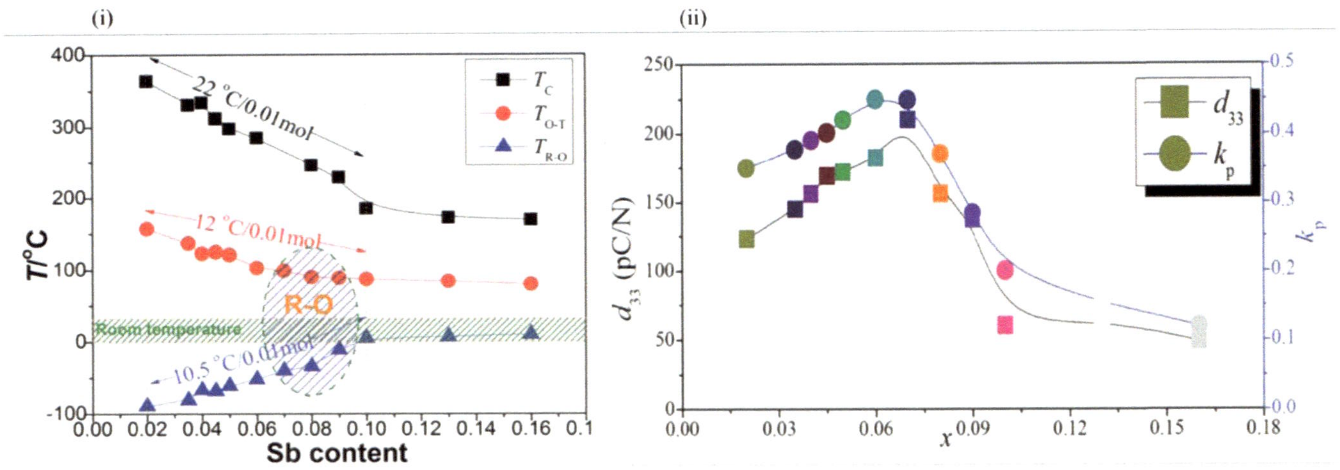

Figure 18.11 (i) Representation of the phase diagrams of Sb-doped KNN piezoceramic. The labels (e.g., 22 °C/0.01 mol, 12 °C/0.01 mol, and 10.5 °C/0.01 mol) indicate the change of T_{R-O}, T_{O-T}, and T_c with the change in Sb content and (ii) piezoelectric coefficients illustrated the effect of constructing the phase boundary by Sb doping.

The shifting of T–C toward low temperatures while keeping the O/M–T transition constant, Zr is often an ideal choice. Furthermore, undoped KNN exhibits T_c with a narrow ε profile. Upon incorporation of Zr^{4+} ions into the crystal lattice, this transition widens and the ε value falls when the Zr^{4+} ions get incorporated into the crystal lattice, which suggests the appearance of a transition of a diffuse phase.

The elimination of the intermediate O phase and adjusting $T_{R\text{-}T}$ closer to ambient temperature can be a new idea and has already been proposed by Wu et al. [107]. Zr^{4+} ions act as the B-site ions in perovskites to shift $T_{R\text{-}O}$ up to RT. Also, $AZrO_3$ (A = Ca, Sr, Ba) and $Bi_{0.5}(Na, K)_{0.5}ZrO_3$ lowers the $T_{O\text{-}T}$ and O-T boundaries are formed. Consequently, it remains unclear how Zr^{4+} contributes to adjusting phase transition boundaries [108]. Temperature independent electric field-induced strains in MnO_2 and $CaZrO_3$ co-modified ceramics ((K, Na, Li)(Nb, Ta)O_3)) have also been realized [109].

18.3.2 Microstructure Engineering: Approach to Control the Grain Size

18.3.2.1 Poling

Control over the domain engineering and crystal orientation of single crystals may enhance the piezoelectric property of KNN. Poling conditions have recently attracted much attention, particularly for LFPP [110]. When the electric field is low, saturated d_{33} can be achieved for the R-T/O-T/R-O phase structure. There could be several mechanisms responsible for that, e.g., domain switching, domain wall motion, domain vibration, electric field-induced phase transition, and lowered energy barrier at the phase boundaries. This reveals that under a low electric field, building a multiphase coexistence phase structure can accomplish high piezoelectricity in KNN ceramics [110, 111]. The innate properties of PPB transition make poling temperatures (T_p), and other poling conditions, very important for KNN ceramics. When it comes to KNN ceramics, poling temperatures near their phase transitions are typically chosen. The d_{33} values of KNN ceramics were improved by tuning the T_P to the T_c. With such a control, d_{33} values for ceramics with phase boundaries can be greatly enhanced [110]. Morozov et al. further investigated how piezoelectricity of KNN ceramics modified with

Mn, Ta, and Li are affected by the T_P. They demonstrated that poling in the tetragonal phase can enhance both the low-signal dielectric and piezoelectric properties in the orthorhombic phase [112]. Initially, unpoled KNN ceramics exhibit solely 90° and 180° domain walls. However, the introduction of Li doping transforms the domain configuration, resulting in the coexistence of various domains (such as 60°, 90°, and 180°) within a single grain.

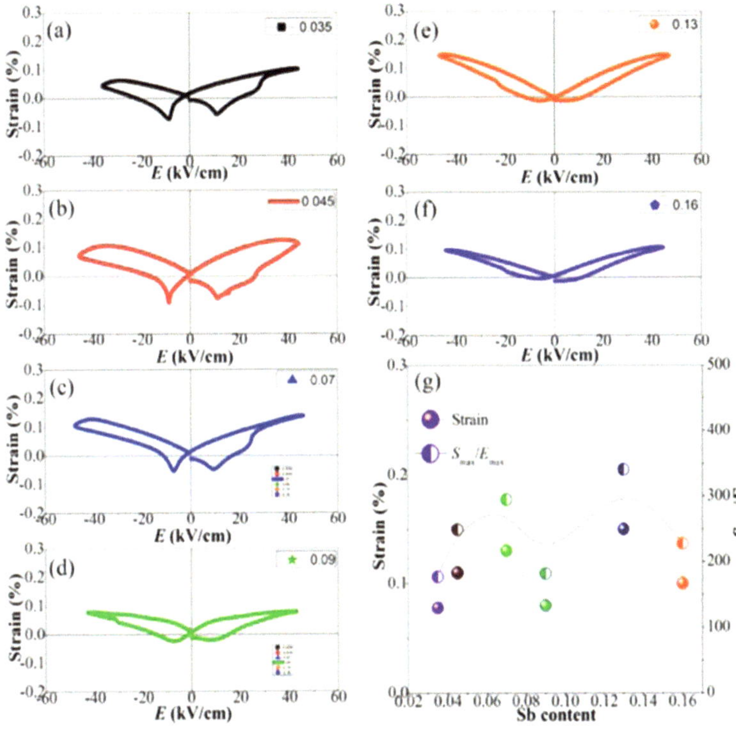

Figure 18.12 (a–f) Strain vs. electric field curves and (g) strain and S_m/E_m values of modified KNN ceramics (doped with Sb) [99]. S_m and E_m refer to S_{max} and E_{max}.

However, the mechanisms behind the better performance of KNN ceramics still remain dubious. The stronger piezoelectric response could be caused by the reduced energy barrier at the polymorphic phase transition (PPT) temperature zone of O and T phases or improved domain wall orientation and lowered strength of the coercive field in the T phase.

Through X-ray diffraction analysis of (K, Na, Li)NbO_3, Zhao et al. investigated the relation of domain reorientations of 90° to poling fields, showing that poling fields (5 kV/mm) can affect 90° domain reorientation, thus enhancing piezoelectricity [113]. Figure 18.12a–f is the representation of the electric field-induced strain of Sb-doped KNN during the poling treatment. The effect on the strain and S_m/E_m values of the ceramics are also shown in Fig. 18.12g. By aging and repoling, piezoelectric properties of KNN ceramics could be enhanced, increasing it from 190 to 324 pC/N due to migration of defect and spontaneous polarization rotation [114]. The study on the effect of poling conditions on the piezoelectric properties of (1−x) $(K_{0.42}Na_{0.58})(Nb_{0.96}Sb_{0.04})O_3$−x$(Bi_{0.5}K_{0.5})0.90Zn_{0.10}ZrO_3$ ceramics was successful in achieving a high d_{33} when T_P and $T_{R\text{-}T}$ coincided. As an electric field above 0.8 kV/mm affected multiple polarization states, complete polarization was ensured [115]. Thus, it is crucial to optimize poling parameters to further augment the piezoelectric activity of KNN.

18.3.2.2 Processing condition

Although the solid-state method has so far been a standard one for preparation of KNN ceramics, modified sintering methods are being explored to overcome a few limitations, such as deliquescence and volatilization of alkali metals [102]. During sintering, K^+ and Na^+ are lost, which makes it difficult to maintain the stoichiometry. Induced oxygen vacancies are assumed to firmly influence the piezoelectric activity. Although an increase in the T_s caused formation of a denser structure, volatilization triggered loss of A-site elements. Using sintering aids, sintering is pursued with liquid phase formation to accelerate diffusion and transformation during densification. This is how grain growth can affect piezoelectric activity. Bimodal distributions of grain size are quite common in oxide-doped KNN ceramics. Large grains can experience noticeable grain expansion during sintering due to differences in surface free energy between small and large grains. Conversely, the addition of some oxides that melt in low temperatures can successfully reduce the T_s and regulate their grain size, e.g., adding MnO_2 and ZrO_2 particles to replace Nb^{5+} ions in the B sites inhibits grain growth to improve packing [116]. As the amount of Zr increases with ZrO_2 addition, transitions from

O-M are apparent for KNN; the final density rises and a glassy phase is formed.

Sinterability may be improved by different sintering aids. For example, Zr as a dopant is commonly used for this purpose. At low concentrations, zirconia inhibits grain growth by restricting grain boundary mobility [117]. Despite sufficient sintering at low temperatures, sintering aids can sometimes degrade piezoelectric features for variations in the T_C. Doping CuO, ZnO, and $K_4CuNb_8O_{23}$ or different oxides [118] is found to lower the T_s. KNN ceramics doped with $LiTaO_3$ and excess Na^+ to optimize T_s and avoid loss of Na^+ to achieve d_{33} ~268 pC/N, much higher than that of the conventionally prepared $LiTaO_3$ ceramics [119]. The dependency of piezoelectric properties on grain growth may be due to interfacial polarization. Generally, the coarser the particles, the easier they are to polarize resulting in a high d_{33}.

Another investigation showed that the microstructure, phase structure, and electrical properties are profoundly influenced by the T_s, which in turn is a critical factor in determining d_{33} [120]. Sb and Zr co-doped ceramics have recently experienced a substantial d_{33} (~465 pC/N), where the $T_{R\text{-}T}$ values and alkali metal content may be controlled by varying the values of T_s [121].

18.3.2.3 Texturing

Along with the routes of compositional engineering to enhance piezoelectric response, crystallographic orientation is also a successful approach for improving piezoelectricity in perovskites since grains have anisotropic contributions to piezoelectricity.

Preferred grain orientation is enabled by texturing and single crystal growth, resulting in improved electromechanical performance by taking advantage of anisotropy [122]. Most studies involving KNN ceramics involve textured systems. By combining grain texturing and composition engineering in the KNN ternary ceramics, some studies improved their piezoelectric properties. Combining texture engineering with compositional design could result in a highly efficient piezoelectricity of about 416 pC/N, akin to the untextured counterpart, even if it is a PZT system. Doping with multivalent ions like Ta^{5+} and Sb^{5+} affects the PPT temperature [123], which may be accountable for the dramatic increase in the piezoelectric response of the ceramics, simultaneously increasing the temperature sensitivity of the material.

TGG and RTGG are popular methods for texturing. One of the biggest breakthroughs was the discovery of extremely textured ceramics ⟨001⟩ synthesized by the RTGG method. Moreover, extremely high piezoelectricity was observed in highly textured ceramics using plate-like $NaNbO_3$ templates. This was due to facilitated polarization rotation and optimized domain configuration with domain wall energy. However, it cannot be ignored that optimization of templates is crucial as both too less and too many templates could result in deviation of composition. Optimal piezoelectric properties in PPB textured ceramics can be for contributions, both intrinsic and extrinsic [124]. Crystallographic orientation and lattice-related contributions constitute the intrinsic part. In textured ceramics, a phase transition induced by electric fields may partially explain their high piezo-response, similar to a PZT configuration near MPB [125]. In polycrystalline FE ceramics, a PPB combined with crystallographic texture can be used to achieve good piezoelectricity. Using this method led to improved piezoelectric characteristics, featuring d_{33} = 550 pC/N and electromechanical coupling factor, k_p = 72% in ceramics with <001> texture and R-O-T phase. This not only presents a novel route for enhancing the electrical properties of piezoceramics but also advances KNN ceramics as a viable substitute for lead-based PEMs [126].

18.4 BNT Ceramics

Earlier in 1960 Smolenskii et al. reported these relaxor-type FE with a rhombohedral structure [127]. BNTs have attracted significant attention because they have comparatively higher P_r (38 μC/cm^2) and maximum permittivity temperature (T_m) of about 335°C. The BNT could be a potential replacement for environmentally toxic PZT. One of the most important features of BNT is that it can form an MPB [128]. In BNT systems, R and T ferroelectric phases are usually accountable for the highest piezoelectricity. Also, the relaxor-to-FE phase (RFE) leads to the maximum strain response. Transition of R to T phase occurs at 300°C, whereas transition from T to C phase occurs at 540°C.

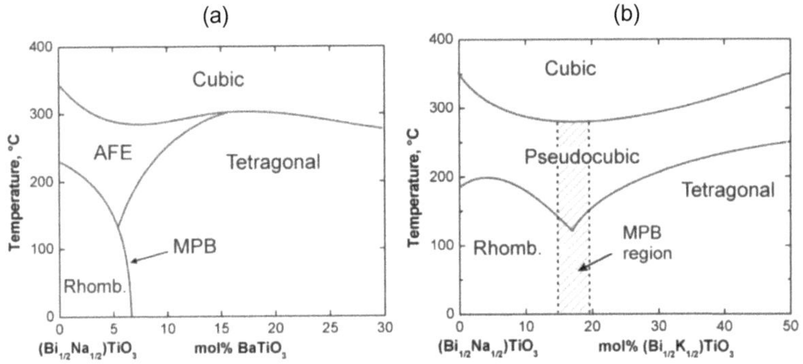

Figure 18.13 MPB constructed by the addition of (a) BT and (b) barium potassium titanate (BKT) into the BNT system [129]. Here, AFE means antiferroelectric phase.

FE to ergodic relaxor (FEER) phase transition is considered to shift to RT by chemical modification with Sr as shown in Fig. 18.13. When the transition occurs amidst a critical electric field, the field-induced strain produced is known as "electrostriction". The reorientation of dipoles within the lattice induced by electrostriction.

High conductivity and E_c of pure BNT ceramic make it difficult to sufficiently pole, which often results in unwanted d_{33} (73–95 pC/N). So, chemical modification (such as solid solutions or ion substitution) is widely used to assist the piezoelectricity of BNT through the fabrication of phase boundaries [130]. Moreover, BNT has significant tanδ due to its narrow range in T_s. Reports also show that during synthesis and sintering, Bi and/or Na are vaporized, resulting in oxygen vacancies [131]. Thus pure BNT ceramics have limited applications. Low T_d brings about another limitation in the working temperature range [132].

To overcome the abovementioned barrier, structure and compositional modification (Fig. 18.14) is considered, which is being done through texture engineering, phase boundary construction by doping, and domain wall modifications to reduce tanδ and improve depolarization and T_c [133].

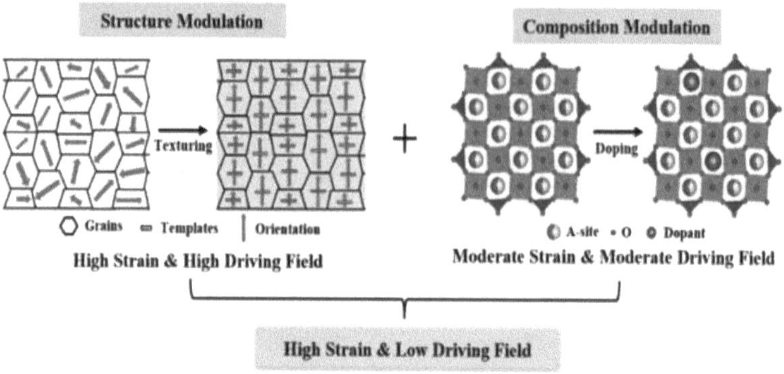

Figure 18.14 Schematic representation of some strategies to improve piezoelectric properties [133].

18.4.1 Chemical Modification: Doping with Different Materials

One of the targets upon which chemical modification is performed is to convert the long-range FE microdomains in non-ergodic relaxors into nanosized domains of an ER state [134]. The increasing ER phase content at small dopant levels produces a piezoelectric response with partial relaxation. As the amount of doping increases, local structure tends to have diminished influence from the electric field. At a point, the material is largely ergodic. Eventually, a greater random field results in dynamic improvement and reduces the size of local polar nanoregions. Generally, Eu, Li, Sm, Pr, and La are used to substitute Bi and Na ions. With increasing mean radii of A-site cation, piezoelectric coefficients are affected in the same way as they should be in addition to the lattice structure parameters. The recently developed system, bismuth sodium titanate modified with Ca (BNCT), has gained significant interest among the several BNT-based solid solution derivative systems investigated to date. The BNCT system revealed that $T_{M\text{-}O}$ occurs at RT at the composition x = 0.10, whereas this transition starts to diminish within the 30–200°C temperature range. The paraelectric T phase might change into a rhombohedral structure if it were to independently become FE as well. According to this scenario, BNCT may display coexistence of FE phases at RT [135]. Among relatively few recent investigations, some revealed a complex evolution and remarkable distortion of this BNCT. To further analyze the structure and the root of the deviation from the usual (*R3c*) constitution of BNCT, Roukos et al. [136] used a technique that explains the certain array of phase transitions. Based on findings at RT, the primary structure of BNCT is R (*R3c*), accompanied by the emergence of T (*P4bm*) domains in a secondary capacity. Structural analysis demonstrates the stabilization of T (*P4bm*) domains within the foundational R (*R3c*) framework. While traces of the elevated-temperature (*P4bm*) phase are entrapped within the R (*R3c*) structure and persist at RT, the initially paraelectric T (*P4bm*) phase tends to shift into a FE state, transitioning to the R (*R3c*) phase. The simultaneous presence of the distinct crystallographic phases *R3c* and *P4bm* in BNCT at RT showcases a property reminiscent of a MPB, a crucial factor for achieving remarkable piezoelectric features.

Ta^{3+}, Nb^{5+}, Sn^{2+}, Zr^{2+}, and Sb^{4+} are popular B-site dopants. Thus, to investigate the relationship between structure and property, and the effect of modification of the B site of piezoelectric ceramic [137], the structural and electrical properties of $(Bi_{0.5}Na_{0.5})(Ti_{1-x}Zr_x)O_3$ (BNTZ) were evaluated. Kulwicki et al. [138] suggested that the conductivity and tanδ could be significantly lowered by isovalent substitution, and therefore Ti^{4+} ion in the B site was substituted by Zr^{4+}. The partial incorporation of Zr^{4+} into BNT crystal is indicated by a secondary phase appearing in the BNZT ceramic when x exceeds 0.1. In compositions where x = 0.4, the secondary phase plays a more significant role in facilitating hopping charges. This phenomenon increases the tanδ. The electric field induced polarization (P-E) P hysteresis loop demonstrates that the highest FE characteristics are achieved when $x = 0.4$. These characteristics include P_r, P_m, and E_c, measuring 40.9 mC/cm^2, 49.5 mC/cm^2, and 51.5 kV/cm, respectively. Additionally, upon application of 80 kV/cm of electric field, this composition yields a moderate strain of 0.268%. Consequently, these observations indicate that introducing appropriate Zr^{4+} into BNT notably influences the dielectric and FE properties of BNTZ ceramics.

This incorporation lowers the number of defect dipoles within the structure, leading to a decrease in internal stress. This reduction contributes to the ceramics exhibiting heightened piezoelectric and FE capabilities, making them potential substitutes for lead-based piezoelectric ceramics. When considering lower levels of doping, the introduction of Y^{3+} (ionic radius: 0.106 nm) in place of Ca^{2+} (ionic radius: 0.134 nm) at the A site of ABO_3 enhances the piezoelectric and FE properties. This effect is achieved by creating interplanar spacing and minimizing oxygen vacancies. However, with excessive yttrium doping, the interplanar spacing diminishes due to the substitution of Zr^{4+} or Ti^{4+} by Y^{3+}.

18.4.2 Microstructure Engineering: Approach to Control the Grain Size

18.4.2.1 Poling

A strain in the LFPP is caused by an electric field-induced phase transformation of 93% $Bi_{0.5}Na_{0.5}TiO_3$–7% $BaTiO_3$, changing its structure from pseudocubic to T. Under the influence of an electric field, the spontaneous polarization of the induced phase aligns predominantly parallel to the field, distinguishing this strain mechanism from the behavior observed in lead-based MPB systems [139]. Hiruma et al. exhibited the determination of T_d and $T_{R\text{-}T}$ for BNT-based solid solutions [140]. Moreover, the dielectric characteristics of xBNT-yBKT-zBT, along with their piezoelectric attributes, were examined at various temperatures. Unlike the unpoled specimen, the poled sample revealed a distinct peak at 177°C with no frequency variation, whereas the unpoled counterpart displayed a wide peak spanning from approximately 150 to 200°C. Beyond the temperature threshold of T_d, a subtle piezoelectric reaction manifested at elevated temperatures, attributed to an alteration in the elastic constant around 265°C [141]. Hence, this implies the $T_{R\text{-}T}$ governing the shift from the R to T structure. Moreover, the ε_r matches the non-polarized ε_r at $T_{R\text{-}T}$. Consequently, both T_d and $T_{R\text{-}T}$ can be ascertained by examining the temperature-dependent behavior of the dielectric properties in polarized and non-polarized states. Additionally, analyses of longitudinal and transverse strain measurements have demonstrated that the phase shift entails a substantial alteration in volume, leading to an elevated piezoelectric coefficient in comparison to the counterpart lacking a phase transition [142].

Defect dipole engineering can also lead to a significant electrostrain. Electrostrain refers to the alteration in the material shape under an electric field; therefore, greater electrostrain values translate to improved piezoelectric performance. Textured piezoceramics of Sr/Nb-doped $Bi_{0.5}(Na_{0.82}K_{0.18})0.5TiO_3$, which are devoid of lead, are designed with defect dipoles. These ceramics showcase an impressive unipolar electro strain of 1.6%, coupled with minimal hysteresis and fatigue resistance. These characteristics are comparable to, or even surpass, those of cutting-edge single-crystal piezoelectric ceramics that contain lead. The augmentation of electrostrain is attributed to defect dipole engineering, reversible electric-field-induced phase transitions, and controlled grain orientation [143]. Regarding this, it is worth mentioning that giant electrostrain results from the introduction of Ta doping, leading to the simultaneous presence of the *R3c* R phase and *P4bm* T phase [144].

18.4.2.2 Texturing

The significant strain induced by an electric field in BNT ceramics, due to the phase transition, presents them as potential substitutes for lead-based materials in actuators. Nonetheless, incorporating BNT-based ceramics into piezoelectric devices poses challenges since achieving such notable strain typically demands a substantial electric field (approximately 70 kV/cm) [145]. To preserve the attributes of the substantial strain, it becomes essential to decrease the required electric field while still upholding the distinct qualities of the significant strain. The electric field-triggered phase transition is intrinsically linked with the crystallographic orientation, and the ⟨001⟩ crystallographic orientation can effectively expedite this process [146]. Consequently, relaxor to ferroelectric phase transition can occur along the $\langle 001 \rangle_c$ orientation [147]. Yet, BNT-based single crystals encounter drawbacks such as expensive synthesis and significant composition, segregation during crystal growth, making it challenging to achieve large-scale production.

Apart from manipulating grain orientation, TGG presents itself as one of the most favorable techniques for attaining a single crystal-like characteristic [148]. Plate-shaped particles of $NaNbO_3$ are frequently employed to promote growth of grain orientation in the morphotropic phase. For further understanding, Liu et al. [133] fabricated a ⟨001⟩-textured BNT-based ceramic using ⟨001⟩-oriented $NaNbO_3$ microplates, where a small amount of CuO and Li_2CO_3 was added to aid sintering to promote growth of template-induced grain. Due to anisotropic behavior, the textured sample achieved a greater reversible phase transition (shown in Fig. 18.15). As a result of synergistic effects between structure and composition, textured ceramic triggered phase transition at 40 kV/cm, which is significantly less than that of the nontextured counterpart. Simultaneously, a large d_{33} = 907 pm/V could be achieved, which is higher than that of the non-textured sample (at 1150°C) and its textured counterpart, respectively. By developing textured BNT-based ceramics, the work paves a new path to improving BNT ceramic near MPB regions.

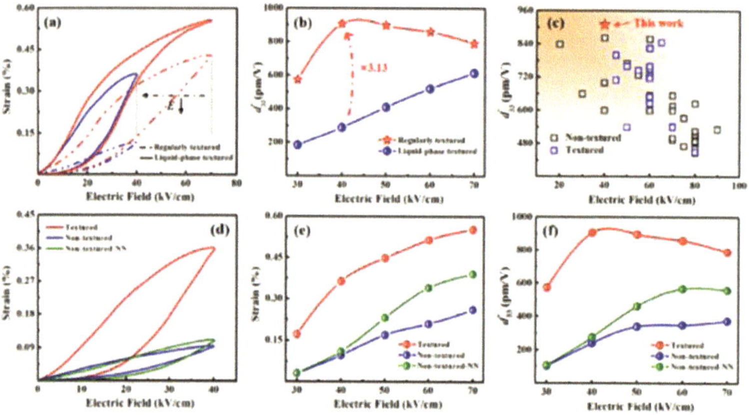

Figure 18.15 A comprehensive comparison between nontextured and textured ceramics. (a) Unipolar *S–E* curves measured at 40 and 70 kV/cm at 10 Hz, (b) normalized d_{33}* vs. electric field for the sample "liquid phase 1000°C-textured" and "regularly 1150°C-textured," (c) representation of d_{33}* values for various BNT LFPP, (d) unipolar *S–E* curves measured at 40 kV/cm at 10 Hz plotted against applied electric field, (e) r strain values, and (f) normalized d_{33}* of the NN-doped nontextured, the nontextured ceramic, "liquid phase 1000°C-textured ceramic" [133].

18.4.2.3 Processing conditions

Despite excellent depolarization behavior and electrical properties, the practical application of BNT is not significant due to the poor thermal stability of electrical properties. The effect of the sintering process is thoroughly investigated to observe the temperature dependency of dielectric, FE, piezoelectric, and pyroelectric parameters. At lower T_s, higher T_d was obtained due to large internal compressive strength which arises from a smaller grain size and vice versa for higher T_S. Although higher T_d is observed when materials are sintered at lower temperatures, poor thermal stability was seen through thermal-induced current. Density, due to the formation of secondary phases, decreases as calcination temperature is increased. Additionally, sintering is carried out to reduce defects and improve crystal quality, enhance grain growth, and increase ceramic density. Based on the optimal piezoelectric and FE values obtained, 1150°C and 850°C can be contemplated as the optimal sintering and calcination temperatures, respectively [149]. Post-heat treatment is also very crucial. The tanδ of BNT coated with bismuth (BNT–10Bi) decreased noteworthily after heating, with enhanced crystallinity and density [150]. Suitably excess Bi promotes the phase formation of BNT perovskite not only by reimbursing the loss of volatile Bi during melting but also by suppressing the melting-induced weight loss. Through a scalable thermal spray process, these results could lead to evolution of compositions that achieve high-quality LFPP coatings [151].

Along with BNT, another promising alkali niobate is bismuth potassium titanate (BKT). Several chemical methods are reported to improve the sinterability of BKT ceramics; however, they have not yet clarified the piezo- and ferroelectricity of the ceramics. Hence, Buhrer et al. sintered BKT ceramics using a hot-pressing (HP) method to achieve dense ceramics. Furthermore, the sinterability of BKT ceramics could be improved by adding BT to the mixture [152]. Aliovalent dopants produce point defects which can promote diffusion during sintering. Sometimes oxygen vacancies are introduced by the acceptor dopant to substitute for cations to enhance their piezo-response.

18.5 BFO Ceramics

BFO is a familiar RT multiferroic material with a distorted R perovskite structure belonging to the space group, *R3c*. The absence of macroscopic magnetization in BFO is due to the spin ordering configuration of G-type in R structure onward $[111]_c$ or $[001]_h$ and around $[110]_h$ axis the cycloid spin structure at RT. The magnetoelectric effect it shows, with the application of electric and magnetic fields, facilitates its magnetization and polarization.

Multiferroic materials are rare because magnetic ordering requires the presence of localized unpaired electrons, whereas distortion in local off-centered symmetry along with empty "*d*" orbital are the most requisite factors. The vacant *d*-subshells of Fe and lone pairs in Bi ion produce off-centered symmetry along the threefold polar axis [111]. It is attractive because of the Neel temperature ($T_N \sim 673$ K), high polarization (i.e., 100 $\mu C/cm^2$), and their high T_c (~ 1100 K). BFO-based ceramics, owing to improved electrical properties and T_c, are considered as benign materials. In the Bi_2O_3–Fe_2O_3 phase systems, two phases in equilibrium were seen: O $Bi_2Fe_4O_9$ *(Pbam)* and R perovskite $BiFeO_3$ (*R3c*). The addition of ABO_3-type compounds, like $BaTiO_3$, has shown drastic change in the phase diagram. In the MPB described by Lee et al., R and T phases are found to coexist [153].

However, because of the high Fe^{2+} ions and oxygen vacancies, large leakage currents were observed, which produces difficulties in obtaining better piezoelectric properties. Synthesis of BFO with only one phase is quite challenging. The magnetoelectric coefficient often deteriorates due to the existence of $Bi_2Fe_4O_9$, $Bi_2O_{2.75}$, and $Bi_{46}Fe_2O_{72}$, which happens for metastable nature [156], peritectic decomposition at relatively lower temperatures, bismuth oxide volatilization, etc. [157]. Although doping can prevent the secondary phase formation, the peculiarity of BFO is not so negligible. The process of annealing and quenching at elevated temperatures makes it even harder to obtain phases with expected stoichiometry. Doping strategies (alkaline and rare-earth ions) were adopted to construct MPBs, reducing leakage current to improve piezo-response.

18.5.1 Chemical Modification: Doping with Different Materials

The coexistence of ferroelectricity and magnetism is mainly initiated by the vacancy of *d* orbitals of ions in A and B sites. Substitution of rare-earth ions at Bi^{3+} sites and transition metal ions at Fe^{3+} sites causes enhancement in magnetic and electric properties. Two distinct purposes are served by doping: firstly, modification of the structural phase to increase its polarization (shown in Fig. 18.16(i)) and, secondly, perturbation of the unit cell. Substitution in Bi site pursued by the addition elements, e.g., Sm [158], La [159], Nd [160], Dy [62], Eu [159], Yb [161], Y [161], Ce [154], and Ho [162], is reported to enhance the formation of a single phase. Despite having an inert character, rare-earth elements influence the physical properties, either by controlling the bond angle or by donating electrons to the B-site ions. Substitution with ions having a small ionic radius at A sites is supposed to produce octahedral tilting of oxygen, accompanied by a smaller tolerance factor. Hence, insulating behavior can also be produced upon doping with the rare-earth element in BFO [163]. Oxygen vacancies produced with excess dopant concentration cause structural distortion, especially during $T_{R\text{-}T}$ for Sr-doped BFO [164], and $T_{R\text{-}O}$ for Dy and Gd-doped BFO [165, 166]. The majority of the works on doping are mainly targeted at reducing the leakage current densities, through either suppression of impurity phases or oxygen vacancies.

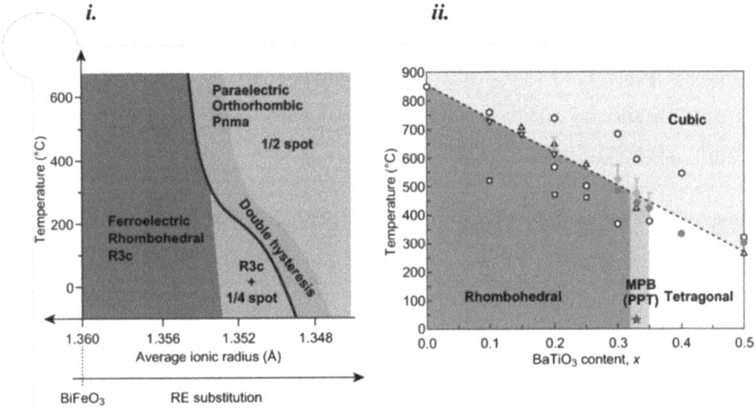

Figure 18.16 Phase diagram of (i) BFO substituted by rare-earth elements [154, 155] and (ii) for the BFO–BTO system [153].

A large piezoelectricity ($d_{33} \approx 50$ pC/N) resulted when leakage current density was reduced [167]. On the other hand, when Dy [62] or Sm is incorporated into BFO [24] reduction of impure phases takes place to result in a high remnant polarization, $P_r \approx 31, 40\ \mu C/cm^2$. La^{3+} substitution at A sites ensures advantages such as oxygen ions stability in the pseudo perovskites layer, which is usually achieved by substitution with ions having small ionic radii like Ce^{3+} or Ce^{4+}. Stabilization of the oxygen octahedron is necessary in the sense that it contributes to the prevention of evaporation and reduction of the amount of leakage current. Even partial doping with Ce can have the same effect on the leakage current, as electron hopping is resisted between Fe divalent cations [168]. A noticeable increase in conductivity, achieved by doping with Mn^{3+}, which prefers to go octahedral site. This modifies the distribution of Fe^{3+} in the lattice, which induces FE behavior [169]. Any deterioration of the FE properties could be due to the transition between non-centrosymmetric and centrosymmetric. Although the lower ionic radius of Mn^{4+} ion shows intense polarization by creating more voids, it reduces the extent of ferroelectricity [170]. Co-doping with large transition metal cations is simply expected to enhance ferroelectricity because of the structural distortion brought out by decreasing the number of free charge carriers during polarization.

However, promoting electrical properties of BFO-based ceramics is rather difficult by ion substitution due to the absence of phase boundaries. Hence, the investigation of BFO ceramics is further put forward using ABO_3-type dopants. The construction of MPB by doping with BT is shown in Fig. 18.16(ii). Typical dopants include $BiFeO_3$-$PbTiO_3$ [171] and $BiFeO_3$-$BaTiO_3$ [172]. Piezoelectricity may be achieved by modifying the R-T phase boundary, but Pb, being toxic, gets relatively less attention [173].

18.5.2 Microstructure Engineering

Microstructural optimization of perovskite materials is considered a better technique for improving piezoelectric behavior.

18.5.2.1 Controlling the grain size

Grain size distribution has an impact on the leakage current density. High grain boundary density resists the mobility of charge carriers, thereby reducing leakage current density. Dopants, irrespective of their ionic radius, induce phase coexistence that can either stimulate or inhibit the growth of grain. The addition of La on the Bi site causes a reduction in grain size [174]. However, the opposite has been observed in another literature [175]. Decreased grain size is caused by the reduced oxygen vacancy achieved by doping with ions, especially those having small ionic radii [176]. The lower the oxygen vacancy, the lower will be the grain growth rate. On the Bi site, doping with ions like La, Sm, Gd, and Y shows suppression of grain growth with increasing dopant concentration. When BFO is doped with Cr at the A site, 90 nm grain size is obtained, along with the coexistence of phases such as T (*P4*) and trigonal (*R3c*); doping at B sites did not produce the same consequences [177]. Growth such as semi-columnar is common in doped perovskites. The reason for enhanced P_r in a doped BFO system is due to charges accumulated in the resistive grain boundaries of the conducting grain, even though the effect is dominant at lower frequencies only [176].

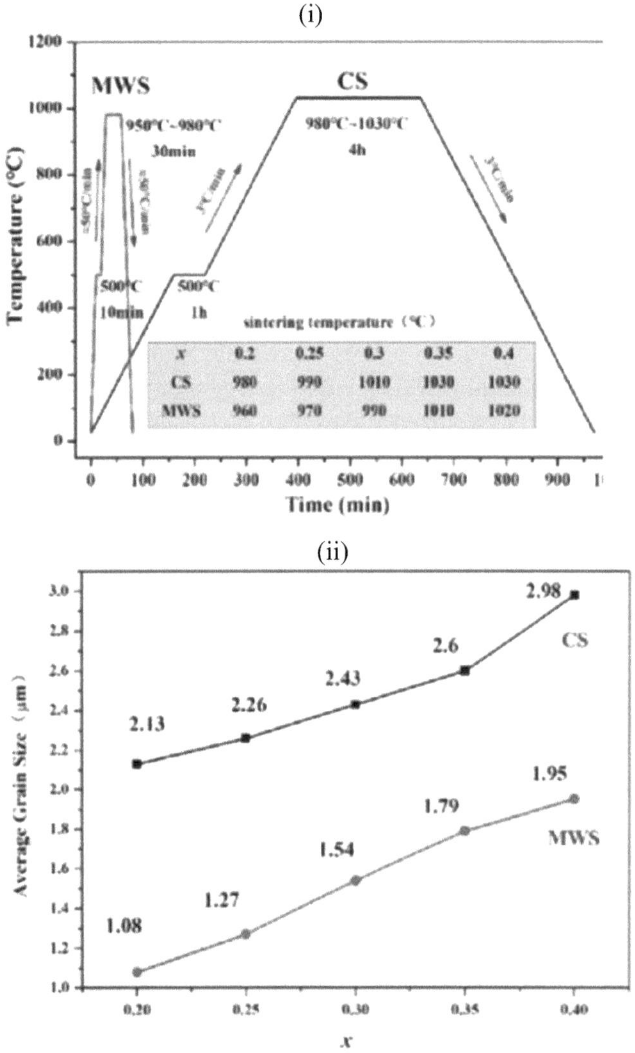

Figure 18.17 (i) Conventional and microwave sintering techniques and (ii) their effects on the average grain size [179].

18.5.2.2 Processing condition

The large grain size of pristine BFO induces volatility by facilitating ion diffusion. BFO in single phase is difficult to produce due to the formation of impurity phases at high temperatures. A method such as sol-gel shows improvement in grain size than conventional solid-state synthesis routes. Mixing at the molecular level, i.e., homogenization, could be achieved by the former technique. The effect of T_s as well as the process is shown in Fig. 18.17. During both synthesis processes, nitric acid is used to leach out the impurity phases obtained after the calcination process, but the reproducibility is found to be low [178].

The sol-gel technique produces soft Ba-doped BFO nanoparticles which were found to be effective in reducing particle size, making it possible to be applied in miniature devices [180]. Oxygen annealing during sintering is found to induce anti-ferromagnetism in Ba-doped BFO [181]. Co-doping by the chemical solution deposition (CSD) method with Ho and Ni shows Bi–O bond dispersion, since Ho–O has higher bond dissociation energy than Bi–O. As a result, the perovskite gets stabilized and oxygen vacancies are lowered.

18.6 Challenges and Prospects

There is a lack of novelty in the design of LFPP materials, and consequently, they still haven't gained comprehensive electrical properties comparable to lead-based materials, which may be remedied by designing new phase boundary (NPB) in LFPP like PZT. Despite these limitations, it must be admitted that great progress has been made in this field, e.g., improved piezoelectricity, strain response, explanation of physical mechanism, further designing electronic devices. More attention should be paid to structure-modified piezo-response improvement under the applications of external fields. Performance evaluation of electronic devices with the use of LFPP should be carried out. LFPP materials are expected to prove their practical uses in the future.

- **Temperature Sensitivity**: For KNN, BNT, BFO, and BT, the temperature dependency in the piezoelectric properties limits its practical applications. Therefore, balancing temperature stability and piezoelectric properties should be focused on in the future. Further attempts may improve MPB formation and domain configuration likewise PZT.

- **Low Density**: A bottleneck problem in textured piezoceramics is their reduced density caused by BT plate-like particle addition. This is regarded as a concerning factor that affects the electrical properties of sintered ceramics. To overcome the abovementioned challenge, the hot pressing or modified hot pressing method could be employed to acquire significant densification of the BCZT.
- **Large Hysteresis**: Although giant strain value determines the applicability of the ceramics especially in actuators, sometimes that restrains their practical purpose.
- **High Leakage Current**: For the long time, development of BFO-based ceramics has been limited due to extreme leakage current that originated from valency variation of Fe. Designing appropriate NPB may result in the optimization of phase boundary and reduction of leakage current. However, BFO, owing to the magnetoelectric effect, is a promising ceramic for future applications.
- **Low T_c and T_d**: BT ceramics having a giant d_{33}, however low $T_c \leq 120°C$ limits its application. Low thermal stability also reduces the piezoelectric properties. Besides, having lower T_d limits the development.

It remains a challenge to achieve simultaneous improvements in piezoelectric properties, T_c, and T_S. Updated phase boundary strategies, new material system designs, and optimized techniques of ceramic preparation are ways to develop applicable BT-based ceramics. Co-fired techniques with multilayer ceramics combined with metal electrodes need to be investigated to improve energy storage, electrocaloric, and electrostrictive applications.

KNN ceramics are still facing several obstacles in their development, for example, bottlenecks in the synthesis methods, mismatches in the improvement of electrical properties, and difficulties meeting device requirements. The easy absorbance of moisture by KNN materials hampers the preparation processes. The electrical properties of KNN depend on mechanical factors (e.g., sintering and calcination temperatures, poling conditions, etc.), and the decrease of densification caused by the volatility of K and Na further complicates the preparation methods. The electrical behavior of KNN is tough to balance, as the improvement of piezoelectricity typically results in low-temperature stability, poor Q_m values, or decreased T_c. Moreover, other electrical parameters, e.g., k_{33}, d_{31}, mechanical properties, and conductivity, have received little attention. In addition, ceramic properties other than piezoelectricity should also be considered. Besides, the temperature stability of piezoelectric parameters, particularly the piezoelectric constant, is crucial. Investigating the effects of manipulating structure (crystal, defect structure, and domain) under external factors (temperature, electric, and magnetic field) to tune performance (dielectric, FE, and piezoelectric behavior) needs to be continued. Studying the service requirements of electric devices and understanding failure mechanisms to develop optimum piezoelectric ceramics is the key research field.

Combined characteristics of strain response, energy storage properties, and piezoelectric performance of BNT have been barely reported. Moreover, the evolution of crystal structure, strain response by external fields, and the energy-storage capacity of these materials have been scarcely reported simultaneously. Therefore, for industrial applications, finding BNT with unique properties is critical. As dielectric capacitors based on BNT can store energy in terms of electric fields, further research on industrial-oriented piezoelectric response is necessary. Bi and Na are volatile elements, which makes it difficult to synthesize BNT solid solutions as the proper stoichiometric ratio in the minimum composition is affected. Additionally, the sensitivity of BNT to A-site non-stoichiometry may produce considerable changes in resistance and tanδ. So, upgrading the synthesis method is important for future researchers. Low P_r and high saturation polarization are vital factors for good energy-storage properties. The strain properties of BNT-based relaxor ferroelectrics are widely investigated as a promising FE material. Yet research on BNT-based antiferroelectric materials lags. To improve energy storage and electromechanical properties, numerous strategies based on solid or chemical doping must be explored (Table 18.3).

Table 18.3 Variation of piezoelectric properties in LFPP with different synthesis methods

Ceramic material	Composition	Method of synthesis	P_r (μC/cm^2)	ε_r	d_{33} (pC/N)	Ref.
BT		Conventional solid state	12.9	4050	546	[182]
		Two-step sintering	10.5	3900	530	[183]
		Sol-gel-hydrothermal	12.56	–	492	[184]
		Hydrothermal	1.19	–	213	[185]
		Spark plasma sintering	~5	~2700	72	[186]
		Sol-gel	12.2	2482	637	[187]
		Plasma activated sintering	12.6	2614	127	[188]
		Microwave sintering	~8	2500	342	[189]
		Conventional solid state	11.69	4500	650	[190]
		Molten salt (50% NaCl-KCl)	9.98	2654	604	[191]

(Continued)

Table 18.3 (*Continued*)

KNN	$K_{0.5}Na_{0.5}NbO_3$	Air sintered			80	[194]
	$0.95KNN-0.05LiNbO_3$	Bridgman	9	185	405	[193]
	$K_{0.5}Na_{0.5}NbO_3$	Hot pressed			160	[192]
	$K_{0.5}Na_{0.5}NbO_3$	Flux method	18	412	148	[195]
	$K_{0.5}Na_{0.5}NbO_3$	Air sintered			80	[195]
	$K_{0.622}Na_{0.378}NbO_3$	Top-seeded solution growth		70	161	[49]
	$Na_{0.5}K_{0.5}Nb0_{.995}Mn_{0.005}O_3$	Flux method	52	424	161	[196]
	$K_{0.5}Na_{0.5}NbO_3$	Air sintered			110	[201]
	$K_{0.14}Na_{0.86}Mn_{0.005}Nb_{0.99}5O_y$	Flux method	37			[194]
	$K_{0.53}Na_{0.47}NbO_3$	Flux method		600	110	[197]
	Mn-doped KNN	Floating zone method	106			[44]
	$K_{0.53}Na_{0.47}Mn_{0.004}Nb_{0.99}6O_y$	Flux method	40			[196]
	$K_{0.5}Na_{0.5}NbO_3$ with B_2O_3 flux	Flux method	32		106	[198]
	$K_{0.5}Na_{0.5}NbO_3$	Solid-state crystal growth	17	1015	80	[199]
	0.5% $Mn-K_{0.5}Na0_{.5}NbO_3$	Flux method		730	270	[50]
	$K_{0.5}Na_{0.5}NbO_3$	Spark plasma sintering			148	[200]
	$(K, Na)(Nb, Ta)O_3$	Top-seeded solution growth		267	200	[41]
	$K_{0.5}Na_{0.5}NbO_3$	Flux method			160	[201]

BNT	BNT-KBT-BT	Reactive template grain growth			314	[202]
	BNT–10Bi	Thermal spray process				[203]
	$Bi_{0.5}Na_{0.5}TiO_3$	Sol-gel		12		[204]
	$Na_{0.5}Bi_{0.5}TiO_3$	Pulse laser deposition			~25	[205]
	BNT-KBT-ST	Reactive template grain growth		28	180	[206]
	BNT-KBT-NN	Reactive template grain growth				[207]
BFO	BFO 0.125 at. % Co	Sintering		70	40	[208]
	$BiFeO_3$–$BaTiO_3$	Solid-state reaction	26.0		134	[209]
	75BF–25BT	Solid solution	10		12.6	[210]
	BF–0.25BT	Solid-state reaction	20.36		225	[211]
	BFO 0.25–0.25 at. % Co–Ti	Sintering		95	40	[208]

Despite extensive research, the novel optical, magnetic, and electronic behaviors of BFO with various dopants (e.g., rare-earth elements) are still in its infancy. Prominent features such as poor quantum efficiency, slow carrier mobility, and limited photocatalytic applications continue to pose challenges. As a result, researchers are directing their efforts toward the regulation of BFO size, crystal structure, and phase transitions to enhance its photocatalytic potential.

Numerous endeavors are underway to substitute rare-earth elements into BFO. However, the introduction of different dopants can lead to the emergence of impurity phases, potentially compromising the intrinsic behavior of pure BFO. Consequently, researchers have undertaken various strategies to address these challenges and have made significant strides in attaining remarkable FE and photocatalytic properties in recent years. Research has opened up new avenues for rare-earth-based BFO applications in communication technology. This is achieved by exploring the fundamental magnetic, electrical, and optical properties associated with the inclusion of rare-earth elements, thus offering a fresh perspective on future research directions.

18.7 Conclusions

In this chapter, developments in LFPP materials have been discussed based on the relationships of doping, texturing, sintering, and phase boundaries to piezoelectric properties. Processing parameters, such as domain configuration, phase boundary, and defects, largely affect the piezoelectric properties of LFPP materials. For enhancement of the electrostrain response, the significant potential of defect dipoles in these materials has been shown. Selecting optimum poling conditions, sintering processes, and texture engineering also play vital roles in improving piezoelectric response. However, the physical mechanisms behind high Q_m and d_{33} values still need to be explored to improve their comprehensive performance. In general, piezoceramics containing multiple phases exhibit MPB behavior; that is, frequent transitions among different phases result in an enhanced piezoelectric response. These materials, however, have some shortcomings, such as insufficient piezoelectric response due to short-range temperature-dependent MPB, which need to be overcome for practical applications to be possible. In the future, a detailed mechanistic understanding of MPB behavior needs to be explored. A well-designed LFPP could potentially replace environmentally toxic PZT ceramics with some further tuning, depending on its structure and properties. This chapter may guide future researchers to study further and develop new LFPP materials. Close attention should also be paid to develop relatively cheaper, high-performance materials and compatible preparation methods to overcome large-scale preparation barriers.

Acknowledgment

The authors acknowledge the Ministry of Science and Technology, Government of Bangladesh for research grant under Special Allocation for Science and Technology. The authors also acknowledge support from the DU-UGC, Bangladesh research grant for the themed research.

References

1. Jaeger RE, Egerton L. Hot Pressing of Potassium-Sodium Niobates. *Journal of the American Ceramic Society.* 45(5):209–13, 1962.
2. Zhou M, Liang R, Zhou Z, Xu C, Nie X, Dong X. Enhanced Curie Temperature and Piezoelectric Properties of $(Ba_{0.85}Ca_{0.15})(Zr_{0.10}Ti_{0.90})O_3$ Lead-Free Ceramics After the Addition of $LiTaO_3$. *Materials Research Bulletin.* 106:213–9, 2018.
3. Hao J, Li W, Zhai J, Chen H. Progress in High-Strain Perovskite Piezoelectric Ceramics. *Materials Science and Engineering R: Reports.* 135:1–57, 2019.
4. Rahman M, Haider J, Akter T, Hashmi MSJ. Techniques for Assessing the Properties of Advanced Ceramic Materials. *Comprehensive Materials Processing.* 1:3–34, 2014.
5. Hu W, Akinc M, Hu H, Beckman S. *Experimental Search for High Curie Temperature Piezoelectric Ceramics with Combinatorial Approaches.* PhD thesis in materials science and engineering submitted to Iowa State University, 5–11, 2011.
6. Liu W, Ren X. Large Piezoelectric Effect in Pb-Free Ceramics. *Physical Review Letter.* 103(25):257602, 2009.

7. Wu J. Perovskite Lead-Free Piezoelectric Ceramics. *Journal of Applied Physics.* 127(19):291–309, 2020.
8. Pandey R, Vats G, Yun J, Bowen CR, Ho-Baillie AWY, Seidel J, et al. Mutual Insight on Ferroelectrics and Hybrid Halide Perovskites: A Platform for Future Multifunctional Energy Conversion. *Advanced Materials.* 31(43):1807376, 2019.
9. Steele JA, Prakasam V, Huang H, Solano E, Chernyshov D, Hofkens J, et al. Trojans That Flip the Black Phase: Impurity-Driven Stabilization and Spontaneous Strain Suppression in γ-$CsPbI_3$ Perovskite. *Journal of the American Ceramic Society.* 143(28):10500–8, 2021.
10. Rödel J, Jo W, Seifert KTP, Anton EM, Granzow T, Damjanovic D. Perspective on the Development of Lead-Free Piezoceramics. *Journal of the American Ceramic Society.* 92(6):1153–77, 2009.
11. Jalalian A. Lead-Free Piezoelectric Materials. *University of Wollongong Thesis Collection 1954-2016.* 2013.
12. Bobic JD, Vijatovic Petrovic MM, Stojanovic BD. Review of the Most Common Relaxor Ferroelectrics and Their Applications. *Magnetic, Ferroelectric, and Multiferroic Metal Oxides.* 1:233–49, 2018.
13. Rowe T, Richtik BN, Dolgos M, Rowe T, Richtik BN, Dolgos M. Synthesis and Characterization of a New Ferroelectric with Low Lead Content, a High Curie Temperature, and a High Piezoelectric Response. *Advanced Electronic Materials.* 9(4):2200910, 2023.
14. Ibn-Mohammed T, Koh SCL, Reaney IM, Acquaye A, Wang D, Taylor S, et al. Integrated Hybrid Life Cycle Assessment and Supply Chain Environmental Profile Evaluations of Lead-Based (Lead Zirconate Titanate) Versus Lead-Free (Potassium Sodium Niobate) Piezoelectric Ceramics. *Energy and Environmental Science.* 9(11):3495–520, 2016.
15. Zheng T, Wu J, Xiao D, Zhu J. Recent Development in Lead-Free Perovskite Piezoelectric Bulk Materials. *Progress in Material Science.* 98:552–624, 2018.
16. Rödel J, Webber KG, Dittmer R, Jo W, Kimura M, Damjanovic D. Transferring Lead-Free Piezoelectric Ceramics into Application. *Journal of the European Ceramic Society.* 35(6):1659–81, 2015.
17. Enomoto M. Structure, Energy and Migration of Phase Boundaries in Steels. *Phase Transformations in Steels.* 1:157–83, 2012.
18. Garcia JE, Rubio-Marcos F. Polymorphic Phase Boundary in Piezoelectric Oxides. *Journal of Applied Physics.* 127(13):131102, 2020.

19. Jaffe B, Cook WRJ, and Jaffe H. *Piezoelectric Ceramics*. Academic Press, London. 1971.

20. Lv X, Zhu J, Xiao D, Zhang XX, Wu J. Emerging New Phase Boundary in Potassium Sodium-Niobate Based Ceramics. *Chemical Society Reviews*. 49(3):671–707, 2020.

21. Wang L, Mao C, Wang G, Du G, Liang R, Dong X. Effect of CuO Addition on the Microstructure and Electric Properties of Low-Temperature Sintered 0.25PMN–0.40PT–0.35PZ Ceramics. *Journal of the American Ceramic Society*. 96(1):24–7, 2013.

22. Damjanovic D. Ferroelectric, Dielectric and Piezoelectric Properties of Ferroelectric Thin Films and Ceramics. *Reports on Progress in Physics*. 61(9):1267, 1998.

23. Safari A, Zhou Q, Zeng Y, Leber JD. Advances in the Development of Pb-Free Piezoelectric Materials for Transducer Applications. *Japanese Journal of Applied Physics*. 62(SJ):SJ0801, 2023.

24. Zheng T, Wu J, Xiao D, Zhu J. Recent Development in Lead-Free Perovskite Piezoelectric Bulk Materials. *Progress in Material Science*. 98:552–624, 2018.

25. Fan P, Liu K, Ma W, Tan H, Zhang Q, Zhang L, et al. Progress and Perspective of High Strain NBT-Based Lead-Free Piezoceramics and Multilayer Actuators. *Journal of Materiomics*. 7:508–44, 2021.

26. Takenaka T, Nagata H. Current Status and Prospects of Lead-Free Piezoelectric Ceramics. *Journal of the European Ceramic Society*. 25(12):2693–700, 2005.

27. Han C, Wu J, Pu C, Qiao S, Wu B, Zhu J, et al. High Piezoelectric Coefficient of Pr_2O_3-Doped $Ba_{0.85}Ca_{0.15}Ti_{0.90}Zr_{0.10}O_3$ Ceramics. *Ceramics International*. 38(8):6359–63, 2012.

28. Ma J, Liu X, Li W. High Piezoelectric Coefficient and Temperature Stability of Ga_2O_3-Doped $(Ba_{0.99}Ca_{0.01})(Zr_{0.02}Ti_{0.98})O_3$ Lead-Free Ceramics by Low-Temperature Sintering. *Journal of Alloys and Compounds*. 581:642–5, 2013.

29. Zhao L, Zhang BP, Zhou PF, Zhao XK, Zhu LF. Phase Structure and Property Evaluation of (Ba, Ca)(Ti, Sn)O_3 Sintered with Li_2CO_3 Addition at Low Temperature. *Journal of the American Ceramic Society*. 97(7):2164–9, 2014.

30. Zhou PF, Zhang BP, Zhao L, Zhao XK, Zhu LF, Cheng LQ, et al. High Piezoelectricity Due to Multiphase Coexistence in Low-Temperature Sintered (Ba, Ca)(Ti, Sn)O_3-CuO_x Ceramics. *Applied Physics Letters*. 103(17):172904, 2013.

31. Sun H, Duan S, Liu X, Wang D, Sui H. Lead-Free $Ba_{0.98}Ca_{0.02}Zr_{0.02}Ti_{0.98}O_3$ Ceramics with Enhanced Electrical Performance by Modifying MnO_2 Doping Content and Sintering Temperature. *Journal of Alloys and Compounds*. 670:262–7, 2016.

32. Chen X, Li Y, Zeng J, Zheng L, Park CH, Li G. Phase Transition and Large Electrostrain in Lead-Free Li-Doped (Ba, Ca)(Ti, Zr)O_3 Ceramics. *Journal of the American Ceramic Society*. 99(6):2170–4, 2016.

33. Wu B, Xiao D, Wu J, Gou Q, Zhu J. Microstructure and Electrical Properties of $(Ba_{0.98}Ca_{0.02})(Ti_{0.94}Sn_{0.06})O_3$–x wt% ZnO Lead-Free Piezoelectric Ceramics Sintered at Lower Temperature. *Journal of Materials Science*. 26(4):2323–8, 2015.

34. Cui Y, Liu X, Jiang M, Zhao X, Shan X, Li W, et al. Lead-Free $(Ba_{0.85}Ca_{0.15})(Ti_{0.9}Zr_{0.1})O_3$–$CeO_2$ Ceramics with High Piezoelectric Coefficient Obtained by Low-Temperature Sintering. *Ceramics International*. 38(6):4761–4, 2012.

35. Zhao L, Zhang BP, Zhou PF, Zhu LF, Wang N. Piezoelectric and Ferroelectric Properties of (Ba, Ca)(Ti, Sn)O_3 Lead-Free Ceramics Sintered with Li_2O Additives: Analysis of Point Defects and Phase Structures. *Ceramics International*. 42(1):1086–93, 2016.

36. Wu J, Wang T, Cheng X, Wang X, Zhang B, Zhu J, et al. Enhanced d_{33} Value in HfO_2-Modified $(Ba_{0.98}Ca_{0.02})(Ti_{0.94}Sn_{0.06})O_3$ Ceramics. *Journal of Alloys and Compounds*. 576:299–301, 2013.

37. Chao X, Wang Z, Tian Y, Zhou Y, Yang Z. $Ba(Cu_{0.5}W_{0.5})O_3$-Induced Sinterability, Electrical and Mechanical Properties of $(Ba_{0.85}Ca_{0.15}Ti_{0.90}Zr_{0.10})O_3$ Ceramics Sintered at Low Temperature. *Materials Research Bulletin*. 66:16–25, 2015.

38. Jie Y, Zhang F, Yang Q, Liu Z, Li Y, Liu Y, et al. Large Piezoelectric Properties in KNN-Based Lead-Free Single Crystals Grown by a Seed-Free Solid-State Crystal Growth Method. *Applied Physics Letters*. 108:182904, 2016.

39. Huo X, Zhang R, Zheng L, Zhang S, Wang R, Wang J, et al. (K, Na, Li)(Nb, Ta)O_3:Mn Lead-Free Single Crystal with High Piezoelectric Properties. *Journal of the American Ceramic Society*. 98(6):1829–35, 2015.

40. Lin D, Li Z, Zhang S, Xu Z, Yao X. Dielectric/Piezoelectric Properties and Temperature Dependence of Domain Structure Evolution in Lead-Free $(K_{0.5}Na_{0.5})NbO_3$ Single Crystal. *Solid State Communications*. 149(39–40):1646–9, 2009.

41. Zheng L, Huo X, Wang R, Wang J, Jiang W, Cao W. Large Size Lead-Free (Na, K)(Nb, Ta)O_3 Piezoelectric Single Crystal: Growth and Full Tensor Properties. *CrystEngComm*. 15(38):7718–22, 2013.

42. Tian H, Hu C, Meng X, Zhou Z, Shi G. Dielectric, Piezoelectric, and Elastic Properties of $K_{0.8}Na_{0.2}NbO_3$ Single Crystals. *Journal of Material Chemistry C.* 3(37):9609–14, 2015.

43. Huo X, Zheng L, Zhang S, Zhang R, Liu G, Wang R, et al. Growth and Properties of Li, Ta Modified (K, Na)NbO_3 Lead-Free Piezoelectric Single Crystals. *Physica Status Solidi.* 8(1):86–90, 2014.

44. Inagaki Y, Kakimoto KI, Kagomiya I. Crystal Growth and Ferroelectric Property of $Na_{0.5}K_{0.5}NbO_3$ and Mn-Doped $Na_{0.5}K_{0.5}NbO_3$ Crystals Grown by Floating Zone Method. *Journal of the European Ceramic Society.* 30(2):301–6, 2010.

45. Gupta S, Priya S. Ferroelectric Properties and Dynamic Scaling of ⟨100⟩ Oriented $(K_{0.5}Na_{0.5})NbO_3$ Single Crystals. *Applied Physics Letters.* 98(24):242906, 2011.

46. Huo X, Zheng L, Zhang R, Wang R, Wang J, Sang S, et al. A High-Quality Lead-Free (Li, Ta) Modified (K, Na)NbO_3 Single Crystal and Its Complete Set of Elastic, Dielectric and Piezoelectric Coefficients with Macroscopic *4mm* Symmetry. *CrystEngComm.* 16(42):9828–33, 2014.

47. Deng H, Zhao X, Zhang H, Chen C, Li X, Lin D, et al. Orientation Dependence of Electrical Properties of Large-Sized Sodium Potassium Niobate Lead-Free Single Crystals. *CrystEngComm.* 16(13):2760–5, 2014.

48. Rafiq MA, Costa ME, Vilarinho PM. Pairing High Piezoelectric Coefficients, d_{33}, with High Curie Temperature (T_c) in Lead-Free (K, Na)NbO_3. *ACS Applied Materials and Interfaces.* 8(49):33755–64, 2016.

49. Tian H, Hu C, Meng X, Tan P, Zhou Z, Li J, et al. Top-Seeded Solution Growth and Properties of $K_{1-x}Na_xNbO_3$ Crystals. *Crystal Growth and Design.* 15(3):1180–5, 2015.

50. Lin D, Li Z, Zhang S, Xu Z, Yao X. Influence of MnO_2 Doping on the Dielectric and Piezoelectric Properties and the Domain Structure in $(K_{0.5}Na_{0.5})NbO_3$ Single Crystals. *Journal of the American Ceramic Society.* 93(4):941–4, 2010.

51. Ge W, Luo C, Zhang Q, Devreugd C, Ren Y, Li J, et al. Ultrahigh Electromechanical Response in $(1-x)(Na_{0.5}Bi_{0.5})TiO_3$-x$BaTiO_3$ Single-Crystals via Polarization Extension. *Journal of Applied Physics.* 111:093508, 2012.

52. Chen CS, Chen PY, Tu CS. Polar Nanoregions and Dielectric Properties in High-Strain Lead-Free 0.93(Bi{sub1/2}Na{sub1/2})TiO{sub3}-0.07BaTiO{sub3} Piezoelectric Single Crystals. *US Department of Energy.* 115, 2014.

53. Zhang Q, Zhang Y, Wang F, Wang Y, Lin D, Zhao X, et al. Enhanced Piezoelectric and Ferroelectric Properties in Mn-Doped $Na_{0.5}Bi_{0.5}TiO_3$–$BaTiO_3$ Single Crystals. *Applied Physics Letters.* 95(10):102904, 2009.

54. Zhang H, Chen C, Deng H, Ren B, Zhao X, Lin D, et al. Ultrahigh Ferroelectric Response in Fe Modified $0.95(Na_{1/2}Bi_{1/2})TiO_3$-$0.05BaTiO_3$ Single Crystals. *Journal of Material Chemistry C.* 2(47):10124–8, 2014.

55. Chen C, Jiang X, Li Y, Wang F, Zhang Q, Luo H. Growth and Electrical Properties of $Na_{1/2}Bi_{1/2}TiO_3$-$BaTiO_3$ Lead-Free Single Crystal with Morphotropic Phase Boundary Composition. *Journal of Applied Physics.* 108(12):124106, 2010.

56. Yamamoto K, Suzuki M, Noguchi Y, Miyayama M. High-Performance $Bi_{0.5}Na_{0.5}TiO_3$ Single Crystals Grown by High-Oxygen-Pressure Flux Method. *Japanese Journal of Applied Physics.* 47:7623–9, 2008.

57. Walker J, Budic B, Bryant P, Kurusingal V, Sorrell C, Bencan A, et al. Robust Polarization and Strain Behavior of Sm-Modified $BiFeO_3$ Piezoelectric Ceramics. *IEEE Transactions on Ultrasonics Ferroelectric Frequency Control.* 62(1):83–7, 2015.

58. Yuan GL, Or SW, Chan HLW. Structural Transformation and Ferroelectric–Paraelectric Phase Transition in $Bi_{1-x}La_xFeO_3$ (x = 0–0.25) Multiferroic Ceramics. *Journal of Physics D.* 40(4):1196, 2007.

59. Jeon N, Rout D, Kim IW, Kang SJL. Enhanced Multiferroic Properties of Single-Phase $BiFeO_3$ Bulk Ceramics by Ho Doping. *Applied Physics Letters.* 98(7):072901, 2011.

60. Yuan GL, Baba-Kishi KZ, Liu JM, Wing Or S, Wang YP, Liu ZG. Multiferroic Properties of Single-Phase $Bi_{0.85}La_{0.15}FeO_3$ Lead-Free Ceramics. *Journal of the American Ceramic Society.* 89(10):3136–9, 2006.

61. Yuan GL, Or SW. Multiferroicity in Polarized Single-Phase $Bi_{0.875}Sm_{0.125}FeO_3$ Ceramics. *Journal of Applied Physics.* 100(2):024109, 2006.

62. Zhang S, Wang L, Chen Y, Wang D, Yao Y, Ma Y. Observation of Room Temperature Saturated Ferroelectric Polarization in Dy Substituted $BiFeO_3$ Ceramics. *Journal of Applied Physics.* 111(7):074105, 2012.

63. Chen X, Wang J, Yuan G, Wu D, Liu J, Yin J, et al. Structure, Ferroelectric, and Piezoelectric Properties of Multiferroic $Bi_{0.875}Sm_{0.125}FeO_3$ Ceramics. *Journal of Alloys and Compounds.* 541:173–6, 2012.

64. Sun C, Wang Y, Yang Y, Yuan G, Yin J, Liu Z. Multiferroic Properties of $Bi_{1-x}Dy_xFeO_3$ (x = 0–0.2) Ceramics at Various Temperatures. *Materials Letter.* 72:160–3, 2012.

65. Baraskar BG, Kadhane PS, Darvade TC, R. James A, Kambale RC, Baraskar BG, et al. $BaTiO_3$-Based Lead-Free Electroceramics with

Their Ferroelectric and Piezoelectric Properties Tuned by Ca^{2+}, Sn^{4+}, and Zr^{4+} Substitution Useful for Electrostrictive Device Application, in (ed. Irzaman H) *Ferroelectrics and Their Applications*. 2018.

66. Panda PK, Sahoo B. PZT to Lead Free Piezo Ceramics: A Review. *Taylor and Francis*. 474(1):128–43, 2015.

67. Kadhane PS, Baraskar BG, Darvade TC, James AR, Kambale RC. Ferroelectric and Piezoelectric Properties of Ca^{2+} and Sn^{4+} Substituted $BaTiO_3$ Lead-Free Electroceramics on the Emphasis of Phase Coexistence. *Solid State Communications*. 306:113797, 2020.

68. Kou Q, Yang B, Sun Y, Yang S, Liu L, Xie H, et al. Tetragonal (Ba, Ca) (Zr, Ti)O_3 Textured Ceramics with Enhanced Piezoelectric Response and Superior Temperature Stability. *Journal of Materiomics*. 8(2):366–74, 2022.

69. Wu J. Perovskite Lead-Free Piezoelectric Ceramics. *Journal of Applied Physics*. 127(19):190901, 2020.

70. Cai W, Zhang Q, Zhou C, Gao R, Zhang S, Li Z, et al. Synergistic Effect of Grain Size and Phase Boundary on Energy Storage Performance and Electric Properties of BCZT Ceramics. *Journal of Materials Science: Materials in Electronics*. 31(12):9167–75, 2020.

71. Zheng T, Wu J, Xiao D, Zhu J. Recent Development in Lead-Free Perovskite Piezoelectric Bulk Materials. *Progress in Materials Science*. 98:552–624, 2018.

72. Wang X, Liang P, Chao X, Yang Z. Dielectric Properties and Impedance Spectroscopy of $MnCO_3$-Modified $(Ba_{0.85}Ca_{0.15})(Zr_{0.1}Ti_{0.9})O_3$ Lead-Free Ceramics. *Journal of the American Ceramic Society*. 98(5):1506–14, 2015.

73. Wu J, Xiao D, Wu W, Chen Q, Zhu J, Yang Z, et al. Composition and Poling Condition-Induced Electrical Behavior of $(Ba_{0.85}Ca_{0.15})(Ti_{1-x}Zr_x)O_3$ Lead-Free Piezoelectric Ceramics. *Journal of the European Ceramic Society*. 32(4):891–8, 2012.

74. Yan X, Zheng M, Gao X, Zhu M, Hou Y. High-Performance Lead-Free Ferroelectric BZT–BCT and Its Application in Energy Fields. *Journal of Material Chemistry C*. 8(39):13530–56, 2020.

75. Tewatia K, Sharma A, Sharma M, Kumar A. Factors Affecting Morphological and Electrical Properties of Barium Titanate: A Brief Review. *Materials Today*. 1(44):4548–56, 2021.

76. Yao Y, Zhou C, Lv D, Wang D, Wu H, Yang Y, et al. Large Piezoelectricity and Dielectric Permittivity in $BaTiO_3$-$xBaSnO_3$ System: The Role of Phase Coexisting. *Europhysics Letters*. 98(2):27008, 2012.

77. Yan Y, Geng LD, Liu H, Leng H, Li X, Wang YU, et al. Near-Ideal Electromechanical Coupling in Textured Piezoelectric Ceramics. *Nature Communications*. 13(1):1–9, 2022.

78. Kadhane PS, Baraskar BG, Darvade TC, James AR, Kambale RC. Ferroelectric and Piezoelectric Properties of Ca^{2+} and Sn^{4+} Substituted $BaTiO_3$ Lead-Free Electroceramics on the Emphasis of Phase Coexistence. *Solid State Communications*. 306:113797, 2020.

79. Hoshina T. Size Effect of Barium Titanate: Fine Particles and Ceramics. *Journal of the Ceramic Society of Japan*. 121(1410):156–61, 2013.

80. Hao M, Fan G, Lou Y, Wen Y, Deng H, Wang F, et al. Regular and Uniform-Shaped $BaTiO_3$ Microplates Prepared Using a Modified Precursor. *Ceramics International*. 45(2):2338–44, 2019.

81. Leontsev SO, Eitel RE. Progress in Engineering High Strain Lead-Free Piezoelectric Ceramics. *Science and Technology of Advanced Materials*. 11:1–13, 2010.

82. Ul R, Marchet P, Pham-Thi M, Tran-Huu-Hue LP. Improved Properties of Doped $BaTiO_3$ Piezoelectric Ceramics. *Physica Status Solidi*. 216(22):1900413, 2019.

83. Liu C, Liu X, Wang D, Chen Z, Fang B, Ding J, et al. Improving Piezoelectric Property of $BaTiO_3$–$CaTiO_3$–$BaZrO_3$ Lead-Free Ceramics by Doping. *Ceramics International*. 40:9881–7, 2014.

84. Cai E, Liu Q, Gu H. Optimization of the Electrical Properties on Lead-Free Barium Titanate-Based Piezoelectric by Combining Grain Growth with Orthorhombic-Tetragonal Phase Boundary. *Journal of Alloys and Compounds*. 948:169775, 2023.

85. Wada S. Domain Wall Engineering in Piezoelectric Crystals with Engineered Domain Configuration. *Handbook of Advanced Dielectric, Piezoelectric, and Ferroelectric Materials*. 266–303, 2008.

86. Hao J, Bai W, Li W, Zhai J. Correlation Between the Microstructure and Electrical Properties in High-Performance $(Ba_{0.85}Ca_{0.15})(Zr_{0.1}Ti_{0.9})O_3$ Lead-Free Piezoelectric Ceramics. *Journal of the American Ceramic Society*. 95(6):1998–2006, 2012.

87. Zheng W, Lin J, Liu X, Yang W, Li Y. Enhanced Ferroelectric and Piezoelectric Performance of $(Ba_{0.85}Ca_{0.15})(Zr_{0.1}Ti_{0.9})O_3$ Lead-Free Ceramics opon Ce and Sb Co-Doping. *RSC Advances*. 11(5):2616–23, 2021.

88. Hao J, Li W, Zhai J, Chen H. Progress in High-Strain Perovskite Piezoelectric Ceramics. *Materials Science and Engineering R: Reports*. 135:1–57, 2019.

89. Tewatia K, Sharma A, Sharma M, Kumar A. Factors Affecting Morphological and Electrical Properties of Barium Titanate: A Brief Review. *Materials Today.* 44:4548–56, 2021.
90. Messing GL, Poterala S, Chang Y, Frueh T, Kupp ER, Watson BH, et al. Texture-Engineered Ceramics – Property Enhancements through Crystallographic Tailoring. *Journal of Materials Research.* 32(17):3219–41, 2017.
91. Du Q, Zhang C, Liu CL, Zhou W, Zeng F, Li K, et al. Highly <001>-Textured $BaTiO_3$ Ceramics with High Piezoelectric Performance Prepared by Vat Photopolymerization. *Additive Manufacturing.* 66:103454, 2023.
92. Seabaugh MM, Kerscht IH, Messing GL. Texture Development by Templated Grain Growth in Liquid-Phase-Sintered A-Alumina. *Journal of the American Ceramic Society.* 80(5):1181–8, 1997.
93. Fu Z, Zhang Z, Lu P, Zhang L, Yao H, Xu F, et al. Mechanistic Interpretation of the Reactive Templated Grain Growth Process of (Li, Ta, Sb) Modified (K, Na)NbO_3 Textured Lead-Free Piezoceramics. *CrystEngComm.* 18(37):7029–35, 2016.
94. Tani T. Crystalline-Oriented Piezoelectric Bulk Ceramics with A Perovskite-Type Structure. *Journal of the Korean Physical Society.* 32(9):1217, 1998.
95. Li F, Zhang S, Xu Z, Chen LQ. The Contributions of Polar Nanoregions to the Dielectric and Piezoelectric Responses in Domain-Engineered Relaxor-$PbTiO_3$ Crystals. *Advanced Functional Materials.* 27(18):1700310, 2017.
96. Wu B, Wu H, Wu J, Xiao D, Zhu J, Pennycook SJ. Giant Piezoelectricity and High Curie Temperature in Nanostructured Alkali Niobate Lead-Free Piezoceramics through Phase Coexistence. *Journal of American Chemical Society.* 138(47):15459–64, 2016.
97. Wu J, Xiao D, Zhu J. Potassium-Sodium Niobate Lead-Free Piezoelectric Materials: Past, Present, and Future of Phase Boundaries. *Chemical Reviews.* 115(7):2559–95, 2015.
98. Zhang N, Zheng T, Wu J. Lead-Free (K, Na)NbO_3-Based Materials: Preparation Techniques and Piezoelectricity. *ACS Omega.* 5(7):3099–107, 2020.
99. Wu J, Tao H, Yuan Y, Lv X, Wang X, Lou X. Role of Antimony in the Phase Structure and Electrical Properties of Potassium–Sodium Niobate Lead-Free Ceramics. *RSC Advances.* 5(19):14575–83, 2015.
100. Ahart M, Somayazulu M, Cohen RE, Ganesh P, Dera P, Mao HK, et al. Origin of Morphotropic Phase Boundaries in Ferroelectrics. *Nature.* 451(7178):545–8, 2008.

101. Shi W, Du J, Zhai Y, Chen C, Wei Y, Li W, et al. Intrinsic and Extrinsic Dielectric Contributions to the Electrical Properties in $CaZrO_3$-Doped KNN-Based Electrical/Optical Multifunctional Ceramics. *Journal of Material Science*. 55(14):5741–9, 2020.

102. Banerjee S, Bairagi S, Wazed Ali S. A Critical Review on Lead-Free Hybrid Materials for Next Generation Piezoelectric Energy Harvesting and Conversion. *Ceramics International*. 47(12):16402–21, 2021.

103. Kosec M, Malič B, Benčan A, Rojac T. KNN-Based Piezoelectric Ceramics. *Piezoelectric and Acoustic Materials for Transducer Applications*. 81–102, 2008.

104. Malic B, Bernard J, Bencan A, Kosec M. Influence of Zirconia Addition on the Microstructure of $K_{0.5}Na_{0.5}NbO_3$ Ceramics. *Journal of the European Ceramic Society*. 28(6):1191–6, 2008.

105. Lim JB, Zhang S, Jeon JH, Shrout TR. (K, Na)NbO_3-Based Ceramics for Piezoelectric "Hard" Lead-Free Materials. *Journal of the American Ceramic Society*. 93(5):1218–20, 2010.

106. Goh PC, Yao K, Chen Z. Titanium Diffusion into $(K_{0.5}Na_{0.5})NbO_3$ Thin Films Deposited on Pt/Ti/SiO_2/Si Substrates and Corresponding Effects. *Journal of the American Ceramic Society*. 92(6):1322–7, 2009.

107. Zhang B, Wu J, Cheng X, Wang X, Xiao D, Zhu J, et al. Lead-Free Piezoelectrics Based on Potassium-Sodium Niobate with Giant d_{33}. *ACS Applied Materials and Interfaces*. 5(16):7718–25, 2013.

108. Qin Y, Zhang J, Yao W, Lu C, Zhang S. Domain Configuration and Thermal Stability of $(K_{0.48}Na_{0.52})(Nb_{0.96}Sb_{0.04})O_3$-$Bi_{0.50}(Na_{0.82}K_{0.18})_{0.50}ZrO_3$ Piezoceramics with High d_{33} Coefficient. *ACS Applied Materials and Interface*. 8(11):7257–65, 2016.

109. Yao FZ, Wang K, Jo W, Webber KG, Comyn TP, Ding JX, et al. Diffused Phase Transition Boosts Thermal Stability of High-Performance Lead-Free Piezoelectrics. *Advanced Functional Materials*. 26(8):1217–24, 2016.

110. Du H, Zhou W, Luo F, Zhu D, Qu S, Pei Z. An Approach to Further Improve Piezoelectric Properties of $(K_{0.5}Na_{0.5})NbO_3$-Based Lead-Free Ceramics. *Applied Physics Letters*. 91(20):202907, 2007.

111. Zheng T, Wu J. Relationship Between Poling Characteristics and Phase Boundaries of Potassium-Sodium Niobate Ceramics. *ACS Applied Materials and Interfaces*. 8(14):9242–6, 2016.

112. Morozov MI, Kungl H, Hoffmann MJ. Effects Of Poling over the Orthorhombic-Tetragonal Phase Transition Temperature in Compositionally Homogeneous (K, Na)NbO_3-Based Ceramics. *Applied Physics Letters*. 98(13):132908, 2011.

113. Zhao Y, Zhao Y, Huang R, Liu R, Zhou H. Poling Field Dependence of Ferroelectric Domains in Tetragonal KNNLN Ceramics. *Ceramics International*. 38:6067–70, 2012.

114. Wang K, Li JF. Domain Engineering of Lead-Free Li-Modified (K, Na) NbO_3 Polycrystals with Highly Enhanced Piezoelectricity. *Advanced Functional Materials*. 20(12):1924–9, 2010.

115. Tao H, Wu J, Wang H. Modification of Strain and Piezoelectricity in $(K,Na)NbO_3$–$(Bi,Na)HfO_3$ Lead-Free Ceramics with High Curie Temperature. *Journal of Alloys and Compounds*. 684:217–23, 2016.

116. Ramajo L, Taub J, Castro MS. Influence of Zirconium Addition on Final Properties of $K_{0.5}Na_{0.5}NbO_3$-Based Ceramics. *Journal of Materials Science: Materials in Electronics*. 25(1):168–73, 2014.

117. Malic B, Bernard J, Bencan A, Kosec M. Influence of Zirconia Addition on the Microstructure of $K_{0.5}Na_{0.5}NbO_3$ Ceramics. *Journal of the European Ceramic Society*. 28(6):1191–6, 2008.

118. Cheng Y, Xing J, Wang T, Wang F, Li R, Sun X, et al. Realization of Densified Microstructure and Large Piezoelectricity in KNN Ceramics via the Addition of Oxide Additives. *Journal of Materials Science: Materials in Electronics*. 32(15):20211–24, 2021.

119. Zheng T, Wu J, Xiao D, Zhu J. Giant d_{33} in Nonstoichiometric (K, Na) NbO_3-Based Lead-Free Ceramics. *Scripta Materialia*. 94:25–7, 2015.

120. Shen ZY, Zhen Y, Wang K, Li JF. Influence of Sintering Temperature on Grain Growth and Phase Structure of Compositionally Optimized High-Performance Li/Ta-Modified $(Na, K)NbO_3$ Ceramics. *Journal of the American Ceramic Society*. 92(8):1748–52, 2009.

121. Zheng T, Wu J, Cheng X, Wang X, Zhang B, Xiao D, et al. High Strain in $(K_{0.40}Na_{0.60})(Nb_{0.955}Sb_{0.045})O_3$–$Bi_{0.50}Na_{0.50}ZrO_3$ Lead-Free Ceramics with Large Piezoelectricity. *Journal of Materials Chemistry C*. 2(41):8796–803, 2014.

122. Du XH, Zheng J, Belegundu U, Uchino K. Crystal Orientation Dependence of Piezoelectric Properties of Lead Zirconate Titanate Near the Morphotropic Phase Boundary. *Applied Physics Letters*. 72(19):2421–3, 1998.

123. Messing GL, Poterala S, Chang Y, Frueh T, Kupp ER, Watson BH, et al. Texture-Engineered Ceramics – Property Enhancements through Crystallographic Tailoring. *Journal of Materials Research*. 32(17):3219–41, 2017.

124. Damjanovic D. Comments on Origins of Enhanced Piezoelectric Properties in Ferroelectrics. *IEEE Transactions on Ultrasonics, Ferroelectrics, and Frequency Control*. 56(8):1574–85, 2009.

125. Guo H, Zhang S, Beckman SP, Tan X. Microstructural Origin for the Piezoelectricity Evolution in $(K_{0.5}Na_{0.5})NbO_3$-Based Lead-Free Ceramics. *Journal of Applied Physics.* 114(15):154102, 2013.

126. Li P, Huan Y, Yang W, Zhu F, Li X, Zhang X, et al. High-Performance Potassium-Sodium Niobate Lead-Free Piezoelectric Ceramics Based on Polymorphic Phase Boundary and Crystallographic Texture. *Acta Materialia.* 165:486–95, 2019.

127. Kitagawa K, Toyoda T, Kitagawa K, Yamamoto T. Mechanical Properties of Small Amount of Mn and Nb Added PZT Piezoelectric Ceramics. *Journal of the Japan Society of Powder and Powder Metallurgy.* 50(5):359–64, 2003.

128. Lin D, Xiao D, Zhu J, Yu P. Piezoelectric and Ferroelectric Properties of $[Bi_{0.5}(Na_{1-x}-yK_xLi_y)_{0.5}]TiO_3$ Lead-Free Piezoelectric Ceramics. *Applied Physics Letters.* 88(6):062901, 2006.

129. Leontsev SO, Eitel RE. Progress in Engineering High Strain Lead-Free Piezoelectric Ceramics. *Science and Technology of Advanced Materials.* 11(4):044302, 2010.

130. Zheng T, Wu J, Xiao D, Zhu J. Recent Development in Lead-Free Perovskite Piezoelectric Bulk Materials. *Progress in Materials Science.* 98:552–624, 2018.

131. Nagata H. Electrical Properties and Tracer Diffusion of Oxygen in Some Bi-Based Lead-Free Piezoelectric Ceramics. *Journal of the Ceramic Society of Japan.* 116(1350):271–7, 2008.

132. Hiruma Y, Yoshii K, Nagata H, Takenaka T. Investigation of Phase Transition Temperatures on $(Bi_{1/2}Na_{1/2})TiO_3$-$(Bi_{1/2}K_{1/2})TiO_3$ and $(Bi_{1/2}Na_{1/2})TiO_3$-$BaTiO_3$ Lead-Free Piezoelectric Ceramics by Electrical Measurements. *Ferroelectrics.* 346(1):114–9, 2007.

133. Liu K, Ma W, Ben N, Zhu H, Xu W, Gao H. Superior Large-Signal Piezoelectric Coefficient in Textured NBT-Based Piezoceramics by Template-Induced Structure and Composition Modulation. *ACS Applied Materials & Interfaces.*15(20):24595–605, 2023.

134. Zhang ST, Kounga AB, Aulbach E, Ehrenberg H, Rödel J. Giant Strain in Lead-Free Piezoceramics $Bi_{0.5}Na_{0.5}TiO_3$–$BaTiO_3$–$K_{0.5}$ $Na_{0.5}NbO_3$ System. *Applied Physics Letters.* 91(11):112906, 2007.

135. Garg R, Senyshyn A, Ranjan R. Complex Structural Phase Transitions in Slightly Ca Modified $Bi_{0.5}Na_{0.5}TiO_3$. *Journal of Physics: Condensed Matter.* 24(45):455902, 2012.

136. Roukos R, Romanos J, Abou Dargham S, Chaumont D. Evidence of a Stable Tetragonal (*P4bm*) Phase in Rhombohedral (*R3c*) Complex

Perovskite $(Na_{1/2}Bi_{1/2})_{1-x}Ca_xTiO_3$ Lead-Free Piezoelectric Materials. *Materials Science and Engineering: B.* 288:116196, 2023.

137. Jin L, Huang Y, Pang J, Zhang L, Jing R, Guo D. High Thermally Stable Dielectric Permittivity, Polarization Enhancement and Electrostrictive Properties in Zr-Substituted Bismuth Sodium Titanate Lead-Free Ferroelectric Ceramics. *Ceramics International.* 46(14), 22889–99, 2020.

138. Haertling GH. Ferroelectric Ceramics: History and Technology. *Journal of the American Ceramic Society.* 82(4):797–818, 1999.

139. Daniels JE, Jo W, Rödel J, Jones JL. Electric-Field-Induced Phase Transformation at a Lead-Free Morphotropic Phase Boundary: Case Study in a 93%$(Bi_{0.5}Na_{0.5})TiO_3$–7% $BaTiO_3$ Piezoelectric Ceramic. *Applied Physics Letters.* 95(3):032904, 2009.

140. Hiruma Y, Nagata H, Takenaka T. Phase Transition Temperatures and Piezoelectric Properties of $(Bi_{1/2}Na_{1/2})TiO_3$–$(Bi_{1/2}K_{1/2})TiO_3$–$BaTiO_3$ Lead-Free Piezoelectric Ceramics. *Japanese Journal of Applied Physics.* 45(9S):7409, 2006.

141. Tian Y, Ma M, Li S, Dong J, Ji X, Wu H. Piezoelectricity and Thermophysical Properties of $Ba_{0.90}Ca_{0.10}Ti_{0.96}Zr_{0.04}O_3$ Ceramics Modified with Amphoteric Nd^{3+} and Y^{3+} Dopants. *Materials.* 16(6):2369, 2023.

142. Zhang MH, Shen C, Zhao C, Dai M, Yao FZ, Wu B, et al. Deciphering the Phase Transition-Induced Ultrahigh Piezoresponse in (K, Na)NbO_3-Based Piezoceramics. *Nature Communications.* 13(1):3434, 2022.

143. Lai L, Li B, Tian S, Zhao Z, Zhang S, Dai Y. Giant Electrostrain in Lead-Free Textured Piezoceramics by Defect Dipole Design. *Advanced Materials.* 35(29):2300519, 2023.

144. Liu X, Tan X. Giant Strain with Low Cycling Degradation in Ta-Doped $[Bi_{1/2}(Na_{0.8}K_{0.2})_{1/2}]TiO_3$ Lead-Free Ceramics. *Journal of Applied Physics.* 120(3):034102, 2016.

145. Liu X, Tan X. Giant Strains in Non-Textured $(Na_{1/2}Bi_{1/2})TiO_3$-Based Lead-Free Ceramics. *Advanced Materials.* 28(3):574–8, 2016.

146. Fancher CM, Iamsasri T, Blundell JE, Bowman KJ. Effect of Crystallographic Texture on the Field-Induced-Phase Transformation Behavior of $Bi_{0.5}Na_{0.5}TiO_3$–$7BaTiO_3$–$2K_{0.5}Na_{0.5}NbO_3$. *Materials Research Letters.* 1(3):156–60, 2013.

147. Park JH, Kang SJL. Solid-State Conversion of (94-x)$(Na_{1/2}Bi_{1/2})TiO_3$-$6BaTiO_3$-x$(K_{1/2}Na_{1/2})NbO_3$ Single Crystals and Their Enhanced Converse Piezoelectric Properties. *AIP Advances.* 6(1):015310, 2016.

148. Li J, Shen Z, Chen X, Yang S, Zhou W, Wang M, et al. Grain-Orientation-Engineered Multilayer Ceramic Capacitors for Energy Storage Applications. *Nature Materials.* 19(9):999–1005, 2020.

149. Turki O, Slimani A, Zina S, Khemakhem H, Abdelmoula N, Lebrun L. Effect of Sintering Temperature of NBT–6BT Lead-free Ceramics on the Structural, Ferroelectric, and Piezoelectric Properties. *Applied Physics A.* 128:186(1–9), 2022.

150. Miriyala K, Ramadurai R. Tunable Ferroelectric Domain Orientation in Polycrystalline and Highly Oriented $Na_{0.5}Bi_{0.5}TiO_3$ Thin Films. *Material Letters.* 178:23–6, 2016.

151. Guo K, Chen S, Tan CKI, Sharifzadeh Mirshekarloo M, Yao K, Tay FEH. Bismuth Sodium Titanate Lead-free Piezoelectric Coatings by Thermal Spray Process. *Journal of the American Ceramic Society.* 100(8):3385–92, 2017.

152. Buhrer CF. Some Properties of Bismuth Perovskites. *Journal of Chemical Physics.* 36(3):798–803, 2004.

153. Hwan Lee M, Jeong Kim D, Su Park J, Wook Kim S, Kwon Song T, Kim MH. High-Performance Lead-Free Piezoceramics with High Curie Temperatures. *Advanced Materials.* 27(43):6976–82, 2015.

154. Pradhan SK, Roul BK. Electrical Behavior of High Resistivity Ce-Doped $BiFeO_3$ Multiferroic. *Physica B: Condensed Matter.* 407(13):2527–32, 2012.

155. Kan D, Pálová L, Anbusathaiah V, Cheng CJ, Fujino S, Nagarajan V. Universal Behavior and Electric-Field-Induced Structural Transition in Rare-Earth-Substituted $BiFeO_3$. *Advanced Functional Materials.* 20(7):1108–15, 2010.

156. Levin EM, Roth RS. Polymorphism of Bismuth Sesquioxide. II. Effect of Oxide Additions on the Polymorphism of Bi_2O_3. *Journal of Research of the National Institute of Standards and Technology.* 68(2):197, 1964.

157. Koizumi H, Niizeki N, Ikeda T. An X-Ray Study on Bi_2O_3-Fe_2O_3 System. *Japanese Journal of Applied Physics.* 3(8):495, 1964.

158. Walker J, Budic B, Bryant P, Kurusingal V, Sorrell CC, Bencan A. Robust Polarization and Strain Behavior of Sm-Modified $BiFeO_3$ Piezoelectric Ceramics. *IEEE Transactions on Ultrasonics, Ferroelectrics, and Frequency Control.* 62(1):83–7, 2015.

159. Tao H, Lv J, Zhang R, Xiang R, Wu J. Lead-Free Rare Earth-Modified $BiFeO_3$ Ceramics: Phase Structure and Electrical Properties. *Materials & Design.* 120:83–9, 2017.

160. Yuan G, Or S. Enhanced Piezoelectric and Pyroelectric Effects in Single-Phase Multiferroic $Bi_{1-x}Nd_xFeO_3$ (x=0–0.15) Ceramics. *Applied Physics Letters.* 88:62905, 2006.

161. Zheng T, Wu J. Effects of Site Engineering and Doped Element Types on Piezoelectric and Dielectric Properties of Bismuth Ferrite Lead-Free Ceramics. *Journal of Materials Chemistry C.* 3(43):11326–34, 2015.

162. Jeon N, Rout D, Kim IW, Kang SJL. Enhanced Multiferroic Properties of Single-Phase $BiFeO_3$ Bulk Ceramics by Ho Doping. *Applied Physics Letters.* 98(7):072901. 2011.

163. Yang CH, Kan D, Takeuchi I, Nagarajan V, Seidel J. Doping $BiFeO_3$: Approaches and Enhanced Functionality. *Physical Chemistry Chemical Physics.* 14(46):15953–62, 2012.

164. Hussain T, Siddiqi SA, Atiq S, Awan MS. Induced Modifications in the Properties of Sr Doped $BiFeO_3$ Multiferroics. *Progress in Natural Science: Materials International.* 23(5):487–92, 2013.

165. Tang YH, Han TC, Liu HL, Lin JG. Doping Effects on the Magnetoelectricity of Bismuth Ferrite Nanoparticles. *Journal of Superconductivity and Novel Magnetism.* 25:2753–5, 2012.

166. Lotey GS, Verma NK. Structural, Magnetic, and Electrical Properties of Gd-Doped $BiFeO_3$ Nanoparticles with Reduced Particle Size. *Journal of Nanoparticle Research.* 14:1–11, 2012.

167. Zheng T, Wu J. Enhanced Piezoelectric Activity in High-Temperature $Bi_{1-x-y}Sm_xLa_y FeO_3$ Lead-Free Ceramics. *Journal of Materials Chemistry C.* 3(15):3684–93, 2015.

168. Quan Z, Liu W, Hu H, Xu S, Sebo B., et al. Microstructure, Electrical and Magnetic Properties of Ce-Doped $BiFeO_3$ Thin Films. *Journal of Applied Physics.* 104(8):084106, 2008.

169. Dhanalakshmi B, Pratap K, Rao BP, Rao PSVS. Effects of Mn Doping on Structural, Dielectric and Multiferroic Properties of $BiFeO_3$ Nanoceramics. *Journal of Alloys and Compounds* 676:193–201, 2016.

170. Chen G, Li Y, Wang X, Wen Z, Lin F, Dai L. Super-Oscillation Far-Field Focusing Lens Based on Ultra-Thin Width-Varied Metallic Slit Array. *IEEE Photonics Technology Letters.* 28(3):335–8, 2015.

171. Fan L, Chen J, Li S, Kang H, Liu L, Fang L. Enhanced Piezoelectric and Ferroelectric Properties in the $BaZrO_3$ Substituted $BiFeO_3$-$PbTiO_3$. *Applied Physics Letters.* 102(2):022905, 2013.

172. Zheng T, Wu J. Quenched Bismuth Ferrite-Barium Titanate Lead-Free Piezoelectric Ceramics. *Journal of Alloys and Compounds.* 676:505–12, 2016.

173. Zhuang J, Wu H, Ren W, Ye ZG. Local Polar Structure and Multiferroic Properties of (1−x)$Bi_{0.9}Dy_{0.1}FeO_{3-x}PbTiO_3$ Solid Solution. *Journal of Applied Physics.*116(6):014903, 2014.

174. Jian Z, Kumar NP, Zhong M, Yemin H, Reddy PV. Structural, Magnetic, and Dielectric Properties of $Bi_{1-x}La_xFeO_3$ (x = 0, 0.1, 0.15 and 0.2). *Journal of Magnetism and Magnetic Materials.* 386:92–7, 2015.

175. Kazhugasalamoorthy S, Jegatheesan P, Mohandoss R, Giridharan N V, Karthikeyan B, Joseyphus RJ. Investigations on the Properties of Pure and Rare Earth Modified Bismuth Ferrite Ceramics. *Journal of Alloys and Compounds.* 493(1–2):569–72, 2010.

176. Barik SK, Ahmed S, Hajra S. Studies of Dielectric Relaxation and Impedance Analysis of New Electronic Material $(Sb_{1/2}Na_{1/2})(Fe_{2/3}Mo_{1/3})O_3$. *Applied Physics A.* 125(3):200, 2019.

177. Kumar A, Yadav KL. Magnetic, Magnetocapacitance and Dielectric Properties of Cr Doped Bismuth Ferrite Nanoceramics. *Materials Science and Engineering.* 176(3):227–30, 2011.

178. Kim JK, Kim SS, Kim WJ. Sol–Gel Synthesis and Properties of Multiferroic $BiFeO_3$. *Materials Letters.* 59(29–30):4006–9, 2005.

179. Ji C, Fan T, Chen G, Bai X, Wang J, He J. Influence of Sintering Method on Microstructure, Electrical and Magnetic Properties of $BiFeO_3$–$BaTiO_3$ Solid Solution Ceramics. *Materials Today Chemistry.* 20:100419, 2021.

180. Hasan M, Hakim MA, Basith MA, Hossain MS, Ahmmad B, Zubair MA. Size Dependent Magnetic and Electrical Properties of Ba-Doped Nanocrystalline $BiFeO_3$. *AIP Advances.* 6(3):035314, 2016.

181. Singh P, Jung JH. Effect of Oxygen Annealing on Magnetic, Electric and Magnetodielectric Properties of Ba-Doped $BiFeO_3$. *Physica B: Condensed Matter.* 405(4):1086–9, 2010.

182. Zheng P, Zhang JL, Tan YQ, Wang CL. Grain-Size Effects on Dielectric and Piezoelectric Properties of Poled $BaTiO_3$ Ceramics. *Acta Materialia.* 60(13–14):5022–30, 2012.

183. Yan X, Zheng M, Sun S, Zhu M, Hou Y. Boosting Energy Harvesting Performance in (Ba,Ca)(Ti,Zr)O_3 Lead-Free Perovskites through Artificial Control of Intermediate Grain Size. *Dalton Transactions.* 47(28):9257–66, 2018.

184. Ji X, Wang C, Zhang S, Tu R, Shen Q, Shi J, et al. Structural and Electrical Properties of BCZT Ceramics Synthesized by Sol–Gel-Hydrothermal Process at Low Temperature. *Journal of Materials Science: Materials in Electronics.* 30(13):12197–203, 2019.

185. Ji W, Fang B, Lu X, Zhang S, Yuan N, Ding J. Tailoring Structure and Performance of BCZT Ceramics Prepared Via Hydrothermal Method. *Physica B: Condensed Matter*. 567:65–78, 2019.

186. Hao J, Bai W, Li W, Zhai J. Correlation between the Microstructure and Electrical Properties in High-Performance $(Ba_{0.85}Ca_{0.15})(Zr_{0.1}Ti_{0.9})O_3$ Lead-Free Piezoelectric Ceramics. *Journal of the American Ceramic Society*. 95(6):1998–2006, 2012.

187. Praveen JP, Karthik T, James AR, Chandrakala E, Asthana S, Das D. Effect of Poling Process on Piezoelectric Properties of Sol–Gel Derived BZT–BCT Ceramics. *Journal of the European Ceramic Society*. 35(6):1785–98, 2015.

188. Li SB, Wang CB, Ji X, Shen Q, Zhang LM. Effect Of Composition Fluctuation on Structural and Electrical Properties of BZT-xBCT Ceramics Prepared by Plasma Activated Sintering. *Journal of the European Ceramic Society*. 37(5):2067–72, 2017.

189. Orlik K, Lorgouilloux Y, Marchet P, Thuault A, Jean F, Rguiti M, et al. Influence of Microwave Sintering on Electrical Properties of BCTZ Lead Free Piezoelectric Ceramics. *Journal of the European Ceramic Society*. 40(4):1212–6, 2020.

190. Wang P, Li Y, Lu Y. Enhanced Piezoelectric Properties of $(Ba_{0.85}Ca_{0.15})(Ti_{0.9}Zr_{0.1})O_3$ Lead-Free Ceramics by Optimizing Calcination and Sintering Temperature. *Journal of the European Ceramic Society*. 31(11):2005–12, 2011.

191. Chen X, Peng Z, Chao X, Yang Z. Structure, Electrical Properties and Reaction Mechanism of $(Ba_{0.85}Ca_{0.15})(Zr_{0.1}Ti_{0.9})O_3$ Ceramics Synthesized by the Molten Salt Method. *Ceramics International*. 43(15):11920–8, 2017.

192. Jaeger RE, Egerton L. Hot Pressing of Potassium-Sodium Niobates. *Journal of the American Ceramic Society*. 45(5):209–13, 1962.

193. Chen K, Xu G, Yang D, Wang X, Li J. Dielectric and Piezoelectric Properties of Lead-Free $0.95(K_{0.5}Na_{0.5})NbO_3$-$0.05LiNbO_3$ Crystals Grown by the Bridgman Method. *Journal of Applied Physics*, 101(4):044103, 2007.

194. Wu J, Xiao D, Zhu J. Potassium-Sodium Niobate Lead-Free Piezoelectric Materials: Past, Present, and Future of Phase Boundaries. *Chemical Reviews*. 115(7):2559–95, 2015.

195. Malic B, Bernard J, Holc J, Jenko D, Kosec M. Alkaline-Earth Doping in $(K, Na)NO_3$ Based Piezoceramics. *Journal of the European Ceramic Society*. 25(12):2707–11, 2005.

196. Inagaki Y, Kakimoto KI. Dielectric and Piezoelectric Properties of Mn-Doped $(Na_{0.5}K_{0.5})NbO_3$ Single Crystals Grown by Flux Method. *Applied Physics Express.* 1(6):0616021–3, 2008.

197. Kizaki Y, Noguchi Y, Miyayama M. Defect Control for Superior Properties in $(K_{0.5}Na_{0.5})NbO_3$ Single Crystals. *Key Engineering Materials.* 350:85–8, 2007.

198. Saravanan R, Rajesh D, Rajasekaran S V, Perumal R, Chitra M, Jayavel R. Crystal Structure, Dielectric Properties of $(K_{0.5}Na_{0.5})NbO_3$ Single Crystal Grown by Flux Method Using B_2O_3 Flux. *Crystal Research and Technology.* 48(1):22–8, 2013.

199. Rafiq MA, Costa MEV, Vilarinho PM. Establishing the Domain Structure of $(K_{0.5}Na_{0.5})NbO_3$ (KNN) Single Crystals by Piezoforce-Response Microscopy. *Science of Advanced Materials.* 6(3):426–33, 2014.

200. Li JF, Wang K, Zhang BP, Zhang LM. Ferroelectric and Piezoelectric Properties of Fine-Grained $(Na_{0.5}K_{0.5})NbO_3$ Lead-Free Piezoelectric Ceramics Prepared by Spark Plasma Sintering. *Journal of the American Ceramic.* 89(2):706–9, 2006.

201. Lin D, Li Z, Zhang S, Xu Z, Yao X. Influence of MnO_2 Doping on the Dielectric and Piezoelectric Properties and the Domain Structure in $(K_{0.5}Na_{0.5})NbO_3$ Single Crystals. *Journal of the American Ceramic Society.* 93(4):941–4, 2010.

202. Zhao ZH, Ye MY, Ji HM, Li XL, Zhang X, Dai Y. Enhanced Piezoelectric Properties and Strain Response in ⟨001⟩ Textured BNT-BKT-BT Ceramics. *Materials & Design.* 137:184–91, 2018.

203. Guo K, Chen S, Tan CKI, Sharifzadeh Mirshekarloo M, Yao K, Tay FEH. Bismuth Sodium Titanate Lead-Free Piezoelectric Coatings by Thermal Spray Process. *Journal of the American Ceramic Society.* 100(8):3385–92, 2017.

204. Dargham SA, Ponchel F, Zaatar Y, Assaad J, Remiens D, Zaouk D. Synthesis and Characterization of BNT Thin Films Prepared by Sol-Gel Method. *Mater. Today* 3(3):810–5, 2016.

205. Miriyala K, Ramadurai R. Tunable Ferroelectric Domain Orientation in Polycrystalline and Highly Oriented $Na_{0.5}Bi_{0.5}TiO_3$ Thin Films. *Mater. Lett.* 178:23–6, 2016.

206. Ahn CW, Kim HS, Woo WS, Won SS, Seog HJ, Chae SA. Low-Temperature Sintering of $Bi_{0.5}(Na, K)_{0.5}TiO_3$ for Multilayer Ceramic Actuators. *Journal of the American Ceramic Society.* 98(6):1877–83, 2015.

207. Frey M, Payne D. Grain-Size Effect on Structure and Phase Transformations for Barium Titanate. *Physical Review B.* 54(5):3158, 1996.

208. Tuluk A, Joshi S, Mahon T, Van Der Zwaag S. Tuning Piezoproperties of $BiFeO_3$ Ceramic by Cobalt and Titanium Dual Doping. *The Journal of Applied Physics.* 131(21):214104, 2022.

209. Wei Y, Wang X, Zhu J, Wang X, Jia J, Damjanovic D. Dielectric, Ferroelectric, and Piezoelectric Properties of $BiFeO_3$–$BaTiO_3$ Ceramics. *Journal of the American Ceramic Society.* 96:3163–3168, 2013.

210. Dai Z, Akishige Y. Electrical Properties of $BiFeO_3$–$BaTiO_3$ Ceramics Fabricated by Mechanochemical Synthesis and Spark Plasma Sintering. *Materials Letters.* 88:36–9, 2012.

211. Xun B, Song A, Yu J, Yin Y, Li JF, Zhang BP. Lead-Free $BiFeO_3$-$BaTiO_3$ Ceramics with High Curie Temperature: Fine Compositional Tuning across the Phase Boundary for High Piezoelectric Charge and Strain Coefficients. *ACS Applied Materials & Interfaces.* 13(3):4192–202, 2021.

Chapter 19

Lead-Free Materials for Next-Generation Piezoelectric Energy Harvesting

R. K. Shukla and Anchal Srivastava
Department of Physics, University of Lucknow, Lucknow-226007, India
rajeshkumarshukla00@gmail.com, srivastava_anchal@lkouniv.ac.in

Piezoelectric energy harvesting has emerged as a promising technology for converting mechanical vibrations into electrical energy, with applications ranging from powering small electronic devices to enhancing the energy efficiency of industrial systems. However, the use of lead-based piezoelectric materials, such as PZT (lead zirconate titanate), has raised environmental and health concerns, necessitating the development of lead-free alternatives. This chapter gives a comprehensive view of different lead-free materials for next-generation piezoelectric energy harvesting. Different aspects of various classes of lead-free materials with challenges and opportunities associated with scalability, reliability,

Piezoelectric Materials: Design and Applications
Edited by N. B. Singh and Dev Kumar Mahato

ISBN 978-981-5129-40-3 (Hardcover), 978-1-003-59897-8 (eBook)
www.jennystanford.com

and compatibility with existing systems have been discussed. This chapter also emphasizes the importance of environmentally friendly alternatives in the pursuit of next-generation energy harvesting technologies. The insights provided in this chapter aim to help researchers, engineers, and policymakers involved in the development and implementation of sustainable energy solutions.

19.1 Introduction

The history of piezoelectricity dates back to the 18th century, although its full scientific understanding and practical applications have developed over time. Initially, in 1707, Pierre Curie observed that certain crystals, like tourmaline, generate an electric charge when subjected to mechanical stress, such as pressure or bending. In 1880, the Curie brothers further explored this phenomenon and coined the term "piezoelectricity," as the word piezein means "squeeze" in Greek. In the mid-19th century, Lord Kelvin and Gabriel Lippmann made significant advancements in understanding the mathematical equations that described the piezoelectric effect. Piezoelectricity found various practical applications during the 20th century. For example, World War I saw the use of piezoelectric quartz crystals in sonar devices for detecting submarines, and World War II saw the use of piezoelectric materials in early radar systems. Barium titanate (BT) ceramics were discovered during World War II for the development of compact capacitors for use in compact radar systems. So, the titania condenser material TiO_2-MgO was doped by Ca/Sr/Ba and the BT perovskite structured piezoelectric material resulted in a ten times higher permittivity. In 1947 RB Gray gave the first experimental verification of piezoelectricity of the electrically poled polycrystalline BT. Later the isomorphic oxides with perovskite structure such as $PbZrO_3$, $PbTiO_3$, and $SrTiO_2$ and their solid solution were studied. The material designing flexibility of binary PZT was enhanced by working on ternary PZT, viz., $Pb(Zn,Nb)O_3$, $Pb(Mg,Nb)O_3$, $Pb(Mn,Sb)O_3$, etc. The extraordinary magnitude of dielectric permittivity was reported in PMZ and PZN in 1985. Piezoelectric in polyvinylidene difluoride (PVDF), discovered in 1969 by Kawai, showed compression when exposed to an electric field. Due to serious environmental problems, lead-free piezoceramic started developing around the year 1999.

The development of piezoelectric transducers allowed for the change of mechanical/acoustical vibrations into electrical signals and vice versa. Resonators, tip movement in scanning probe microscopy and atomic force microscopy, sensors for vibrations in air and beneath water, various other sensors, medical imaging, precision positioning systems, microphones, quartz watches, and inkjet printers all bank on the piezoelectricity inherent in several crystals and ceramics. Piezoelectricity is characteristically present even in bones. Later PZT became widely used due to their strong piezoelectric properties. Researchers have been working on finding environmentally friendly alternatives, such as lead-free piezoelectric materials to mitigate environmental pollution. In recent years, piezoelectric materials have been explored for their potential in energy harvesting, where mechanical vibrations from various sources can be converted into electricity for powering small electronic devices and sensors. Piezoelectricity is also being studied for applications in self-powered sensors, wearable technology, and even harvesting energy from everyday movements like walking.

The property of piezoelectricity is due to polarization happening at the atomic scale. Crystals show piezoelectric behavior only if they are non-centrosymmetric. Piezoelectricity is a consequence of either naturally occurring dipoles or those induced by artificial means in the crystal structure of these materials. The piezoelectricity can remain preserved even if the performance of the nanomaterial is enhanced owing to the free motion of the domain boundary, thereby leading to expansion or contraction in material-volume. Piezoelectric properties of metal oxide nanostructures have been demonstrated to enhance the photoresponsivity, etc. which may lead to improved device performances of related optoelectronic devices. For many sensing applications, soft piezo materials are needed whereas for many high-power applications, hard piezo materials are required [1]. A comparison of different parameters for soft and hard piezoelectric materials is given in Table 19.1.

Table 19.1 Comparison of different parameters for soft and hard piezoelectric materials

Type	Piezoelectric constants	Permittivity	Dielectric constants	Dielectric losses	Electromechanical coupling factors	Mechanical quality factors	Coercive field	Depolarization
Soft piezo material	Larger	Higher	Larger	Higher	Larger	Low	Lower	Easier
Hard piezo material	Smaller	Lower	Smaller	Lower	Smaller	High	Higher	Harder

The magnitude of piezoelectric energy serves as an indicator of the coupling of elastic properties with that of dielectric. These coupling constants are defined as the electromechanical coupling coefficients.

When a force is applied to any material, at the microscopic level, a potential is established within the internal structure due to the piezoelectric property of the material. This process results in the redistribution of surface charges and the generation of a polarization phenomenon. When the piezoelectric material is put between two electrodes, the external load induces a transient current of electrons driven by the established potential. In such materials, the direction of the spontaneous polarization depends on the state of the ferroelectric domain and the maximum magnitude of the spontaneous polarization unlike in single crystals where only the magnitude matters [2].

For harvesting the energy, the important parameters are piezoelectric strain constant d, piezoelectric voltage constant g, square root of the mechanical-electrical energy conversion efficiency, mechanical quality factor Q, and dielectric constant ε. However, besides material properties, the frequency, the available volume, and the form of the mechanical energy available, corresponding to the specific energy harvesting unit, also should be taken into consideration [3].

Owing to this phenomenon, cutting-edge fields known as "Piezotronics and Piezophototronics" have emerged. The application of piezophototronics to light-emitting diodes, photocells, and solar cells enhances their performance. Thus, piezoelectricity, piezotronics, and piezophototronics represent interdisciplinary fields that explore the interplay between mechanical, electrical, and optical properties in various materials, leading to a range of innovative applications in electronics and photonics.

19.2 Energy Harvesting

"Energy harvesting" encompasses a comprehensive array of approaches and methodologies aimed at harnessing energy sources that are typically insufficient. Generally compact, often on a centimeter scale, energy harvesters primarily generate electrical energy. They find utility in devices with minimal energy requirements, such as sensors and microcircuitry. Common energy sources for these systems include waste heat, ambient electromagnetic radiation, and vibrations. The manner in which energy is captured relies on a microgenerator, designed according to the specific energy source and operational mode. For example, vibrations can be captured using either mechanical or solid-state devices, viz., devices using piezoelectric materials. Sensors in moving vehicles for preventing accidents, for example, can provide earlier warnings and status updates, enabling timely repairs and maintenance, as opposed to fixed schedules or post-breakdown scenarios. For this, batteries are needed which have to be changed periodically. Energy harvesting offers a solution to this predicament by eliminating the need for frequent battery replacements.

The available technologies are piezoelectricity, thermoelectricity, electromagnetic radiation, solar, and wind. Piezoelectricity is the process through which mechanical energy gets transformed into electrical energy, evident when a slab experiences mechanical stress and generates a voltage across it. This phenomenon is inherent to specific materials and occurs within the solid state. Consequently, the conversion of mechanical forces and motions into electric energy becomes achievable without the need for mobile components or complex apparatus. Conversely, the inverse holds true as well: applying a voltage to the slab results in its deformation referred to as the inverse piezoelectric effect.

In any crystal, the charge distribution is such that the total charge is balanced, rendering the crystal electrically neutral. However, for piezoelectric crystals, mechanical force – elongating or compressive – has the capacity to disrupt this equilibrium and create electric polarisation to induce a voltage across the crystal. The piezoelectric generator harvests electrical energy by converting the mechanical stresses occurring in the material due to vibrations. Due to the depletion of conventional fossil fuels, the energy crisis is looming large, and therefore the research interest in materials exhibiting piezoelectricity has surged.

19.3 Next-Generation Energy Harvesting Materials

Characteristics necessary for a piezoelectric energy harvesting material include a high piezoelectric coefficient, good mechanical flexibility, high Curie temperature (T_c), low dielectric loss, wide bandwidth response to effectively capture a range of frequencies from various mechanical inputs, and high energy density to produce significant electrical energy output. Lead-based materials, the first one being PZT, show useful properties but lead is a toxic element.

Extensive research efforts have been dedicated to advancing lead-free piezoceramics as a viable alternative to PZT in the realm of electromechanical devices like actuators, sensors, and transducers. In the quest for developing Pb-free piezoelectric materials, the initial selection of constituent chemical elements takes into account various factors, including cost, toxicity, and ionic polarizability. Following this, various crystal structures are evaluated based on fundamental principles and multiple phase diagrams with a particular emphasis on identifying morphotropic phase boundaries that offer favorable piezoelectric properties. Insights gained from density functional theory can effectively bridge the gap between atomic-level considerations and strategies based on phase diagrams. The morphotropic phase boundary (MPB), where different ferroelectric phases coexist in ferroelectric systems, facilitates a decrease in domain wall energy, thereby facilitating domain mobility. Consequently, piezoelectric solid solutions with perovskite structures, which are crucial for applications like sensors and actuators, demonstrate superior piezoelectric responses near the MPB.

The assessment of the performance of these materials relies on several critical parameters. Among these, the piezoelectric coefficient stands out as a key indicator, representing the degree of coupling between the mechanical and dielectric properties within piezoelectric structures. The piezoelectric coefficient is commonly represented as d_{xy}, where x and y denote the directions of the electric field and that of stress/strain. Thus, the coefficients in the longitudinal, transverse, and tangential directions are denoted by d_{33}, d_{31}, and d_{15}, respectively. The significant parameters, along with the electromechanical coupling coefficient k signifying the degree of energy conversion, encompass Curie temperature (T_c), depolarization temperature (T_d), remanent polarization (P_r), spontaneous polarization (P_s) (the inherent polarization of each electric domain), resistance (R), and dielectric dissipation factor tan d. For superior piezoelectric materials, elevated values of d_{33}, k, ε_r, T_c, T_d, and P_r are sought after, while minimizing R and tan δ.

Research on lead-free piezoelectric materials started nearly five decades back resulting in BT based-, sodium potassium niobate (KNN) based-, and sodium bismuth titanate (BNT) based-ceramic systems, bismuth layer-structured ferroelectrics (BLSFs) and ZnO nanostructures. Their combinations $(Bi,Na)TiO_3$–$BaTiO_3$ (BNT–BT) systems [4, 5] and $(Bi,Na)TiO_3$–$(Bi,K)TiO_3$ (BNT–BKT) systems have also been reported [6]. The BNT–BT–KNN ternary system [7, 8] showed a piezoelectric response comparable to that of the PZT system [9, 10]. This invited worldwide research on BNT-based systems.

19.3.1 Barium Titanate-Based Lead-Free Piezoelectric Systems

The ceramic system based on BT, possessing a perovskite structure of ABO_3 (where A includes elements like calcium, strontium, and barium, and B encompasses elements like niobium, tantalum, and titanium), marks a significant advancement as the first documented lead-free piezoelectric ceramic. This BT-based ceramic exhibits favorable electrical stability, effective electromechanical coupling, and minimal dielectric loss at room temperature. Nevertheless, its performance is constrained by a lower T_c and d_{33} coefficient. $BaTiO_3$, (BT) is a typical lead-free piezoelectric material that has remained a field of continuous study [11–22]. Ca and Zr co-doped BT (BZT–BCT) can replace the PZT system [23]. The disappearing polarization anisotropy is the primary cause for the huge piezoelectricity in $Ba(Zr_{0.2}Ti_{0.8})O_3$–$(Ba_{0.7}Ca_{0.3})TiO_3$ (BZT–BCT) ceramics [24] but their low Curie temperature may be an impediment in their application [25].

Enhanced piezoelectric characteristics have been achieved through innovative structural designs and refined preparation methods. A sol-gel approach to finely control the phase composition of BZT–BCT has been reported [26]. This led to reduced processing temperatures, smaller particle sizes, and a resulting product with notable properties: a high P_r of 12.2 mC/cm^2, an elevated d_{33} coefficient of 637 pC/N, and a substantial electromechanical coupling coefficient (k) of 0.596.

In a sandwich-structured piezoelectric BT configuration, positioning a porous BT layer between two compact outer BT layers created varying electric field concentrations in the porous layer. The outcome was diverse piezoelectric properties dependent on factors such as the porous structure of the interlayer and its relative thickness in relation to the overall structure. Notably, an optimal d_{33} coefficient of 124.5 pC/N was achieved with a 60% volumetric porosity of the interlayer and a relative thickness ratio of 0.52 [27].

19.3.2 Bismuth Ferrite-Based Lead-Free Piezoelectric Systems

Bismuth ferrite (BF)-based systems are one of the Pb-free piezoelectric solid solution systems. These systems, a possible alternative for Pb-based piezoelectric materials, exhibit interesting behavior that is yet to be understood. These systems are potential candidates for applications requiring piezoelectricity at high temperatures. The various compositions studied are BF–$SrTiO_3$ (BF–ST), BNT–BF, BNT–BKT–BF, etc. [28–33]. BF–ST is reported to exhibit a very high energy density of 9.3 J/cm^3 and outstanding efficiency of 90% [29] which is comparatively quite favorable. The piezoelectric coefficient of 69 pC/N in 0.625BF–0.375ST has been reported [28]. However, the piezoelectric response of BF–ST systems was found to be unstable over time.

In the pursuit of elevated performance in piezoelectric, doping modification has emerged as a pivotal strategy. This involves the introduction of suitable donor and acceptor elements into the ceramic structure. The resulting improvement in piezoelectric properties has been attributed to the formation of MPB within the doped ceramics. The electric domains of this MPB contribute to the presence of numerous polarization patterns, further enhancing the ceramic's piezoelectric behavior.

Various BF-based systems were subjected to evaluation for their piezoelectric properties. In the case of the 0.625BF–0.375ST composition, a piezoelectric coefficient of 69 pC/N was determined [28]. Across a spectrum of BF concentrations, the significant piezoelectric response (d_{33}) ranged from 32 to 182 pm/V [33]. Nevertheless, a marked asymmetric piezoelectric response was observed over a broad composition range, attributed to the potent internal bias caused by defect dipoles.

Later the breakthrough, unveiling of remarkable piezoelectric performance (402 pC/N) within a $BiFeO_3$–$BaTiO_3$ (BF–BT)-based system [34], triggered a series of studies and advancements in such systems. Consequently, BF–BT-based systems emerged as a potential contender for the subsequent next generation of Pb-free piezoelectric systems.

BF has garnered attention as a multiferroic substance that showcases both ferromagnetic traits and ferroelectricity at room temperature. Notably, BF boasts an impressive T_c of 825°C. Despite these qualities, BF's piezoelectric attributes remain lackluster, impeding its viability as a promising Pb-free piezoelectric material. To address this, a novel approach involving the amalgamation of the high piezoelectricity of BT and the elevated T_c of BF has emerged in the form of the BF–BT system. This proposition aims to supplant lead-based piezoelectric setups, and recent studies [35–37] have shed light on this innovative avenue.

A core-shell structure in the grains of $0.75Bi(Fe_{0.99}Ti_{0.01})O_3$–$0.25BaTiO_3$ lead-free piezoelectric ceramic has been observed by high-energy synchrotron X-ray diffraction [38]. The structures of the core were rhombohedral for both slow cooling and quenching whereas the shell is found to be pseudo-cubic in the slow cooling and rhombohedral in the quenching method. The core-shell structure in other doped BF–BT lead-free systems has also been reported [39]. Multiple lead-free systems based on BF and characterized by a pseudo-cubic structure demonstrate substantial piezoelectric response and ferroelectricity [40–43]. Several methods and meticulous techniques are suggested to elucidate the precise crystal structure and the primary factors contributing to the heightened ferroelectricity and piezoelectric performance [39].

The electrical characteristics of piezoelectric materials, encompassing dielectric, ferroelectric, pyroelectric, and piezoelectric properties, are intricately linked to the arrangement of domains within them. The deliberate manipulation of the configuration of the domains to alter the piezoelectric response of a material is termed domain engineering, a captivating subject within the realm of piezoelectric materials. A variety of approaches for such engineering for creating textures, for the growth of single crystals, and for prudently varying the compositional design have been put forth [44–49]. Of particular note is the ability to regulate piezoelectricity by influencing the movement of domain walls through doping. Notably, the piezoelectric characteristics of BT exhibit significant enhancement with an increase in domain wall density, attributed to the reduction in domain size [50, 51]. Exploring the domain structure holds significance in comprehending the magnitude of ferroelectricity and piezoelectricity [52, 53]. At the juncture where rhombohedral and cubic phases coexist, accompanied by prevalent nano-sized domain structures, the peak of piezoelectric behavior is attained.

The transition of domain structure from macro to nano size within a polar nano region can be ascertained through piezoresponse force microscopy (PFM), a valuable technique for characterizing domain arrangements. The variability in domain size based on the ratio in the BF–BT system underscores the practicality of domain engineering within BF–BT-based ceramics. For 0.75BF–0.25BT composition, the density of domain walls increases as the domain size diminishes. The domain size is governed by the proportion of BF, decreasing proportionally as BF content is reduced [54]. Within the BT–BMT–BF system (where M signifies magnesium), one composition demonstrates the coexistence of both macro- and nano-sized domains, while in another, macro-domain patterns vanish, leaving only nano-sized domains. Interestingly, in contrast to BT-based ceramics, the BT–BMT–BF system experiences a decline in ferroelectric and piezoelectric properties due to the reduced domain size. Nonetheless, the exploration of domain engineering within BF-based systems is still at its nascent stage. Consequently, the expectation arises that domain engineering will assume a pivotal role in augmenting piezoelectric capabilities, prompting the initiation of numerous investigations employing domain engineering techniques.

The enhancement of performance in piezoelectric ceramics through doping modification has emerged as a crucial strategy. This involves the introduction of suitable donors and acceptors into the ceramic structure. The augmented piezoelectric characteristics are linked to the emergence of a new MPB within the doped ceramics. Within this MPB, electric domains exhibit numerous polarization orientations, facilitating their alignment when subjected to an electric field. This phenomenon was observed by Zhu et al. in the context of BF–BT ceramics doped with $Bi(Mg_{0.5}Ti_{0.5})O_3$ (BMT), where an MPB was identified between the pseudo-cubic (PC) and rhombohedral (R) phases. The optimal composition occurred at a BF:BT:BMT ratio of 1:1:0.04, resulting in d_{33} and T_c values of 154 pC/N and 482°C, respectively.

Various challenges, notably high conductivity, hinder the enhancement of piezoelectric properties within the BF–BT system. Incorporating dopants and constructing ternary systems are standard techniques for tailoring the piezoelectric characteristics of Pb-free piezoelectric materials. Currently, a significant portion of research in the BF–BT domain revolves around these methodologies. Notable piezoelectric coefficient, d_{33}, reaching approximately 150 pC/N, and T_c, surpassing 400°C, have been obtained through the incorporation of successful ternary dopants, including (Mg,Ti), (Zn,Ti), Ga, (Ni,Ti), (Mg,Nb), and (Zn,Nb) [55–59].

Recently, enhancement in the piezoelectric properties of Pb-free BF-based systems has been obtained through the process of quenching and remarkably high k of 324, 402, and 454 pC/N for water-quenched BF–BT with (Zn,Ti) and Ga doped systems are achieved [34, 60]. This breakthrough has spurred further investigations into the effects of various quenching parameters, including quenching temperature and cooling rate [61, 62].

19.3.3 Potassium Sodium Niobate (KNN)-Based Lead-Free Piezoelectric Systems

The KNN-based ceramic system has emerged as a highly promising alternative in the realm of lead-free piezoelectric materials. Textured KNN systems exhibit strong piezoelectric properties [63]. These include a notably high T_c of 420°C, a substantial P_r of 33 mC/cm^2, and a considerable electromechanical coupling coefficient exceeding 0.454 [64, 65]. The fabrication of the KNN-based ceramic system is primarily accomplished through a solid-state method [66]. Nonetheless, this approach requires exceedingly high temperatures within the range of 1000–1500°C, and it often falls short in terms of product purity and performance. Consequently, alternative techniques such as normal sintering, a wet chemical method, and a seed-free solid-state crystal growth process [67] have been used to address these limitations. In the context of normal sintering, researchers successfully synthesized $0.96(K_{0.5}Na_{0.5})0.95Li_{0.05}Nb_{1-x}$ $SbxO_3$–$0.04BaZrO_3$ yielding products of higher purity (approximately 99.5%) compared to those obtained via the solid-state method (which typically yields products with purity around 90%).

Recent advancements have been made to counter challenges related to the hygroscopic nature of alkaline oxides associated with KNN-based materials synthesized by conventional sintering [68, 69]. Innovative techniques like spark plasma-, sealed-, and microwave-sintering have been introduced [70, 71]. A comparative study between microwave sintering and traditional sintering methods for $(K_{0.5}Na_{0.5})NbO_3$ ceramics has been made [72]. While there was negligible variation in piezoelectric properties, stability was enhanced by the microwave sintering. Notably, microwave sintering saved time by 97% as compared to the traditional method.

To further enhance the piezoelectric attributes of the KNN-based ceramic system, a post-processing technique "domain engineering" has been employed. This involves manipulating the domain structure of the ceramics by applying an electric field to the fabricated material [73]. The augmented density of domain walls subsequently reduces the potential barrier for polarization switching, thereby contributing to the augmentation of piezoelectric characteristics [74]. For example, the piezoelectric properties of 0.5 wt% Mn-doped KNN improved through field cooling – a process that entails applying an electric field and then gradually cooling the sample at a specific rate [75]. By adjusting the electric field strength, temperature, and cooling rate, samples with varying domain sizes and piezoelectric properties were achieved. As a result, the product exhibited d_{33} and T_c values of up to 350 pC/N and 416°C, respectively.

However, domain engineering does not suit ceramics with small single crystals [76]. Therefore, tantalum has been frequently employed to support the growth of single crystals in KNN-based ceramics [77], and thus a large, high-quality single crystal of tantalum-modified KNN (K, Na)(Nb, Ta)O_3 (KNNT) was produced, through top-seeded solution growth, of size 12 × 11 × 11 mm, which is much bigger than that of unmodified KNN, which had a diameter of 4 mm. Notably, the k_{33} and k_t values reached to 0.827 (0.68) and 0.646 (0.507), respectively. These values even surpassed those of PZT, i.e., k_{33} = 0.68 and k_t = 0.507. The values in the bracket are the corresponding values for the unmodified KNN.

Three categories of phase boundaries have been explored and implemented in KNN-based ceramics: rhombohedral-orthorhombic (R-O), orthorhombic-tetragonal (O-T), and rhombohedral-tetragonal (R-T) phase boundaries [78]. The O-T phase boundary gained prominence in the research and development of KNN-based materials. This is primarily attributed to the initial significant advancement in d_{33}, which is attributed to the emergence of this phase boundary. However, the R-O boundary induces a poor piezoelectric effect. Such effect could be mitigated or done away with by the R-T phase boundary. A novel material composition denoted as $(1-x)$-$(K_{0.48}Na_{0.52})(Nb_{0.95}Sb_{0.05})O_3$-$xBi_{0.5}Ag_{0.5}ZrO_3$, referred to as (1-x)KNNSxBAZ, demonstrates the mediation of the R-T phase boundary through controlled adjustments of BAZ concentrations. In this study, the incorporation of both Sb5+ and BAZ elements enables the modulation of their respective transitions from rhombohedral to orthorhombic and from tetragonal to orthorhombic, ultimately facilitating the creation of the R-T phase boundary. The resulting ceramic with x = 0.0425 yields a giant d_{33} of approximately 490 pC/N [78] surpassing the results reported earlier in the literature in the related ceramics.

19.3.4 Barium Sodium Titanate-Based Pb-Free Piezoelectric Systems

The BNT-based material is known for its efficacy as a lead-free piezoelectric material, distinguished by its remarkable P_r of 38 mC/cm^2 and a high T_c of 320°C. This has led to extensive investigation and exploration. Particularly, the modified BT as $(1\text{-}x)(Na_{0.5}Bi_{0.5})TiO_3$–$xBaTiO_3$ (NBBT) has been an attraction. Nevertheless, the application of NBBT monocrystals has encountered limitations arising from pronounced leakage current, substantial dielectric losses, and relatively low domain wall mobility (Td) [76]. Efforts to enhance NBBT's performance have focussed on larger single crystals, which have shown improved piezoelectric properties attributed to a reduction in oxygen vacancies and the associated leakage current. Good progress by employing the Templated Seed Surface Growth (TSSG) technique to produce a sizable and high-quality $0.95(Na_{0.5}Bi_{0.5})TiO_3$–$0.05BaTiO_3$ (NBBT95/5) monocrystal in a rhombohedral phase has been made [79]. This [001]c-poled NBBT95/5 multidomain single crystal exhibited a remarkable d_{33} = 360 pC N^{-1}, d_{15} = 162 pC N^{-1}, and d_{31} = −113 pC N^{-1}, k_{33} = 0.720 and k_t = 0.540, low dielectric loss (tan δ = 1.1%), and a high T_c of 135°C. These attributes position NBBT95/5 as a prospective next-generation substitute for PZT ceramics across various applications.

Further, iron (Fe) has been doped into NBBT single crystals to augment their piezoelectric properties. The incorporation of 0.2 at% Fe into the monocrystal material resulted in a notable enhancement of the d_{33} coefficient, i.e., up to 590 pC N^{-1} [80]. The introduction of Fe ions induced a transition from macro- to nano-domains, significantly bolstering the piezoelectric capabilities of NBBT.

In tandem with NBBT, various other iterations of BNT-based ceramics have attracted attention. For example, the synthesis of a [001]-textured $0.83Na_{0.5}Bi_{0.5}TiO_3$–$0.17Bi_{0.5}K_{0.5}TiO_3$ (BNT–BKT) ceramic through the Template Grain Growth (TGG) method employing anisotropic particles such as $BaTiO_3$, $SrTiO_3$, and $NaNbO_3$ as templates [81] led to notably improved thermal stability at elevated temperatures.

19.3.5 Bismuth Layer-Structured Ferroelectrics (BLSFs)

In recent times, the Aurivillius-phase BLSFs have been spotlighted as potential materials for high-temperature piezoelectric applications. They stand out owing to their impressive characteristics like elevated Curie temperatures and superior resistance against thermal depoling. Structurally, BLSFs have a general formula of $(Bi_2O_2)(A_{m-1}B_mO_{3m+1})$. These compounds blend alternating perovskite blocks $(A_{m-1}B_mO_{3m+1})$ with a Bi_2O_2 layer. Here, "m" indicates the count of stacked oxygen octahedra along the layer's perpendicular axis. Within this formation, "A" and "B" signify two distinct cations located at the dodecahedra and octahedra centers, respectively [82–84]. The discovery of BLSFs' ferroelectric properties was first made by Smolenskii, Isupov, and Agranovskaya [82] published in 1961. In this general formula, A could be K^+, Na^+, Ca^{2+}, Sr^{2+}, Pb^{2+}, Ba^{2+}, Ln^{3+}, Bi^{3+}, Ce^{4+}, etc., and B could be Fe^{3+}, Cr^{3+}, Ti^{4+}, Nb^{5+}, Ta^{5+}, W^{6+}, Mo^{6+}, etc. [84]. BLSFs have merits of high T_c and low tan d but on the other hand (i) their coercive electric field is relatively high resulting in difficulty to polarize and (ii) their piezoelectric activity is low as the spontaneous polarization encounters two-dimensional confinement owing to the crystal structure.

$BaBi_4Ti_4O_{15}$ (BBT), a BLSF, modified with 0.5 wt% Ce showed T_c and d_{33} as 407°C and 24.2 pC N^{-1}. Another BLSF $CaBi_4Ti_4O_{15}$ (CBT) showed a high T_c (790°C) but low d_{33} (8 pC N^{-1}). Nb-Mn co-doped CBT ferroelectric (CBTN–Mn) prepared by the solid-state method show good values (d_{33} = 23 pC N^{-1} and T_c = 790°C), thereby becoming a promising piezoelectric material for use at high temperatures.

ZnO, a versatile compound semiconductor, is also an important lead-free piezoelectric material as it is non-toxic and lends itself easily to several types of nanostructures.

19.3.6 Piezoelectric Polymers and Polymer Composites

Lead-free piezoelectric materials can be organic also, viz., PVDF and its copolymers, and their composites, and some biopolymers [4, 84]. Currently, there is a growing trend toward addressing the issues of fragility and incompatibility with irregular mechanical deformations. This has led to the widespread adoption of a strategy involving the incorporation of piezoelectric ceramics into flexible and pliable polymer matrices. For instance, one well-known biopolymer, BC (bacterial cellulose), is recognized for its robust mechanical strength, chemical stability, and strong crystalline structure. Leveraging the textured nano-fibrillated architecture of BC, it is frequently employed either as a reinforcing element or as a base matrix for crafting functional composite materials. Researchers have taken advantage of this concept by creating piezoelectric BT/BC composites. Within this design, BT nanoparticles exhibiting piezoelectric properties were evenly dispersed throughout the matrix formed by interconnected networks of BC nanofibers.

The resulting nanogenerator built upon this distinctive BT/BC structure showcased a remarkable peak power density (P) of 0.64 mW/cm^{-2}, a value tenfold greater than that achieved with BT/polydimethylsiloxane composites.

Piezoelectric ceramics are generally brittle, and therefore polymers, which are flexible, are preferred in many applications. Although the d_{33} coefficient of polymers has been quite less than the PZT [85], for the past two decades extensive research has been done on polymers like PVDF, its copolymers P(VDF-TrFE), polypropylene (PP), etc. for the purpose of energy harvesting besides sensors, actuators, and medical devices because of their good voltage to strain ratio. The piezoelectric coefficient of PVDF has been reported as 20–30 pC/N [86] and the piezoelectric property is due to its crystal structure ($-CH_2-CF_2-$) Piezoelectricity can also be induced in some polymers, viz., vinylidene cyanide and vinyl acetate (PVDCN-VAc), a combination of PVDF and PMMA, polyuria, etc. by poling and mechanical stretching. Electrospinning can also lead to piezoelectricity via inducing β-phase in PVDF and researchers are working on electro-spun PVDF fibers. Screen-printing method, sol-gel method, and chemical vapor deposition method have been used for preparing piezoelectric polymers. Energy harvesting up to 0.97 mW at 33 Hz from a small area of 2.4 cm^2 has been reported from a pile of ten layers of P(VDF-TrFE) obtained by the screen-printing method [87]. This energy harvester depending on a cantilever design is capable of working for around 107 cycles and can actuate a microcontroller to exchange messages with a mobile phone, in a wireless mode. Such structures are capable of harvesting ~100 mW for more than 10^7 cycles. Structures like rugby-ball can allow for higher stress and can therefore lead to higher power generation.

The piezoelectric polymers can be categorized as bulk, composite, and voided charged polymers (VCP). Polymer composites involve piezoelectric ceramic nanoparticles with polymers. As the name suggests, VCP has gas voids introduced which are charged by applying an electrical field that creates internal dipoles, thereby leading to the piezoelectric effect.

Piezocomposites are continuously being studied as they have moderate piezoelectric coupling coefficients and can lead to self-powered wearable devices, nanogenerators, etc. [88]. Several composites for energy harvesting have come up, by blending polymers like PVDF, PDMS, etc. with inorganic piezoelectric materials like ZnO, $ZnSO_3$, CdS, $NaNbO_3$, $BaTiO_3$, and $BiFeO_3$ [88]. Sheet, rod, bar, or nanoparticle powder of piezoelectric materials are embedded in an organic polymer matrix material. A superior advantage of such composites is their flexibility in space as well as in operation. Spray coating and RF sputtering have been used to prepare films of these composites as well as composites with multi-walled carbon nanotubes. A composite of mesoporous PVDF made with ZnO NPs has also been reported [89]. ZnO is a non-perovskite structure with piezoelectric properties and has also been used as filler along with polyaniline (PANi). Additionally, such composites possess pliability and biocompatibility [90]. They can convert very small vibrational responses in the body or can profile rough surfaces of the tissues into noticeable signals after amplification via electronics [91].

The nanoconfinement effect is another efficient option for enhancing the piezoelectric response of PVDF and thereby the energy harvesting from various piezoelectric generators can also be enhanced [92]. Moreover, the use of piezoelectric biodegradable polymers – polysaccharides, polypeptides, and polynucleotides – is an important step toward building eco-friendly sustainable components and devices for electronics. These polymers are stabilized in the form of nanofibers [93]. Further, successful fabrication of quasi-2d, α-phase P(VDF-TrFE) films on the SiO_2/Si substrates coated with AlO_x/Al has been reported at room temperature with a strong piezoelectric response greater than that of the β-phase. The piezoelectric properties of two polarization states are also studied which throws light on the fundamental dependence of piezoelectric properties on the molecular structures and which enables novel opportunities for future energy harvesters [94].

Using the piezo-triboelectric effect, resonance energy harvesting has been obtained via an acoustic core-shell configuration, which finds a potential application in artificial cochleae [95]. Here $BaTiO_3$ particles and porous PVDF-TrFE act as the core and the shell, respectively.

Some of the polymers and inorganic-organic hybrids showing piezoelectricity are enumerated in Table 19.2.

Table 19.2 Polymers and inorganic-organic hybrids showing piezoelectricity

Polymer-based piezoelectric materials	References
PVDF and polyurethane	[96]
PEDOT:PSS core and P(VDF-TrFE) shell nanofiber 13.2 V for two harvesters in parallel	[97]
Graphene/PVDF/graphene device induced voltage of 7.16 V (peak to peak) with a tensile stress of 1.75 MPa Metal/PVDF-TrFE/metal device induced voltage of 9.1 V (peak to peak) with a stress of 2.18 MPa	[98]
Diphenylalanine (FF) peptide nanotubes	[99–104]
Polycrystalline films of amino acids	[105]
PNTS (self-assembled diphenylalanine peptide nanotubes) for developing green piezo devices and piezoelectric materials	[100, 106]
Transparent piezoelectric energy harvesting device with screen-printed electrodes on PVDF sheets	[107]
PVDF-HFP (poly(vinylidene fluoride-hexafluoropropylene) (piezoelectricity coefficient d_{31} is comparable to PVDF)	[108, 109]
CuO nanorods used in designing polyacrylonitrile electro-spun fibrous materials	[110]
CNT-reinforced PVDF/KNN based electro-spun nanogenerator	[111]
$BaTiO_3$/PDMS composites for piezo nanogenerators	[112]
Fully sustainable, self-poled, biowaste piezo nanogenerator using pomelo fruit membrane	[113]
Fish skin (collagen) – a natural biodegradable piezoelectric material	[114]
Silk fibroins	[115–118]
Piezoelectric M13 bacteriophage	[119, 120]
Synthetic biodegradable piezoelectric polymer PLA and PLLA	[121, 122]

19.4 Conclusion and Future Strategies

Over the course of this review, it's evident that the global scientific community is pivoting away from traditional lead-based piezoelectric materials in search of environmentally friendly and efficient alternatives. The necessity to transition is driven by environmental concerns as well as by regulatory impositions and public health considerations. The exploration of KNN, BT, BT-BF, BNT, BLSFs, and ZnO, among others, showcases the vast potential and the strides that have already been made in obtaining next-generation lead-free piezoelectric materials for energy harvesting. Each of these materials brings unique properties and potential applications to the table. For instance, KNN exhibits promising piezoelectric coefficients competitive with traditional PZT, while BLSFs are notable for their high-temperature applications, and ZnO, especially in nanoforms, has displayed versatility. Moreover, the underlying mechanisms of these next-generation materials, from morphotropic phase boundaries to intricate crystal structures, have provided insights into how piezoelectricity can be optimized without the use of lead. However, challenges remain. Many of these materials require further research to enhance their piezoelectric responses, optimize their processing methods, and improve their overall performance to meet or surpass their lead-based counterparts consistently. PVDF and its copolymers P(VDF-TrFE), biodegradable polymers, etc. have opened a new gateway for energy harvesting besides sensors, actuators, and medical devices because of their good voltage-to-strain ratio. In summary, the journey toward a sustainable and efficient lead-free piezoelectric future is well underway paving the way for a greener, safer, and more efficient technological landscape.

Further improvement in the piezoelectric properties may be achieved (i) by optimizing the preparation process, viz., proper sintering temperature to obtain denser microstructure and (ii) by ion doping. Doping facilitates the formation of MPB leading to better piezoelectric performance (iii) by exploiting the anisotropy-based piezoelectric response and (iv) by the use of polymers and polymer composites. Fabrication of oriented single crystals can lead to enhanced piezoelectricity. Domain engineering via poling along a nonpolar axis is also seen to enhance piezoelectricity. With the help of advanced computational techniques, predictive modeling can identify potential lead-free materials, thus reducing the time and resources needed for experimental validation.

References

1. APC, Americanpiezo.com
2. Meng, Y; Chen, G; Huang, M. Piezoelectric materials: Properties, advancements, and design strategies for high-temperature applications. *Nanomaterials (Basel)*, 12(7), 1171, 2022.
3. Gosavi, S.K.; Balpande, S.S. A comprehensive review of micro and nano scale piezoelectric energy harvesters. *Sens. Lett.*, 17(3), 180–195, 2019.
4. Kim, S.; Nam, H.; Calisir, I. Lead-free $BiFeO_3$-based piezoelectrics: A review of controversial issues and current research state. *Materials*, 15, 4388, 2022
5. Takenaka, T.; Maruyama, K.I.; Sakata, K. $(Bi_{1/2}Na_{1/2})TiO_3$-$BaTiO_3$ system for lead-free piezoelectric ceramics. *Jpn. J. Appl. Phys.*, 30, 2236, 1991.
6. Sasaki, A.; Chiba, T.; Mamiya, Y.; Otsuki, E. Dielectric and piezoelectric properties of $(Bi_{0.5}Na_{0.5})TiO_3$-$(Bi_{0.5}K_{0.5})TiO_3$ systems. *Jpn. J. Appl. Phys.*, 38, 5564, 1999.
7. Zhang, S.T.; Kounga, A.B.; Aulbach, E.; Ehrenberg, H.; Rodel, J. Giant strain in lead-free piezoceramics $Bi_{0.5}Na_{0.5}TiO_3$-$BaTiO_3$-$K_{0.5}Na_{0.5}NbO_3$ system. *Appl. Phys. Lett.*, 91, 112906, 2007.
8. Zhang, S.T.; Kounga, A.B.; Aulbach, E.; Granzow, T.; Jo, W.; Kleebe, H.J.; Rodel, J. Lead-free piezoceramics with giant strain in the system $Bi_{0.5}Na_{0.5}TiO_3$-$BaTiO_3$-$K_{0.5}Na_{0.5}NbO_3$, structure and room temperature properties. *J. Appl. Phys.*, 103, 034107, 2008.
9. Jo, W.; Dittmer, R.; Acosta, M.; Zang, J.; Groh, C.; Sapper, E.; Wang, K.; Rodel, J. Giant electric-field-induced strains in lead-free ceramics for actuator applications—Status and perspective. *J. Electroceram.*, 29, 71–93, 2012.
10. Rodel, J.; Jo, W.; Seifert, K.T.P.; Anton, E.M.; Granzow, T.; Damjanovic, D. Perspective on the development of lead-free piezoceramics. *J. Am. Ceram. Soc.*, 92, 1153–1177, 2009.
11. Khanal, G.P.; Kim, S.; Kim, M.; Fujii, I.; Ueno, S.; Wada, S. Grain-size dependence of piezoelectric properties in thermally annealed $BaTiO_3$ ceramics. *J. Ceram. Soc. Jpn.*, 126, 536–541, 2018.

12. Hoshina, T.; Hatta, S.; Takeda, H.; Tsurumi, T. Grain size effect on piezoelectric properties of $BaTiO_3$ ceramics. *Jpn. J. Appl. Phys.*, 57, 0902BB, 2018.

13. Huan, Y.; Wang, X.; Fang, J.; Li, L. Grain size effect on piezoelectric and ferroelectric properties of $BaTiO_3$ ceramics. *J. Eur. Ceram. Soc.*, 34, 1445–1448, 2014.

14. Kinoshita, K.; Yamaji, A. Grain-size effects on dielectric properties in barium titanate ceramics. *J. Appl. Phys.*, 47, 371–373, 1976.

15. Mangaiyarkkarasi, J.; Saravanan, R.; Ismail, M.M. Chemical bonding and charge density distribution analysis of undoped and lanthanum doped barium titanate ceramics. *J. Chem. Sci.*, 128, 1913–1921, 2016.

16. Kim, S.; Noda, S.; Abe, T.; Yokoi, Y.; Nakahira, Y.; Moriyoshi, C.; Kuroiwa, Y. Electric-field-induced structural changes for cubic system of lead-free and lead-based perovskite-type oxides. *Jpn. J. Appl. Phys.*, 59, SPPA05, 2020.

17. Abe, T.; Kim, S.; Moriyoshi, C.; Kitanaka, Y.; Noguchi, Y.; Tanaka, H.; Kuroiwa, Y. Visualization of spontaneous electronic polarization in Pb ion of ferroelectric $PbTiO_3$ by synchrotron-radiation X-ray diffraction. *Appl. Phys. Lett.*, 117, 252905, 2020.

18. Aoyagi, S.; Kuroiwa, Y.; Sawada, A.; Yamashita, I.; Atake, T. Composite structure of $BaTiO_3$ nanoparticle investigated by X-ray diffraction. *J. Phys. Soc. Jpn.*, 71, 1218–1221, 2002.

19. Oshime, N.; Ohwada, K.; Sugawara, K.; Abe, T.; Yamauchi, R.; Ueno, T.; Machida, A.; Watanuki, T.; Ueno, S.; Fujii, I.; et al. Bragg coherent diffraction imaging allowing simultaneous retrieval of three-dimensional shape and strain distribution for 40–500 nm particles. *Jpn. J. Appl. Phys.*, 60, SFFA07, 2021.

20. Kim, S.; Hwang, G.; Song, K.; Jeong, C.K.; Park, K.; Jang, J.; Kim, K.; Ryu, J.; Choi, S. Inverse size-dependence of piezoelectricity in single $BaTiO_3$ nanoparticles. *Nano Energy*, 58, 78–84, 2019.

21. Vriami, D.; Damjanovic, D.; Vleugels, J.; Biest, O.V. Textured $BaTiO_3$ by templated grain growth and electrophoretic deposition. *J. Mater. Sci.*, 50, 7896–7907, 2015.

22. Mallik, H.S.; Fujii, I.; Khanal, G.P.; Kim, S.; Ueno, S.; Suzuki, T.S.; Wada, S. Fabrication of <111>-oriented $BaTiO_3$ ceramics by high magnetic field electrophoretic deposition using hexagonal-tetragonal coexisting $BaTiO_3$ powder. *J. Ceram. Soc. Jpn.*, 128, 469–474, 2020.

23. Liu, W.; Ren, X. Large piezoelectric effect in Pb-free ceramics. *Phys. Rev. Lett.*, 103, 257602, 2009.

24. Gao, J.; Hu, X.; Wang, Y.; Liu, Y.; Zhang, L.; Ke, X.; Zhong, L.; Zhao, H.; Ren, X. Understanding the mechanism of large dielectric response in Pb-free (1-x)$Ba(Zr_{0.2}Ti_{0.8})O_3$–x$(Ba_{0.7}Ca_{0.3})TiO_3$ ferroelectric ceramics. *Acta Mater.*, 125, 177–186, 2017.

25. Kim, S.W.; Choi, H.I.; Lee, M.H.; Park, J.S.; Do, D.; Kim, M.H.; Song, T.K.; Kim, W.J. Electrical properties and phase of $BaTiO_3$-$SrTiO_3$ solid solution. *Ceram. Int.*, 39, S487–S490, 2012.

26. Praveen, J. P.; Karthik, T.; James, A.; Chandrakala, E.; Asthana, S.; Das, D. Effect of poling process on piezoelectric properties of sol–gel derived BZT–BCT ceramics. *J. Eur. Ceram. Soc.*, 35, 1785–1798, 2015.

27. Roscow, J.I.; Lewis, R.W.C.; Taylor, J.; Bowen, C.R. Modelling and fabrication of porous sandwich layer barium titanate with improved piezoelectric energy harvesting figures of merit. *Acta Mater.*,128, 207–217, 2017.

28. Makarovic, M.; Bencan, A.; Walker, J.; Malic, B.; Rojac, T. Processing, piezoelectric and ferroelectric properties of (x)$BiFeO_3$-(1-x)$SrTiO_3$ ceramics. *J. Eur. Ceram. Soc.*, 39, 3693–3702, 2019.

29. Yan, F.; Bai, H.; Ge, G.; Lin, J.; Shi, C.; Zhu, K.; Shen, B.; Zhai, J.; Zhang, S. Composition and structure optimized $BiFeO_3$-$SrTiO_3$ lead-free ceramics with ultrahigh energy storage performance. *Small*, 18, 2106515, 2022.

30. Lu, Z.; Wang, G.; Li, L.; Huang, Y.; Feteira, A.; Bao, W.; Kleppe, A.K.; Xu, F.; Wang, D.; Reaney, I.M. In situ poling X-ray diffraction studies of lead-free $BiFeO_3$-$SrTiO_3$ ceramics. *Mater. Today Phys.*, 19, 100426, 2021.

31. Wu, M.; Fang, L.; Li, G.; Elouadi, B. Dielectric and ferroelectric properties of (1-x)$BiFeO_3$-x$Bi_{0.5}Na_{0.5}TiO_3$ solid solution. *Ferroelectrics*, 478, 18–25, 2015.

32. Zhou, C.; Liu, X.; Li, W.; Yuan, C. Structure and piezoelectric properties of $Bi_{0.5}Na_{0.5}TiO_3$- $Bi_{0.5}Ka_{0.5}TiO_3$-$BiFeO_3$ lead-free piezoelectric ceramics. *Mater. Chem. Phys.*, 114, 832–836, 2009.

33. Fujii, I.; Ito, Y.; Suzuki, T.; Wada, T. Ferroelectric and piezoelectric properties of $(Bi_{1/2}Na_{1/2})TiO_3$-$BiFeO_3$ ceramics. *J. Mater. Res.*, 31, 28–35, 2016.

34. Lee, M.H.; Kim, D.J.; Park, J.S.; Kim, S.W.; Song, T.K.; Kim, M.H.; Kim, W.J.; Do, D.; Jeong, I.K. High-performance lead-free piezoceramics with high Curie temperatures. *Adv. Mater.*, 27, 6976–6982, 2015.

35. Catalan, G.; Scott, J.F. Physics and applications of bismuth ferrite. *Adv. Mater.*, 21, 2463–2485, 2009.

36. Kim, S.; Khanal, G.P.; Nam, H.; Fujii, I.; Ueno, S.; Moriyoshi, C.; Kuroiwa, Y.; Wada, S. Structural and electrical characteristics of potential candidate lead-free $BiFeO_3$-$BaTiO_3$ piezoelectric ceramics. *J. Appl. Phys.*, 122, 164105, 2017.

37. Ismailzade, I.H.; Ismailov, R.M.; Alekberov, A.I.; Salaev, F.M. Investigation of the magnetoelectric $(ME)_H$ effect in solid solutions of the systems $BiFeO_3$-$BaTiO_3$ and $BiFeO_3$-$PbTiO_3$. *Phys. Status Solidi A*, 68, K81–K85, 1981.

38. Calisir, I.; Kleppe, A.K.; Feteir, A.; Hall, D.A. Quenching-assisted actuation mechanisms in core-shell structured $BiFeO_3$-$BaTiO_3$ piezoceramics. *J. Mater. Chem. C*, 7, 10218–10230, 2019.

39. Wang, D.; Fan, Z.; Li, W.; Zhou, D.; Feteira, A.; Wang, G.; Murakami, S.; Sun, S.; Zhao, Q.; Tan, X.; et al. High energy storage density and large strain in $Bi(Zn_{2/3}Nb_{1/3})O_3$-doped $BiFeO_3$-$BaTiO_3$ ceramics. *ACS Appl. Energy Mater.*, 1, 4403–4412, 2018.

40. Chen, B.; Ma, J.; Wu, S.; Wu, B. Phase structure and electrical properties of $(1\text{-}x)K_{0.48}Na_{0.52}NbO_3$-$xBi_{0.46}La_{0.04}$ $(Na_{0.82}K_{0.18})_{0.5}ZrO_3$ lead-free piezoceramics. *J. Mater. Sci. Mater. Electron.*, 28, 3299–3308, 2017.

41. Cheng, S.; Zhang, B.P.; Zhao, L.; Wang, K.K. Enhanced insulating and piezoelectric properties of $BiFeO_3$-$BaTiO_3$-$0.5Na_{0.5}TiO_3$ ceramics with high Curie temperature. *J. Am. Ceram. Soc.*, 102, 7355–7365, 2019.

42. Wei, Y.X.; Wang, X.T.; Jia, J.J.; Wang, X.L. Multiferroic and piezoelectric properties of $0.65BiFeO_3$-$0.35BaTiO_3$ ceramics with pseudo-cubic symmetry. *Ceram. Int.*, 38, 3499–3502, 2012.

43. Chen, J.G.; Cheng, J.R.; Guo, J.; Cheng, Z.X.; Wang, J.L.; Liu, H.B.; Zhang, S.J. Excellent thermal stability and aging behaviors in $BiFeO_3$-$BaTiO_3$ piezoelectric ceramics with rhombohedral phase. *J. Am. Ceram. Soc.*, 103, 374–381, 2020.

44. Nakamura, K.; Ando, H.; Shimizu, H. Partial domain inversion in $LiNbO_3$ plates and its applications to piezoelectric devices. In *Proceedings of the IEEE 1986 Ultrasonics Symposium*, Williamsburg, VA, USA, pp. 527–530, 1986.

45. Nakamura, K.; Shimizu, H. Poling of ferroelectric crystals by using interdigital electrodes and its application to bulk-wave transformer. In *Proceedings of the IEEE 1983 Ultrasonics Symposium*, Atlanta, GA, USA, pp. 719–722, 1983.

46. Webjorn, J.; Laurell, F.; Arvidsson, G. Blue light generated by frequency doubling of laser diode light in a lithium niobate channel waveguide. *IEEE Photonics Technol. Lett.*, 1, 316–318, 1989.

47. Terabe, K.; Higuchi, S.; Takekawa, S.; Nakamura, M.; Goto, T.; Kitamura, K. Nanoscale domain engineering of a $Sr_{0.61}Ba_{0.39}Nb_2O_6$ single crystal using a scanning force microscope. *Ferroelectrics*, 292, 83–89, 2003.

48. Cho, Y.; Hiranaga, T.; Fujimoto, K.; Wagatsuma, Y.; Ones, A.; Terabe, K.; Kitamura, K. Tbit/ferroelectric data storage based on scanning nonlinear dielectric microscopy. *Trans. Mater. Res. Soc. Jpn.*, 28, 109–112, 2003.

49. Hiranaga, Y.; Fujimoto, K.; Cho, Y.; Wagatsuma, Y.; Onoe, A.; Terabe, K.; Kitamura, K. Nano-sized inverted domain formation in stoichiometric $LiTaO_3$ single crystal using scanning nonlinear dielectric microscopy. *Integr. Ferroelectr.*, 49, 203–209, 2002.

50. Wada, S.; Yako, K.; Kakemoto, H.; Tsurumi, T.; Erhart, J. Enhanced piezoelectric property of barium titanate single crystals with the different domain sizes. *Key Eng. Mater.*, 269, 19–22, 2004.

51. Wada, S.; Tsurumi, T. Enhanced piezoelectric property of barium titanate single crystals with engineered domain configurations Br. *Ceram. Trans.*, 103, 93–96, 2004.

52. Kitagawa, S.; Ozaki, T.; Horibe, Y.; Yoshii, K.; Mori, S. Ferroelectric domain structures in $BiFeO_3$-$BaTiO_3$. *Ferroelectrics*, 376, 122–128, 2008.

53. Ozaki, T.; Kitagawa, S.; Nishihara, S.; Hosokoshi, Y.; Suzuki, M.; Noguchi, Y.; Miyayama, M.; Mori, S. Ferroelectric properties and nano-scaled domain structures in $(1\text{-}x)BiFeO_{3\text{-}x}BaTiO_3$ $(0.33 < x < 0.5)$. *Ferroelectrics*, 385, 155–161, 2009.

54. Mitsui, R.; Fujii, I.; Nakashima, K.; Kumada, N.; Kuroiwa, Y.; Wada, S. Enhancement in the piezoelectric properties of $BaTiO_3$-$Bi(Mg_{1/2}Ti_{1/2})O_3$-$BiFeO_3$ system ceramics by nanodomain. *Ceram. Int.*, 39, S695–S699, 2013.

55. Zhou, Q.; Zhou, C.; Yang, H.; Chen, G.; Li, W.; Wang, H. Dielectric, ferroelectric, and piezoelectric properties of $Bi(Ni_{1/2}Ti_{1/2})O_3$-modified $BiFeO_3$–$BaTiO_3$ ceramics with high Curie temperature. *J. Am. Ceram. Soc.*, 95, 3889–3893, 2012.

56. Murakami, S.; Ahmed, N.T.A.F.; Wand, D.; Feteira, A.; Sinclair, D.C.; Reaney, M.I. Optimising dopants and properties in BiMeO3 (Me = Al, Ga, Sc, Y, $Mg_{2/3}Nb_{1/3}$, $Zn_{2/3}Nb_{1/3}$, $Zn_{1/2}Ti_{1/2}$) lead-free BaTiO3-BiFeO3 based ceramics for actuator applications. *J. Eur. Ceram. Soc.*, 38, 4220–4231, 2018.

57. Lin, Y.; Zhang, L.; Yu, J. Stable piezoelectric property of modified $BiFeO_3$-$BaTiO_3$ lead-free piezoceramics. *J. Mater. Sci. Mater. Electron.*, 26, 8432–8441, 2015.

58. Fujii, I.; Mitsui, R.; Nakashima, K.; Kumada, N.; Shimada, M.; Watanabe, T.; Hayashi, J.; Yabuta, H.; Kubota, M.; Fukui, T.; et al. Structural, dielectric, and piezoelectric properties of Mn-doped $BaTiO_3$–$Bi(Mg_{1/2}Ti_{1/2})O_3$–$BiFeO_3$ ceramics. *Jpn. J. Appl. Phys.*, 50, 09ND07, 2011.

59. Murakami, S.; Wang, D.; Mostaed, A.; Khesro, A.; Feteira, A.; Sinclair, D.C.; Fan, Z.; Tan, X.; Reaney, I.M. High strain (0.4%) $Bi(Mg_{2/3}Nb_{1/3})O_3$-$BaTiO_3$-$BiFeO_3$ lead-free piezoelectric ceramics and multilayers. *J. Am. Ceram. Soc.*, 101, 5428–5442, 2018.

60. Lee, M.H.; Kim, D.J.; Choi, H.I.; Kim, M.H.; Song, T.K.; Kim, W.J.; Do, D. Thermal quenching effects on the ferroelectric and piezoelectric properties of $BiFeO_3$-$BaTiO_3$ ceramics. *ACS Appl. Electron. Mater.*, 1, 1772–1780, 2019.

61. Nam, H.; Kim, S.; Aizawa, T.; Fujii, I.; Ueno, S.; Wada, S. Influence of quenching temperature on piezoelectric and ferroelectrics properties in $BaTiO_3$-$Bi(Mg_{1/2}Ti_{1/2})O_3$-$BiFeO_3$ ceramics. *Ceram. Int.*, 44, S199–S202, 2018.

62. Kim, D.S.; Cheon, C.I.; Lee, S.S.; Kim, J.S. Effect of cooling rate on phase transitions and ferroelectric properties in $0.75BiFeO_3$-$0.25BaTiO_3$ ceramics. *Appl. Phys. Lett.*, 109, 202902, 2016.

63. Saito, Y.; Takao, H.; Tani, T.; Nonoyama, T.; Takatori, K.; Homma, T.; Nagaya, T.; Nakamura, M. Lead-free piezoceramics. *Nature*, 432, 84–87, 2004.

64. Wang, X.; Wu, J.; Xiao, D.; Zhu, J.; Cheng, X.; Zheng, T.; Zhang, B.; Lou X.; Wang, X. Giant piezoelectricity in potassium–sodium niobate lead-free ceramics. *J. Am. Chem. Soc.*, 136, 2905–2910, 2014.

65. Egerton, L; Dillon, DM. Piezoelectric and dielectric properties of ceramics in the system potassium—sodium niobate. *J. Am. Ceram. Soc.*, 42(9), 438–442, 1959.

66. Hresˇcˇak, J.; Malicˇ, B. Cilensˇek, J.; Bencˇan, A. *J. Therm. Anal. Calorim.*, 127, 129–136, 2017.

67. Yang, J.; Zhang, F.; Yang, Q.; Liu, Z.; Li, Y.; Liu, Y.; Zhang, Q. Large piezoelectric properties in KNN-based lead-free single crystals grown by a seed-free solid-state crystal growth method. *Appl. Phys. Lett.*, 108, 182904, 2016.

68. Lin, D.; Kwok, K.W.; Chan, H.L. Piezoelectric and ferroelectric properties of $K_xNa_{1-x}NbO_3$ lead-free ceramics with MnO_2 and CuO doping. *J. Alloys Compd.* 461(1–2), 273–178, 2008.

69. Rubio-Marcos, F.; Romero, J.; Martin-Gonzalez, M.; Fernandez, J. Lead-free piezoceramics: Revealing the role of the rhombohedral-tetragonal

phase coexistence in enhancement of the piezoelectric properties. *J. Eur. Ceram. Soc.*, 30, 2763–2771, 2010.

70. Men, T.; Yao, F.; Zhu, Z.; Wang, K.; Li., J. Piezoelectric properties of $(K_{0.5}Na_{0.5})NbO_3$-$BaTiO_3$ lead-free ceramics prepared by spark plasma sintering. *J. Adv. Dielectr.*, 6, 1650013, 2016.

71. Li, J.F.; Wang, K.; Zhang, B.P.; Zhang, L.M. Ferroelectric and piezoelectric properties of fine-grained $Na_{0.5}K_{0.5}NbO_3$ lead-free piezoelectric ceramics prepared by spark plasma sintering. *J. Am. Ceram. Soc.*, 89, 706–709, 2006.

72. Feizpour, M.; Barzegar Bafrooei, H.; Hayati, R.; Ebadzadeh, T. Microwave-assisted synthesis and sintering of potassium sodium niobate lead-free piezoelectric ceramics. *Ceram. Int.*, 40, 871–877, 2014.

73. Zheng, T.; Wu, J.; Xiao, D.; Zhu, J. Recent development in lead-free perovskite piezoelectric bulk materials. *Prog. Mater. Sci.*, 98, 552–624, 2018.

74. Xu, F.; Trolier-Mckinstry, S.; Ren, W.; Xu, B.; Xie, Z.L., Hemker, K.J. Domain wall motion and its contribution to the dielectric and piezoelectric properties of lead zirconate titanate films. *J. Appl. Phys.*, 89, 1336–1348, 2001.

75. Lin, D.; Zhang, S.; Cai, C.; Liu, W. Domain size engineering in 0.5%MnO_2-$(K_{0.5}Na_{0.5})NbO_3$ lead-free piezoelectric crystals. *J. Appl. Phys.*, 117, 113–126, 2015.

76. Zheng, L.; Huo, X.; Wang, R.; Wang, J.; Jiang, W.; Cao, W. Large size lead-free $(Na,K)(Nb,Ta)O_3$ piezoelectric single crystal: growth and full tensor properties. *Cryst. Eng Comm*, 15, 7718–7722, 2013.

77. Hollenstein, E.; Davis, M.; Damjanovic, D.; Setter, N. Piezoelectric properties of Li-and Ta-modified $(K_{0.5}Na_{0.5})NbO_3$ ceramics. *Appl. Phys. Lett.*, 87, 182905, 2005.

78. Wang, X.; Wu, J.; Xiao, D.; Cheng, X.; Zheng, T.; Lou, X.; Binyu Zhang, B.; Zhu, J. New Potassium–Sodium Niobate Ceramics with a Giant d_{33}. *ACS Appl. Mater. Interfaces*, 6, 6177–6180, 2014.

79. Zheng, L.; Yi, X.; Zhang, S.; Jiang, W.; Yang, B.; Zhang, R.; Cao, W. Complete set of material constants of $0.95(Na_{0.5}Bi_{0.5})TiO_3$-$0.05BaTiO_3$ lead-free piezoelectric single crystal and the delineation of extrinsic contributions. *Appl. Phys. Lett.*, 103, 3683–3685, 2013.

80. Xue, S.; Ma, J.; Zhao, X.; Wang, F.; Sun, D.; Wang, T.; Shi, W.; Fu, Z.; Zhou, H.; Luo, H. Unraveling delocalized electrons in metal-induced gap states from second harmonics. *Appl. Phys. Lett.*, 111, 162902, 2017.

81. Bai, W.; Xi, J.; Zhang, J.; Shen, B.; Zhai, J.; Yan, H. Effect of different templates on structure evolution and large strain response under a low electric field in < 0 0 1>-textured lead-free BNT-based piezoelectric ceramics. *J. Eur. Ceram. Soc.*, 35, 2489–2499, 2015.

82. Smolenskii, G.A.; Isupov, V.A.; and Agranovskaya, A.I. Ferroelectrics of the oxygen-octahedral type with layered structure. *Sov. Phys. Solid State*, 3, 651–655, 1961.

83. Tan, Z.; Xi, J.; Xing, J.; Wu, B.; Zhang, Q.; Chen, Q.; Zhu, J. Understanding the piezoelectric origin of bismuth layer-structured ferroelectric polycrystal using first-principle method. *J. Eur. Ceram. Soc.*, 42(9), 3865–3876, 2022.

84. Wei, H.; Wang, H.; Xia, Y.; Cui, D.; Shi, Y.; Dong, M.; Liu, C.; Ding, T.; Zhang, J.; Ma, Y.; Wang, N.; Wang, Z.; Sun, Y.; Wei, R.; Guo, Z. An overview of lead-free piezoelectric materials and devices. *Mater. Chem. C*, 6, 12446, 2018.

85. Schlaberg, H.; Duffy, J. Piezoelectric polymer composite arrays for ultrasonic medical imaging applications, *Sens. Actuators A Phys.*, 44, 111–117, 1994.

86. Sun, W.; Zhao, D.; Tan, T.; Yan, Z.; Guo, P.; Luo, X. Low-velocity water flowenergy harvesting using vortex-induced vibration and galloping. *Appl. Energy*, 251, 113392, 2019.

87. Godard, N.; Allirol, L.; Latour, A.; Glinsek, S.; Gerard, M.; Polesel, J.; Dos Santos, F.D.; Defay, E. 1-mW vibration energy harvester based on a cantilever with printed polymer multilayers. *Cell Rep. Phys. Sci.*, 1, 100068, 2020.

88. Wang, Z.L., Towards self-powered nanosystems: From nanogenerators to nanopiezotronics. *Adv. Funct. Mater.*, 18, 3553–3567, 2008.

89. Mao, Y.; Zhao, P.; McConohy, G.; Yang, H.; Tong, Y.; Wang, X., Sponge-like piezoelectric polymer films for scalable and integratable nanogenerators and self-powered electronic systems. *Adv. Energy Mater.*, 4, 1301624, 2014.

90. Dagdeviren, C.; Joe, P.; Tuzman, O.L.; Park, K.-I.; Lee, K.J.; Shi, Y.; Huang, Y.; Rogers, J.A. Recent progress in flexible and stretchable piezoelectric devices for mechanical energy harvesting, sensing and actuation. *Extreme Mech. Lett.*, 9, 269–281, 2016.

91. Xin, Y.; Sun, H.; Tian, H.; Guo, C.; Li, X.; Wang, S.; Wang, C. The use of polyvinylidene fluoride (PVDF) films as sensors for vibration measurement: A brief review. *Ferroelectrics*, 502, 28–42, 2016.

92. Surmenev, R.A.; Chernozem, R.V.; Pariy, I.O.; Surmeneva, M.A. A review on piezo- and pyroelectric responses of flexible nano- and

micropatterned polymer surfaces for biomedical sensing and energy harvesting applications. *Nanomater. Energy*, 79, 105442, 2021.

93. Persano, L.; Ghosh, S.K.; Pisignano, D. Enhancement and function of the piezoelectric effect in polymer nanofibers. *Acc. Mater. Res.*, 2022(3), 900–912, 2022.

94. Qian, J.; Jiang, S.; Wang, Q.; Zheng, S.; et al. Unveiling the piezoelectric nature of polar α-phase P(VDF-TrFE) at quasi-two-dimensional limit. *Sci. Rep.*, 8,532, 2018.

95. Zheng, J.; Yu, Z.; Wang, Y.; Fu, Y.; Chen, D.; Zhou, H. Acoustic core–shell resonance harvester for application of artificial cochlea based on the piezo-triboelectric effect. *ACS Nano*, 15(11), 17499–17507, 2021.

96. Chowdhury, A.; et al. Leveraging shape memory coupled piezoelectric properties in melt-extruded composite filament based on polyvinylidene fluoride and polyurethane. *Macromol. Mater. Eng.*, 305.12: 2000296 (2020).

97. Han, J.; Kim, J.H.; Choi, H.J.; et al. Origin of enhanced piezoelectric energy harvesting in all-polymer-based core-shell nanofibers with controlled shell-thickness. *Compos. B Eng.*, 223, 109141, 2021.

98. Park, S.; Kim, Y.; Jung, H.; et al. Energy harvesting efficiency of piezoelectric polymer film with graphene and metal electrodes. *Sci. Rep.* 7, 17290, 2017.

99. Park, H.; Kim, Y.; Kim, Y.; Lee, C.; et al. Self-assembly of unidirectionally polarized piezoelectric peptide nanotubes using environmentally friendly solvents. *Appl. Surf. Sci.*, 618, 156588, 2023.

100. Kholkin, A.; Amdursky, N.; Bdikin, I.; Gazit, E.; Rosenman, G. Strong piezoelectricity in bioinspired peptide nanotubes. *ACS Nano*, 4, 610, 2010.

101. Safaryan, S.; et al. Diphenylalanine-based microribbons for piezoelectric applications via inkjet printing. *ACS Appl. Mater. Interfaces*, 10(12), 10543–10551, 2018.

102. Nguyen, V.; Jenkins, K.; Yang, R. Epitaxial growth of vertically aligned piezoelectric diphenylalanine peptide microrods with uniform polarization. *Nano Energy*, 17, 323–329, 2015.

103. Nguyen, V.; Zhu, R.; Jenkins, K.; Yang, R. Self-assembly of diphenylalanine peptide with controlled polarization for power generation. *Nat. Commun.*, 7(1), 13566, 2016.

104. Vasilev, S.; Zelenovskiy, P.; Vasileva, D.; Nuraeva, A.; Shur, V.Y.; Kholkin, A.L. Piezoelectric properties of diphenylalanine microtubes prepared from the solution. *J. Phys. Chem. Solids*, 93, 68, 2016.

105. Guerin, S.; Tofail, S.A.M.; Thompson, D. Longitudinal piezoelectricity in orthorhombic amino acid crystal films. *Cryst. Growth Des.*, 18(9), 4844–4848, 2018.

106. Bosne, E.D.; Heredia, A.; Kopyl, S.; Karpinsky, D.V.; Pinto, A.G.; Kholkin, A.L. Piezoelectric resonators based on self-assembled diphenylalanine microtubes. *Appl. Phys. Lett.*, 102, 073504, 2013.

107. Rodrigues-Marinho, T.; Pereira, N.; Correia, V.; Miranda, D.; Lanceros-Méndez, S.; Costa, P.; Transparent piezoelectric polymer-based materials for energy harvesting and multitouch detection devices, *ACS Appl. Electron. Mater.*, 4(1), 287–296, 2022.

108. Huan, Y.; Liu, Y.; Yang, Y.; Wu, Y. Influence of extrusion, stretching and poling on the structural and piezoelectric properties of poly(vinylidene fluoride-hexafluoropylene) copolymer films. *J. Appl. Polym. Sci.*, 104, 858–862, 2007.

109. Künstler, W.; Wegener, M.;, Seiβ, M.; Gerhard-Multhaupt, R. Preparation and assessment of piezo- and pyroelectric poly(vinylidene fluoride-hexafluoropropylene) copolymer films. *Appl Phys A.*, 73, 641–645, 2001.

110. Bairagi, S.; Chowdhury, A.; Banerjee, S.; et al. Investigating the role of copper oxide (CuO) nanorods in designing flexible piezoelectric nanogenerator composed of polyacrylonitrile (PAN) electrospun web-based fibrous material. *J Mater Sci: Mater Electron*, 33, 13152–13165, 2022.

111. Bairagi, S.; Ali, W. Investigating the role of carbon nanotubes (CNTs) in the piezoelectric performance of a PVDF/KNN-based electrospun nanogenerator. *Soft Matter*, 16(20), 4876–4886, 2020.

112. Meisak, D.; Kinka, M.; Plyushch, A.; Macutkevič, J.; et al. Piezoelectric nanogenerators based on $BaTiO_3$/PDMS composites for high-frequency applications. *ACS Omega*, 8(15), 13911–13919, 2023.

113. Bairagi, S.; Ghosh,S.; Ali, S.W. A fully sustainable, self-poled, bio-waste based piezoelectric nanogenerator: electricity generation from pomelo fruit membrane. *Sci. Rep.*, 10, 12121, 2020.

114. Ghosh, S.K.; Mandal, D. Sustainable energy generation from piezoelectric biomaterial for noninvasive physiological signal monitoring. *ACS Sustain. Chem. Eng.*, 5(10), 8836–8843, 2017.

115. Liu, J.-H.; Li, W.-D.; Jia, J.; Tang, C.-Y.; et al. Structure-regenerated silk fibroin with boosted piezoelectricity for disposable and biodegradable oral healthcare device. *Nano Energy*, 103(A), 107787, 2022.

116. Wen, D.-L.; Sun, D.-H.; Huang, P.; Huang, W.; Su, M.; Wang, Y.; Han, M.-D.; Kim, B.; Brugger, J.; Zhang, H.-X. Recent progress in silk fibroin-based flexible electronics. *Microsyst. Nanoeng.*, 7, 35, 2021.

117. Yucel, T.; Peggy Cebe, and David L. Kaplan. Structural origins of silk piezoelectricity. *Adv. Funct. Mater.*, 21(4), 779–785, 2011.

118. Joseph, J.; Singh, S.G.; Vanjari, S.R.K. Leveraging innate piezoelectricity of ultra-smooth silk thin films for flexible and wearable sensor applications. *IEEE Sens. J.*, 17(24), 8306–8313, 2017.

119. Park, I.W.; Kim, K.W.; Hong, Y.; Yoon, H.J.; Lee, Y.; Gwak, D.; Heo, K. Recent developments and prospects of M13-bacteriophage based piezoelectric energy harvesting devices. *Nanomaterials*, 2020(10), 93, 2020.

120. Lee, B.Y.; Zhang, J.; Zueger, C.; Chung, W.J.; Yoo, S.Y.; Wang, E.; Meyer, J.; Ramesh, R.; Lee, S.W. Virus-based piezoelectric energy generation. *Nat Nanotechnol.*, 7(6): 351–356, 2012.

121. Ma, X.; Hu, Q.; Dai, Y.; He, P.; Zhang, X. Disposable sensors based on biodegradable polylactic acid piezoelectric films and their application in wearable electronics. *Sens. Actuators A Phys.*, 346, 113834, 2022.

122. Farahani, A.; Zarei-Hanzaki, A.; Abedi, H.R.; Haririan, I.; Akrami, M.; Aalipour, Z.; Tayebi, L.; An investigation into the polylactic acid texturization through thermomechanical processing and the improved d33 piezoelectric outcome of the fabricated scaffolds. *J. Mater. Res. Technol.*, 15, 6356–6366, 2021.

Chapter 20

3D Printed Piezoelectric Energy Harvesters: Materials, 3D Printing Techniques, and Applications

Neksumi Musa,[a,b] Mubarak Dahiru,[b] and N. B. Singh[c]
[a]*Department of Environmental Sciences, Sharda University, Greater Noida, India*
[b]*School of Science and Technology, State Polytechnic, Yola, Nigeria*
[c]*Department of Chemistry, Sharda University, Greater Noida, India*
nbsingh43@gmail.com

This chapter provides a comprehensive review of 3D printed piezoelectric energy harvesters, covering the materials used, 3D printing techniques employed, and various applications of the technology. Piezoelectric energy harvesting, the conversion of mechanical energy into electrical energy, has received significant attention in recent years due to its potential to power small-scale electronics and devices. With the emergence of 3D printing technology, the fabrication of piezoelectric energy harvesters has become more accessible, allowing for the development of

Piezoelectric Materials: Design and Applications
Edited by N. B. Singh and Dev Kumar Mahato

ISBN 978-981-5129-40-3 (Hardcover), 978-1-003-59897-8 (eBook)
www.jennystanford.com

unique designs and improved performance. This chapter explores the various materials used for 3D printed piezoelectric energy harvesters, including ceramics, polymers, and composites. It also discusses different 3D printing techniques employed to create these devices, such as fused deposition modeling (FDM), stereolithography (SLA), and selective laser sintering (SLS). Furthermore, various applications of 3D printed piezoelectric energy harvesters, including wearable devices, sensors, and structural health monitoring systems, have been discussed. Valuable insights into the materials, techniques, and applications of 3D printed piezoelectric energy harvesters, have been discussed in detail in this chapter.

20.1 Introduction

Piezoelectric energy harvesting is a technology that involves the conversion of mechanical energy into electrical energy using piezoelectric materials. It has gained significant attention in recent years due to its potential to power small-scale electronics and devices in various applications. Piezoelectric energy harvesting is based on the principle of the piezoelectric effect, which refers to the ability of certain materials to generate an electric charge in response to mechanical stress or deformation. Piezoelectric materials, such as certain ceramics, polymers, and composites, exhibit this effect and can be used to convert mechanical energy into electrical energy [1]. The piezoelectric effect occurs due to the arrangement of atoms or molecules within the material. When an external force or pressure is applied to a piezoelectric material, it causes a deformation in its crystal structure, displacing charged particles. This displacement separates positive and negative charges, creating an electric potential difference or voltage across the material.

20.2 Importance and Applications of Piezoelectric Energy Harvesters

Piezoelectric energy harvesters have gained importance due to their ability to scavenge energy from ambient vibrations, mechanical movements, or environmental sources. They offer several advantages over traditional power sources, such as batteries or wired connections, including their small size, lightweight, and self-sustainability [2]. This makes them particularly suitable for low-power electronic devices and wireless sensor networks.

The applications of piezoelectric energy harvesters are diverse and continue to expand in various fields. Some notable applications include the following [3]:

1. Wireless Sensor Networks: Piezoelectric energy harvesters can power wireless sensor nodes in environmental monitoring systems, smart buildings, and infrastructure. They can harvest energy from ambient vibrations or mechanical vibrations induced by human activities, eliminating the need for frequent battery replacements or wired connections.
2. Wearable Electronics: Piezoelectric energy harvesters are used in wearable devices, such as fitness trackers, smartwatches, and medical sensors. They can convert the mechanical energy generated by human motion or body movements into electrical energy to power wearable electronics, reducing the reliance on conventional batteries.
3. Structural Health Monitoring: Piezoelectric energy harvesters integrated into civil structures, such as bridges, buildings, or aircraft, can capture the mechanical vibrations induced by the structural dynamics. This harvested energy can be used for powering sensors in structural health monitoring (SHM) systems, providing continuous monitoring and early detection of potential damage or structural flaws.
4. Internet of Things (IoT) Devices: The proliferation of IoT devices relies on efficient and sustainable power sources. Piezoelectric energy harvesters can be employed to power low-power IoT devices, enabling them to operate autonomously without the need for frequent battery replacement or wired connections.
5. Human Body Energy Harvesting: The human body generates mechanical energy through movements, such as walking, running, or typing. Piezoelectric energy harvesters can be integrated into wearable devices or clothing to capture and convert this mechanical energy into electrical energy for powering wearable electronics or medical devices.

6. Environmental Energy Harvesting: Piezoelectric energy harvesters can also harness energy from ambient vibrations or environmental sources, such as wind, water, or acoustic waves. They can be used in remote or off-grid locations where conventional power sources are not readily available.

Piezoelectric energy harvesting is a promising technology that enables the conversion of mechanical energy into electrical energy using piezoelectric materials. Its importance lies in its potential to power small-scale electronics and devices in various applications, offering advantages such as self-sustainability, small size, lightweight, and compatibility [4].

20.3 Overview of 3D Printing Technology

20.3.1 3D Printing Techniques and Their Working Principles

3D printing, also known as additive manufacturing, is a revolutionary manufacturing process that enables the creation of three-dimensional objects from digital models by layer-by-layer deposition of materials [5]. Several 3D printing techniques are available, each with its working principles and suitability for different applications. Applications in energy sector are given in Table 20.1.

Fused Deposition Modeling (FDM): FDM is one of the most widely used 3D printing techniques. It involves the extrusion of a thermoplastic filament through a heated nozzle [6]. The nozzle moves along a predetermined path, depositing the molten material layer by layer to form the desired object. Once the material cools and solidifies, the layers fuse, creating a solid structure. The working principle of FDM is based on the controlled heating and deposition of the thermoplastic material. The printer software slices the digital model into thin layers, and the nozzle deposits the material according to the layer-by-layer instructions. FDM offers good accuracy, and ease of use, and is suitable for prototyping and low-volume production [7].

Table 20.1 3D printing techniques for piezoelectric energy harvesters and their applications

3D printing technique	Description	Application in piezoelectric energy harvesters
Fused deposition modeling (FDM)	FDM is a 3D printing technique that uses a thermoplastic filament, which is melted and extruded layer by layer to create the desired object	FDM is widely used for fabricating piezoelectric energy harvesters due to its simplicity, low cost, and ability to create complex geometries. It allows for the integration of piezoelectric materials and conductive traces to form the necessary structure for energy conversion
Stereolithography (SLA)	SLA is a 3D printing technique that uses a liquid resin that is solidified layer by layer using a UV laser	SLA is particularly suitable for producing intricate and high-resolution piezoelectric energy harvesters. It enables the fabrication of detailed structures and precise electrode patterns, resulting in improved energy conversion efficiency
Selective laser sintering (SLS)	SLS is a 3D printing technique that uses a high-powered laser to selectively fuse powdered materials, such as ceramics or polymers, layer by layer	SLS offers advantages in manufacturing piezoelectric energy harvesters by allowing the use of a wide range of materials, including ceramics and composites. It enables the production of complex geometries and can achieve high mechanical strength and thermal stability

Stereolithography (SLA): SLA is an additive manufacturing technique that uses photopolymerization to create 3D objects. It employs a vat of liquid photopolymer resin and a UV laser [8]. The UV laser selectively solidifies the resin layer by layer, following the digital model's cross-sectional data. SLA works based on the principle of polymerization. The UV laser scans the surface of the liquid resin, causing it to solidify and adhere to the previously solidified layers. The build platform gradually moves down as each layer is completed. SLA offers high resolution and can produce intricate and detailed parts, making it suitable for applications where precision is crucial [9].

Selective Laser Sintering (SLS): SLS is a 3D printing technique that utilizes a high-power laser to selectively fuse powdered materials. The process begins with a thin layer of powder material being spread over the build platform [10]. The laser then scans the surface, selectively melting and fusing the powder particles according to the digital model. SLS operates on the principle of thermal fusion. The laser's high intensity raises the temperature of the powder particles to the point of sintering, causing them to fuse. As each layer is completed, a new layer of powder is spread, and the process is repeated. SLS offers the advantage of using a wide range of materials, including plastics, metals, and ceramics [11].

20.4 Advantages and Limitations of 3D Printing in Energy Harvester Fabrication

20.4.1 Advantages of 3D Printing in Energy Harvester Fabrication

(a) *Design Freedom*: 3D printing allows for the creation of complex geometries and intricate structures that would be challenging or impossible to achieve using traditional manufacturing methods. This design freedom enables the optimization of energy harvester shapes for improved performance and efficiency [12].

(b) *Customization and Personalization*: With 3D printing, energy harvesters can be easily customized and tailored to specific requirements. This flexibility allows for the adaptation of energy harvesting devices to different applications and environments, maximizing their effectiveness [13].

(c) *Rapid Prototyping and Iteration*: 3D printing enables quick prototyping and iterative design processes. Energy harvester designs can be rapidly produced and tested, facilitating the optimization of device performance and reducing time to market.

(d) *Material Selection*: 3D printing offers a wide range of compatible materials for energy harvester fabrication, including polymers, composites, and even some metals. This material versatility allows for the selection of materials with suitable properties for piezoelectric energy harvesters, such as high piezoelectric coefficients, mechanical strength, and durability [14].

(e) *Reduced Waste and Cost*: 3D printing is an additive manufacturing process that minimizes material waste compared to traditional subtractive manufacturing methods. The ability to print complex structures with precise material deposition leads to the efficient use of materials [15]. Additionally, 3D printing can eliminate the need for costly tooling or molds, reducing production costs, especially for small-scale or customized energy harvesters.

20.4.2 Limitations of 3D Printing in Energy Harvester Fabrication

(a) *Limited Material Selection*: While 3D printing offers a wide range of compatible materials, the selection of materials with piezoelectric properties may be limited. Piezoelectric materials suitable for 3D printing may have different properties or performances compared to traditional bulk materials [16].

(b) *Lower Mechanical Strength*: Some 3D printing materials, particularly polymers, may have lower mechanical strength compared to traditional manufacturing materials. This can impact the overall durability and long-term performance of the energy harvester.

(c) *Resolution and Accuracy*: The resolution and accuracy of 3D printing can vary depending on the chosen technique and printer capabilities. Fine details or intricate features may not be reproduced with high precision, affecting the performance or efficiency of the energy harvester.

(d) *Post-Processing Requirements*: After 3D printing, energy harvesters may require additional post-processing steps, such as cleaning, surface treatment, or assembly. These steps can add complexity, time, and cost to the fabrication process [17].

(e) *Scalability and Production Volume*: 3D printing is well-suited for low to medium production volumes. However, it may not be as efficient or cost-effective for large-scale production compared to traditional manufacturing methods, such as injection molding or casting.

Despite these limitations, ongoing advancements in 3D printing technologies, materials, and processes are continuously addressing these challenges. Researchers and engineers are developing new techniques and optimizing existing ones to improve the resolution, mechanical strength, and material compatibility of 3D printed energy harvesters, expanding their potential in various applications [18].

20.4.3 Additional Advantages of 3D Printing in Energy Harvester Fabrication

(a) *Integration of Multiple Components*: 3D printing allows for the seamless integration of multiple components into a single printed structure. This capability enables the incorporation of additional functionalities, such as electronic components or sensors, directly into the energy harvester design, streamlining the assembly process and enhancing the overall performance.

(b) *Rapid Design Iteration*: 3D printing enables rapid design iteration by quickly modifying and printing new prototypes. This iterative process allows for the exploration of different design variations, optimizing the energy harvester's performance, efficiency, and manufacturability [19].

(c) *Accessibility and Democratization*: 3D printing technology has become more accessible and affordable, making it easier for researchers, engineers, and even enthusiasts to experiment with energy harvester designs. This accessibility promotes innovation and knowledge-sharing within the field, accelerating advancements in energy harvesting technology [20].

20.4.4 Additional Limitations of 3D Printing in Energy Harvester Fabrication

(a) *Size and Scale Limitations*: 3D printers have specific build volumes and size limitations, restricting the size and scalability of the printed energy harvesters. Large-scale energy harvesting applications may require alternative manufacturing methods or the assembly of multiple printed components [21].

(b) *Surface Finish and Quality*: Depending on the 3D printing technique used, the surface finish of printed energy harvesters may not be as smooth or polished compared to traditional manufacturing methods. Surface roughness can affect the overall performance and aesthetics of the device.

(c) *Process Complexity*: 3D printing, especially with more advanced techniques, may involve complex setup, calibration, and maintenance procedures. Proper printer calibration, material handling, and post-processing steps are necessary to achieve consistent and reliable results, requiring skilled operators or technicians.

(d) *Cost of 3D Printing Equipment and Materials*: While 3D printing has become more affordable, the initial investment in high-quality 3D printers and compatible materials can still be a significant cost factor, particularly for research groups or organizations with limited resources.

It is important to note that the advantages and limitations of 3D printing in energy harvester fabrication are continuously evolving as technology advances and new techniques are developed. Researchers and manufacturers are actively working to overcome these limitations and optimize the use of 3D printing for energy harvesting applications [22]. Advantages and limitations of 3D printing in energy harvester fabrication are given in Table 20.2.

Table 20.2 Advantages and limitations of 3D printing in energy harvester fabrication

Aspect	Advantages	Limitations
Design freedom	Enables the creation of complex geometries and intricate structures	Limited material selection for piezoelectric properties
	Optimizes energy harvester shapes for improved performance	Lower mechanical strength, especially in polymers
Customization and personalization	Allows easy customization and tailoring to specific requirements	Varying resolution and accuracy, impacting performance
Rapid prototyping and iteration	Facilitates quick prototyping and iterative design processes	Additional post-processing requirements
Material selection	Offers a wide range of compatible materials for energy harvester fabrication	Scalability and production volume limitations
Reduced waste and cost	Minimizes material waste and efficient material usage	Requires expertise and proper maintenance for equipment
	Eliminates the need for costly tooling or molds	
	Reduces production costs, especially for small-scale or customized energy harvesters	
Integration of multiple components	Enables seamless integration of multiple components into a single printed structure	Complex design and assembly processes
Rapid design iteration	Facilitates rapid modification and printing of new prototypes	An iterative process may extend development timelines
Accessibility and democratization	Increased accessibility and affordability of 3D printing technology	Limited expertise and knowledge in utilizing 3D printing technology

Aspect	Advantages	Limitations
Size and scale limitations		Restrictions on the size and scalability of printed energy harvesters
Surface finish and quality		Surface roughness may affect performance and aesthetics
Process complexity		Requires complex setup, calibration, and maintenance procedures
Cost of 3D printing equipment and materials		The initial investment in equipment and materials can be significant

20.5 Materials for 3D Printed Piezoelectric Energy Harvesters

The following are the various materials for 3D printed piezoelectric energy harvesters:

- Ceramics as piezoelectric materials for energy harvesting
- Polymers with piezoelectric properties suitable for 3D printing
- Composite materials for enhanced performance in energy harvesters

20.5.1 Ceramics as Piezoelectric Materials for Energy Harvesting

Ceramic materials have long been recognized for their excellent piezoelectric properties, making them a popular choice for piezoelectric energy harvesters. They offer high piezoelectric coefficients, good mechanical stability, and temperature resistance, making them suitable for a wide range of applications [23]. With the advent of 3D printing technology, the fabrication of ceramic-based piezoelectric energy harvesters has become more accessible and customizable.

20.5.2 Piezoelectric Properties of Ceramics

Piezoelectricity in ceramics arises from the crystal structure and arrangement of their constituent atoms or ions. When subjected to mechanical stress or pressure, these materials undergo a deformation, causing a shift in their electric charges and resulting in the generation of an electric potential difference [24]. This inherent piezoelectric effect enables the conversion of mechanical energy into electrical energy, making ceramics ideal for energy harvesting applications. Ceramic materials exhibit high piezoelectric coefficients, which determine the amount of electrical charge generated in response to mechanical stress. The choice of ceramic material is crucial as it directly impacts energy harvesting efficiency. Some commonly used ceramic materials for 3D printed piezoelectric energy harvesters include lead zirconate titanate (PZT), aluminum nitride (AlN), and zinc oxide (ZnO) [25].

20.5.3 Advantages of Ceramics in Energy Harvesting

Various advantages of ceramics in energy harvesting are shown in Fig. 20.1.

(a) *High Piezoelectric Coefficients*: Ceramics, such as PZT, exhibit exceptionally high piezoelectric coefficients, allowing for efficient conversion of mechanical energy into electrical energy. This high piezoelectric response enhances the energy harvesting performance of the device [26].

(b) *Mechanical Stability*: Ceramics possess excellent mechanical stability and stiffness, enabling them to withstand high levels of mechanical stress and deformation without significant degradation in performance. This property is crucial for energy harvesters operating in demanding environments.

(c) *Temperature Resistance*: Ceramic-based piezoelectric materials exhibit good thermal stability and can operate over a wide temperature range. This characteristic makes them suitable for energy harvesting applications in extreme temperatures or environments.

(d) *Broad Frequency Range*: Ceramics can exhibit piezoelectric response across a wide frequency range, allowing for the capture of mechanical vibrations or oscillations occurring at different frequencies. This versatility enables the utilization of ceramic-based energy harvesters in diverse applications with varying vibration frequencies [27].

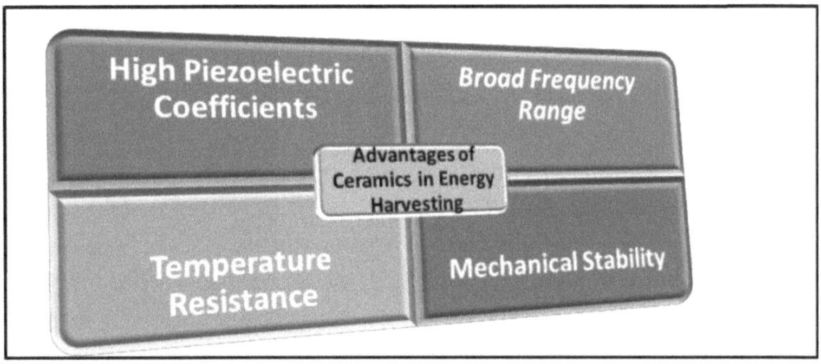

Figure 20.1 Advantages of ceramics in energy harvesting.

20.5.4 Challenges and Considerations in 3D Printing Ceramics

(a) *Material Selection and Compatibility*: The selection of suitable ceramic materials for 3D printing can be challenging due to factors such as powder characteristics, sintering behavior, and compatibility with printing techniques. It is important to choose ceramics that exhibit both desirable piezoelectric properties and good printability [28].

(b) *Sintering and Post-Processing*: 3D printed ceramic structures require post-processing steps, such as sintering, to achieve the desired material properties and density. Sintering conditions, including temperature and time, need to be carefully controlled to prevent deformation or cracking during the process.

(c) *Printing Resolution and Accuracy*: Achieving high-resolution and accurate 3D printed ceramic structures can be demanding. The choice of printing technique, printer parameters, and material formulation significantly impacts the final printed part's quality, surface finish, and dimensional accuracy [29].

(d) *Material Homogeneity and Microstructure*: The microstructure and homogeneity of the printed ceramic material play a vital role in determining its piezoelectric properties. The uniform distribution of piezoelectric crystals or domains within the printed ceramic structure is crucial for optimal energy harvesting performance.

Researchers and manufacturers are actively exploring and optimizing 3D printing processes and techniques for ceramics to address the challenges associated with 3D printing. Some notable advancements include the development of ceramic-based 3D printing methods, such as ceramic inkjet printing and stereolithography, specifically tailored for piezoelectric applications [30]. Ceramic inkjet printing utilizes ceramic suspensions or inks containing ceramic particles dispersed in a liquid medium. The ink is deposited layer by layer through inkjet nozzles, guided by computer-controlled movements. This technique offers precise control over material deposition, enabling the creation of intricate ceramic structures with high resolution. Furthermore, it allows for combining different ceramic materials or integrating multiple layers with varying compositions, optimizing the piezoelectric properties of the printed structures [31].

20.5.5 Stereolithography and Its Application in Piezoelectric Energy Harvester Production

SLA is another promising technique for 3D printing ceramic-based piezoelectric energy harvesters. It uses a UV laser to selectively solidify a ceramic nanoparticle liquid resin [32]. The laser scans the liquid resin, solidifying it layer by layer according to the digital model. SLA offers high resolution and the ability to create complex geometries, facilitating the fabrication of intricate piezoelectric structures. In addition to refining printing techniques, efforts are being made to enhance the properties and performance of 3D printed ceramic materials for energy harvesting applications. This includes the optimization of ceramic powder formulations, surface treatments, and post-processing methods to achieve higher density, improved piezoelectric response, and enhanced mechanical stability [33].

Furthermore, research is focused on developing novel ceramic composite materials by incorporating additives or functionalizing the ceramics to enhance their piezoelectric properties or tailor their response to specific environmental conditions. These advancements aim to overcome the inherent limitations of 3D printing ceramics and improve the overall performance and efficiency of 3D printed piezoelectric energy harvesters. It is worth noting that while ceramics offer excellent piezoelectric properties, they may have certain drawbacks compared to other materials. Ceramics are typically more brittle and prone to fracture under high mechanical stress or impact, limiting their use in applications that require exceptional durability or resistance to mechanical shocks. In such cases, alternative materials, such as polymers or composites, may be more suitable for 3D printed energy harvesters [34].

The use of ceramics as piezoelectric materials for 3D printed energy harvesters holds great promise. With continued advancements in material formulations, printing techniques, and post-processing methods, ceramic-based piezoelectric energy harvesters can contribute significantly to the development of efficient and customizable energy harvesting solutions in various fields, including wearable electronics, SHM, and IoT devices [35].

20.6 Polymers with Piezoelectric Properties Suitable for 3D Printing

Polymers are a versatile class of materials widely used in 3D printing due to their processability, flexibility, and ability to tailor their properties for specific applications. While most polymers do not exhibit inherent piezoelectric properties, recent advancements in material engineering have led to the development of piezoelectric polymers that can be utilized for 3D printing piezoelectric energy harvesters. These piezoelectric polymers offer unique advantages, such as improved flexibility, lightweight characteristics, and the ability to conform to complex shapes, making them suitable for a range of applications [36].

20.6.1 Piezoelectric Mechanisms in Polymers

Polymers can exhibit piezoelectricity through various mechanisms, including molecular orientation, charge transfer, and the presence of polar groups. When subjected to mechanical stress, these polymers undergo a deformation, resulting in the generation of an electric charge or voltage across their surfaces. This piezoelectric effect in polymers is influenced by factors such as polymer structure, molecular alignment, and the presence of functional groups.

20.6.2 Piezoelectric Polymers for 3D Printing

Several piezoelectric polymers have been developed and successfully utilized for the 3D printing of energy harvesters. Some of the commonly used piezoelectric polymers include the following:

(a) Polyvinylidene fluoride (PVDF): PVDF is a widely studied and commercially available piezoelectric polymer with excellent piezoelectric properties. PVDF-based materials, such as PVDF-TrFE (TrFE:trifluoroethylene), PVDF-HFP (HFP: hexafluoropropylene), and their copolymers, demonstrate good piezoelectric response and mechanical flexibility, making them suitable for 3D printing applications. PVDF-based materials can be processed using various 3D printing techniques, including FDM and inkjet printing [37].

(b) Polyurethane (PU): Polyurethane-based piezoelectric polymers have gained attention for their excellent mechanical properties and high flexibility. By incorporating piezoelectric nanoparticles, such as PZT or barium titanate ($BaTiO_3$), into the polymer matrix, piezoelectric properties can be imparted to polyurethane. These composite materials can be 3D printed using techniques such as FDM or SLA [38].

(c) Poly(L-lactic acid) (PLLA): PLLA is a biocompatible and biodegradable polymer that can be modified to exhibit piezoelectric properties. By incorporating piezoelectric fillers, such as piezoelectric ceramic nanoparticles or polymers with inherent piezoelectricity, into PLLA, piezoelectric behavior can be induced. 3D printing techniques like FDM and SLA have been utilized to fabricate PLLA-based piezoelectric structures [39].

(d) Poly(vinylidene fluoride-co-tetrafluoroethylene) (P(VDF-TrFE)): P(VDF-TrFE) is a copolymer that combines the desirable properties of PVDF and TrFE. It exhibits enhanced piezoelectric properties compared to PVDF and improved mechanical flexibility compared to pure TrFE. P(VDF-TrFE) can be 3D printed using techniques such as inkjet printing or FDM, allowing for the fabrication of complex piezoelectric structures [39].

20.6.3 Advantages of Polymer-Based Piezoelectric Materials for 3D Printing

Polymer-based piezoelectric materials offer several advantages for the 3D printing of energy harvesters.

(a) Flexibility and Conformability: Polymer-based piezoelectric materials are inherently flexible, allowing them to conform to various shapes and deformations. This flexibility is particularly advantageous in applications where the energy harvester needs to be integrated into curved or irregular surfaces, such as wearable devices or biomedical sensors. Additionally, the ability to print flexible and stretchable piezoelectric materials opens up new possibilities for energy harvesting from dynamic or non-planar sources.

(b) Lightweight: Polymer-based piezoelectric materials are generally lightweight, which is beneficial for applications where weight reduction is a critical factor, such as in wearable devices or unmanned aerial vehicles (UAVs). The lightweight nature of these materials also contributes to the overall comfort and usability of energy harvesting devices.

(c) Processability and Customizability: Polymers are highly processable materials, allowing for precise control over the fabrication process. 3D printing techniques enable the creation of complex geometries and intricate designs with ease. This level of customization is advantageous for tailoring energy harvester designs to specific applications, optimizing their performance, and integrating additional functionalities, such as sensors or electronics, into the structure.

(d) Cost-Effectiveness: Polymer-based piezoelectric materials are generally more cost-effective compared to traditional piezoelectric ceramics. Polymers are widely available and can be produced at a lower cost, making them suitable for large-scale production and commercial applications [40].

20.6.4 Challenges and Considerations in 3D Printing Polymer-Based Piezoelectric Materials

While polymer-based piezoelectric materials offer unique advantages, there are certain challenges and considerations associated with their 3D printing. Challenges and considerations in 3D printing polymer-based piezoelectric materials are given in Fig. 20.2.

(a) Piezoelectric Performance: The piezoelectric properties of polymers are typically lower compared to ceramic-based materials. Achieving high piezoelectric coefficients and maintaining good mechanical properties in polymer composites require careful material selection, processing parameters, and optimization of the printing techniques. Improvements in the piezoelectric performance of polymer-based materials remain an active area of research.

(b) Stability and Durability: Polymer-based materials may experience degradation over time due to environmental factors, such as temperature, humidity, and UV radiation. Ensuring the stability and durability of 3D printed piezoelectric polymers is crucial for long-term reliability and performance.

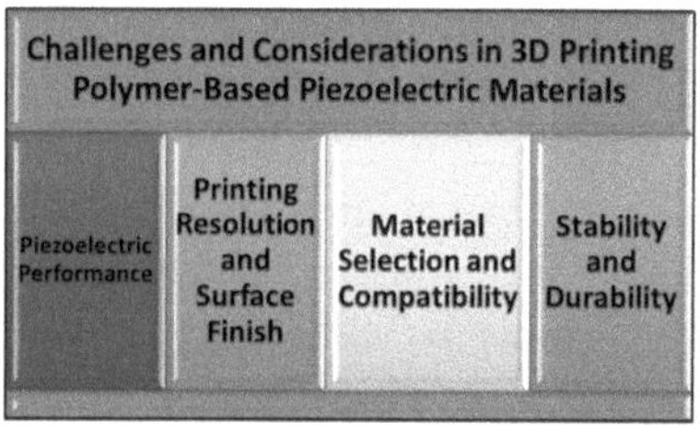

Figure 20.2 Challenges and considerations in 3D printing polymer-based piezoelectric materials.

(c) Material Selection and Compatibility: The choice of polymer matrix, piezoelectric fillers, and their compatibility is essential for achieving the desired piezoelectric properties and printability. Material selection should consider factors such as adhesion between layers, interfacial compatibility, and the ability to uniformly disperse piezoelectric fillers throughout the polymer matrix.

(d) Printing Resolution and Surface Finish: Achieving high-resolution and smooth surface finish in 3D printed polymer-based piezoelectric structures can be challenging. The choice of printing technique, printer parameters, and post-processing steps should be optimized to achieve the desired level of detail, accuracy, and surface quality.

Researchers and manufacturers are actively addressing these challenges by developing novel polymer composites, optimizing printing techniques, and exploring new approaches, such as hybrid printing methods and the integration of functional additives to enhance the piezoelectric performance of 3D printed polymer-based energy harvesters [41].

Polymer-based piezoelectric materials for 3D printing offer significant potential for the fabrication of lightweight, flexible, and customizable energy harvesters. Continued advancements in material formulations, printing techniques, and post-processing methods will further enhance the performance and expand the applications of 3D printed polymer-based piezoelectric energy harvesting devices.

20.7 Composite Materials for Enhanced Performance in Energy Harvesters

Composite materials offer unique opportunities to enhance the performance of energy harvesters by combining different materials with complementary properties. These materials can be engineered to exhibit improved piezoelectric response, mechanical strength, durability, and other desired characteristics. The development of composite materials for energy harvesting applications has gained significant attention due to their ability to overcome the limitations of individual materials and achieve enhanced performance. This section explores the use of composite materials in energy harvesters and discusses their benefits and applications.

20.7.1 Types of Composite Materials for Energy Harvesters

Types of composite materials for energy harvesters are shown in Fig. 20.3.

(a) Polymer Matrix Composites (PMCs): Polymer matrix composites consist of a polymer matrix reinforced with fillers such as nanoparticles, fibers, or flakes. These composites can exhibit improved mechanical properties, thermal stability, and piezoelectric response compared to pure polymers.

By incorporating piezoelectric fillers, such as piezoelectric ceramics or polymers, into the polymer matrix, the overall piezoelectric performance of the composite can be enhanced. PMCs can be tailored to meet specific requirements, such as flexibility, lightweight, or high stiffness, making them suitable for a wide range of energy harvesting applications [42].

(b) Ceramic Matrix Composites (CMCs): Ceramic matrix composites comprise a ceramic matrix reinforced with ceramic or non-ceramic fibers. CMCs offer exceptional mechanical properties, high-temperature stability, and improved piezoelectric response compared to pure ceramics. The addition of piezoelectric fibers or nanoparticles to the ceramic matrix enhances the overall piezoelectric properties of the composite. CMCs are suitable for applications where high mechanical strength, thermal resistance, and durability are required [43].

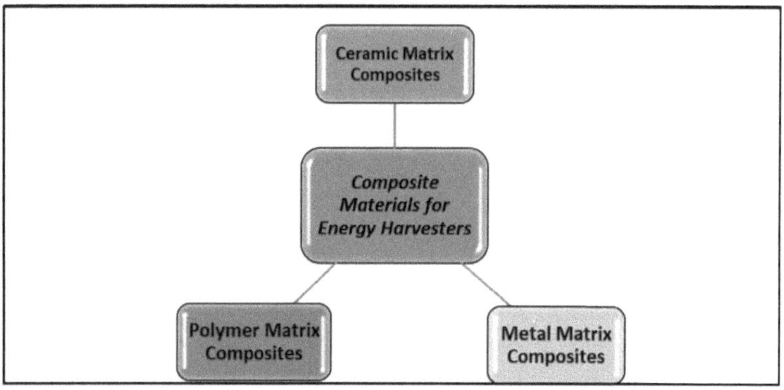

Figure 20.3 Types of composite materials for energy harvesters.

(c) Metal Matrix Composites (MMCs): Metal matrix composites are composed of a metal matrix reinforced with ceramic or carbon fibers. MMCs provide excellent mechanical properties, such as high strength, stiffness, and toughness. By incorporating piezoelectric materials into the metal matrix, the composite can exhibit enhanced piezoelectric response. MMCs are well-suited for applications where high mechanical loads or harsh operating conditions are present.

20.7.2 Benefits of Composite Materials in Energy Harvesters

(a) Enhanced Piezoelectric Response: Composite materials can exhibit improved piezoelectric response compared to individual materials. By combining materials with high piezoelectric coefficients, the overall piezoelectric performance of the composite can be increased. This enables higher energy conversion efficiency in energy harvesters [44].
(b) Tailored Mechanical Properties: Composite materials can be engineered to possess specific mechanical properties, such as high stiffness, flexibility, or toughness, depending on the application requirements. This allows for the design of energy harvesters that can withstand mechanical stresses, deformations, or vibrations encountered in real-world environments.
(c) Lightweight and Flexible Designs: Composite materials often offer lightweight and flexible characteristics, making them ideal for applications that demand low weight or require conformability to curved or irregular surfaces. This is particularly advantageous for wearable devices or energy harvesters integrated into flexible structures.
(d) Improved Durability and Reliability: The combination of different materials in composites can enhance the overall durability and reliability of energy harvesters. By reinforcing a less durable material with a stronger or more resilient one, the composite can withstand harsh operating conditions, including temperature variations, humidity, or mechanical shocks.
(e) Design Freedom and Customizability: Composite materials provide design freedom, enabling the creation of complex geometries and customized energy harvester architectures. 3D printing technologies can be employed to fabricate intricate composite structures with precise control over material distribution, leading to improved performance and efficiency [45].

20.7.3 Applications of Composite Materials in Energy Harvesters

Composite materials find applications in various energy harvesting domains, including:

(a) Structural Health Monitoring: Composite-based energy harvesters can be integrated into structural components to continuously monitor the health and integrity of structures, such as bridges, buildings, or aircraft. The enhanced mechanical properties of composite materials make them suitable for embedding energy harvesters in load-bearing components without compromising structural integrity. These energy harvesters can convert mechanical vibrations or strains induced by structural deformations into electrical energy, enabling real-time monitoring and diagnostics of the structure's health [46].

(b) Wearable and Biomedical Devices: Composite materials offer lightweight, flexible, and biocompatible properties, making them ideal for wearable energy harvesting devices and biomedical applications. Energy harvesters integrated into wearable devices, such as smartwatches or fitness trackers, can capture ambient mechanical energy from body movements and convert it into electrical power to extend the device's battery life. In biomedical applications, composite-based energy harvesters can be used to power implantable or wearable medical devices, eliminating the need for frequent battery replacements or external power sources.

(c) Internet of Things (IoT) Devices: Composite materials are well-suited for energy harvesters in IoT devices, where small-scale, self-powered sensors and actuators are deployed in various environments. These energy harvesters can scavenge energy from ambient vibrations, thermal gradients, or other available energy sources, providing a sustainable power source for IoT devices. The lightweight and customizable nature of composite materials allows for the integration of energy harvesting capabilities into compact IoT devices without adding significant weight or size.

(d) Automotive and Aerospace Applications: Composite materials are extensively used in the automotive and aerospace industries due to their high strength-to-weight ratio. In these applications, composite-based energy harvesters can be incorporated into vehicle components or aircraft structures to convert mechanical vibrations or airflow-induced vibrations into electrical energy. This harvested energy can be utilized to power vehicle systems, such as wireless sensors, tire pressure monitoring systems, or onboard electronics, reducing the reliance on external power sources and increasing overall energy efficiency.

20.7.4 Challenges and Considerations in Composite-Based Energy Harvesters

Despite the numerous benefits, there are certain challenges and considerations associated with the use of composite materials in energy harvesters.

(a) Material Compatibility: The compatibility between different materials within the composite, such as the matrix and reinforcement, is crucial to ensure strong bonding, uniform distribution of piezoelectric fillers, and efficient energy transfer. Incompatibility issues can lead to delamination, reduced mechanical strength, or decreased piezoelectric performance.

(b) Manufacturing Complexity: The fabrication of composite materials and the integration of energy harvesting components can be more complex compared to individual materials. Careful consideration of manufacturing processes, such as layering techniques, curing conditions, or post-processing steps, is required to achieve a uniform and defect-free composite structure.

(c) Cost Considerations: Composite materials can be costlier compared to individual materials, primarily due to the inclusion of reinforcing elements or the need for specialized manufacturing techniques. The cost-effectiveness of composite-based energy harvesters should be evaluated against their performance benefits and specific application requirements.

(d) Environmental Factors: Composite materials may exhibit sensitivity to environmental conditions, such as temperature, humidity, or UV radiation. These factors can potentially degrade the composite structure, affecting its long-term performance and reliability. Proper material selection, protective coatings, or encapsulation techniques should be employed to mitigate environmental degradation [47].

(e) Recycling and Sustainability: Composite materials pose challenges in terms of recycling and sustainability due to the difficulty of separating the different components and the potential environmental impact of composite disposal. Considerations should be given to developing eco-friendly and recyclable composite materials to minimize their environmental footprint.

Researchers and industry stakeholders continue to address these challenges through advanced material engineering, manufacturing process optimization, and sustainability initiatives. Composite-based energy harvesters hold great promise for advancing the field of energy harvesting, enabling self-powered systems, and contributing to the development of sustainable technologies in various sectors.

20.8 Fused Deposition Modeling (FDM) for Fabricating Piezoelectric Energy Harvesters

FDM is a widely used 3D printing technique for fabricating piezoelectric energy harvesters. FDM is known for its simplicity, accessibility, and versatility, making it a popular choice for both research and industrial applications [48]. This section provides an extensive overview of FDM as a 3D printing technique for piezoelectric energy harvesters, including its working principles, advantages, limitations, and notable applications.

20.8.1 Working Principles of Fused Deposition Modeling

FDM works by extruding a thermoplastic filament through a heated nozzle, which deposits the material layer by layer to create a three-dimensional object. The process involves the following steps [49]:

(a) Filament Feed: The thermoplastic filament, typically in a spool form, is fed into the 3D printer.
(b) Filament Melting: The filament passes through a heated nozzle, where it is melted to a semi-liquid state.
(c) Layer Deposition: The molten filament is extruded through the nozzle and deposited onto the build platform in a controlled manner, following the predetermined design pattern.
(d) Layer Adhesion: As each layer is deposited, it fuses with the previous layer, creating a strong bond between the layers.
(e) Build Platform Advancement: After the completion of each layer, the build platform is lowered or moved to accommodate the deposition of the subsequent layer.
(f) Cooling and Solidification: Once the object is fully printed, it undergoes cooling and solidification, resulting in a solid, fully functional structure.

Working principles of fused deposition modeling are shown in Fig. 20.4.

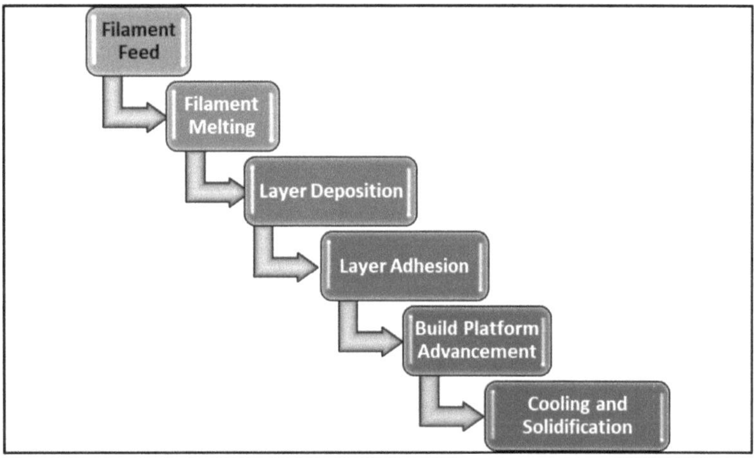

Figure 20.4 Working principles of fused deposition modeling.

20.8.2 Advantages of Fused Deposition Modeling for Piezoelectric Energy Harvesters

(a) Material Compatibility: FDM supports a wide range of thermoplastic materials, including those with piezoelectric properties. This flexibility allows for the selection of suitable materials for piezoelectric energy harvesters based on their desired characteristics, such as piezoelectric coefficients, mechanical strength, or flexibility.

(b) Cost-Effectiveness: FDM is known for its affordability, as the thermoplastic filaments used in the process are relatively inexpensive compared to other 3D printing materials. This makes FDM an economically viable option for prototyping and small-scale production of piezoelectric energy harvesters.

(c) Design Flexibility: FDM offers design freedom, enabling the fabrication of complex geometries, intricate patterns, and customized energy harvester designs. This allows for the optimization of the device's shape, size, and internal structure to maximize energy conversion efficiency and performance.

(d) Rapid Prototyping: FDM enables rapid prototyping, allowing researchers and engineers to quickly iterate and refine their designs. This accelerates the development process and reduces time-to-market for new piezoelectric energy harvester concepts.

(e) Ease of Use: FDM is a user-friendly 3D printing technique that requires minimal training and expertise. Its simplicity and accessibility make it suitable for researchers, engineers, and students interested in fabricating piezoelectric energy harvesters [50].

20.8.3 Limitations and Considerations of Fused Deposition Modeling for Piezoelectric Energy Harvesters

(a) Resolution and Surface Finish: FDM typically exhibits lower resolution and surface finish compared to other high-precision 3D printing techniques. The layering effect can result in visible layer lines and reduced detail in the printed object. However, advancements in FDM technology, such as smaller nozzle sizes and improved slicing algorithms, have somewhat mitigated this limitation.

(b) Anisotropic Mechanical Properties: FDM-produced parts often exhibit anisotropic mechanical properties, meaning that their strength and stiffness can vary depending on the direction in which the layers are deposited. This anisotropy can affect the performance and reliability of piezoelectric energy harvesters, particularly in applications that require uniform mechanical properties in all directions. Design considerations, such as optimizing the layer orientation or incorporating support structures, can help address this limitation to some extent.

(c) Limited Material Selection: While FDM offers a wide range of thermoplastic materials, the selection of piezoelectric materials for energy harvesters may be somewhat limited compared to other 3D printing techniques. However, advancements in material development and filament formulations are expanding the range of piezoelectric materials compatible with FDM.

(d) Temperature Sensitivity: FDM relies on heat to melt and extrude the thermoplastic filament. This temperature sensitivity can affect printed parts' dimensional accuracy and stability, especially when working with materials that exhibit significant thermal expansion or contraction. Proper calibration, temperature control, and material selection are important considerations to ensure the accurate and reliable fabrication of piezoelectric energy harvesters [51].

20.8.4 Applications of Fused Deposition Modeling in Piezoelectric Energy Harvesters

FDM has found applications in various areas of piezoelectric energy harvesting, including [52]:

(a) Wearable Electronics: FDM enables the fabrication of flexible, lightweight, and customizable energy harvesters that can be seamlessly integrated into wearable devices. These energy harvesters can scavenge energy from human motions, such as body movements or vibrations, and convert it into electrical power to sustain the operation of wearable electronics, such as fitness trackers, smartwatches, or health monitoring devices.

(b) Internet of Things (IoT) Sensors: FDM-produced piezoelectric energy harvesters can be utilized in small-scale IoT sensors that are deployed in various environments. These sensors can harvest energy from ambient vibrations, such as machinery vibrations or vibrations induced by human activities, and power the sensing and communication components of the IoT devices. FDM allows for the fabrication of compact and self-powered sensors, reducing the need for frequent battery replacements or external power sources.

(c) Structural Health Monitoring: FDM can be employed to fabricate piezoelectric energy harvesters integrated into structural components, such as bridges, buildings, or aerospace structures. These energy harvesters can capture mechanical vibrations or strains induced by structural deformations, enabling continuous monitoring of the structural health. The self-powered nature of these energy harvesters eliminates the need for external power sources, providing a cost-effective and sustainable solution for SHM applications.

(d) Smart Packaging and Consumer Electronics: FDM-produced piezoelectric energy harvesters can be integrated into smart packaging systems or consumer electronics to power low-power devices or provide energy for wireless communication. The ability to customize the shape, size, and internal structure of energy harvesters using FDM allows for seamless integration into the packaging or electronics, providing a self-sustaining power source.

In summary, FDM is a versatile and accessible 3D printing technique for fabricating piezoelectric energy harvesters. It offers advantages such as material compatibility, cost-effectiveness, design flexibility, and rapid prototyping. However, it also has limitations, including resolution and surface finish, anisotropic mechanical properties, limited material selection, and temperature sensitivity. Despite these limitations, FDM has found applications in wearable electronics, IoT sensors, SHM, smart packaging, and consumer electronics. With ongoing advancements in FDM technology and material development, the potential for fabricating high-performance piezoelectric energy harvesters using FDM continues to expand [53].

20.9 Stereolithography and Its Application in Piezoelectric Energy Harvester Production

SLA is an additive manufacturing technique that uses photopolymerization to create three-dimensional objects. SLA has gained significant attention in various industries, including the production of piezoelectric energy harvesters. This section provides an extensive overview of SLA as a 3D printing technique and explores its applications in piezoelectric energy harvester production, including working principles, advantages, limitations, and notable examples.

20.9.1 Working Principles of Stereolithography

SLA operates based on the principle of photopolymerization, where a liquid resin is selectively cured layer by layer using a light source, typically a UV laser. The SLA process involves the following steps [54]:

(a) Preparation of Build Platform: A build platform is positioned inside a vat containing a liquid photosensitive resin.

(b) Photopolymerization: The UV laser scans the surface of the resin, selectively solidifying it according to the desired pattern or cross-sectional shape of the object.

(c) Layer-by-Layer Build: After the completion of each layer, the build platform moves down, allowing a new layer of resin to flow over the previously solidified layer. The UV laser then selectively cures the resin to create the next layer.

(d) Post-Processing: Once the entire object is printed, it undergoes post-processing, which typically involves removing excess resin, cleaning, and curing the object under UV light to ensure full polymerization and optimal mechanical properties.

20.9.2 Advantages of Stereolithography for Piezoelectric Energy Harvester Production

SLA offers several advantages for the production of piezoelectric energy harvesters [55].

(a) High Resolution and Surface Quality: SLA is known for its exceptional resolution, enabling the fabrication of intricate details and fine features in piezoelectric energy harvesters. The ability to produce high-quality, smooth surfaces is beneficial for achieving optimal performance in energy harvesting applications.

(b) Material Selection: SLA supports a wide range of photopolymer resins, including those with piezoelectric properties. This allows for the fabrication of energy harvesters using specialized piezoelectric materials tailored for specific performance requirements.

(c) Design Freedom: SLA offers design freedom, enabling the creation of complex geometries, internal structures, and customized energy harvester designs. This flexibility allows for the optimization of the device's shape, size, and internal architecture to maximize energy conversion efficiency.

(d) Rapid Prototyping: SLA enables rapid prototyping, allowing for quick iterations and refinement of piezoelectric energy harvester designs. This accelerates the development process, reduces time-to-market, and facilitates the exploration of various design iterations.

(e) Accuracy and Precision: SLA ensures high accuracy and dimensional precision, resulting in the consistent and reliable fabrication of piezoelectric energy harvesters. This is crucial for maintaining the desired performance and functionality of the devices.

20.9.3 Limitations and Considerations of Stereolithography for Piezoelectric Energy Harvester Production

Despite its advantages, SLA also has certain limitations and considerations [55].

(a) Material Limitations: While SLA offers a wide selection of photopolymer resins, the range of commercially available piezoelectric resins specifically tailored for energy harvesting applications may be relatively limited. However, ongoing research and development efforts are expanding the range of available materials.

(b) Limited Build Volume: SLA machines typically have a smaller build volume compared to other 3D printing technologies. This can restrict the size of piezoelectric energy harvesters that can be produced in a single print. Large-scale energy harvesters may require printing multiple components and subsequent assembly.

(c) Post-Processing Requirements: SLA-printed objects require post-processing to remove excess resin and cure the printed parts fully. Post-curing is essential to ensure the piezoelectric energy harvesters' mechanical strength, stability, and optimal performance. Depending on the specific resin used, post-curing may involve exposing the printed parts to UV light or other curing methods, which can add time and complexity to the fabrication process.

(d) Cost: SLA printers and photopolymer resins can be relatively expensive compared to other 3D printing technologies. The cost of equipment, materials, and post-processing requirements should be taken into consideration when evaluating the feasibility and scalability of using SLA for piezoelectric energy harvester production.

20.9.4 Applications of Stereolithography in Piezoelectric Energy Harvester Production

SLA has found various applications in the production of piezoelectric energy harvesters, including [56]:

(a) Micro-Energy Harvesters: SLA's high resolution and precision make it suitable for fabricating micro-scale piezoelectric energy harvesters. These devices can be used to scavenge energy from ambient vibrations or mechanical deformations in small-scale systems, such as micro-electromechanical systems (MEMS), wearable electronics, or biomedical devices.

(b) Customized Energy Harvesters: SLA's design freedom allows for the fabrication of customized piezoelectric energy harvesters tailored to specific applications. This includes designing the shape, size, and internal structure of the energy harvester to optimize its performance in a given environment or system.

(c) Integration with Complex Systems: SLA enables the integration of piezoelectric energy harvesters into complex systems or components. For example, SLA can fabricate energy harvesters seamlessly integrated into 3D printed structural components or electronic devices, enabling self-powered smart systems.
(d) Rapid Prototyping and Concept Validation: SLA's ability to quickly produce high-resolution prototypes facilitates the rapid prototyping and concept validation of new piezoelectric energy harvester designs. Engineers and researchers can iterate and test different design iterations efficiently, enabling faster development cycles and optimization of the energy harvester's performance.

20.9.5 Future Developments and Considerations in SLA

As SLA technology continues to advance, several areas of development and consideration can be expected.

(a) Material Development: Research efforts are focused on developing new photopolymer resins with enhanced piezoelectric properties, enabling the fabrication of high-performance piezoelectric energy harvesters using SLA. The availability of a broader range of materials specifically tailored for energy harvesting applications would expand the design possibilities and improve overall performance.
(b) Scalability and Cost Optimization: Future developments in SLA technology aim to improve scalability and reduce costs. This includes increasing the build volume, enhancing printing speed, and optimizing material usage to make SLA more viable for large-scale production of piezoelectric energy harvesters.
(c) Multi-Material Printing: SLA machines capable of printing with multiple materials simultaneously or sequentially could enable the fabrication of complex, multi-functional piezoelectric energy harvesters with integrated components or sensors. This would further enhance the performance and versatility of the energy harvesting devices.

In conclusion, SLA is a powerful 3D printing technique for the production of piezoelectric energy harvesters. Its high resolution, design flexibility, and rapid prototyping capabilities make it suitable for creating customized energy harvesters and exploring new design concepts. While SLA does have limitations such as material availability and post-processing requirements, ongoing advancements in SLA technology and material development are expanding its potential in the field of piezoelectric energy harvesting. With further research and innovation, SLA has the potential to drive the development of high-performance, customized, and scalable piezoelectric energy harvesters, enabling their integration into various applications such as micro-energy harvesters, customized devices, and complex systems. As SLA technology continues to evolve, addressing material limitations, improving scalability, and optimizing cost-effectiveness will be essential for its widespread adoption in the field.

Furthermore, the integration of SLA with other 3D printing techniques, such as multi-material printing or hybrid approaches, holds promise for enhancing the performance and functionality of piezoelectric energy harvesters. By combining SLA with other additive manufacturing methods, it becomes possible to create hybrid structures that incorporate different materials or components with varying properties. This opens up opportunities for developing innovative piezoelectric energy harvesting devices that can meet specific application requirements and achieve higher energy conversion efficiencies. Additionally, advancements in SLA post-processing techniques, such as automated resin removal and curing processes, will contribute to streamlining the production workflow and reducing the time and effort required for finishing printed parts. Automation and optimization of post-processing steps will enhance the overall efficiency and scalability of SLA-based piezoelectric energy harvester fabrication.

SLA is a valuable 3D printing technique for the production of piezoelectric energy harvesters. Its high resolution, design flexibility, and rapid prototyping capabilities make it well-suited for creating customized and complex energy harvesting devices. With ongoing developments in materials, post-processing techniques, and hybrid approaches, SLA is poised to play a significant role in advancing the field of piezoelectric energy harvesting, enabling the development of efficient, customizable, and scalable solutions for a wide range of applications.

20.10 Selective Laser Sintering and Its Advantages in Energy Harvester Manufacturing

SLS is a popular additive manufacturing technique used for the production of various components, including piezoelectric energy harvesters [57]. This section provides an extensive overview of SLS as a 3D printing technique and explores its advantages in energy harvester manufacturing, including working principles, material versatility, design flexibility, production scalability, and notable examples.

20.10.1 Working Principles of Selective Laser Sintering

SLS is an additive manufacturing process that utilizes a high-power laser to selectively fuse powdered materials, typically polymers or composites, layer by layer. The SLS process involves the following steps [58]:

(a) Powder Bed Preparation: A thin layer of powdered material is spread evenly across the build platform.
(b) Laser Scanning and Fusion: A high-power laser scans the cross section of the part, selectively melting and fusing the powder particles to form a solid layer.
(c) Layer-by-Layer Build: After each layer is fused, a new layer of powder is applied, and the laser repeats the scanning and fusion process until the complete part is fabricated.
(d) Cooling and Post-Processing: Once the build is complete, the part is cooled, the excess powder is removed, and post-processing steps, such as cleaning and surface finishing, may be performed to achieve the desired final properties.

20.10.2 Applications of Selective Laser Sintering in Energy Harvester Manufacturing

SLS has found numerous applications in the manufacturing of piezoelectric energy harvesters, including [59]:

(a) Customized Energy Harvesters: SLS enables the fabrication of customized energy harvesters tailored to specific applications.

The design flexibility of SLS allows for the creation of unique energy harvester geometries and internal architectures that can be optimized for specific energy harvesting requirements. This includes designing energy harvesters with complex shapes, varying thicknesses, and internal structures that enhance the device's performance and efficiency.

(b) Integration with Complex Systems: SLS-produced energy harvesters can be seamlessly integrated into complex systems or components. The ability to directly manufacture energy harvesters as part of a larger assembly or system allows for tighter integration, reduced assembly steps, and improved overall system performance. This is particularly beneficial for applications where space is limited or the energy harvester must be integrated into existing structures or devices.

(c) Functional Prototyping and Concept Validation: SLS's rapid prototyping capabilities enable the efficient fabrication of functional prototypes for energy harvesters. Engineers and researchers can quickly iterate and test different design concepts, optimizing the energy harvester's performance and functionality. This facilitates the validation of design ideas, accelerates the development process, and reduces the time to market for new energy harvesting technologies.

(d) Education and Research: SLS is widely used in educational institutions and research laboratories for studying and exploring new energy harvesting concepts. Its ability to produce complex geometries and customized designs allows researchers and students to investigate the fundamental principles of energy harvesting, evaluate different materials and configurations, and advance the understanding of piezoelectric energy harvesting technology.

20.10.3 Notable Examples of SLS-Printed Piezoelectric Energy Harvesters

Several notable examples demonstrate the successful application of SLS in the manufacturing of piezoelectric energy harvesters [60].

(a) Wearable Energy Harvesters: SLS has been used to fabricate flexible, wearable energy harvesters that can scavenge energy from body movements or ambient vibrations. These energy harvesters can power small electronic devices or sensors, making them suitable for healthcare monitoring, sports, and wearable technology applications.

(b) Structural Health Monitoring Systems: SLS-printed piezoelectric energy harvesters have been integrated into SHM systems for various industries, such as aerospace, civil engineering, and transportation. These energy harvesters can detect and convert mechanical vibrations or strains into electrical energy, enabling continuous monitoring of structural integrity and providing valuable data for maintenance and safety purposes.

(c) Internet of Things (IoT) Devices: SLS has been used to manufacture miniaturized piezoelectric energy harvesters for IoT devices. These devices can scavenge energy from environmental vibrations or user interactions to power wireless sensors or low-power IoT devices, eliminating the need for battery replacement or external power sources.

Advantages of SLS in energy harvester manufacturing are given in Table 20.3.

Table 20.3 Advantages of selective laser sintering (SLS) in energy harvester manufacturing

S. No.	Advantages
1	Material Versatility: SLS is compatible with a wide range of materials, including polymers, composites, and even some ceramics. This versatility enables the fabrication of energy harvesters using different piezoelectric materials, allowing for material selection based on specific performance requirements.
2	Design Flexibility: SLS provides excellent design freedom, allowing the production of complex geometries, intricate internal structures, and customized energy harvester designs. This flexibility enables the optimization of the energy harvester's shape, size, and internal architecture to enhance its performance and efficiency.
3	Scalability and Production Efficiency: SLS is highly scalable, making it suitable for both prototyping and large-scale production. The powder bed-based manufacturing process allows for the simultaneous fabrication of multiple parts, maximizing production efficiency. Moreover, SLS can produce fully functional, ready-to-use energy harvesters in a single manufacturing step, eliminating the need for complex assembly processes.

S. No.	Advantages
4	High Mechanical Strength: SLS-produced energy harvesters exhibit excellent mechanical strength and structural integrity due to the fusion of the powder particles at high temperatures. This ensures robustness and durability, which are crucial for energy harvesters operating in challenging environments.
5	Multi-Material Capabilities: SLS can enable the fabrication of energy harvesters with multi-material properties by using different powdered materials in a single build. This capability allows for the integration of various functional components, such as electrodes or support structures, directly into the energy harvester, enhancing its overall performance.
6	Cost-Effectiveness: SLS can be a cost-effective manufacturing method for energy harvesters, particularly for low- to medium-production volumes. The ability to utilize powdered materials efficiently, minimize waste, and eliminate the need for dedicated tooling or molds contributes to cost savings and shortens the overall production cycle.

As SLS technology continues to advance, several areas of development and consideration can be expected:

(a) Material Development: Further research and development efforts are focused on expanding the range of piezoelectric materials compatible with SLS. This includes exploring new polymer composites, ceramics, and hybrid materials that exhibit enhanced piezoelectric properties and can be efficiently processed using SLS.

(b) Enhanced Printing Parameters: Ongoing advancements aim to optimize the SLS printing parameters for energy harvester fabrication, such as laser power, scan speed, and layer thickness. These optimizations can lead to improved printing accuracy, resolution, and overall energy harvester performance.

(c) Post-Processing Automation: Automation of post-processing steps, such as powder removal, cleaning, and surface finishing, can streamline the SLS manufacturing process. This includes the development of advanced post-processing equipment or techniques that reduce manual intervention, decrease labor costs, and improve the overall efficiency of energy harvester production.

(d) Integration with Other Manufacturing Techniques: Exploring the integration of SLS with other manufacturing techniques, such as post-print machining or assembly processes, can expand the possibilities for fabricating complex and multi-functional energy harvesters. This integration can enhance the performance, reliability, and functionality of the final devices.

SLS is a highly versatile and advantageous 3D printing technique for the manufacturing of piezoelectric energy harvesters. Its working principles, material versatility, design flexibility, scalability, and notable examples demonstrate its effectiveness in this field. SLS enables the fabrication of customized energy harvesters with complex geometries and internal structures that optimize their performance. It allows for the integration of energy harvesters into complex systems and components, reducing assembly steps and improving overall system performance. The rapid prototyping capabilities of SLS facilitate functional prototyping and concept validation, accelerating the development process for new energy harvesting technologies. Moreover, SLS finds applications in education and research, enabling the exploration of new concepts and advancing the understanding of piezoelectric energy harvesting.

SLS offers significant advantages in energy harvester manufacturing, making it a valuable tool for producing efficient and customized piezoelectric energy harvesters. Its capabilities and ongoing advancements contribute to the progress of the field, paving the way for innovative energy harvesting solutions in various applications.

20.11 Design Considerations for 3D Printed Piezoelectric Energy Harvesters

Designing efficient and effective piezoelectric energy harvesters involves careful consideration of various factors. This section focuses on design considerations specifically related to 3D printed piezoelectric energy harvesters. It covers the optimization of geometry and shape, integration of electronics and connections, as well as design challenges and solutions unique to 3D printed energy harvesters.

1. Optimization of Geometry and Shape for Enhanced Energy Conversion
   (a) Geometric Configuration: The geometric configuration of an energy harvester plays a critical role in maximizing energy conversion. The shape and size of the piezoelectric material, as well as the positioning of electrodes, need to be optimized for efficient energy harvesting. 3D printing allows for the fabrication of complex geometries, enabling the design of intricate structures that enhance energy conversion capabilities. The use of lattice structures, variable thicknesses, or tailored shapes can improve mechanical deformation and strain distribution, resulting in enhanced energy harvesting performance [61].
   (b) Resonance Frequency: Energy harvesters are typically designed to resonate at specific frequencies to match the excitation source. The geometry and dimensions of the energy harvester can be adjusted to achieve the desired resonance frequency. 3D printing enables the customization of energy harvester designs, allowing for the optimization of the structure's dimensions to target specific resonance frequencies and enhance energy conversion efficiency.
   (c) Flexibility and Conformability: In certain applications, such as wearable devices or integration into curved surfaces, the flexibility and conformability of the energy harvester are crucial. 3D printing techniques, such as flexible filament printing or multi-material printing, can be employed to fabricate energy harvesters with flexible and conformable characteristics. This enables their integration into various form factors and ensures efficient energy harvesting even in dynamic or irregular environments.

20.11.1 Integration of Electronics and Connections within the 3D Printed Structure

(a) Electrode Integration: The integration of electrodes within the 3D printed structure is essential for effective energy harvesting. It is important to ensure proper electrical contact between the piezoelectric material and the electrodes. Design considerations include the positioning of electrodes to maximize the piezoelectric effect and the selection of suitable materials for the electrodes to minimize electrical resistance. 3D printing allows for the direct integration of electrodes during the fabrication process, simplifying the assembly and improving the overall reliability of the energy harvester.

(b) Wiring and Connection Design: Efficient energy transfer from the energy harvester to external devices or storage systems requires appropriate wiring and connection design. The routing of wiring and the connection points should be carefully planned to minimize signal loss and ensure reliable electrical connections. Integration of wiring channels or conduits within the 3D printed structure can facilitate the routing of wires and protect them from external influences.

20.11.2 Design Challenges and Solutions Specific to 3D Printed Energy Harvesters

(a) Material Selection: The choice of materials for 3D printed energy harvesters is critical to achieving the desired performance. Piezoelectric materials suitable for 3D printing, such as certain polymers or ceramics, may have specific mechanical or electrical properties that need to be considered during the design process. Material compatibility, including adhesion between layers, thermal expansion, and mechanical stability, should be carefully evaluated to ensure the integrity and functionality of the energy harvester [62].

(b) Manufacturing Constraints: 3D printing has certain manufacturing constraints that influence the design of energy harvesters. These include layer thickness, printing resolution, and build orientation limitations. Designing energy harvesters that adhere to these constraints while maintaining optimal performance requires careful consideration and iterative design optimization. Collaboration between design engineers and 3D printing experts is crucial to overcoming these challenges and achieving the successful fabrication of energy harvesters [62].

20.11.3 Design for Performance Optimization

(a) Structural Integrity: The structural integrity of 3D printed energy harvesters is essential for their long-term functionality. Design considerations should include the appropriate thickness, infill density, and support structures to ensure mechanical strength and stability. Finite element analysis (FEA) or simulation tools can aid in evaluating the structural performance of the design and identifying areas prone to stress concentration or failure.

(b) Vibration Isolation: Energy harvesters can be subjected to various external vibrations or mechanical forces that are not related to the desired energy source. Designing effective vibration isolation mechanisms, such as compliant structures or damping materials, can help reduce the impact of unwanted vibrations on energy harvesting performance. 3D printing allows for the integration of such isolation features directly into the energy harvester design.

(c) Environmental Considerations: Depending on the intended application, 3D printed energy harvesters may be exposed to different environmental conditions, such as temperature variations, moisture, or chemical exposure. Designing for environmental robustness includes selecting appropriate materials, coatings, or encapsulation methods to protect the energy harvester and maintain its performance and reliability over time [63].

20.11.4 Testing and Validation

Designing 3D printed piezoelectric energy harvesters involves iterative testing and validation to ensure their performance matches the design objectives. This includes experimental characterization of the energy harvester's electrical output, mechanical properties, and resonant frequency. Additionally, simulation tools can aid in predicting and optimizing the energy harvesting performance based on the design parameters. Designers should also consider real-world conditions and evaluate the energy harvester's performance under different excitation sources or environmental scenarios [42].

The design considerations for 3D printed piezoelectric energy harvesters encompass optimizing geometry and shape, integrating electronics and connections, addressing specific design challenges, and ensuring performance optimization. Through careful design, engineers can harness the capabilities of 3D printing to create customized energy harvesters that are efficient, reliable, and tailored to meet the requirements of various applications [33].

20.12 Applications of 3D Printed Piezoelectric Energy Harvesters

20.12.1 Wearable Devices Powered by 3D Printed Energy Harvesters

One prominent application of 3D printed piezoelectric energy harvesters is in wearable devices. Wearable technology has gained significant attention in recent years, offering a wide range of applications in the healthcare, sports, and lifestyle sectors. The integration of 3D printed energy harvesters in wearables enables the generation of electrical power from ambient mechanical vibrations or movements, thereby reducing or eliminating the need for conventional batteries.

By embedding 3D printed piezoelectric energy harvesters in clothing, accessories, or wearable devices, users can power various functionalities, such as health monitoring sensors, activity trackers, or communication modules. The design flexibility of 3D printing allows for the creation of energy harvesters that conform to the shape and size requirements of wearable devices, ensuring a seamless integration without compromising comfort or aesthetics. The advantages of 3D printed energy harvesters in wearable devices include their lightweight nature, customization options, and the ability to harvest energy from a wide range of movements or vibrations, including body motion, breathing, or environmental vibrations. This opens up new possibilities for self-powered wearable technology that enhances user experience and extends battery life, making wearables more practical and sustainable.

20.12.2 Sensors and Wireless Sensor Networks Utilizing 3D Printed Energy Harvesters

In the realm of sensing technology, 3D printed piezoelectric energy harvesters find applications in powering sensors and wireless sensor networks (WSNs). Sensors play a crucial role in monitoring and collecting data in various domains, including environmental monitoring, SHM, and industrial automation.

The integration of 3D printed energy harvesters with sensors enables self-powered sensing systems that eliminate the need for frequent battery replacement or wired power sources. This is particularly advantageous in remote or inaccessible locations where sensor deployment and maintenance can be challenging. 3D printed piezoelectric energy harvesters can power a wide range of sensors, including temperature sensors, humidity sensors, pressure sensors, and vibration sensors. These energy harvesters convert ambient mechanical energy, such as vibrations or deformations, into electrical energy, providing a sustainable power source for sensor operation.

Furthermore, 3D printing allows for the fabrication of complex sensor housing designs that can accommodate both the sensor and the energy harvester, optimizing the overall system size and integration. The customization options provided by 3D printing enable the design of sensor nodes that are specifically tailored to the application requirements, making them more efficient and effective. WSNs consist of multiple interconnected sensor nodes that collaborate to monitor and transmit data. The incorporation of 3D printed piezoelectric energy harvesters in WSNs enhances their autonomy and reduces the dependency on external power sources. This leads to the development of self-sustainable networks that can operate for extended periods without human intervention.

20.12.3 Structural Health Monitoring Systems Enhanced by 3D Printed Energy Harvesters

Structural health monitoring (SHM) systems are employed to assess the integrity and performance of structures, such as bridges, buildings, or aircraft, in real time. These systems rely on sensors to measure various parameters, including strain, deformation, vibration, or temperature, to detect any deviations from normal operating conditions. The integration of 3D printed piezoelectric energy harvesters within SHM systems allows for self-powered sensing capabilities. The energy harvesters convert mechanical energy resulting from structural vibrations or deformations into electrical energy to power the sensors, eliminating the need for external power sources or battery replacement [64].

3D printing enables the fabrication of energy harvesters with customized shapes and sizes, enabling their seamless integration into structural components or sensor nodes within the SHM system. The energy harvesters can be strategically placed in areas of interest to capture the mechanical energy and provide a sustainable power source for the sensors. The use of 3D printed piezoelectric energy harvesters in SHM systems offers several advantages. Firstly, it eliminates the reliance on wired power sources, reducing the complexity of the system and making it more flexible for installation and deployment. Additionally, the energy harvesters can be designed to be lightweight and low-profile, minimizing their impact on the structural behavior and ensuring accurate monitoring [65].

Moreover, 3D printing allows for the integration of multiple components within the energy harvester, such as sensors, signal conditioning circuitry, or wireless communication modules. This integration not only simplifies the system architecture but also improves the overall efficiency and reliability of the SHM system. The self-powered nature of 3D printed energy harvesters in SHM systems ensures continuous monitoring without interruptions caused by battery replacement or power supply issues. It enables real-time data acquisition and analysis, facilitating early detection of structural anomalies, preventing catastrophic failures, and enhancing maintenance and repair strategies. Furthermore, the customization capabilities of 3D printing enable the adaptation of energy harvesters to specific structural geometries, materials, and monitoring requirements. The design can be optimized to maximize energy conversion efficiency, ensuring that even low-level vibrations or deformations can be effectively harnessed for power generation.

The application of 3D printed energy harvesters in SHM systems has the potential to revolutionize the field of structural monitoring by providing self-powered, autonomous, and scalable solutions. It enables continuous data collection, facilitates remote monitoring capabilities, and reduces operational costs associated with power supply and maintenance. 3D printed piezoelectric energy harvesters find valuable applications in wearable devices, sensors, and WSNs, as well as SHM systems. The integration of these energy harvesters enables self-powered operation, reducing dependency on external power sources and enhancing the efficiency, reliability, and autonomy of these systems. As 3D printing technology continues to advance, the design and fabrication of energy harvesters will further evolve, opening up new possibilities for sustainable and innovative applications in various fields.

20.13 Performance Evaluation and Optimization of 3D Printed Energy Harvesters

Performance evaluation and optimization are critical steps in the development of 3D printed energy harvesters. This process involves assessing the energy conversion efficiency, identifying areas for improvement, and optimizing the design and materials to enhance overall performance. The following aspects are crucial in evaluating and optimizing the performance of 3D printed energy harvesters [66].

20.13.1 Measurement Techniques for Evaluating Energy Conversion Efficiency

Accurate measurement techniques are essential for evaluating the energy conversion efficiency of 3D printed energy harvesters. Several parameters need to be assessed, including output voltage, current, power, and electrical impedance. These measurements help determine the effectiveness of the energy harvester in converting mechanical energy into electrical energy. Standardized evaluation methods such as impedance spectroscopy, power-voltage curves, and dynamic response analysis can provide valuable insights into the performance characteristics of the energy harvester. These techniques enable the determination of key performance indicators such as power density, voltage sensitivity, and resonant frequency. Comparative studies can be conducted to analyze the performance of different 3D printed designs and materials.

20.13.2 Strategies for Improving the Performance of 3D Printed Energy Harvesters

To optimize the performance of 3D printed energy harvesters, various strategies can be employed [67]:

(a) Material Selection: The choice of materials plays a crucial role in determining the energy conversion efficiency of the harvester. Different materials have varying piezoelectric properties and mechanical characteristics that directly impact performance. Through material selection and optimization, the energy harvester can be tailored to the specific requirements of the application, maximizing energy conversion efficiency.

(b) Design Optimization: The geometric design of the energy harvester significantly influences its performance. Factors such as shape, size, thickness, and electrode configuration can be optimized to enhance energy conversion efficiency. Techniques like FEA and computational modeling can assist in optimizing the design by simulating and analyzing the mechanical and electrical behavior of the harvester.

(c) Resonant Frequency Tuning: Resonant frequency plays a crucial role in maximizing energy harvesting efficiency. By adjusting the design parameters or employing additional tuning mechanisms, such as adding mass or altering the shape, the resonant frequency can be matched to the operating frequency of the ambient vibrations, thereby optimizing energy conversion efficiency.

(d) Integration of Multiple Harvesters: In some cases, integrating multiple 3D printed energy harvesters in an array or network configuration can enhance performance. This allows for increased power generation and improved energy harvesting efficiency by capturing energy from multiple sources or increasing the effective surface area for energy conversion.

Strategies and approaches for enhancing the performance of 3D printed energy harvesters are given in Fig. 20.5.

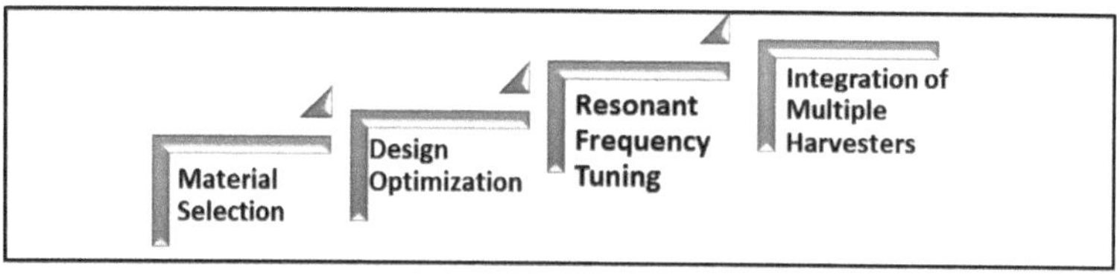

Figure 20.5 Enhancing the performance of 3D printed energy harvesters: Strategies and approaches.

20.14 Comparative Analysis of Different 3D Printed Designs and Materials

Comparative analysis plays a vital role in evaluating the performance of different 3D printed designs and materials. By fabricating and testing multiple prototypes, researchers can identify the strengths and weaknesses of different design approaches and materials. Comparative studies provide insights into the influence of design parameters, such as shape, size, and electrode configuration, on energy conversion efficiency. Evaluating the performance of various materials, including ceramics, polymers, and composites, helps determine their suitability for specific applications. The comparative analysis enables researchers to identify the trade-offs between material properties, manufacturing considerations, and energy harvesting performance [67]. Performance evaluation and optimization of 3D printed energy harvesters involve the use of measurement techniques to assess energy conversion efficiency, strategies for enhancing performance through material selection and design optimization, and comparative analysis of different designs and materials. These steps are crucial in advancing the field of 3D printed energy harvesters, enabling the development of efficient and tailored solutions for various applications.

20.15 Challenges and Future Directions in 3D Printed Piezoelectric Energy Harvesters

20.15.1 Limitations and Potential Issues in Current 3D Printing Technologies

While 3D printing has shown great promise in the fabrication of piezoelectric energy harvesters, there are still some limitations and potential issues that need to be addressed. These include [68]:

(a) Material Selection and Properties: The availability of suitable piezoelectric materials that can be effectively 3D printed remains a challenge. The performance of the energy harvester heavily relies on the piezoelectric properties of the materials used. Therefore, advancements in material development and characterization techniques are needed to optimize the performance of 3D printed energy harvesters.

(b) Resolution and Accuracy: The resolution and accuracy of current 3D printing technologies may limit the intricacy and complexity of the energy harvester designs. Higher resolution printing techniques are required to achieve finer details, such as smaller feature sizes, thin layers, and precise electrode patterns, to enhance the overall performance of the energy harvesters.

(c) Manufacturing Process Control: Ensuring consistency and repeatability in the manufacturing process is crucial for producing reliable and high-performance energy harvesters. Controlling factors such as print parameters, material properties, and post-processing steps is essential to minimize variations and achieve consistent results.

(d) Scalability: Scaling up the production of 3D printed energy harvesters to meet commercial demands is another challenge. The current printing speed and build volume limitations of 3D printers may hinder large-scale manufacturing. Advancements in printing technologies, such as faster printing techniques and larger build platforms, are necessary to address this challenge.

20.15.2 Emerging Trends and Advancements in the Field of 3D Printed Energy Harvesters

The field of 3D printed energy harvesters is constantly evolving, driven by emerging trends and advancements. Some notable trends and advancements include [68]:

(a) Multimaterial Printing: Multimaterial 3D printing techniques enable the integration of different materials within a single energy harvester, allowing for the combination of piezoelectric materials, conductive materials, and structural materials. This opens up new possibilities for enhanced energy conversion efficiency, improved device integration, and increased functionality.

(b) Hybrid Approaches: Hybrid approaches that combine traditional manufacturing techniques with 3D printing are being explored. For instance, incorporating post-processing steps such as sintering, annealing, or electrode deposition on 3D printed structures can enhance the piezoelectric properties and overall performance of the energy harvesters.

(c) Flexible and Stretchable Energy Harvesters: Advances in flexible and stretchable materials, along with 3D printing techniques, enable the fabrication of energy harvesters that can conform to curved surfaces or withstand mechanical deformations. These flexible and stretchable energy harvesters offer new opportunities for applications in wearable electronics, smart textiles, and biomedical devices.

(d) Advanced Design Optimization: Computational design optimization techniques, coupled with 3D printing, allow for the exploration of complex geometries and optimization of energy harvester performance. Computer-aided design (CAD) software, FEA, and optimization algorithms enable the development of energy harvesters with improved energy conversion efficiency, structural integrity, and durability.

20.15.3 Prospects and Research Directions for 3D Printed Piezoelectric Energy Harvesters

The future of 3D printed piezoelectric energy harvesters holds great prospects for further advancements and novel applications. Several research directions can be pursued to push the boundaries of this field.

(a) Enhanced Piezoelectric Materials: Continued research and development of piezoelectric materials with improved properties, such as higher piezoelectric coefficients, broader operating temperature ranges, and enhanced mechanical durability, will significantly impact the performance of 3D printed energy harvesters.

(b) Integration of Functionalities: Exploring the integration of additional functionalities within 3D printed energy harvesters, such as energy storage capabilities or wireless communication modules, can expand their applications and create more self-sufficient energy systems.

(c) Environmental Adaptability: Developing 3D printed energy harvesters that can operate in diverse environmental conditions, such as extreme temperatures, humidity, or corrosive environments, will open doors for applications in harsh and demanding settings.

(d) Biocompatible Energy Harvesters: Research efforts focused on developing biocompatible and biodegradable materials for 3D printed energy harvesters can enable their use in biomedical applications, such as implantable devices and wearable health monitors.

(e) Manufacturing Process Improvements: Advancements in 3D printing technologies, including higher-resolution printing, faster printing speeds, and improved process control, will enhance the manufacturing capabilities and scalability of 3D printed energy harvesters.

Addressing the limitations and potential issues in current 3D printing technologies, exploring emerging trends and advancements, and pursuing research directions can drive the field of 3D printed piezoelectric energy harvesters toward greater performance, versatility, and applicability in various industries and domains. Continued collaboration between researchers, engineers, and material scientists is vital for unlocking the full potential of this technology and realizing its impact on energy harvesting and beyond.

20.16 Conclusion and Recommendations

A comprehensive review of 3D printed piezoelectric energy harvesters, focusing on the materials used, 3D printing techniques employed, and various applications of the technology have been discussed. Key findings and insights from the chapter include:

(a) Piezoelectric energy harvesting is a promising technology that converts mechanical energy into electrical energy, with potential applications in powering small-scale electronics and devices.

(b) The emergence of 3D printing technology has made the fabrication of piezoelectric energy harvesters more accessible, enabling the development of unique designs and improved performance.

(c) Various materials, including ceramics, polymers, and composites, can be 3D printed with piezoelectric properties suitable for energy harvesting applications.
(d) Different 3D printing techniques, such as FDM, SLA, and SLS, offer flexibility and precision in fabricating piezoelectric energy harvesters.
(e) 3D printed piezoelectric energy harvesters have found applications in wearable devices, sensors, WSNs, and SHM systems, demonstrating their versatility and potential impact in various fields.

20.17 Recommendations for Further Research and Development in the Field

Based on the current status, several recommendations can be made for further research and development in the field of 3D printed piezoelectric energy harvesters.

(a) Material Development: Further research is needed to explore and optimize the piezoelectric properties of existing materials and develop new materials with improved performance characteristics, such as enhanced piezoelectric coefficients, mechanical durability, and environmental adaptability.
(b) Design Optimization: Continued research efforts should focus on optimizing the design of 3D printed energy harvesters to maximize energy conversion efficiency, mechanical robustness, and integration of additional functionalities.
(c) Manufacturing Process Improvements: Advancements in 3D printing technologies, such as higher resolution printing, faster printing speeds, and improved process control, are necessary to enhance the manufacturing capabilities and scalability of 3D printed energy harvesters.
(d) System Integration: Exploring the integration of 3D printed energy harvesters into larger energy systems, such as self-powered devices and IoT networks, can open up new opportunities for energy harvesting and utilization.
(e) Performance Characterization: Further research should focus on developing standardized measurement techniques and

evaluation methods to accurately assess the performance of 3D printed energy harvesters, allowing for comparative studies and benchmarking of different designs and materials.

20.18 Closing Remarks on the Significance of 3D Printed Piezoelectric Energy Harvesters

In conclusion, 3D printed piezoelectric energy harvesters hold significant promise in the field of energy harvesting. The accessibility and flexibility offered by 3D printing technology have enabled the development of unique designs, optimized materials, and enhanced performance. These energy harvesters have already found applications in wearable devices, sensors, and SHM systems, demonstrating their potential to power a wide range of electronic devices and contribute to the advancement of various industries.

As research and development in the field continue, it is essential to address the limitations and challenges associated with 3D printing technologies, materials, and manufacturing processes. This requires collaborative efforts between researchers, engineers, material scientists, and industry stakeholders to drive innovation, optimize performance, and realize the full potential of 3D printed piezoelectric energy harvesters. With further advancements in material science, design optimization, manufacturing processes, and system integration, 3D printed piezoelectric energy harvesters are poised to play a crucial role in the development of self-powered electronics, IoT devices, and sustainable energy solutions. The continued exploration and adoption of this technology will contribute to a greener and more energy-efficient future.

References

1. Cámara-Molina J. C., Moliner E., Martínez-Rodrigo M. D., Connolly D. P., Yurchenko D., Galvín P., and Romero A. "3D printed energy harvesters for railway bridges - Design optimisation," *Mech. Syst. Signal Process.*, **190**, 110133, 2022, doi: 10.1016/j.ymssp.2023.110133.
2. Hossain S. S., Jang S., Park S., Bae C. J. "Understanding ink design and printing dynamics of extrusion-based 3D printing: Defect-free

dense piezoelectric ceramics," *J. Manuf. Process.*, **92**, 1–11, 2023, doi: 10.1016/j.jmapro. 2023.02.018.

3. Zhou X., Parida K., Halevi O., Liu Y., Xiong J., Magdassi S., and Lee-Pooi S., "All 3D printed stretchable piezoelectric nanogenerator with non-protruding kirigami structure," *Nano Energy*, **72**, 104676, 2020, doi: 10.1016/j.nanoen. 2020.104676.
4. Godard N., Allirol L., Latour A., Glinsek S., Gérard M, Polesel J., Santos F.D.D., Defay E., "1-mW vibration energy harvester based on a cantilever with printed polymer multilayers," *Cell Reports Phys. Sci.*, **1**(6), 2020, doi: 10.1016/j.xcrp.2020.100068.
5. Wang Y., Hong M., Venezuela J., Liu T., and Dargusch M., "Expedient secondary functions of flexible piezoelectrics for biomedical energy harvesting," *Bioact. Mater.*, **22**, 291–311, 2023, doi: 10.1016/j.bioactmat.2022.10.003.
6. Zong H. et al., "Rheological and printability evaluation of melt-cast explosives for fused deposition modeling (FDM) 3D printing," *FirePhysChem*, **4**(1), 34–41, 2024. doi: 10.1016/j.fpc.2023.05.007.
7. Winarso R., Anggoro P. W., Ismail R., Jamari J., and Bayuseno A. P., "Application of fused deposition modeling (FDM) on bone scaffold manufacturing process: A review," *Heliyon*, **8**(11), e11701, 2022, doi: 10.1016/j.heliyon.2022.e11701.
8. Jnr M. H. et al., "Stereolithography (SLA) for the evaluation of thermal and mechanical properties of commercial polypropylene," *Procedia Manuf.*, **55**(C), 205–212, 2021, doi: 10.1016/j.promfg.2021.10.029.
9. Rosado E. and Moreno R., "Mullite-silica scaffolds obtained by stereolithography and reaction sintering," *Open Ceram.*, **14**(March), 100361, 2023, doi: 10.1016/j.oceram.2023.100361.
10. Lekurwale S., Karanwad T., and Banerjee S., "Selective laser sintering (SLS) of 3D printlets using a 3D printer comprised of IR/red-diode laser," *Ann. 3D Print. Med.*, **6**, 100054, 2022, doi: 10.1016/j.stlm.2022.100054.
11. Stoia D. I. and Marsavina L., "Effect of aluminum particles on the fracture toughness of polyamide-based parts obtained by selective laser sintering (SLS)," *Procedia Struct. Integr.*, **18**, 163–169, 2019, doi: 10.1016/j.prostr.2019.08.150.
12. Hunde B. R. and Woldeyohannes A. D., "3D printing and solar cell fabrication methods: A review of challenges, opportunities, and future prospects," *Results Opt.*, **11**(January), 100385, 2023, doi: 10.1016/j.rio.2023.100385.

13. Nazir A. et al., "Multi-material additive manufacturing: A systematic review of design, properties, applications, challenges, and 3D printing of materials and cellular metamaterials," *Mater. Des.*, **226**, 111661, 2023, doi: 10.1016/j.matdes.2023.111661.

14. Cui Y., Zhu J., Tong H., and Zou R., "Advanced perspectives on MXene composite nanomaterials: Types synthetic methods, thermal energy utilization and 3D printed techniques," *iScience*, **26**(1), 105824, 2023, doi: 10.1016/j.isci.2022.105824.

15. Bairagi S., Shahid-ul-Islam, Shahadat M., Mulvihill D. M., and Ali W., "Mechanical energy harvesting and self-powered electronic applications of textile-based piezoelectric nanogenerators: A systematic review," *Nano Energy*, **111**(April), 108414, 2023, doi: 10.1016/j.nanoen.2023.108414.

16. Babu A., Aazem I., Walden, S. Bairagi, Mulvihill D. M., and Pillai S. C., "Electrospun nanofiber based TENGs for wearable electronics and self-powered sensing," *Chem. Eng. J.*, **452**(P1), 139060, 2023, doi: 10.1016/j.cej.2022.139060.

17. Wu Y., Ma Y., Zheng H., and Ramakrishna S., "Piezoelectric materials for flexible and wearable electronics: A review," *Mater. Des.*, **211**, 110164, 2021, doi: 10.1016/j.matdes.2021.110164.

18. Shah M. A., Pirzada B. M., Price G., Shibiru A. L., and Qurashi A., "Applications of nanotechnology in smart textile industry: A critical review," *J. Adv. Res.*, **38**, 55–75, 2022, doi: 10.1016/j.jare.2022.01.008.

19. Pang Y., Cao Y., Derakhshani M., Fang Y., Wang Z. L., and Cao C., "Hybrid energy-harvesting systems based on triboelectric nanogenerators," *Matter*, **4**(1), 116–143, 2021, doi: 10.1016/j.matt.2020.10.018.

20. Manjakkal L., Núñez C. G., Dang W., and Dahiya R., "Flexible self-charging supercapacitor based on graphene-Ag-3D graphene foam electrodes," *Nano Energy*, **51**(March), 604–612, 2018, doi: 10.1016/j.nanoen.2018.06.072.

21. Li W., Cao Y., Wang C., and Sepúlveda N., "Ferroelectret nanogenerators for the development of bioengineering systems," *Cell Reports Phys. Sci.*, **4**(5), 1–35, 2023, doi: 10.1016/j.xcrp.2023.101388.

22. Liang H., Hao G., and Olszewski O. Z., "A review on vibration-based piezoelectric energy harvesting from the aspect of compliant mechanisms," *Sensors Actuators A Phys.*, **331**, 112743, 2021, doi: 10.1016/j.sna.2021.112743.

23. Yu Y., Luo C., Chiba H., Y. Shi, and Narita F., "Energy harvesting and wireless communication by carbon fiber-reinforced polymer-

enhanced piezoelectric nanocomposites," *Nano Energy*, **113**(May), 108588, 2023, doi: 10.1016/j.nanoen.2023.108588.

24. Cheng S. et al., "Enhanced piezoelectric properties and thermal stability of $Bi_{0.5}Na_{0.5}TiO_3$ modified $BiFeO_3BaTiO_3$ ceramics with morphotropic phase boundary," *J. Mater.*, **9**(3), 464–471, 2023, doi: 10.1016/j.jmat.2023.01.001.

25. Mesrar M., Elbasset A., Echatoui N. S., Abdi F., and Lamcharfi T., "Microstructural and high-temperature dielectric, piezoelectric and complex impedance spectroscopic properties of $K_{0.5}Bi_{0.5}TiO_3$ modified NBT-BT lead-free ferroelectric ceramics," *Heliyon*, **9**(4), e14761, 2023, doi: 10.1016/j.heliyon.2023.e14761.

26. Jain Ruth D. E., "Study on vibrational Raman modes and electric properties of lead-free sodium bismuth titanate – bismuth magnesium zirconate piezoelectric ceramics," *Chem. Phys. Impact*, **6**, 100172, 2023, doi: 10.1016/j.chphi.2023.100172.

27. Ferrero G. et al., "Piezoelectric properties of mechanochemically processed $0.67BiFeO_3$-$0.33BaTiO_3$ ceramics," *J. Eur. Ceram. Soc.*, **43**(2), 350–361, 2023, doi: 10.1016/j.jeurceramsoc.2022.10.024.

28. Sun D., Lu Y., and Karaki T., "Review of the applications of 3D printing technology in the field of piezoelectric ceramics," *Resour. Chem. Mater.*, **2**(2), 128–142, 2023, doi: 10.1016/j.recm.2023.02.001.

29. Donate R., Paz R., Moriche R., Jesús M., Alemán-domínguez M. E., and Monzón M., "An overview of polymeric composite scaffolds with piezoelectric properties for improved bone regeneration," *Mater. Des.*, **231**, 112085, 2023, doi: 10.1016/j.matdes.2023.112085.

30. Panda P. K., Sahoo B., and Thejas T. S., "High strain lead-free piezo ceramics for sensor and actuator applications: A review," *Sensors Int.*, **4**, 100226, 2023, doi: 10.1016/j.sintl.2022.100226.

31. Shaukat H. et al., "Piezoelectric materials: Advanced applications in electro-chemical processes," *Energy Reports*, **9**, 4306–4324, 2023, doi: 10.1016/j.egyr.2023.03.077.

32. Gao Y. et al., "Printable electrode materials for supercapacitors," *ChemPhysMater*, **1**(1), 17–38, 2022, doi: 10.1016/j.chphma.2021.09.002.

33. Cheng M. et al., "A review of flexible force sensors for human health monitoring," *J. Adv. Res.*, **26**, 53–68, 2020, doi: 10.1016/j.jare.2020.07.001.

34. Kumar C. et al., "Multiscale in-situ quantification of the role of surface roughness and contact area using a novel Mica-PVS triboelectric

nanogenerator," *Nano Energy*, **107**, 108122, 2023, doi: 10.1016/j.nanoen.2022.108122.

35. Yan Y. et al., "Implantable nerve guidance conduits: Material combinations, multi-functional strategies and advanced engineering innovations," *Bioact. Mater.*, **11**, 57–76, 2022, doi: 10.1016/j.bioactmat.2021.09.030.

36. Shahbazi M., Jäger H., Ettelaie R., Mohammadi A., and Asghartabar Kashi P., "Multimaterial 3D printing of self-assembling smart thermo-responsive polymers into 4D printed objects: A review," *Addit. Manuf.*, **71**, doi: 10.1016/j.addma.2023.103598.

37. Magazine R., van Bochove B., Borandeh S., and Seppälä J., "3D inkjet-printing of photo-crosslinkable resins for microlens fabrication," *Addit. Manuf.*, **50**, doi: 10.1016/j.addma.2021.102534.

38. Heck J., Goding J., Portillo Lara R., and Green R., "The influence of physicochemical properties on the processibility of conducting polymers: A bioelectronics perspective," *Acta Biomater.*, **139**, 259–279, 2022, doi: 10.1016/j.actbio.2021.05.052.

39. Li J., Wu C., Chu P. K., and Gelinsky M., "3D printing of hydrogels: Rational design strategies and emerging biomedical applications," *Mater. Sci. Eng. R Rep.*, **140**, 100543, 2020, doi: 10.1016/j.mser.2020.100543.

40. Tabriz A. G. and Douroumis D., "Recent advances in 3D printing for wound healing: A systematic review," *J. Drug Deliv. Sci. Technol.*, **74**, 103564, 2022, doi: 10.1016/j.jddst.2022.103564.

41. Melnyk L. A. and Oyewumi M. O., "Integration of 3D printing technology in pharmaceutical compounding: Progress, prospects, and challenges," *Ann. 3D Print. Med.*, **4**, 100035, 2021, doi: 10.1016/j.stlm.2021.100035.

42. Hamidah I., Pawinanto R. E., Mulyanti B., and Yunas J., "A bibliometric analysis of micro electro mechanical system energy harvester research," *Heliyon*, **7**(3), e06406, 2021, doi: 10.1016/j.heliyon.2021.e06406.

43. Rana S., Singh V., and Singh B., "Recent trends in 2D materials and their polymer composites for effectively harnessing mechanical energy," *iScience*, 25(2), 103748, 2022, doi: 10.1016/j.isci.2022.103748.

44. He T., Guo X., and Lee C., "Flourishing energy harvesters for future body sensor network: From single to multiple energy sources," *iScience*, 24(1), 101934, 2021, doi: 10.1016/j.isci.2020.101934.

45. Yang Z., Zhou S., Zu J., and Inman D., "High-performance piezoelectric energy harvesters and their applications," *Joule*, **2**(4), 642–697, 2018, doi: 10.1016/j.joule.2018.03.011.

46. Zabek D. and Morini F., "Solid state generators and energy harvesters for waste heat recovery and thermal energy harvesting," *Therm. Sci. Eng. Prog.*, **9**, 235–247, 2019, doi: 10.1016/j.tsep.2018.11.011.

47. Jabri M., Masoumi S., Sajadirad F., West R. P., and Pakdel A., "Thermoelectric energy conversion in buildings," *Mater. Today Energy*, **32**, 101257, 2023, doi: 10.1016/j.mtener.2023.101257.

48. Lee C. and Tarbutton J. A., "Electric poling-assisted additive manufacturing process for lead-free piezoelectric device fabrication," *Procedia Manuf.*, **1**, 320–326, 2015, doi: 10.1016/j.promfg.2015.09.035.

49. Zhang H. et al., "A novel method to combine fused deposition modelling and inkjet printing in manufacturing multifunctional parts for aerospace application," *J. Mater. Res. Technol.*, **24**, 4405–4426, 2023, doi: 10.1016/j.jmrt.2023.04.059.

50. Glatt M., Greco S., Yi L., Kirsch B., and Aurich J. C., "Development and implementation of a system for the automated removal of parts produced by Fused Deposition Modeling," *Procedia CIRP*, **103**(March), 109–114, 2021, doi: 10.1016/j.procir.2021.10.017.

51. Xu Z., Fostervold R., and Razavi S. M. J., "Thickness effect on the mechanical behavior of PLA specimens fabricated via Fused Deposition Modeling," *Procedia Struct. Integr.*, **33**(C), 571–577, 2021, doi: 10.1016/j.prostr.2021.10.063.

52. Fekete I., Ronkay F., and Lendvai L., "Highly toughened blends of poly(lactic acid) (PLA) and natural rubber (NR) for FDM-based 3D printing applications: The effect of composition and infill pattern," *Polym. Test.*, **99**, 2021, doi: 10.1016/j.polymertesting.2021.107205.

53. Rathee S., Srivastava M., Maheshwari S., and Siddiquee A. N., "Effect of varying spatial orientations on build time requirements for FDM process: A case study," *Def. Technol.*, **13**(2), 92–100, 2017, doi: 10.1016/j.dt.2016.11.006.

54. Lakkala P., Munnangi S. R., Bandari S., and Repka M., "Additive manufacturing technologies with emphasis on stereolithography 3D printing in pharmaceutical and medical applications: A review," *Int. J. Pharm. X*, **5**, 100159, 2023, doi: 10.1016/j.ijpx.2023.100159.

55. Naula E. A., Rodríguez B. L., Garza-Castañon L. E., and Martínez-López J. I., "Manufacturing of stereolithography enabled soft tools for point of care micromixing and sensing chambers for underwater vehicles," *Procedia Manuf.*, **53**, 443–449, 2021, doi: 10.1016/j.promfg.2021.06.047.

56. Hoffmann P., Köllner D., Simon S., Ichi-Kakimoto K., and Fey T, "Modular lead-free piezoceramic/polymer composites with locally adjustable piezoelectric properties," *Open Ceram.*, **13**, 0–3, 2023, doi: 10.1016/j.oceram.2022.100320.

57. Zhou M., Zhu W., Yu S., Tian Y., and K. Zhou, "Selective laser sintering of carbon nanotube–coated thermoplastic polyurethane: Mechanical, electrical, and piezoresistive properties," *Compos. Part C Open Access*, **7**, 100212, 2022, doi: 10.1016/j.jcomc.2021.100212.

58. Westphal E. and Seitz H., "A machine learning method for defect detection and visualization in selective laser sintering based on convolutional neural networks," *Addit. Manuf.*, **41**, 101965, 2021, doi: 10.1016/j.addma.2021.101965.

59. Agrawaal H. and Thompson J. E., "Additive manufacturing (3D printing) for analytical chemistry," *Talanta Open*, **3**(January), 100036, 2021, doi: 10.1016/j.talo.2021.100036.

60. Grossin D. et al., "A review of additive manufacturing of ceramics by powder bed selective laser processing (sintering/melting): Calcium phosphate, silicon carbide, zirconia, alumina, and their composites," *Open Ceram.*, **5**, 2021, doi: 10.1016/j.oceram.2021.100073.

61. Mariello M. et al., "Flexible piezoelectric AlN transducers buckled through package-induced preloading for mechanical energy harvesting," *Nano Energy*, **85**, 105986, 2021, doi: 10.1016/j.nanoen.2021.105986.

62. Shi J., Li H., Xu F., and Tao X., "Materials in advanced design of personal protective equipment: a review," *Mater. Today Adv.*, **12**, 2021, doi: 10.1016/j.mtadv.2021.100171.

63. Liu S., Rao Y., Jang H., Tan P., and Lu N., "Strategies for body-conformable electronics," *Matter*, **5**(4), 1104–1136, 2022, doi: 10.1016/j.matt.2022.02.006.

64. Kumbhakar P. et al., "Prospective applications of two-dimensional materials beyond laboratory frontiers: A review," *iScience*, **26**(5), 106671, 2023, doi: 10.1016/j.isci.2023.106671.

65. Quan Y., Wu X., Zhu S., Zeng X., Zeng Z., and Zheng Q., "Triboelectric nanogenerators for clinical diagnosis and therapy: A report of recent progress," *Med. Nov. Technol. Devices*, **16**, 100195, 2022, doi: 10.1016/j.medntd.2022.100195.

66. Lin Y., Tseng K., and Ma C., "Investigation of resonant and energy harvesting characteristics of piezoelectric fiber composite bimorphs," *Mater. Des.*, **197**, 109267, 2021, doi: 10.1016/j.matdes.2020.109267.

67. Zhao B. et al., "A graded metamaterial for broadband and high-capability piezoelectric energy harvesting," *Energy Convers. Manag.*, **269**, 116056, 2022, doi: 10.1016/j.enconman.2022.116056.

68. Elshaekh D. N., Mohamed H. A., Shawkey H. A., and Kayed S. I., "Printed circularly polarized spilt ring resonator monopole antenna for energy harvesting," *Ain Shams Eng. J.*, **14**(6), 102182, 2023, doi: 10.1016/j.asej.2023.102182.

Index